Joint IMEKO TC1-TC7-TC13 Symposium: Measurement Science Behind Safety and Security 2014

AA008858

Journal of Physics: Conference Series Volume 588

Madeira, Portugal
3 – 5 September 2014

Editors:

J. A. Sousa
A. S. Ribeiro
E. Filipe

ISBN: 978-1-5108-0291-9
ISSN: 1742-6588

Printed from e-media with permission by:

Curran Associates, Inc.
57 Morehouse Lane
Red Hook, NY 12571

proceedings.com

Some format issues inherent in the e-media version may also appear in this print version.

Copyright© (2014) by the Institute of Physics
All rights reserved. The material featured in this book is subject to
IOP copyright protection, unless otherwise indicated.

Printed by Curran Associates, Inc. (2015)

For permission requests, please contact the Institute of Physics
at the address below.

Institute of Physics
Dirac House, Temple Back
Bristol BS1 6BE UK

Phone: 44 1 17 929 7481
Fax: 44 1 17 920 0979

techtracking@iop.org

Additional copies of this publication are available from:

Curran Associates, Inc.
57 Morehouse Lane
Red Hook, NY 12571 USA
Phone: 845-758-0400
Fax: 845-758-2634
Email: curran@proceedings.com
Web: www.proceedings.com

TABLE OF CONTENTS

PREFACE

The 2014 Joint IMEKO (International Measurement Confederation) TC1-TC7-TC13 Symposium was organized by RELACRE – Portuguese Association of Accredited Laboratories and the Portuguese Society for Metrology, on 3-5 September 2014. The work of this symposium is reported in this volume.

The scope of the symposium includes the main topics covered by the above Technical Committees:

TC1 Education and Training in measurement and Instrumentation
TC7 Measurement Science
TC13 Measurements in Biology and Medicine

The effort towards excellence of previous events, in this well established series, is maintained. There has been a special focus on measurement science behind safety and security, with the aim of highlighting the interdisciplinary character of measurement science and the importance of metrology in our daily lives. The discussion was introduced by keynote lectures on measurement challenges in biometrics, health monitoring and social sciences, to promote useful interactions with scientists from different disciplines. The Symposium was attended by experts working in these areas from 18 countries, including USA, Japan and China, and provided a useful forum for them to share and exchange their work and ideas.

In total over fifty papers are included in the volume, organized according to the presentation sessions. Each paper was independently peer-reviewed by two reviewers from a distinguished international panel.

The Symposium was held in Funchal, capital of Madeira Islands, known as the Atlantic Pearl. This wonderful Atlantic archipelago, formed by Madeira and Porto Santo islands, discovered in the 14th century, was chosen to host the 2014 IMEKO TC1-TC7-TC13 Joint Symposium "Measurement Science behind Safety and Security". It was the first territory discovered by the Portuguese sailors, when set out to discover a new world, in an epic journey where instrumentation and quality of measurement played a central role in the success of the enterprise, and gave an important technological edge to those men. A fortunate analogy to the advances sought by the research undertaken in the areas covered by this event.

We would like to take this opportunity to thank the members of the Symposium Steering and International Programme Committees, many of whom acted as reviewers of the papers presented here under a very short timescale, as well as our partner, SPMet – Portuguese Society for Metrology.

The editors hope that this volume will provide a useful contribution to enhancing the science, technology, education, and training in measurement and instrumentation.

João A. Sousa, Álvaro S. Ribeiro and Eduarda Filipe
Editors

Content from this work may be used under the terms of the Creative Commons Attribution 3.0 licence. Any further distribution of this work must maintain attribution to the author(s) and the title of the work, journal citation and DOI.
Published under licence by IOP Publishing Ltd

Peer review statement

All papers published in this volume of *Journal of Physics: Conference Series* have been peer reviewed through processes administered by the proceedings Editors. Reviews were conducted by expert referees to the professional and scientific standards expected of a proceedings journal published by IOP Publishing.

Content from this work may be used under the terms of the Creative Commons Attribution 3.0 licence. Any further distribution of this work must maintain attribution to the author(s) and the title of the work, journal citation and DOI.
Published under licence by IOP Publishing Ltd

IMEKO IOP Publishing
Journal of Physics: Conference Series **588** (2015) 012001 doi:10.1088/1742-6596/588/1/012001

Load Error in Metrology – A General Approach

Karl H. Ruhm[1]
Senior Lecturer, *Institute of Machine Tools and Manufacturing*
Swiss Federal Institute of Technology (ETH), Zurich, Switzerland

E-mail: ruhm@ethz.ch

Abstract. The behaviour of a measurement processes is shaped by three types of measurement errors, by *transfer errors*, *disturbance errors* and *load errors*. The following sections will provide for the first time a general and consistent treatment of the load error in Metrology: An expanded model of the measurement process will reveal the significance of this rarely treated error. Besides, there are many comparable loading phenomena in different fields outside Metrology; some concepts may with good reason be transferred and adopted. It will be shown that the main concern has to be focused on backward structures because of physical loading effects, which interdict the commonly assumed nonreactive relations in metrological structures. These backward relations recommend the use of *linear fractional representations* (LFR) as models, with the *Redheffer star product* as a valuable operator. Given such general structures, load error corrections can be derived for dedicated applications.

1. Introduction

Important topics in Metrology are measurement errors and uncertainties. While *measurement uncertainties* keep being covered in numerous publications, *measurement errors* are widely ignored. This situation seems to change. Measurement errors experience some sort of rehabilitation as well as an alignment with self-evident error concepts in other fields: "The approach based on true value and error was questioned as being based on unknowable quantities, i.e. idealized concepts. The very terms were almost banned from the literature, and whoever dared to use them was considered suspiciously as a supporter of old ideas." [1].

On the other hand, a large number of error types have survived in workshops, labs, offices, institutions and lecture halls, though often without being based on sound theoretical definitions. It has been shown recently that for physical measurement processes, three and only three types of measurement errors exist, *transfer errors*, *disturbance errors* and *load errors* [2]. All other common errors are subtypes of these three main error types. Of course, model errors are active too, but they reside on a different analytical level.

Considering these three error types, the load errors are the least familiar ones. Usually they are covered casually, if at all, and only in order to avoid momentary, specific inconveniences. A general consideration and treatment does not exist yet.

As a start the following sections will repeat the definition of the three main error types and concentrate on *load errors* later on. In a *forward analysis* we describe errors and error processes as distinct items.

[1] To whom any correspondence should be addressed.

Content from this work may be used under the terms of the Creative Commons Attribution 3.0 licence. Any further distribution of this work must maintain attribution to the author(s) and the title of the work, journal citation and DOI.
Published under licence by IOP Publishing Ltd

The strategic tools will be taken from Signal and System Theory, which is supported by versatile software for analysis and simulation.

One of several notable features will be the avoidance of the usual simplifying nonreactive behaviour between quantities of interest. This means that the single conventional *block* in a *signal relation diagram* will not be appropriate anymore to describe the relation between two signals. Note that the topic does not just concern *signals*, but also *systems* as *sources of signals* due to their insufficient *loadability* or inappropriate generalised *impedance* respectively. If we consider the domain of processes [3], the *removal of mass, energy, impulse and information* loads sources and thus changes their *state*. At first, the whole issue seems to become needlessly complicated. But as soon as the structure of all relations is compiled systematically, in a signal relation diagram for example, the result will turn out to be clear, comprehensible and implementable. In some fields the concept of generalised quadrupole and multipole representations is predominant.

So, the recurrent demand for a coherent consideration of *process* P and *measurement process* M, both linked in the *process under measurement* PUM (Figure 1), gets a mandatory relevance in the analysis and discussion of load errors and their compensation.

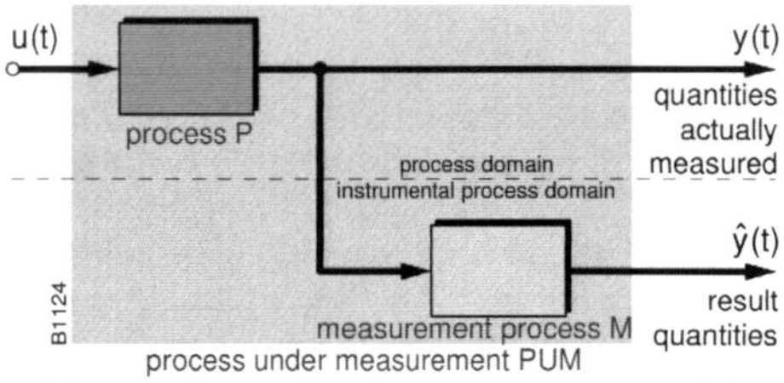

Figure 1. Process P and measurement process M linked in the process under measurement PUM

An additional abnormality arises from the fact that the so called load error does not appear in a measurement result $\hat{\mathbf{y}}$(t) directly, but only indirectly by the impact on the quantity to be measured *within the process* P, which means, on the measurand **y**(t) itself. This inconvenience makes this error type almost untraceable. Whether a load error is acceptable or unacceptable, whether it has to be corrected eventually, this does not concern the following investigations and statements [4]. They tend to develop the concept of a holistic framework by creating a common structure and thus fostering strategies for the correction of load errors.

It is worth mentioning here that we only talk about *errors* and not about *uncertainties* in the result quantities. Of course, once we have defined and analysed the load errors in a *forward analysis*, further investigations on our *uncertain knowledge* about the errors, as well as the parameters in question, may follow in a *backward analysis* concerning *error uncertainties*. They will then be included in the general uncertainty budget for a final measurement result according to the Guide to the Expression of Uncertainty in Measurement (GUM) [5]. *Errors are independent of uncertainties, but uncertainties are not independent of errors.*

In the following sections we assume multivariate, linear dynamic systems. Without loss of generality, we express our statements in the time domain. They are valid in the frequency domain as well. So, the describing transfer function matrices **G** may be *temporal transfer function matrices* **G**(t) or *spectral transfer function matrices* **G**(s) [6].

2. General Error Structure

How does nonideality actually arise, or more precisely, which deviation from a stated ideal behaviour may occur? This question is treated by means of Signal and System Theory [2].

Firstly, what does the model of an ideal, multivariable measurement process M look like at all? Normally, we state the *ideal (nominal) transfer function matrix* $\mathbf{G}_{nom}$ of a measurement system, which relates one set (vector) of output signals to one set (vector) of input signals, to be equal the identity matrix **I**. Of course, it may adopt any other reasonable structure and parameter set.
Secondly, if we have got the *nonideal transfer function matrix* **G**, the deviation between ideal and nonideal behaviour leads directly to the *error definitions* of interest.
But, what makes a transfer function matrix **G** nonideal? We assume that a nonideal system will be disturbed via additional inputs and will disturb its surroundings by additional outputs. So the nonideal system will have two sets (vectors) of input signals and two sets (vectors) of output signals (Figure 2).

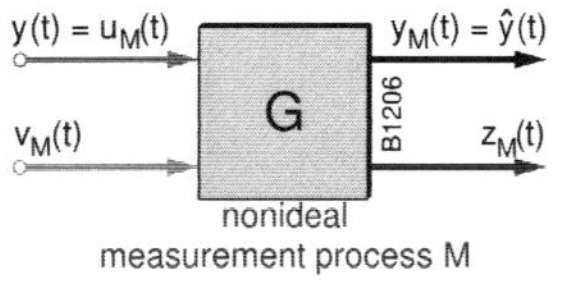

Figure 2. Generalised model of the nonideal measurement process M with disturbance quantities $v_M(t)$ and load quantities $z_M(t)$.

This structure reveals three error sources at the utmost:

- Irregularities *within* the measurement process M between the input quantities $\mathbf{u}_M(t)$ and the output quantities $\mathbf{y}_M(t)$. This leads to *transfer errors.*
- Influences *from* the environment via the disturbance quantities $\mathbf{v}_M(t)$ at the input *on* the measurement process M. This leads to *disturbance errors.*
- Influences *from* the measurement process M via the *load quantities* $z_M(t)$ at the output *on* any environment. This leads to *load errors.*

This generalised model of the nonideal measurement process M is a multivariate system, which is normally assumed to be linear, time-invariant (LTI) and dynamic (Figure 3), but which can readily become nonlinear and time-variant (NLTV).

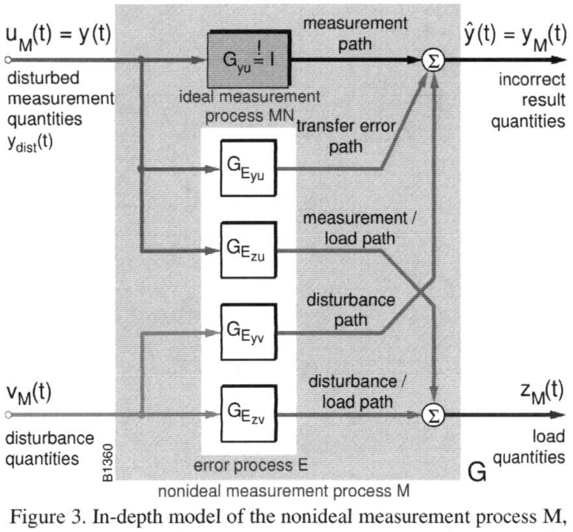

Figure 3. In-depth model of the nonideal measurement process M, consisting of the defined ideal (nominal) measurement process MN and the summarising error process E.

In this detailed signal relation diagram the internal *decomposition* assumes a parallel connection of the *nominal, ideal measurement process* MN and the *error process* E, The error process is the source of the three defined error types. Here already, it is obvious that the contribution to load errors does not affect the measurement result $\hat{\mathbf{y}}(t)$ directly.

It has to be mentioned, that besides this *parallel connection* of ideal system and error system, there is also the possibility of the *series connection* and the *feedback connection* [2]. However, the two latter show certain disadvantages.

3. Thermal Example

Load quantities $\mathbf{z}_M(t)$ impact the environment. In measurement procedures, especially the process P of interest is loaded. A plastic integrated circuit (IC) as process P may serve as a simple example (Figure 4), whose temperature $\vartheta_2(t)$ should be measured by a metallic temperature sensor process S as part of the measurement process M. Under measurement, thermal energy will flow from the hot IC to the cold sensor process S. The temperature of interest $\vartheta_2(t)$ will decrease accordingly and the sensor acquires a temperature, which is lower than it would be under normal, unconnected conditions.

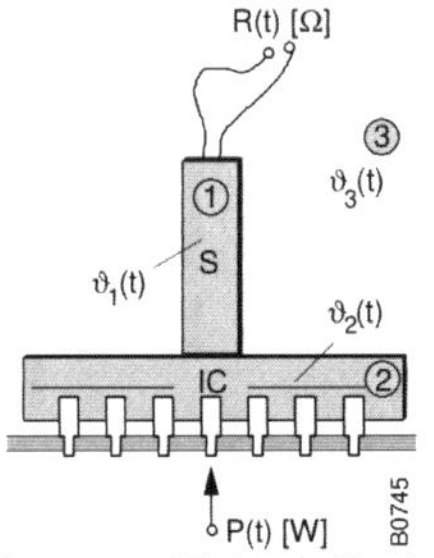

Figure 4. Sensor process S loads process IC thermally during a measurement procedure.

4. Load Error Structure

The dominant background is the physical *connection on*, or *insertion in*, a process P of a given measurement process M (Figure 5). Thereby the load quantities $\mathbf{z}_M(t)$ of the measurement process M become the disturbance quantities $\mathbf{v}(t)$ of process P, which corrupt the quantities to be measured $\mathbf{y}_{dist}(t)$ more or less.

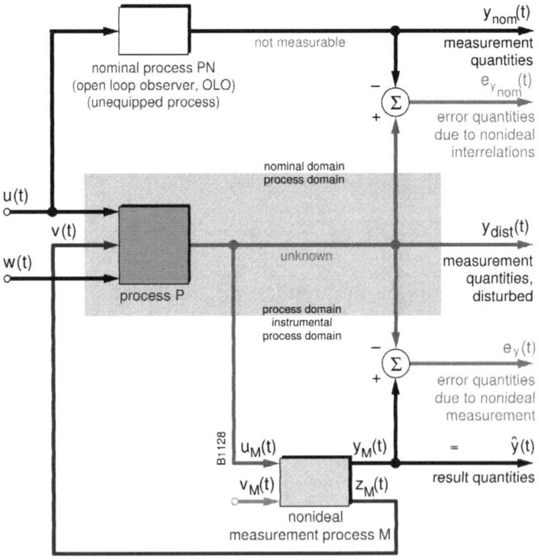

Figure 5. General structure of the loading impact
of a nonideal measurement process M on a process P to be measured.

This impact leads to systematic load errors. Again, they often remain undetected, because they obviously become effective outside the actual measurement process M.

The main reason for load errors is mass and / or energy removal from process P, which goes parallel with the withdrawal and transfer of information during the measurement procedure. This means that impedance mismatch, or generally speaking, the match of *capability* and *necessity* between *sources* and *sinks*, is an issue.

Besides, resisting elements (impedances) combined with storing elements (capacities) become responsible for dynamic effects and therefore for *dynamic load errors*. Mathematical modelling of the interconnected processes with load quantities and load errors reveals and quantifies such effects: The measurement path (chain) is not nonreactive anymore. Feedback paths appear in different configurations. We get extended series connections (Figure 6).

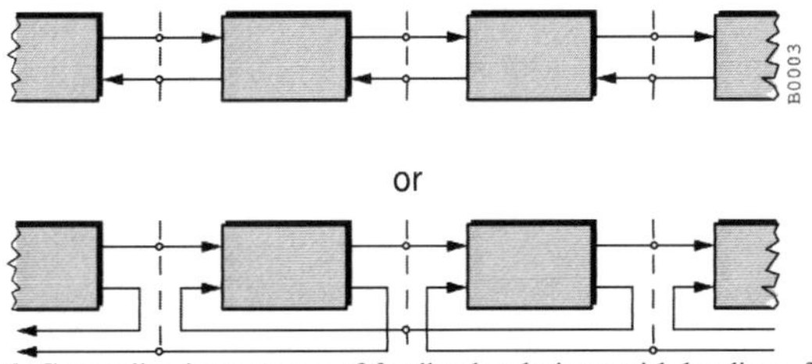

Figure 6. Generalised structures of feedback relations with loading effects.

It has to be amended that the product of *forward* (intensive) *quantities* and *backward* (extensive) *quantities* between two blocks mark *generalised* (signal) *power*. Signal and System Theory handles such structures elegantly by *Linear Fractional Representation* (LFR), in which the so-called *Redheffer Star Product* [7] plays a particular role.

5. Electrical Example

The connection of an electrical source Q to an electrical sink L serves as another obvious and impressive example. In measurement these two processes could represent a thermocouple and an indicating millivolt meter, which is described by the following model (Figure 7).

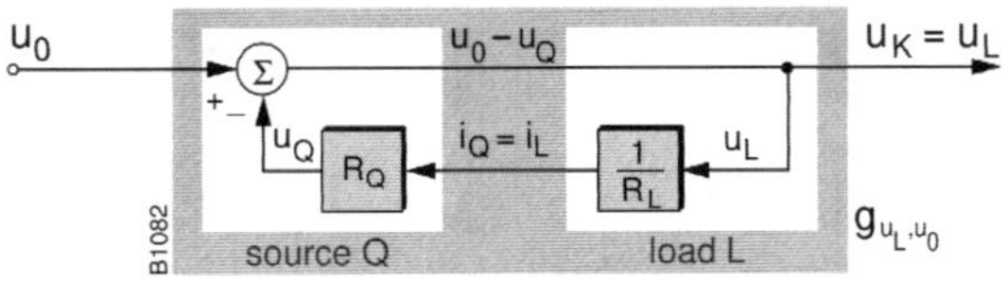

Figure 7. An unmatched series connection of two electrical systems with back-load effect.

The transfer equation of this series connection, which describes the relation between input voltage u_0 and output voltage u_K is well known:

$$u_K = u_L = \frac{1}{1 + \frac{R_Q}{R_L}} u_0 = g_{u_L,u_0} u_0 \quad [V]$$

Apparently, the quantity u_0 to be measured, drops to u_0 u_Q because of the load (sink) impedance R_L and the output (source) impedance R_Q. Unfortunately, the primary measure for an improvement of the nonideal measurement process, a source impedance $R_Q = 0$ and / or a load impedance $R_L = \infty$, is not realisable. However, the second measure, the implementation of a reconstruction process R will be *the* solution, as soon as the model is given. Another similar process is the voltage divider with its nonlinear characteristic caused by loading due to the similar impedance mismatch.

6. General Structure

The following general structure provides a basis for the analysis of loading effects. Unfortunately, the main obstacle of a widespread application is the usual lack of a suitable qualitative and quantitative model (structure, parameter) for the process P as well as for the nonideal measurement process M. Therefore trial and error strategies usually prevail.

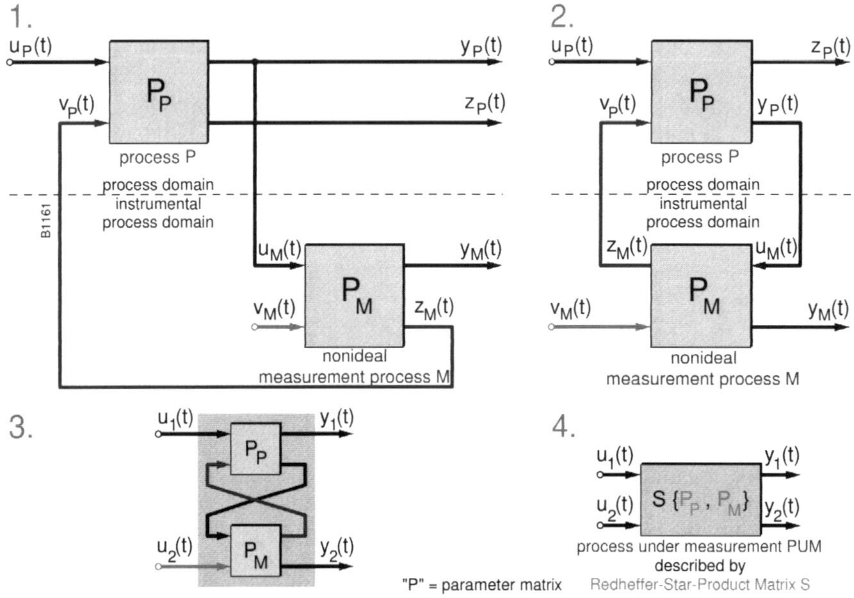

Figure 8. Reduction from step 1. to step 4. of the specific measurement structure to a general, universally valid structure.

This general structure of the *process under measurement* PUM, here a linear, dynamic system of two sets of input signals and two sets of output signals, can be implemented, analysed and discussed conveniently by common software products. The so-called Redheffer Star Product in the Linear Fractional Representation framework [7] serves as a ready-to-use function.

7. Reconstruction Process

Since we are normally unable to improve given processes and measurement processes according to specific demands physically, we design a correction of load errors by implementing a reconstruction process R, connected in series with the nonideal process under measurement. The model of this reconstruction process will be the inverse model of the nonideal process, as can be imagined intuitively. This concept is generally valid, though several conditions concerning the inversion operations must be fulfilled.

Preconditions for a successful realisation of a reconstruction process is the availability of models. On the one hand, we need the individual models of the process P and the measurement process M, each unconnected. On the other hand, we need the model of the process under measurement (PUM), both sub-processes interconnected. They all have to be developed by analytical and / or empirical identification (calibration).

8. Summary

The main reasons for unwanted load effects are physical dependencies between processes and measurement processes. They are seldom recognised and therefore not taken into consideration. Here, sources and structures of load errors have been investigated from the point of view of Signal and System Theory. Patterns of relevant relations become simple when approached systematically. They can be converted into the presented, universally valid structure, supported by dedicated algorithms.

References

[1] Bich, W.: "From Errors to Probability Functions. Evaluation of the Concept of Measurement Uncertainty". IEEE Transactions on Instrumentation and Measurement (TIM), 8 (2012), p. 2153
[2] Ruhm, K. H.: "The Significant Role of Errors in Metrology – Justification by System Theory" XX IMEKO World Congress (2012), Busan, Korea; http://imeko2012.kriss.re.kr/sub02/tc/down/SW_Error%20in%20Metrology.pdf
[3] Ruhm, K. H.: "Process and System – A Dual Definition, Revisited with Consequences in Metrology" 13th IMEKO TC1-TC7 Joint Symposium (2010), London http://iopscience.iop.org/1742-6596/238/1/012037/pdf/1742-6596_238_1_012037.pdf
[4] Profos, P.: "Messfehler", Teubner (1983)
[5] BIPM: Guide to the expression of uncertainty in measurement (GUM) (2008) http://www.bipm.org/utils/common/documents/jcgm/JCGM_100_2008_E.pdf
[6] Ruhm, K. H.: "Concept and Term «Transfer» Concerning Systems" http://www.mmm.ethz.ch/dok01/d0000536.pdf
[7] Redheffer, R. M.: "On a Certain Linear Fractional Transformation" Journal of Mathematical Physics, 39 (1960), 269 – 286

Quality Assessment of Vertical Angular Deviations for Photometer Calibration Benches

A. Silva Ribeiro[1], A. Costa Santos[1], J. Alves e Sousa[2], A. B. Forbes[3]
[1] National Laboratory of Civil Engineering (LNEC), Lisbon, Portugal
[2] Regional Laboratory of Civil Engineering (LREC-RAM), Funchal, Madeira, Portugal
[3] National Physical Laboratory (NPL), Teddington, United Kingdom

E-mail: asribeiro@lnec.pt

Abstract. Lighting, both natural and electric, constitutes one of the most important aspects of the life of human beings, allowing us to see and perform our daily tasks in outdoor and indoor environments. The safety aspects of lighting are self-evident in areas such as road lighting, urban lighting and also indoor lighting. The use of photometers to measure lighting levels requires traceability obtained in accredited laboratories, which must provide an associated uncertainty. It is therefore relevant to study the impact of known uncertainty sources like the vertical angular deviation of photometer calibration benches, in order to define criteria to its quality assessment.

1. Introduction and Motivation

Lighting, both natural and electric, constitutes one of the most important aspects of the life of human beings, allowing us to see and perform our daily tasks in outdoor and indoor environments. The interior spaces of buildings must be designed so as to provide adequate lighting conditions, concerning illumination levels, uniformity and absence of glare, for occupants to perform adequately all visual tasks and activities in conditions of safety and comfort, without eyestrain, and in an efficient way. The safety aspects of lighting are self-evident in areas such as road lighting, including tunnels, urban lighting and also indoor lighting.

The basic quantity used to quantify the lighting environment is the illuminance (E_V), which is defined as the quotient of the luminous flux (lm) incident on a given surface by the area (m^2) of that surface. The S.I. unit of illuminance is the lux (lm/m^2) [1].

Illuminances are usually measured with photosensors, commonly referred as photometers or illuminance meters. Using the photoelectric effect, those photosensors, detect light and convert it into electrical signals obtained from photocurrents developed by sets of photoresistors, photodiodes and photomultipliers connected to other optical elements.

The traceability of these type of instruments can be obtained through calibration in laboratory using different testing techniques, namely, absolute, substitution and monitor approaches (as described in [2]). All these methods use a setup that includes a reference stable light source and a photometric bench and, in some methods, a reference photometer (having properties similar to the photometer to be calibrated).

The setup formed by the geometric bench intends to establish a 1-dimensional line between the centre of a standard light source and the photometer being calibrated, allowing the establishment of a relation between the distance of both elements and the photometer measurements. The experimental setup has

Content from this work may be used under the terms of the Creative Commons Attribution 3.0 licence. Any further distribution of this work must maintain attribution to the author(s) and the title of the work, journal citation and DOI.
Published under licence by IOP Publishing Ltd

different sources of uncertainty, one of which is the important vertical angular deviation associated with the vertical profile of the bench, creating an angular deviation effect influencing the measurement results. The aim of the proposed work is to study this influence in the uncertainty of the measurement, and use that contribution to establish criteria that can be applied to the quality assessment of a photometer calibration bench at LNEC's Photometry and Colorimetry Unit, with respect to the measurement accuracy required.

The study is based on the example 1.2 of the CIE guide [3] related to the measurement of the effective distance. The experimental example given illustrates the uncertainty evaluation applying the ISO-GUM [4] conventional approach, and its validation with a Monte Carlo alternative approach, considering the existing nonlinearity effects in the mathematical model, for a discussion on the adequacy of current practices in this type of testing.

2. Experimental Procedure and Measurement Model

LNEC´s photometry and colorimetry laboratory has a photometric bench where the calibration of photometers is carried out. The method of calibration can be considered an absolute method since it relies on a traceable luminous intensity standard lamp. The basic method of calibration is based on the *inverse square law of illumination* through the variation of the distance to the standard lamp of the photometer to be calibrated. The measured distance differences and the respective illuminance values of the photometer, a "corrected" value for the illuminances (calibration factor) can then be determined (see Figures 1 and 2)

Figure 1 – Illustration of the Photometric Bench used in the calibration of photocells

The experimental procedure intends to measure the effective distance between the filament of the reference luminous intensity standard lamp and the aperture of a photometer. It considers a minimum of two positions (d_0, $d_0 + \Delta d$) on a linear scale of the photometer bench, measuring the photocurrent (or the illuminances) in each of these points (y_0 at d_0 and y_1 at $d_0+\Delta d$).

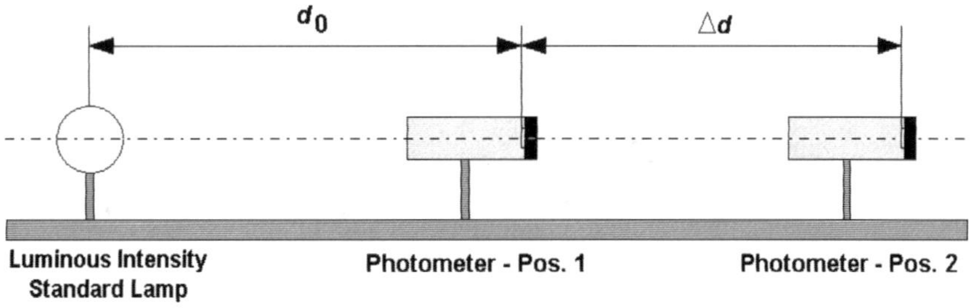

Figure 2 – Setup of the effective distance measurement for photometer calibration purposes.

The theoretical approach considers that the luminous intensity I (in cd), is constant, thus leading to following equation:

$$I = E_0 \cdot d_0^2 = E_1 \cdot (d_0 + \Delta d)^2 \qquad (1)$$

with the photocurrent values y_0 and y_1 the output quantities related to E at d_0 and ($d_0 + \Delta d$). The mathematical model that yields the relation between input and output quantities is:

$$d_0 = \Delta d \Big/ \left(\sqrt{\frac{y_0}{y_1}} - 1 \right) \qquad (2)$$

This simplified model, however, does not state clearly some of the quantities that are known sources of uncertainty, due to the modelling (stray-light effect, non-lambertian angular distribution, instability of the source and "field of view" limitation of the photometer aperture) and due to the experimental setup conditions (vertical and horizontal angular deviations due to the photometer bench and temperature and humidity effects during testing).
The focus of this study is on the impact that the vertical angular deviation of the photometer bench, θ, has in measurement results, especially those related to uncertainty, with the intention of defining criteria for the assessment of the bench experimental setup.

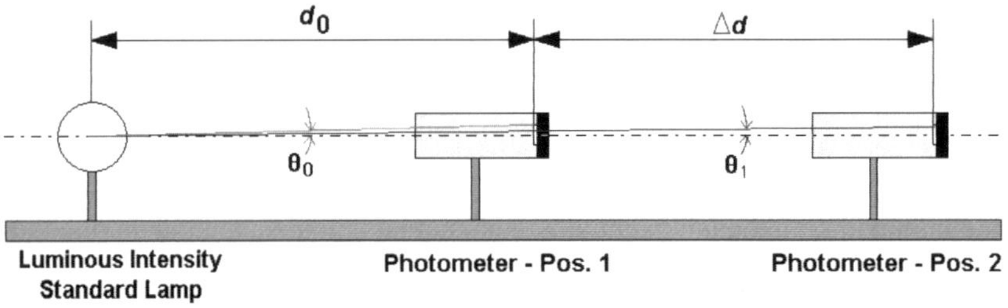

Figure 3 – Angular deviations of the calibration setup.

To study this problem, equation (2) is modified to include the vertical angle variation, θ, between the two distances (θ is the difference between θ_0 and θ_1) yielding the following uncertainty mathematical model:

$$d_0 = \left(\frac{\Delta d}{cos(\theta)} \right) \Big/ \left(\sqrt{\frac{y_0}{y_1}} - 1 \right) \qquad (3)$$

The uncertainty analysis (without vertical angular deviation effect) is provided by [3,5] for a case study, using the ISO-GUM approach. The evaluation of uncertainty performed in this study use the data of that example and evaluates the measurement uncertainty using also a Monte Carlo Method (MCM), considering that the non-linear effects due to the mathematical model might be significant.

3. Measurement Uncertainty of Effective Distance using GUM and Monte Carlo Methods

The case studied was based on an experimental example proposed in [3], including the measurement of the effective distance between the filament of a luminous intensity lamp and the aperture of a photometer and the evaluation of its measurement uncertainty using classical GUM approach, being the results compared with those resulting from the use of a Monte Carlo method approach first using the same mathematical model (2) and secondly using model (3) which includes the contribution of the vertical angular deviation effect, being evaluated its impact for different angular uncertainty values aiming to establish basic knowledge to develop quality control criteria.

The experimental procedure described in [3] considers a source as luminous intensity standard (having nearly Lambertian angular distribution) illuminating a photometer at a distance d_0. In a setup similar to the one shown in Fig. 2, both the offset-corrected photometer signal, y_0, and y_1, obtained after the distance dislocation of Δd, are the average values of sets of 10 independent readings. In these base conditions, equation (2) gives estimates of the effective distance d_0. The experimental data is presented in Table 1.

Journal of Physics: Conference Series **588** (2015) 012002 doi:10.1088/1742-6596/588/1/012002

Table 1 – Input quantities estimates and measurement uncertainty probabilistic information

Quantity	Estimate	Standard uncertainty	PDF* info
Δd	0,500 0 m	0,000 5 m	Type B - Gaussian, $\nu^{**} = \infty$, no correlation.
y_0	8,000 0 V	0,000 8 V	Type A - Gaussian, $\nu^{**} = 9$, no correlation.
y_1	6,000 0 V	0,000 6 V	Type A - Gaussian, $\nu^{**} = 9$, no correlation.

* Probability Distribution Function ** Degrees of Freedom (DoF)

In order to apply GUM approach, again considering the mathematical model (2), sensitivity coefficients need also to be considered, with expressions and estimated values listed below.

$$c_1 = \frac{\partial d_0}{\partial \Delta d} = \frac{1}{\sqrt{\frac{y_0}{y_1}} - 1} = 6{,}464 \tag{4}$$

$$c_2 = \frac{\partial d_0}{\partial y_0} = \frac{\Delta d}{2 \cdot \sqrt{\frac{y_0}{y_1}} \cdot (y_0 + y_1) - 4y_0} = -1{,}508 \, \frac{\mathrm{m}}{\mathrm{V}} \tag{5}$$

$$c_3 = \frac{\partial d_0}{\partial y_1} = \frac{\Delta d \cdot \sqrt{\frac{y_0}{y_1}}}{2y_1 \cdot \left(\sqrt{\frac{y_0}{y_1}} - 1 \right)} = 2{,}010 \, \frac{\mathrm{m}}{\mathrm{V}} \tag{6}$$

Table 2 summarizes all this information as proposed by GUM [4] leading to the standard measurement uncertainty of the effective distance, also found in the same Table.

Table 2 – Uncertainty budget table related to the effective distance measurement

Quantity X_i	Symbol	Value	Absolute Standard uncertainty	Type	DoF	Absolute sensitivity c_i	Absolute contribution $u_i(y)$	Relative contribution $w_i(y)$
Dislocation / m	Δd	0,500 0	0,000 5	B	∞	+ 6,464 1	+ 0,003 2	+ 0,001 0
Initial photometer signal / V	y_0	8,000 0	0,000 8	A	9	- 1,507 8	- 0,001 2	- 0,000 4
Dislocated photometer signal / V	y_1	6,000 0	0,000 6	A	9	+ 2,010 4	+ 0,001 2	+ 0,000 4
Effective distance / m	d_0	**3,232 1**			365		**0,003 6**	0,001 1

The values of the degrees of freedom for the several input quantities were considered using the Welch-Satterthwaite formula [6,7], to obtain the related coverage factor [2] that should multiply the standard measurement uncertainty in order to evaluate the expanded measurement uncertainty of the effective distance estimate.

$$v_{eff} = \frac{u^4(y)}{\sum_{i=1}^{n} \frac{u_i^4(y)}{v_i}} = \frac{0{,}0036^4}{\frac{0{,}0012^4}{9} + \frac{0{,}0012^4}{9}} = 365 \tag{7}$$

$$k(v_{eff} = 365) = 2{,}00 \tag{8}$$

$$U_{95}(d_0) = k \cdot u(d_0) = 2{,}00 \cdot 0{,}0036 \text{ m} = 0{,}0072 \text{ m} \tag{8}$$

To evaluate the effect in uncertainty due to non-linear behaviour of the mathematical model used (2) a Monte Carlo method simulation was performed, considering that it would allow to validate the results obtained using the GUM approach. The numerical simulation developed used the same experimental estimates of the input quantities and the probabilistic information shown in Table 1, for numerical sequences obtained from 10^6 runs each, using MatLab software validated routines (namely to generate pseudo-random numbers sequences with known PDFs and to order the output numerical sequences) and also using the procedure described in [8] to obtain probability intervals from the output PDF.

Table 3 – GUM vs. MCM comparative results

Quantity X_i	Symbol	Approach	Estimated value / m	Expanded measurement uncertainty (95% confidence) / m
Effective distance / m	d_0	GUM	3,232 1	0,007 2
		MCM	3,232 0	0,007 2

The values presented shows that GUM approach is suitable to evaluate effective distance measurement uncertainty in the conditions of the example [3] (all input PDFs are Gaussian), which cannot be generalized if input quantities with different PDF had been considered. The comparison study also showed that the output PDF obtained using MCM has a Gaussian shape (Fig. 4), which differs from the trapezoidal PDF presented in [3]

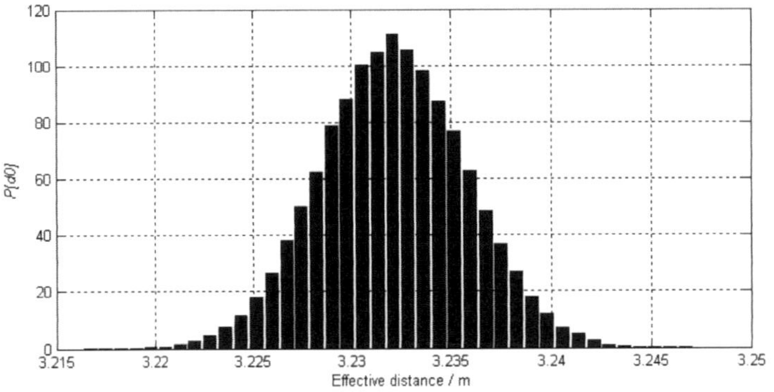

Figure 4 – Effective distance output PDF obtained from MCM study.

4. Analysis of the Effect of Vertical Angular Deviation in the Effective Distance Uncertainty

The second part of the study was intended to evaluate the effect of vertical angular deviation input quantity, in the effective distance output combined uncertainty. Different intervals of uncertainty of vertical angular deviation (from 1° to 5° – using Gaussian PDFs) were tried, now using equation (3). The evaluation of the measurement uncertainty intervals was based on a MCM approach using the same data of the previous analysis. The following table resumes the conditions and results obtained.

Table 4 – Measurement estimates of effective distance and related uncertainty considering different vertical angular deviation uncertainty intervals

Quantity X_i	Symbol	$u(\theta)$ / °	Estimated value / m	Expanded measurement uncertainty (95% confidence) / m
Effective distance / m	d_0	1	3,233	0,007
		3	3,236	0,014
		5	3,245	0,034

One main advantage of using MCM is the access to the output PDF information which, in this case, is particularly informative. In fact, looking for the PDFs presented in the following Figures, is possible to see that although the vertical angle deviation (θ) has Gaussian PDF, the distributions of cos(θ) – Figures on the left – are highly asymmetric leading to the need of knowing how they influence output PDFs (Figures on the right).

In fact, Table 4 shows a deviation of the estimates and an increase of uncertainty due to this quantity, reaching about 1% of the output estimate. This evaluation allows the user to establish tolerance limits for the vertical angular deviation observed on the experimental bench metrological characterization, according to the tolerances required for the experimental setup (usually, from 2% to 5 %).

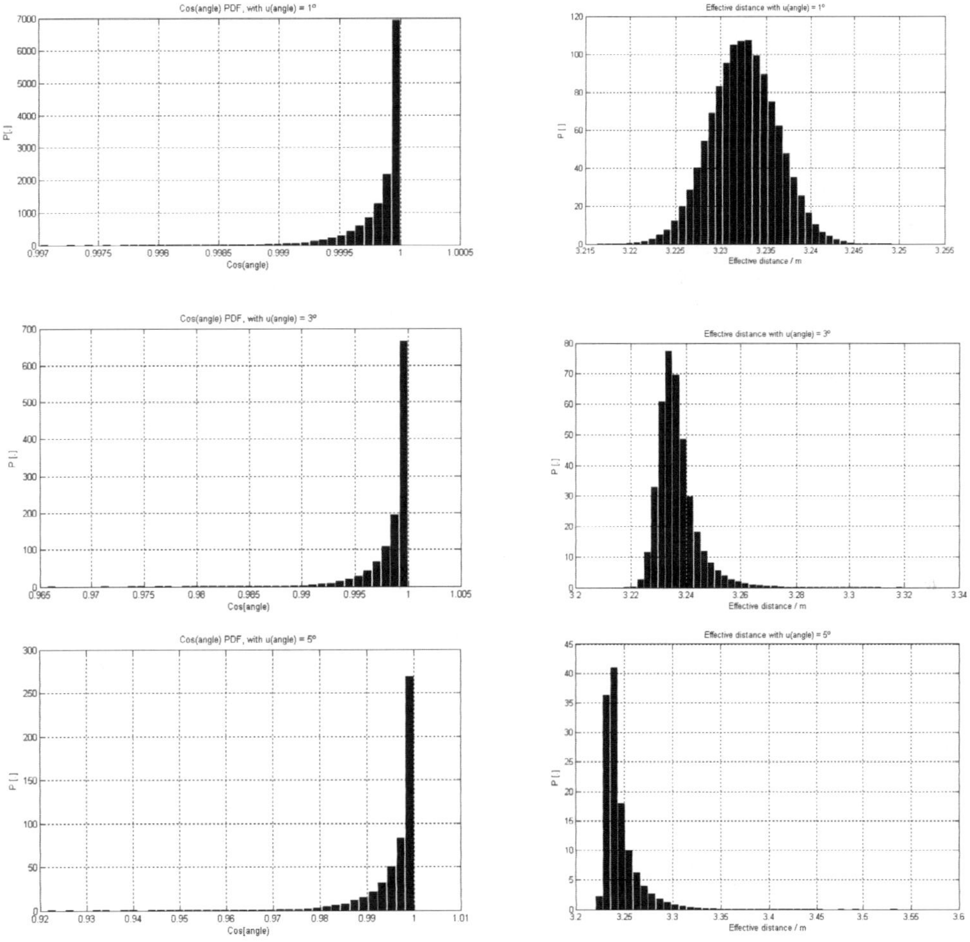

Figure 5 – Input PDFs of cos(θ) and related output PDFs (effective distance) obtained for vertical angle deviation (θ) measurement uncertainties of 1°, 3° and 5°.

5. Conclusions

The progress of metrology techniques to study known measurement problems permits the access to relevant information that improves our knowledge of the quality of measurement. In the present case, regarding the calibration of photometers by the evaluation of effective distance, the output measurement quantity can be strongly affected by the setup that supports the measurement - the experimental bench - being, therefore, relevant to establish quality assessment criteria suitable to the requirements and tolerances to perform this task with confidence.

From the several geometric influence quantities that must be considered, the vertical angular deviation is certainly a very important one, which motivated this work. The first study aimed to use an MCM approach to validate measurement uncertainty results obtained using the GUM conventional approach, considering that the mathematical model used could be affected by its non-linear nature, with the conclusion of suitability on the latter for input quantities having Gaussian PDFs (according with the example given in [3]).

The second study aimed to evaluate how measurement uncertainty due to vertical angular deviation could affect both estimate and measurement uncertainty of the output quantity. In this case, the results shows an asymmetric influence not relevant for lower uncertainty values of this input quantity ($u(\theta)=1°$) but to be considered for higher uncertainty values (reaching 1% of uncertainty of the output quantity). These results imply the need for the user to promote a geometrical characterization of the calibration bench in order to evaluate deviations along the longitudinal support for the displacement of the photometer to be calibrated.

Finally, it is advisable that, for calibration purposes, the user should establish criteria for the quality assessment of vertical angular deviations for photometer calibration benches, based on the uncertainty evaluation and the expected accuracy of illuminance and effective distance measurements.

This study is intended to be further developed in order to establish a broader analysis of the uncertainty contributions due to other geometrical and environmental quantities aiming to establish a generalized quality assessment of this type of testing setups.

References and Bibliography

[1] International Commission of Illumination (CIE) - *International Lighting Vocabulary*. Vienna: CIE, 2011. Standard CIE S 017/E:2011.

[2] International Commission of Illumination (CIE) - *Characterization of the Performance of Illuminance Meters and Luminance Meters.* Vienna: CIE, 2013. Standard CIE S 023/E:2013

[3] International Commission of Illumination (CIE) - *Determination of Measurement Uncertainties in Photometry- Supplement 1: Modules and Examples for the Determination of Measurement Uncertainties. Part 1: Modules for the Construction of Measurement Equations and Part 2: Examples for Models with Individual Inputs.* Vienna: CIE, 2011. CIE 198:2011; CIE 198-SP1.1:2011; CIE 198-SP1.2:2011.

[4] BIPM, IEC, IFCC, ISO, IUPAC, IUPAP and OIML. Guide to the Expression of Uncertainty in Measurement (Geneva, Switzerland: International Organization for Standardization, 1995), ISBN 92-67-10188-9.

[5] International Commission of Illumination (CIE) - *Determination of Measurement Uncertainties in Photometry*. Vienna: CIE, 2011. CIE 198:2011

[6] Satterthwaite, F. E., (1946) *Biometrics Bull.* **2**(6), 110-114.

[7] Welch, B. L., (1947) *Biometrika*. **34**, 28-35.

[8] M. G. Cox and P. M. Harris. Software Support for Metrology Best Practice Guide No. 6: Uncertainty Evaluation. NPL Report DEM-ES-011, National Physical Laboratory, Teddington, UK (2006).

Pressure coefficient evaluation on the surface of the SONDA III model tested in the TTP Pilot Transonic Wind Tunnel

M L C C Reis[1], J B P Falcao Filho[1], E Basso[1] and V R Caldas[2]

[1]Institute of Aeronautics and Space
Pr. Mal. Eduardo Gomes, 50, CEP 12228904, São José dos Campos, SP, Brazil
[2]University of Taubate
Av. 9 de Julho, 199, CEP 12020-200, Taubate, SP, Brasil

E-mail: marialuisamlccr@iae.cta.br

Abstract. A test campaign of the Brazilian sounding rocket Sonda III was carried out at the Pilot Transonic Wind Tunnel, TTP. The aim of the campaign was to investigate aerodynamic phenomena taking place at the connection region of the first and second stages. Shock and expansion waves are expected at this location causing high gradients in airflow properties around the vehicle. Pressure taps located on the surface of a Sonda III half model measure local static pressures. Other measured parameters were freestream static and total pressures of the airflow. Estimated parameters were pressure coefficients and Mach numbers. Uncertainties associated with the estimated parameters were calculated by employing the Law of Propagation of Uncertainty and the Monte Carlo method. It was found that both uncertainty evaluation methods resulted in similar values. A Computational Fluid Dynamics simulation code was elaborated to help understand the changes in the flow field properties caused by the disturbances.

1. Introduction

Scale models of aerospace vehicles are tested in wind tunnels in order to predict the performance of their full scale counterparts during actual flight.

This study presents the results of a test campaign of a half-model of the SONDA III sounding vehicle carried out at the Brazilian Pilot Transonic Facility, TTP.

The purpose of the tests is to evaluate changes in the airflow field properties due to the presence of shock and expansion waves in the inter-stage region of the vehicle.

The evaluated parameter is the pressure coefficient ($-C_p$). Measured parameters are the total airflow temperature, T_0, total and static pressures, p_0 and p_∞, and static pressure, p, taken at pressure taps located at 22 stations distributed longitudinally along the model surface.

Uncertainties in the measured parameters are evaluated and propagated to the pressure coefficient by using both the Law of Propagation of Uncertainty and the Monte Carlo method. Mach number of the airflow, M, and associated uncertainty are also evaluated.

Experimental data are compared to numerical data obtained throughout Computational Fluid Dynamics, CFD.

[1] To whom any correspondence should be addressed.

Content from this work may be used under the terms of the Creative Commons Attribution 3.0 licence. Any further distribution of this work must maintain attribution to the author(s) and the title of the work, journal citation and DOI.
Published under licence by IOP Publishing Ltd

1.1.1. The sounding rocket SONDA III

The aerospace vehicle SONDA III (figure 1) is a bi-stage vehicle developed by the Institute of Aeronautics and Space, IAE. It is one of the sounding rocket family named Sonda, which started with Sonda I, first launched in 1965. In 1996, the first stage of Sonda III was evolved to receive European experiments onboard. This new single stage vehicle was known as VS-30. A boosted version, the VSB-30, which contains the S31 booster motor, was developed in 2001.

This bi-stage geometric conception has been utilized in many aerospace designs and a better understanding of the aerodynamic phenomena taking place around the vehicle is important for future design initiatives.

Figure 1: The Sounding Rocket Sonda III.

With this purpose, a half-model of the vehicle was tested in the TTP wind tunnel in order to study the flow field in the inter-stage region, where shock and expansion waves are expected. Shock waves arise on the vehicle surface when the airflow becomes locally supersonic. The flow properties change through the shock wave and undesirable effects which compromise the flight performance may occur. Therefore, an investigation of the behaviour of shock and expansion waves in wind tunnels is important to characterize their interactions with the model. The nominal Mach number values covered in the tests varied from $M = 0.20$ to 1.10.

1.2. Shock and expansion waves

When a rocket vehicle is in a free flight condition at high velocity, the flow may become locally supersonic over the surface of the vehicle [1].

The freestream Mach number at which $M = 1$ is first achieved on the surface is called the critical Mach number.

In supersonic regimes, oblique shock waves occur when supersonic flow encounters a concave corner and it is attached to the corner as long as the concave corner angle is below a defined limit. Above this limit a more complex detached shock wave occurs. In contrast, if the supersonic flow encounters a convex corner, expansion waves occur. Flow field properties change across shock and expansion waves, as shown in figures 2a and 2b.

Tables and diagrams relating the angle of the corner (θ), the angle of the shock (β), and the Mach number (M) can be found in [2].

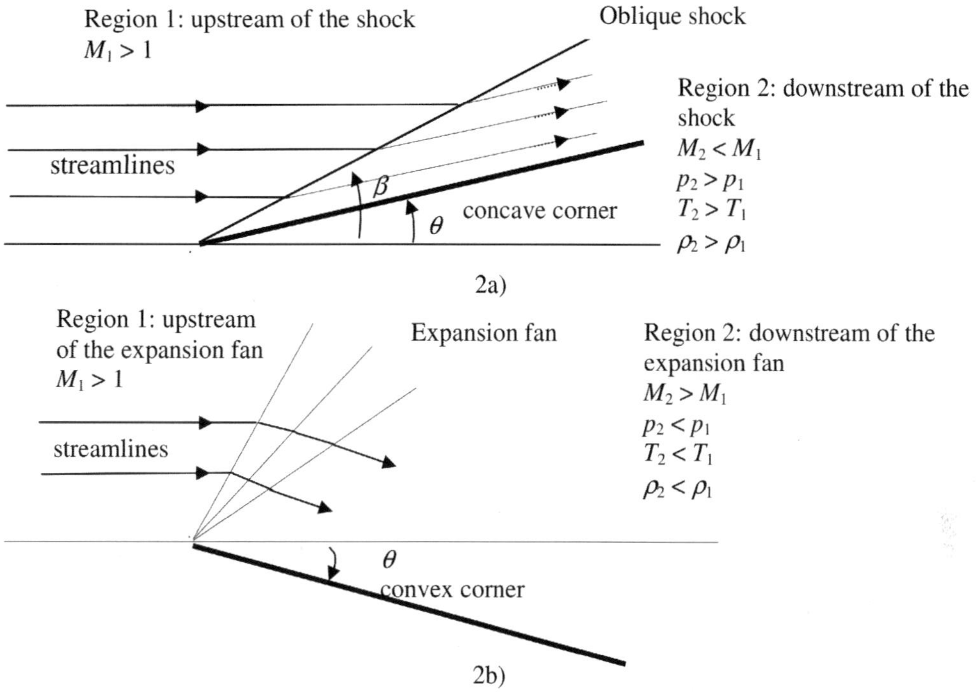

Figure 2: Deflection of a supersonic flow in the presence of: (a) concave corner, (b) convex corner. *M*: Mach number. *p*, *T*, and *ρ*: static pressure, temperature and density, respectively.

1.3. The TTP wind tunnel

The Pilot Transonic Wind Tunnel, TTP, is located at the Aerodynamics Division of the Institute of Aeronautics and Space, Brazil (figure 3). The tunnel is continuously driven by an 830 kW main axial compressor and can also be operated intermittently by means of an injection system, which supplies airflow for around 30 seconds. It is a variable-pressure wind tunnel with control capability to independently vary Mach number, stagnation pressure, stagnation temperature and humidity. The test section is 0.25 m high and 0.30 m wide and has longitudinally slotted walls to favor the uniformity of the airflow. At TTP, the Mach number range can vary from 0.20 to 1.3. Some configurations can be changed in order to allow better flow control such as reentry flaps position and the rate of forced mass extraction, determined by the Plenum Evacuation System, PES.

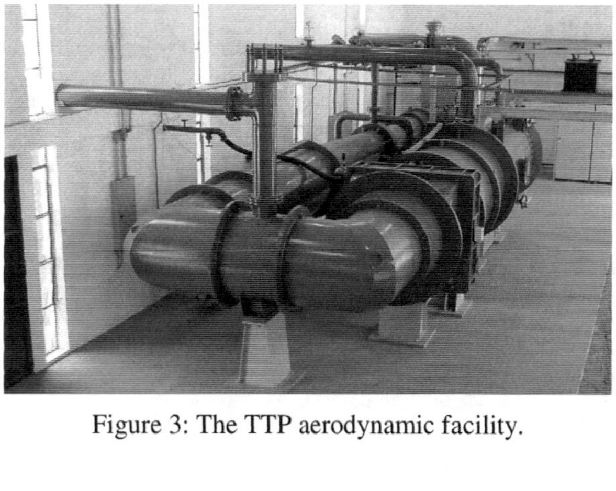

Figure 3: The TTP aerodynamic facility.

1.4. Description of the Model

Figure 4 shows a 1:8 scale model conception, mounted on an Aluminum platform which will be fixed to the lateral wall of the tunnel test section. In figure 4 the model is configured with one fin perpendicular to the tunnel wall. Threads made in special places allow the model to change the direction in relation to the platform in order to adjust angles of attack of 0°, 5.5°, 8°, 12°, 13° and 15.5°. Between the half-model and the platform special spacers can be installed to create a gap between the tunnel wall and the model. This way the possible boundary layer formed on the tunnel wall can be eliminated and thus better represents the free flight situation.
The total length of the model is 0.778 m and the diameter of the first stage is 0.070 m.

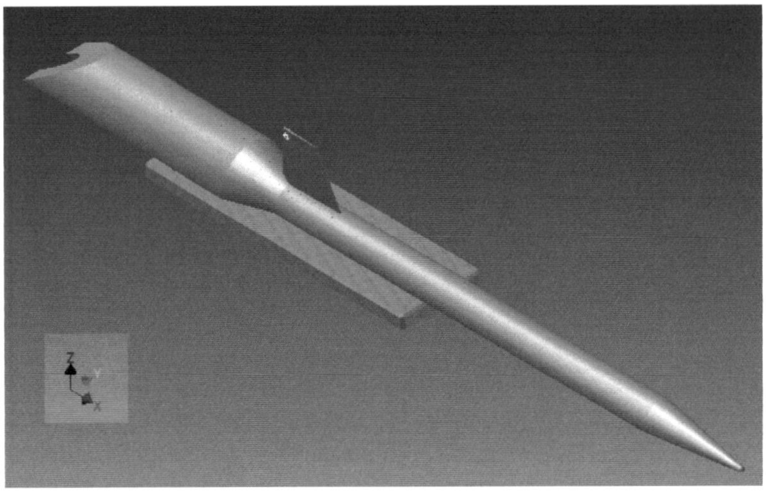

Figure 4: Model conception of Sonda III with scale 1:8 mounted on an Aluminum platform.

The Sonda III model with details of plastic tubes is presented in figure 5. The tubing connects the pressure taps and pressure instrumentation. There are a total of 154 measuring pressure taps located in 22 positions along the model surface. The positions are identified by letters **A** to **V** and are distributed in a region encompassing the inter-stage frustum cone (figure 6). The classification is in alphabetic order, departing from the base of the first stage in the direction of the nose of the second stage. Taps **A** to **H** belong to the first stage, this latter defining the limit border of the connection between both stages. The connection region has a frustum cone shape, with taps **I**, **J**, **K** and **L**. The remaining pressure taps, **M** to **V** are located in the second stage of the model.

Figure 5: Sonda III model with detail of the tube connections.

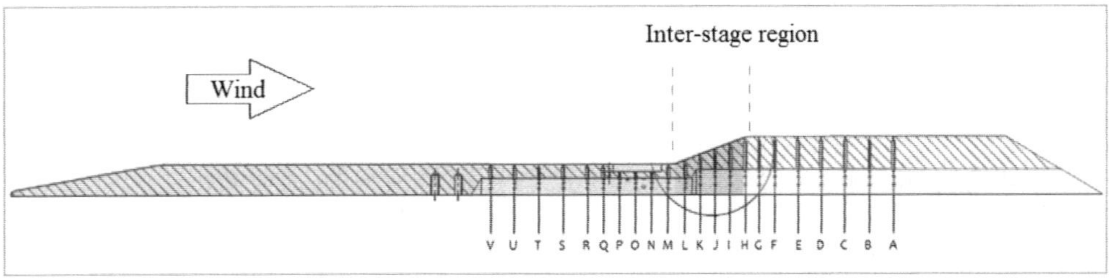

Figure 6: Pressure tap positions around the inter-stage region.

2. Metodology

This study describes results obtained with the half-model installed on the test section lateral wall of the tunnel. The aim of the wind tunnel tests is to determine the pressure distribution on the model surface. This section describes the data reduction of the estimated parameters. Firstly, the mathematical models of the estimated quantities, which are the negative pressure coefficient, $-C_p$, and the freestream Mach number of the airflow, M_∞, are presented. Afterwards, the Law of Propagation of Uncertainty is applied to the mathematical models of both parameters, to estimate the associated uncertainties. Information about the Monte Carlo implementation is then given, such as the number of trials and probability distributions of the measured quantities. The section ends with a description of some properties of the Computational Fluid Dynamics code employed in the numerical simulation.

2.1. Pressure coefficient

The pressure distribution on the model surface can be represented in terms of the pressure coefficient defined by [2]:

$$C_p \equiv \frac{p - p_\infty}{q_\infty} \tag{1}$$

where p is the local pressure measured on each pressure tap station of the model, p_∞ is the freestream static pressure and q_∞ is the freestream dynamic pressure. The static pressure sensor used to measure p_∞ is positioned on the upper part of the wall at the beginning of the wind tunnel test section. The freestream dynamic pressure, q_∞, is defined by:

$$q_\infty \equiv \frac{1}{2}\rho_\infty V_\infty^2 \tag{2}$$

where ρ_∞ and V_∞ are the freestream density and velocity, respectively.
In this study, the air is considered as a perfect gas. Its density is calculated by:

$$\rho_\infty = \frac{p_\infty}{RT_\infty} \tag{3}$$

p_∞: freestream pressure in pascal;
R: perfect gas constant, equal to 287 J/(kg.K) for normal air; and
T_∞: freestream temperature expressed in kelvin.

The airflow Mach number, M, is the ratio between velocity, V, and the speed of sound, a:

$$M = V/a \tag{4}$$

In equation (4), a represents the speed of sound travelling through the air, considered in this study as a perfect gas. Its value is estimated by:

$$a = \sqrt{\gamma RT} \tag{5}$$

where
T: temperature expressed in kelvin; and
γ: ratio of specific heats, equal to 1.4 for air considered as a perfect gas.

Rearranging equation (1) by using (2), (3), (4) and (5), results in:

$$-C_P = \frac{p_\infty - p}{\frac{1}{2}\gamma p_\infty M_\infty^2} \tag{6}$$

In regions of the flow where $p > p_\infty$, the pressure coefficient $-C_p$ will be a negative value.
Considering the flow in the test section as compressible flow, the freestream Mach number, *i. e.*, the Mach number at undisturbed condition (before reaching the model) is given by:

$$M_\infty^2 = \frac{2}{\gamma - 1}\left[\left(\frac{p_0}{p_\infty}\right)^{\frac{(\gamma-1)}{\gamma}} - 1\right] \tag{7}$$

where p_0 is the freestream total pressure.

As already mentioned, a static pressure tap is located on the upper wall of the test section and connected to a pressure sensor to measure p_∞. The total pressure sensor used to measure p_0 is located in the stilling chamber of the circuit.

2.2. Law of propagation of uncertainty
According to reference [3], the value of the uncertainty in measurement is the positive square root of equation (8):

$$u_c^2 = \sum_{i=1}^{N}\left(\frac{\partial y}{\partial x_i}\right)^2 u^2(x_i) \tag{8}$$

where y is the output quantity and $x_{i's}$ are the input quantity. Equation (8) is called the Law of Propagation of Uncertainty and its evaluation is known as the GUM uncertainty framework [4].
Applying equation (8) to (6) results in the values of the uncertainty in the pressure coefficient:

$$u_{-C_p}^2 = \left(\frac{\partial C_p}{\partial p}\right)^2 u_p^2 + \left(\frac{\partial C_p}{\partial p_\infty}\right)^2 u_{p_\infty}^2 + \left(\frac{\partial C_p}{\partial M_\infty}\right)^2 u_{M_\infty}^2 \tag{9}$$

which leads to:

$$u_{-C_p}^2 = \left(\frac{-1}{\frac{\gamma}{2} p_\infty M_\infty^2} \right)^2 u_p^2 + \left(\frac{p}{\frac{\gamma}{2} p_\infty^2 M_\infty^2} \right)^2 u_{p_\infty}^2 + \left(\frac{2(p_\infty - p)}{\frac{\gamma}{2} p_\infty M_\infty^3} \right)^2 u_{M_\infty}^2 \qquad (10)$$

In equation (10), the uncertainties in the static pressure, u_p, and in the freestream pressure, $u_{p\infty}$, are estimated combining two components: the uncertainty declared in the calibration certificate of pressure instrumentation and the uncertainty evaluated as type A, quantified by calculating the standard deviation of the temporal pressure signal. To evaluate the uncertainty in the freestream Mach number, one applies the Law of Propagation of Uncertainty to equation (7):

$$u_{M_\infty}^2 = \left(\frac{\partial M_\infty}{\partial p_0} \right)^2 u_{p_0}^2 + \left(\frac{\partial M_\infty}{\partial p_\infty} \right)^2 u_{p_\infty}^2 \qquad (11)$$

which results in:

$$u_{M_\infty}^2 = \left(\frac{0.714285}{M_\infty} \right)^2 \left[\frac{1}{p_\infty^2} \left(\frac{p_\infty}{p_0} \right)^{1.428571} u_{p_0}^2 + \left(\frac{-p_0}{p_\infty^2} \right)^2 \left(\frac{p_\infty}{p_0} \right)^{1.428571} u_{p_\infty}^2 \right] \qquad (12)$$

2.3. Monte Carlo method

Reference [4] was used to supply Mach number and pressure coefficients values and associated uncertainties by using the Monte Carlo method.

The input quantities for the freestream Mach number, M_∞, are the total pressure, p_0, and the freestream static pressure, p_∞ (equation 7).

The input quantities for the pressure coefficient, $-C_p$, at each pressure station on the half SONDA III model, include M_∞ itself and the local static pressure p taken at pressure taps **A** to **V**, along with p_0 and p_∞ of the airflow (equation 6).

The mean value and standard deviation of the pressure signals supplied by the instruments during the runs of the wind tunnel tests are computed and this information is used to evaluate the measurement uncertainty by using the Monte Carlo method. The probability distributions for the input quantities are propagated through the measurement models expressed by equations (6) and (7), resulting in the estimated output quantities $-C_p$ and M_∞ and the associated standard uncertainties, u_{-Cp} and $u_{M\infty}$ (equations 10 and 12).

Codes in MatLab® were developed for the implementation of the Monte Carlo method. Gaussian distributions were assigned for all input quantities. The selected number of trials was 50,000.

2.4. Computational Fluid Dynamics code

The numerical code used to solve Reynolds-Average Navier-Stokes equations was the CFD++® solver. The wind tunnel walls were not considered in the simulation therefore the vehicle is in free flight condition. In this study, the model representation is confined to a dihedral angle of 10°, with origin at the symmetric axis of the model fuselage. This is possible because the simulated model is without fins and is positioned at null angle of attack. These conditions lead to axisymmetric flow and it is possible to limit the computational field. Pre-processing was carried out by using ANSYS ICEM® CFD meshing software. The airflow is considered viscous and the turbulence model is the Spalart Allmaras [5]. The resulting y+ value is less than 1, adequate for the investigation of boundary layer effects. Meshing is composed of hexahedrons and prisms. The code application for this kind of rocket geometry was validated by using a cone-cylinder model.

3. Results and discussion

The test campaign to analyse the pressure coefficient distribution on the Sonda III half-model surface carried out at the transonic wind tunnel TTP considered the nominal Mach numbers 0.20, 0.30, 0.40, 0.50, 0.60, 0.70, 0.80, 0.85, 0.90, 0.95, 1.00, 1.05 and 1.10. The angle of attack of the model is 0°.

The pressure signals obtained for total pressure, p_0, and freestream static pressure, p_∞, as a function of time, are shown in figures 7a and 7b respectively. The presented signals were obtained for nominal Mach number, M, equal to 0.95.

When the desired regime settles down, an interval of the pressure signals is chosen and the average and standard deviation are calculated. The interval between 40 s and 120 s was chosen for $M = 0.95$, and the transient of the signal discarded. The transient is caused by the wind tunnel drier system, which controls the humidity of the airflow. The ratio p_0/p_∞ of average values of total and static pressures is computed. This ratio is included in equation (7) to supply the freestream Mach number value, M_∞. The same procedure is performed for static pressure values in taps **A** to **V** in order to obtain pressure coefficients (equation 6).

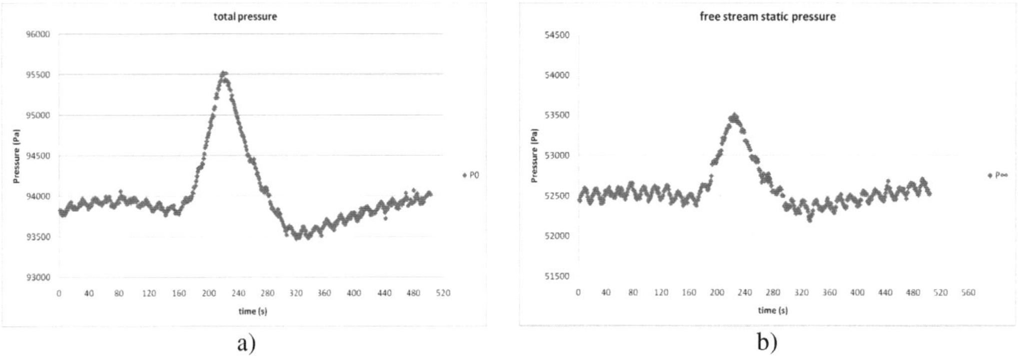

Figure 7: Temporal pressure signals for: (a) total pressure, (b) freestream static pressure.

In Figure 8 one can see the local Mach number at each pressure tap of the model, for all regimes covered by the wind tunnel tests. The origin of the x-axis coincides with the beginning of the frustum cone and is located between taps **L** and **M**. For nominal Mach number 0.80 there is a region on the model surface where the local Mach number is above 1, indicating that the critical Mach number is between regimes 0.70 and 0.80.

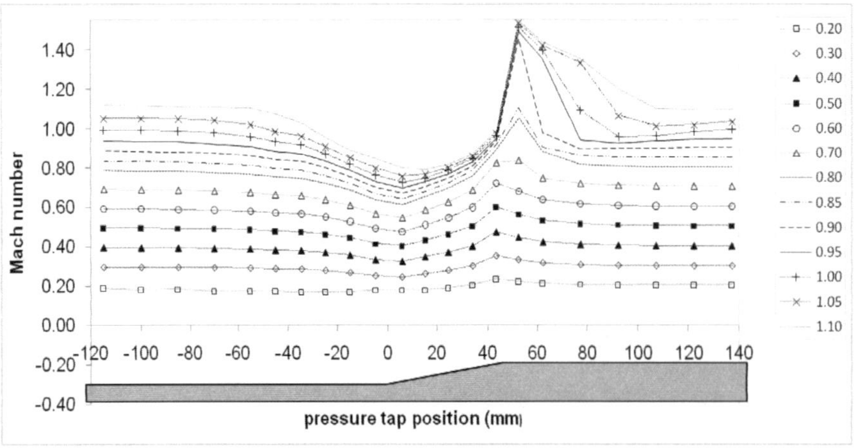

Figure 8: Mach number at pressure taps **A** to **V**.

Figure 9 and table 1 present the pressure coefficient values for nominal Mach number regime 0.90, obtained at pressure taps positions **A** to **V** of the SONDA III model, by using equation (6). The airflow direction is from left to right in the picture. Uncertainties estimated by using the GUM approach (equation 10) are compared to the Monte Carlo results in table 1. No differences were found when comparing both methods.
Pressure coefficient data are presented in table 2 for regimes 0.60 and 1.10. Only Monte Carlo results are shown.

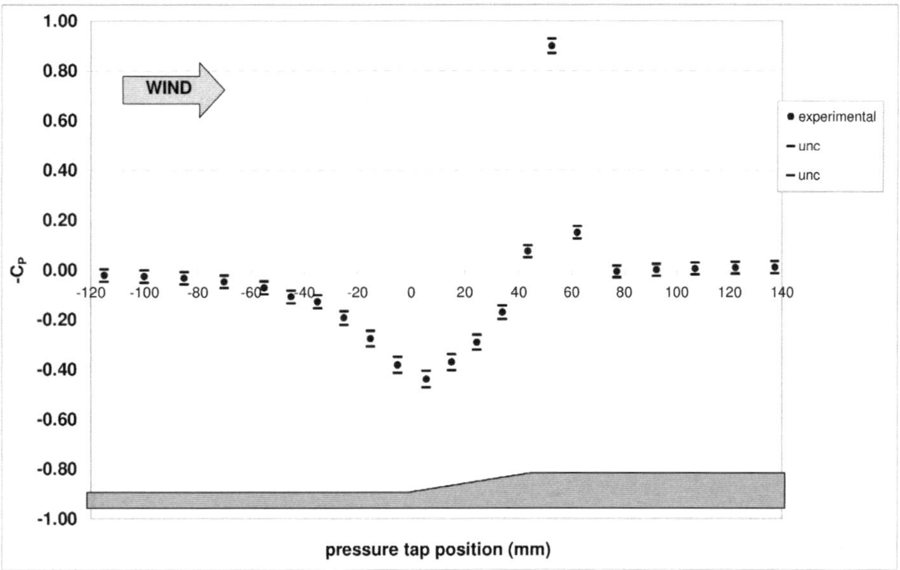

Figure 9: Pressure coefficient and associated uncertainty at pressure tap locations. $M_\infty = 0.90$.

Table 1: Pressure coefficients and associated uncertainties (values were multiplied by10). $M_\infty = 0.90$.

Pressure tap	**A**	**B**	**C**	**D**	**E**	**F**	**G**	**H**	**I**	**J**	**K**
$-C_p$ GUM	0.11	0.09	0.05	0.00	-0.07	1.50	8.99	0.75	-1.72	-2.92	-3.72
u_{-C_p} GUM	0.24	0.24	0.24	0.24	0.24	0.25	0.29	0.24	0.28	0.30	0.32
$-C_p$ MC	0.11	0.09	0.05	0.00	-0.07	1.50	8.99	0.75	-1.72	-2.92	-3.72
u_{-C_p} MC	0.24	0.24	0.24	0.24	0.24	0.25	0.29	0.24	0.28	0.30	0.32
Pressure tap	**L**	**M**	**N**	**O**	**P**	**Q**	**R**	**S**	**T**	**U**	**V**
$-C_p$ GUM	-4.39	-3.82	-2.77	-1.94	-1.29	-1.09	-0.72	-0.47	-0.33	-0.26	-0.22
u_{-C_p} GUM	0.34	0.32	0.31	0.27	0.26	0.26	0.26	0.25	0.25	0.25	0.25
$-C_p$ MC	-4.39	-3.82	-2.77	-1.94	-1.29	-1.09	-0.72	-0.47	-0.33	-0.26	-0.22
u_{-C_p} MC	0.34	0.32	0.31	0.27	0.26	0.26	0.26	0.25	0.25	0.25	0.25

Table 2: Comparison of pressure coefficients and associated uncertainties for subsonic and supersonic regimes.

Pressure tap	**A**	**B**	**C**	**D**	**E**	**F**	**G**	**H**	**I**	**J**	**K**
					$M = 0.60$						
$-C_p$	0.16	0.18	0.21	0.30	0.54	1.19	2.54	3.89	-0.05	-1.66	-2.69
u_{-C_p}	0.19	0.19	0.19	0.19	0.19	0.18	0.18	0.19	0.19	0.20	0.21
Pressure tap	**L**	**M**	**N**	**O**	**P**	**Q**	**R**	**S**	**T**	**U**	**V**
$-C_p$	-3.69	-3.27	-2.21	-1.53	-1.03	-0.92	-0.63	-0.45	-0.34	-0.29	-0.25
u_{-C_p}	0.23	0.22	0.23	0.20	0.19	0.19	0.19	0.19	0.19	0.19	0.19
					$M = 1.10$						
Pressure tap	**A**	**B**	**C**	**D**	**E**	**F**	**G**	**H**	**I**	**J**	**K**
$-C_p$	0.00	-0.05	0.02	-0.62	3.45	4.36	5.45	-1.66	-3.55	-4.45	-4.86
u_{-C_p}	0.06	0.06	0.05	0.19	0.04	0.04	0.05	0.07	0.08	0.08	0.15
Pressure tap	**L**	**M**	**N**	**O**	**P**	**Q**	**R**	**S**	**T**	**U**	**V**
$-C_p$	-4.70	-3.95	-3.22	-2.19	-1.04	-0.35	0.13	0.16	0.18	0.27	0.35
u_{-C_p}	0.17	0.13	0.13	0.08	0.09	0.08	0.06	0.05	0.07	0.06	0.06

A quantitative comparison between Computation Fluid Dynamics results and experimental data is plotted in figure 10 for nominal Mach number 0.90. The numerical and experimental pressure distributions agree to a certain extent, but considerable differences occur at points **E** and **F**. The departure from experimental data reveals that the simulation was not adequately capable of capturing details of aerodynamic phenomena occurring in this region. Uncertainty limits are not sufficient to explain the discrepancies between experimental and simulation data.

Differences may have been caused due to the fact that the simulation considers the entire model, not the half-model attached to the wind tunnel wall. Therefore, the simulation is for a vehicle traveling in free flight. Half-model computational simulation would be desirable.

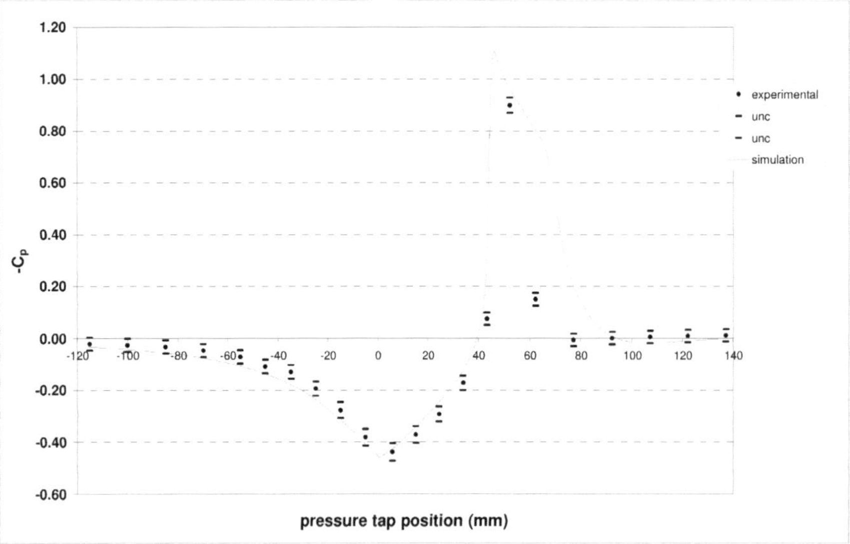

Figure 10: CFD and experimental results for freestream Mach number equal to 0.90.

Pressure coefficients obtained for all Mach number regimes are plotted together in figure 11. Uncertainty values were omitted for clarity. Vertical axis represents $-C_p$. A schematic representation of the model region where the measurements were taken was also drawn at the bottom of the graph, highlighting the initial and the final sections of the frustum cone.

Following the wind direction in figure 11, in the region of pressure taps of the second stage identified by **T**, **U** and **V**, the static pressure is very close to p_∞, resulting in near zero values of $-C_p$ (equation 6). This fact reveals that aerodynamic disturbances were not relevant at these points. Nevertheless, one can note a different behavior between subsonic and supersonic airflows, *i. e.*, supersonic Mach number pressure distributions are apart from the subsonic ones. This also occurs after the frustum cone, in positions **A**, **B** and **C**.

Enlarged versions of these regions are shown in figure 12.

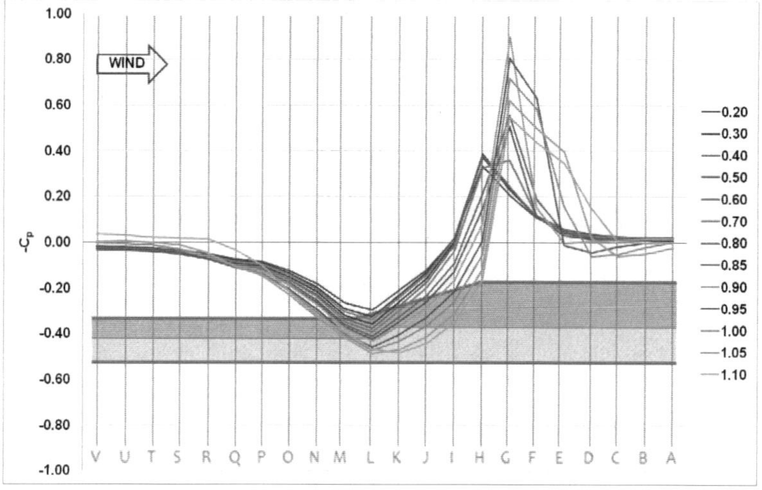

Figure 11: Pressure coefficient distribution on model surface. M_∞= 0.20 to 1.10.

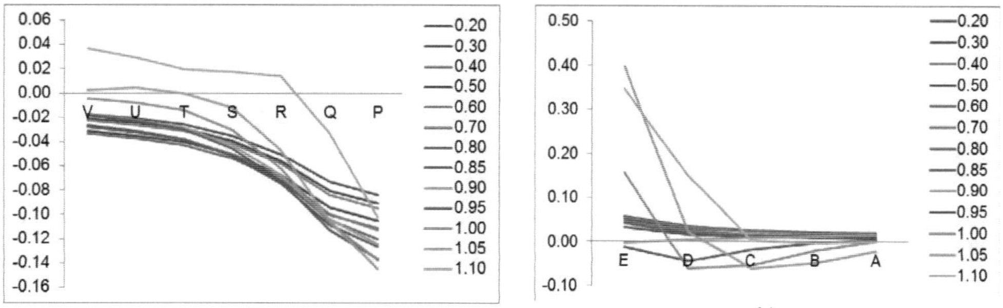

a) b)

Figure 12: Regions presenting pressure coefficient values near zero: (a) second stage, (b) first stage.

To better visualize the effects of the concave and convex corners in the inter-stage region and the presence of shock and expansion waves, graphics are ordered in increasing values of Mach number in figure 13.

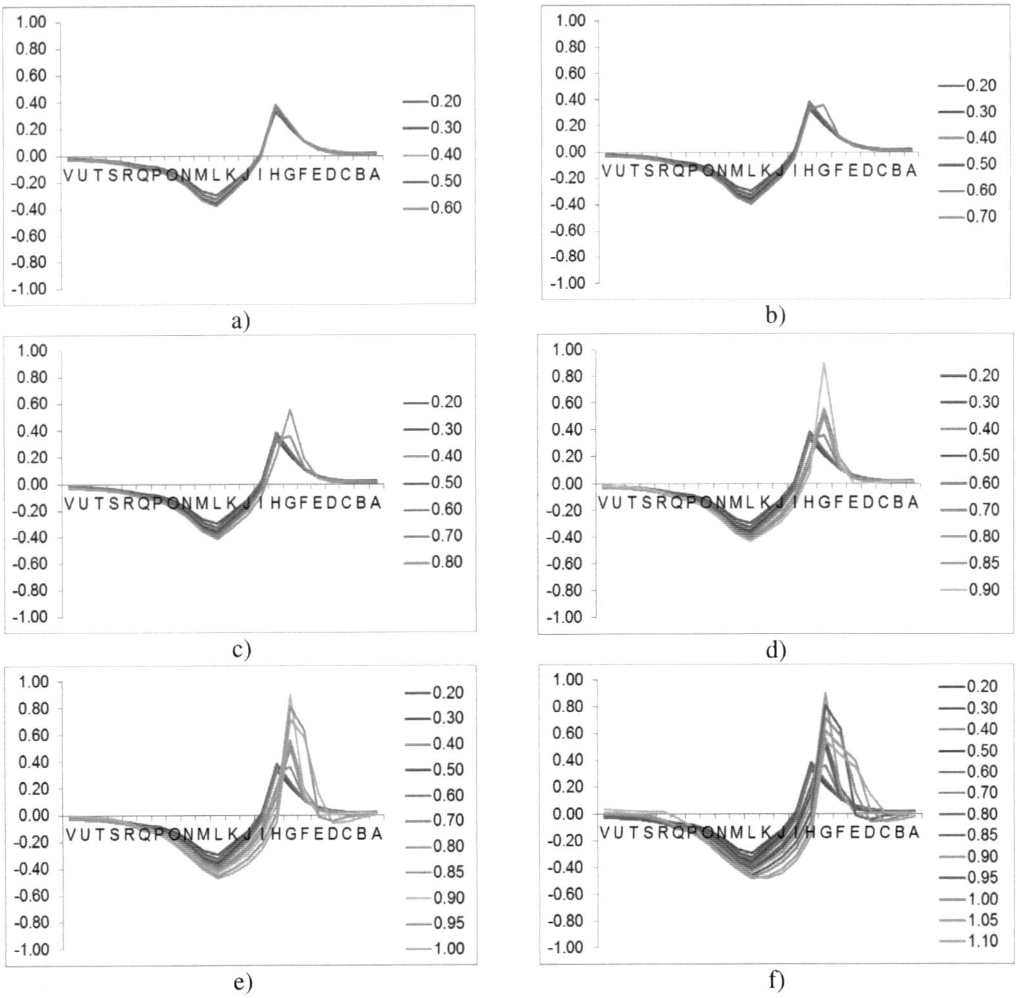

Figure 13: Comparison of $-C_p$ distributions for increasing Mach number.

One can see similar pressure distribution patterns for the subsonic ranges in figure 13a for Mach number up to 0.60. As a result of the disturbance caused by the concave corner around point **L**, the airflow static pressure starts to increase and reaches its maximum value at this point. From point **L** to **I**, which are located in the frustum region, the airflow local static pressure continuously decreases. This tendance remains up to pressure tap **H**, giving the minimum pressure value. From this point on, the pressure starts to recuperate the freestream condition.

The pattern is modified from figure 13b on. The local pressures after the area expansion change in such a way that the pressure coefficient distribution becomes more complex. From figure 13c on, the regime is supercritical. In figures 13c and 13d, it is possible to observe the successive growth of $-C_p$ due to the expansion region around the convex corner at point **H**. The difference (p_∞-p) becomes more positive for Mach number equal to 0.90 and $-C_p$ achieves maximum value. The $-C_p$ did not continue to increase, as shown in figure 13e.

At freestream supersonic flow condition, a shock wave would be expected in the region of the pressure tap **L** due to the concave corner if the relation θ-β-M (angle of the corner, angle of the shock and Mach number) was adequate. The local static pressure behind the shock increases in relation to p_∞, resulting in negative values for $-C_p$. For Mach numbers 1.05 and 1.10, the airflow presents this

behavior even before reaching the concave region (see figures 12a and 13f). Apparently, the shock wave is detached due to the high angle of the frustum cone, and negative values start to occur earlier, characterizing a local subsonic regime.
Simulation by employing computational fluid dynamics was done for the Mach number regime 0.90. Some properties of the numerical code have already been mentioned in section 2.4. A diagram of the whole computational field defined by a dihedral angle of 10° is shown in figure 14, highlighting the central plan. The model is positioned at the botton right and the distance is defined from the begining of the first stage of the model.

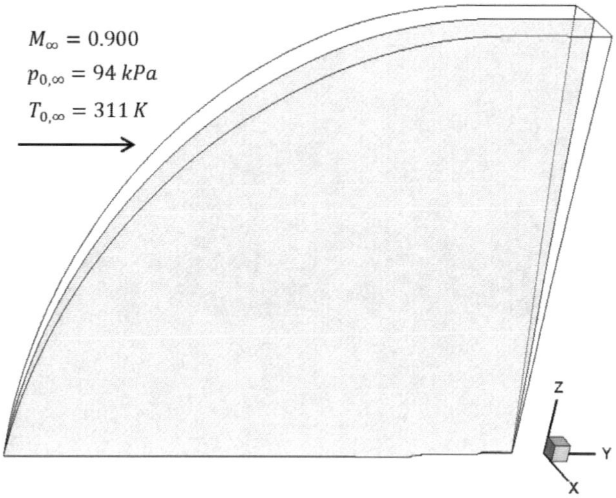

Figure 14: Diagram of the airflow field; x-axis unit: meter.

The computational grid topology around the vehicle for the central plane is shown in figure 15a. Note the grid clustering where a better anlysis of areodynamic phenomena is desired, *i. e.*, inter-stage region, nose and next to the surface. A detailed view of the point clustering at the inter-stage part can be seen in figure 15b. The very closely spaced grid in the vicinity of the model surface is necessary in order to accuratelly capture the boundary layer effects, according to the requirement ($y+ < 1$) of the Spalart Allmaras turbulence model [5]. There are 10 circumferentially and uniformly spaced planes (1° between planes). The whole field has 482,760 calculation cells.

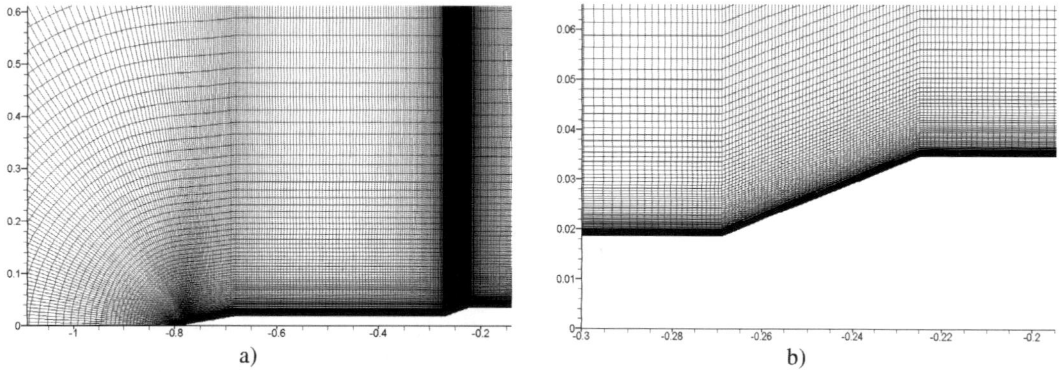

Figure 15: a) Computational meshing around the vehicle. b) A zoom view of the inter-stage region.

Figure 16 summarizes the phenomena which occur around the model at an angle of attack equal to 0° in Mach number regime 0.90. The airflow parameter shown is static pressure. The freestream pressure is equal to 55,580 Pa. The flow is from left to right. One can observe the static pressure increasing before the frustum cone and the pressure decreasing after this region. After passing through the cone, expansion waves are present followed by a shock wave which is identified by the coalescing of isobaric curves. This shock wave interacts with the boundary layer in the vicinity of the model surface resulting in shock thickening. The boundary layer grows after the passage of the shock wave, because of the adverse pressure gradient. This fact is better seen in the Mach number field shown in figure 17. The supersonic region defined by the line corresponding to Mach number 1.00 starts almost at the end of the frustum cone and has an extension of 0.37 times the diameter of the first stage.

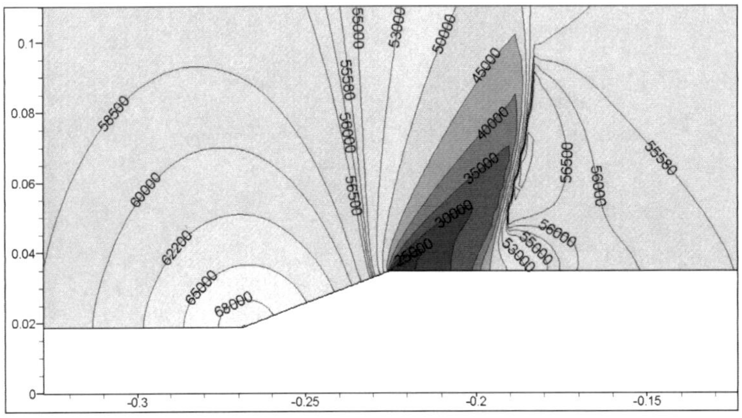

Figure 16: CFD static pressure in the inter-stage region. $M_{\infty} = 0.90$

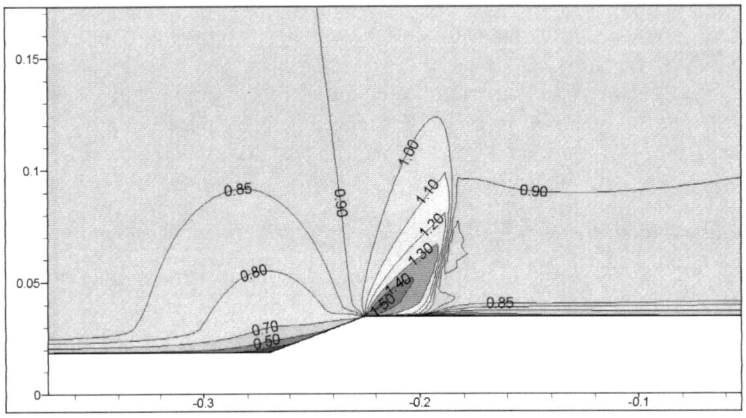

Figure 17: Mach number field. $M_{\infty} = 0.90$.

Figure 18 is a detailed view of the shock-wave boundary-layer interaction region with velocity vectors. One can see the reduction of velocity profile (region in blue), caused by the shock wave impinging over the boundary layer. As a consequence, the boundary-layer external frontier is enlarged.

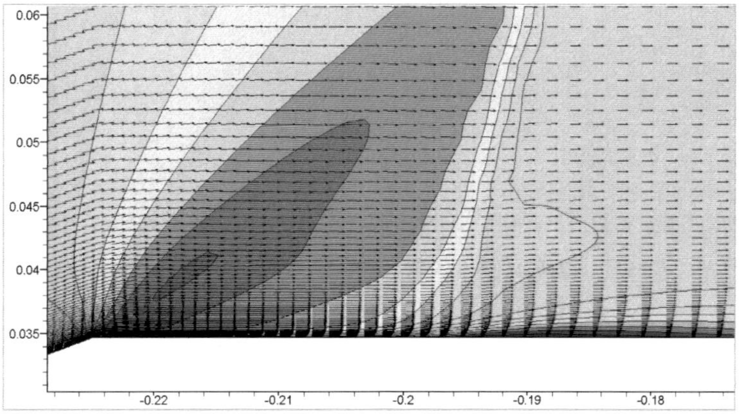

Figure 18: A detail of the velocity profile.

4. Conclusions

The paper has presented the results for a test campaign carried out at the Pilot Transonic Wind Tunnel. A half-model of the sounding rocket SONDA III, equipped with pressure taps around the inter-stage region, was tested in the subsonic and transonic regimes. Pressure coefficient distributions around the vehicle were estimated, as well as the associated uncertainties, to supply information about aerodynamic phenomena taking place in the airflow around the vehicle. Uncertainties were evaluated by using the Law of Propagation of Uncertainty and Monte Carlo method.

It was found that both methods resulted in similar uncertainty values and these were not significant to indicate the presence of measurement disturbances caused by the aerodynamic circuit.

A Computational Fluid Dynamics numerical code was used to better understand the changes in airflow properties caused by the presence of the model. This paper detailed the simulation results for Mach number equal to 0.90. Experimental and simulation data did not agree completely and causes will be better investigated in future studies.

5. References

[1] Délery J and Dussauge J P 2009 Some physical aspects of shock wave/boundary layer interactions *Shock waves* (Springer-Verlag) **19** pp 453–68. DOI 10.1007/s00193-009-0220-z

[2] Anderson Jr J D 2007 *Fundamentals of Aerodynamics* (Mc Graw Hill) New York p 985

[3] BIPM/JCGM 100:2008 *Evaluation of measurement data* – Guide to the expression of uncertainty in measurement (GUM 1995 with minor corrections) p 134

[4] BIPM/JCGM 101:2008 *Evaluation of measurement data* – Supplement 1 to the Guide to the expression of uncertainty in measurement – Propagation of distributions using a Monte Carlo method p 82

[5] Wilcox D C 1993 *Turbulence Modeling for CFD* La Cañada DCW Industries p 460

IMEKO IOP Publishing
Journal of Physics: Conference Series **588** (2015) 012004 doi:10.1088/1742-6596/588/1/012004

Structural observation of long-span suspension bridges for safety assessment: implementation of an optical displacement measurement system

L Lages Martins[1], J M Rebordão[2] and A Silva Ribeiro[1]

[1]National Laboratory for Civil Engineering (LNEC), Lisbon, Portugal
[2]Universidade de Lisboa, Faculdade de Ciências, Laboratory of Optics, Lasers and Systems (LOLS), Lisbon, Portugal

E-mail: lfmartins@lnec.pt

Abstract. This paper addresses the implementation of an optical displacement measurement system in the observation scenario of a long-span suspension bridge and its contribution for structural safety assessment. The metrological background required for quality assurance of the measurements is described, namely, the system's intrinsic parameterization and integration in the SI dimensional traceability chain by calibration, including its measurement uncertainty assessment.

1. Introduction

Observation is an important activity in safety assessment of large structures (such as bridges, viaducts, buildings, dams, among others), contributing for their structural condition monitoring and allowing a reliable analysis of its operational performance in accordance with safety requirements and historical records, thus providing relevant information for management entities.

In the long-span suspension bridge observation context, research efforts have been geared towards the development of non-contact measurement systems, capable of determining the 3D displacement of critical regions, namely, the bridge's main span central section. Optical systems are an attractive solution for the mentioned measurement problem, particularly, for the case of metallic bridges observation, where the accuracy of microwave interferometric radar systems [1] and global navigation satellite systems [2] can be compromised, namely, by the multi-path effect resulting from electromagnetic wave reflections in the bridge's structural components.

Recently, an optical methodology suitable for long-distance measurement of suspension bridges dynamic displacement has been studied and proposed by the authors [3]. This approach is supported by the use of a digital camera rigidly installed beneath the bridge's stiffness girder, oriented towards a set of four active targets placed at a tower foundation, materializing the world referential 3D system. Provided both the camera's intrinsic parameters and the targets relative coordinates are accurately known (by previous testing), non-linear optimization methods can be used to determine the position of the camera's projection center. The temporal evolution of this quantity is considered representative of the bridge's dynamic displacement at the location of the camera.

Following the design and development stages detailed in [3], this paper describes the implementation of the system in a real observation scenario: the 25th of April long-span suspension bridge (P25A) in Lisbon (Portugal). The experimental apparatus is described and displacement results

Content from this work may be used under the terms of the Creative Commons Attribution 3.0 licence. Any further distribution of this work must maintain attribution to the author(s) and the title of the work, journal citation and DOI.
Published under licence by IOP Publishing Ltd

are also presented and discussed. Special focus is given to the assurance of the metrological quality process: (i) the camera's intrinsic parameterization by the diffractive optical element method [4]; (ii) the system's field calibration, aiming its integration in the SI traceability chain; (iii) the measurement uncertainty evaluation, supported in the use of a Monte Carlo method and following the main guidelines given in the GUM-S1 supplement [5].

2. Displacement measurement in a long-span suspension bridge

In order to implement a displacement measurement system in the P25A bridge, a solution based on off-the-shelf optical components (listed in Table 1) was developed for a observation distance near 500 m. To obtain suitable 3D displacement measurement sensitivity, a 600 mm high focal length lens (composed by a 300 mm telephoto lens and a 2x teleconverter) was used.

Table 1 – Main optical components of the displacement measurement system installed in the P25A bridge

Component	Brand	Model	Technical specifications
Digital camera	IMPERX	IPX 2M30H-LM	Monochromatic CCD sensor; 1080 x 1920 pixels; 7,4 μm squared pixel
Telephoto lens	SIGMA	300 mm F2.8 EX APO DG	300 mm focal length
Teleconverter	SIGMA	2x EX APO DG	2x focal length
Optical filter	Schneider Kreuznach	B+W infrared filter 092	100% transmissivity above 730 nm wavelength
Leds	Avago Technologies	HSDL 4260	875 nm near-infrared wavelength

Figure 1 shows the set of four active targets placed in the P25A bridge south tower foundation, facing the bridge's main span where the camera was installed (Fig. 2). Each of the four targets was composed by 16 leds, distributed in a circular geometrical pattern and capable of emitting a narrow near-infrared beam (875 nm wavelength) compatible with the camera's spectral sensitivity. An optical filter on the camera reduces the environment visible irradiance from the many other elements of the observation scenario, therefore improving contrasts in the targets image (see Fig. 3).

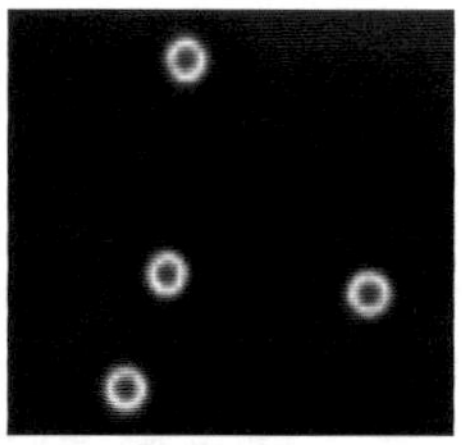

Figures 1 to 3 – Active targets on the south tower foundation (left); digital camera installed in the stiffness girder (center); detail of the image of the targets (right)

High frequency vibrations in the camera were mitigated by rubber washers and expanded polystyrene foam mounted between the camera's support structure and the stiffness girder lower surface. Other operational concerns were related to: (i) the acquisition frequency – a reference value of 15 Hz was defined, based on the image acquisition maximum capacity of the applied digital camera; (ii) the exposure time – a value of 2,5 ms was established in order to avoid image blurring due to the bridge's dynamic displacement; (iii) the lens f-number or relative aperture – this operational parameter was set to *f*/5,6, allowing a minimum amount of light in the camera's sensor (avoiding saturation) but keeping all the active targets visible. The images were processed fitting an ellipse to each visible target on the image and estimating its center.

Since the P25A bridge has two main decks (an upper road deck and a lower train deck), two types of displacement records – with and without train circulation – were obtained during field testing of the displacement measurement system. Table 2 presents the maximum (peak-to-peak) displacements for each of the studied structural scenarios. Figure 4 illustrates the vertical displacement of the bridge's main span central section. Close to t = 120 s, the peak detail of the steps related to the four carriage connections of the train is also noticed.

Table 2 – Maximum displacement values recorded for the P25A bridge central section during field testing

Load condition	Transverse displacement	Vertical displacement	Longitudinal displacement
With passengers train	0,39 m	1,69 m	0,07 m
Without passengers train	0,29 m	0,53 m	0,02 m

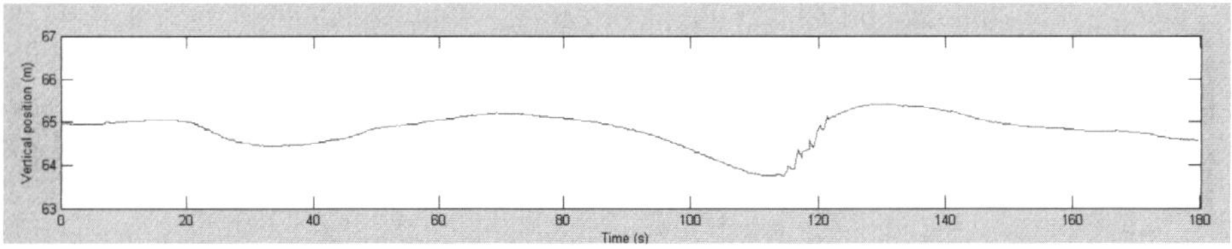

Figure 4 – P25A bridge central section vertical displacement during passengers train circulation

As expected, the train circulation on the P25A bridge increases the 3D displacement magnitudes, namely in the vertical direction where a maximum (peak-to-peak) displacement of 1,69 m was recorded. This value is comparable to previous known vertical displacement measurements obtained during the P25A bridge static loading test performed in 1999 (after its structural reinforcement for train circulation), where a maximum vertical displacement of 3,15 m (2,37 m downward plus 0,78 upward) was recorded for a static distributed load of 77,5 kN/m. Passengers train distributed loads are usually between 20,7 kN/m (empty train situation) and 28,8 kN/m (overload train), thus originating lower vertical displacements.

3. Metrological quality assessment

3.1. Introduction
In order to evaluate the accuracy level of the displacement estimates, a metrological quality assessment was performed, following three stages: camera's intrinsic parameterization, system's field calibration and measurement uncertainty evaluation – described in detail in the following sections.

3.2. Intrinsic parameterization
Before installation in the P25A bridge, the digital camera was subjected to laboratorial testing to obtain its intrinsic parameters, namely, focal length, principal point coordinates and distortion coefficients. Being a high focal length camera (close to 600 mm), the diffractive optical element (DOE) method [4] was applied since conventional parameterization approaches, usually applied for reduced focal lengths, are not suitable because they generate ill-conditioned matrices in least squares estimation.

A collimated laser beam and a DOE were used on an optical bench to create a regular spatial distribution of diffraction dots in the camera's focal plane. Based on the accurate knowledge of the laser wavelength, DOE grating period, observed diffraction orders and centroid locations of the corresponding diffraction dots, the intrinsic parameters can be determined using only one image in a controlled laboratorial environment, considering the pinhole and 1st order radial distortion models and using non-linear optimization [6].

In order to evaluate the measurement uncertainties related to the intrinsic parameters, a Monte Carlo method study was performed following the main guidelines mentioned in [5] due to the use of a non-linear optimization procedure. In the input probabilistic formulation, the laser wavelength estimate ($632,8 \times 10^{-9}$ m) was considered constant and a Gaussian probability function was adopted for the DOE grating period quantity, centered at $152,4 \times 10^{-6}$ m with a standard measurement uncertainty of $0,15 \times 10^{-6}$ m. Based on the digital image processing performances, a standard measurement uncertainty of $p/4$ (where p is pixel size) was assigned to the diffraction image coordinates estimates. Results are presented in Table 3. They are supported in 10 000 Monte Carlo runs, for which a computational accuracy level of 0,1 mm (for the focal length quantity) and 0,01 pixel (for principal point coordinates) were achieved.

Table 3 – Estimates, measurement expanded uncertainties and correlation coefficients obtained from Monte Carlo simulations

Focal length		Principal point x-coordinate		Principal point y-coordinate	
Average value (mm)	95% expanded uncertainty (mm)	Average value (pixel)	95% expanded uncertainty (pixel)	Average value (pixel)	95% expanded uncertainty (pixel)
599,95	0,75	545,61	0,06	960,41	0,09
Correlation coefficients					
Focal length / Principal point x-coordinate: -0,28					
Focal length / Principal point y-coordinate: -0,35					
Principal point x-coordinate / Principal point y-coordinate: -0,25					

The intrinsic parameters estimates for focal length and principal point coordinates are close to the corresponding nominal values (600 mm, 540 pixel and 960 pixel). As expected for high focal length lenses, the estimate for the first-order coefficient of radial distortion is residual and, therefore, is considered to be insufficient to affect image coordinates due to image processing performances. Radial distortion is, consequently, irrelevant and the distortion model can be removed from optimization, both in the parameterization and in displacement measurement procedures.

3.3. Dimensional traceability

Dimensional traceability of the optical measurement system was achieved by a field calibration test performed at the P25A bridge, where both the camera and the set of targets were placed, respectively, in the bridge's south anchorage and in the south tower foundation (Figure 5). When compared with the bridge's central section, these two regions have shown reduced dynamics and can be considered nearly static regions, allowing establishing a line of sight symmetric to the measurement line of sight between the central section and the south tower foundation – the camera remaining, approximately, at the same altitude for both geometrical configurations.

Figures 5 and 6 – Calibration field test and laboratorial dimensional testing of the calibration device prototype

The calibration method proposed implies the acquisition of images of the set of targets supported by known 3D reference positions in the bridge's main displacement directions (vertical, transverse and longitudinal). The system's measurement model can be applied to each image, allowing to know the camera's projection center virtual displacement between two images (the initial and final targets position) which can be compared with the reference displacement initially applied to the set of targets. In order to implement this calibration method, a dedicated device prototype (shown in Figures 5 and 6) was developed for the installation of the set of targets and application of reference displacement in the bridge's main displacement directions.

A SI traceable 3D coordinate measuring machine[1] was used to determine the spatial coordinates of the targets LED's in all four reference positions (initial position, 250 mm in the longitudinal direction,

[1] The same standard equipment was also used for dimensional testing of the targets world coordinates (relative to one of the targets) required for the displacement measurement process.

350 mm in the transverse direction and 250 mm in the vertical direction) in order to obtain reference displacement estimates. Table 4 shows the reference and measured displacement values and corresponding deviations obtained in the field calibration test.

Table 4 – Field calibration results

	Transverse displacement			Vertical displacement			Longitudinal displacement		
	Measured value (mm)	Reference value (mm)	Deviation (mm)	Measured value (mm)	Reference value (mm)	Deviation (mm)	Measured value (mm)	Reference value (mm)	Deviation (mm)
X	-349,9	-350,6	0,7	-2,0	-0,8	-1,2	0,3	-0,4	0,7
Y	1,5	0,1	1,4	-251,0	251,3	0,3	1,7	0,5	1,2
Z	0,0	-0,1	0,1	0,0	0,0	0,0	250,0	251,5	-1,5

The analysis of the values in the previous Table shows that the maximum calibration deviations in each displacement direction is between -1,5 mm and 1,4 mm. It should also be mentioned that, if vertical refraction corrections are applied to targets world coordinates [7], the vertical deviation presented in the vertical displacement test (0,3 mm) becomes null, thus improving its accuracy.

3.4. Measurement uncertainty evaluation

The measurement uncertainty evaluation was performed in order to estimate the dispersion of values related to the calibration deviations mentioned in section 3.3, aiming to quantify the accuracy level of the optical displacement measurement system.

In a probabilistic approach, the measurement uncertainty related to the calibration deviation combines the dispersions of values assigned to the reference displacement and the initial and final positions of the camera projection center.

Regarding the reference displacement, the following uncertainty components were quantified: (i) the measurement of the initial and final target position, which combines both the instrumental uncertainty of the 3D coordinate measurement machine and the circularity deviation obtained in the measurement of the geometrical pattern of the targets; (ii) the return to zero deviation when the referential returns to the initial position; (iii) the differential displacement between pairs of targets; (iv) the repeatability related to the operational method of mounting the set of targets in the calibration device.

The initial and final position of the camera projection center combines the measurement uncertainties of the following input quantities: (i) the focal length and principal point quantities, as mentioned in 3.2; (ii) the targets image coordinates which, in turn, combine the measurement uncertainties related to the digital image processing and turbulence due to thermal vertical gradients; beam wandering tests performed at the P25A bridge (in a similar way as described for the field calibration test in section 3.3 but without moving the set of targets) revealed that this uncertainty component changes between 0,13 pixel (in the winter, with shadow over the targets) and 0,56 pixel (in the summer, without shadows over the targets); (iii) the targets world coordinates which include the uncertainty contributions related to: the laboratorial dimensional measurement; the target circularity; the environmental temperature influence to the set of targets on the P25A bridge; the transport and assembly of the set of targets in the tower foundation; the vertical refraction correction.

Due to the non-linear optimization procedure when determining the camera's projection center position, a Monte Carlo method was used for the propagation of measurement uncertainties of the input quantities previously referred to the output quantities measurement uncertainties. This method is particularly suitable since this multivariable process implies performing several iterations, making difficult to apply other approaches, namely, the conventional ISO-GUM method. In addition, it allows knowing the shape of the output probability density functions and the optimization process stability. Table 5 summarizes the results based on 10000 simulation runs.

Table 5 –Measurement standard uncertainties of input, intermediate and output quantities

INPUT QUANTITIES	**Camera's intrinsic parameters**		**Target world coordinates**	
	Focal length, f	$0{,}38\cdot10^{-3}$ m	Transverse direction (X)	$0{,}57\cdot10^{-3}$ m
	Principal point x-coordinate, u_0	0,030 pixel	Vertical direction (Y)	$0{,}69\cdot10^{-3}$ m
	Principal point y-coordinate, v_0	0,045 pixel	Longitudinal direction (Z)	$0{,}36\cdot10^{-3}$ m
	Correlation coefficient (f, u_0)	-0,28	**Target image coordinates**	
	Correlation coefficient (f, v_0)	-0,35	x-coordinate	0,13 pixel
	Correlation coefficient (u_0, v_0)	-0,25	y-coordinate	0,13 pixel
	Reference displacement	**Transverse displacement**	**Vertical displacement**	**Longitudinal displacement**
	Transverse direction (X)	$0{,}97\cdot10^{-3}$ m	$0{,}58\cdot10^{-3}$ m	$0{,}59\cdot10^{-3}$ m
	Vertical direction (Y)	$0{,}91\cdot10^{-3}$ m	$0{,}58\cdot10^{-3}$ m	$0{,}51\cdot10^{-3}$ m
	Longitudinal direction (Z)	$0{,}90\cdot10^{-3}$ m	$0{,}47\cdot10^{-3}$ m	$0{,}45\cdot10^{-3}$ m
INTERMEDIATE QUANTITIES	**Camera initial position**	**Transverse displacement**	**Vertical displacement**	**Longitudinal displacement**
	Transverse direction (X)	$0{,}95\cdot10^{-3}$ m	$0{,}95\cdot10^{-3}$ m	$0{,}95\cdot10^{-3}$ m
	Vertical direction (Y)	$0{,}95\cdot10^{-3}$ m	$0{,}95\cdot10^{-3}$ m	$0{,}95\cdot10^{-3}$ m
	Longitudinal direction (Z)	$8{,}5\cdot10^{-3}$ m	$8{,}5\cdot10^{-3}$ m	$8{,}5\cdot10^{-3}$ m
	Camera final position	**Transverse displacement**	**Vertical displacement**	**Longitudinal displacement**
	Transverse direction (X)	$1{,}0\cdot10^{-3}$ m	$1{,}0\cdot10^{-3}$ m	$1{,}0\cdot10^{-3}$ m
	Vertical direction (Y)	$0{,}95\cdot10^{-3}$ m	$0{,}95\cdot10^{-3}$ m	$0{,}95\cdot10^{-3}$ m
	Longitudinal direction (Z)	$9{,}0\cdot10^{-3}$ m	$9{,}0\cdot10^{-3}$ m	$9{,}0\cdot10^{-3}$ m
OUTPUT QUANTITIES	**Calibration deviation**	**Transverse displacement**	**Vertical displacement**	**Longitudinal displacement**
	Transverse direction (X)	$1{,}7\cdot10^{-3}$ m	$1{,}5\cdot10^{-3}$ m	$1{,}5\cdot10^{-3}$ m
	Vertical direction (Y)	$1{,}7\cdot10^{-3}$ m	$1{,}4\cdot10^{-3}$ m	$1{,}4\cdot10^{-3}$ m
	Longitudinal direction (Z)	$13\cdot10^{-3}$ m	$13\cdot10^{-3}$ m	$13\cdot10^{-3}$ m

The analysis of the results shows that, for all the calibration directions, the measurement standard uncertainty related to the longitudinal direction (13 mm) is significantly higher than for the other two orthogonal directions (vertical and transverse) where a maximum value of 1,7 mm is reached. This difference can be justified by the geometrical configuration of the optical measurement system which has reduced sensitivity for the longitudinal direction.

It should be noticed that the system's field calibration was performed during winter, being the standard uncertainty of the targets image coordinates relatively low (about 0,13 pixel) when comparing with summer conditions where previous field tests showed a higher dispersion (close to 0,6 pixel). In this case, the measurement uncertainty can be increased by a multiplicative factor of 1,5 up to 2,2 in all displacement directions.

4. Conclusions

The system's implementation on the P25A bridge allowed to estimate its central section 3D displacement considering different loading scenarios, namely, due to train circulation where a maximum (peak-to-peak) vertical displacement value of 1,69 m was recorded, being consistent with historical values obtained during static loading tests performed in 1999.

Quality assurance of the obtained displacement estimates was demonstrated by uncertainty evaluation which encompasses system's intrinsic parameters, dimensional testing and field calibration. A 1,7 mm standard measurement uncertainty was calculated for the vertical and transverse calibration deviations while, in the longitudinal direction, a value of 13 mm was obtained. The Monte Carlo method was particularly suitable for the measurement uncertainty evaluation considering that the measurement procedure includes an iterative multivariable non-linear optimization.

References

[1] Pasquale G, Bernardini G, Ricci P, Gentile C, 2010 *IEEE Aerospace and Electronics Systems Magazine* **25**, 19-26
[2] Nakamura S, 2000 *Journal of Structural Engineering* **126**, 1413-1419
[3] Martins L, Rebordão J M, Ribeiro A, 2013 *Journal of Physics: Conference Series* **459**, 012055
[4] Bauer M, Grießbach D, Hermerschmidt A, Krüger S, Scheele M, Schischmanow A, 2008 *Optics Express* **16**, 20241-20248
[5] *Supplement 1 to the Guide to the expression of Uncertainty in Measurement (S1 GUM)*, Joint Committee for Guides in Metrology, 2008
[6] Lagarias J, Reeds J, Wright M, Wright P, 1998 *SIAM Journal of Optimization* **9,** 112-147
[7] Martins L, Rebordão J M, Ribeiro A, 2014 *International Journal of Thermophysics* **34**, 5

The advantage of aiding mechanisms in measurements with callipers

Emanuel Paquete Correia[1§] and João A. Sousa[2]
[1]*LREC*
Edifício LREC, Rua de S. Gonçalo, 9500-343, Ponta Delgada, Azores
[2]LREC, Rua Agostinho Pereira de Oliveira, 9000-264 Funchal, Portugal
E-mail: emanuel.p.correia@azores.gov.pt

Abstract. Making accurate measurements with a calliper is far from a trivial task. It requires control of the positioning and force application during measurement, which can only be achieved by experienced users. Research has been made in order to help the inexperienced user to make accurate measurements and minimize the training that is normally required. The developed techniques consist on the application of a structure to support the calliper, thus minimizing positioning errors, and a force control accessory for the calliper's movable jaw, resulting in an estimated operator's uncertainty of about 40% of the calliper's resolution value.

1. Introduction

The calliper is an instrument for dimensional measuring, frequently used in Civil Engineering tests, and in a wide variety of other applications. In the performance of my job as a calibration technician, I have found out that the calliper is not an easy to handle instrument because its accuracy depends a lot on the user's ability and experience. Considering the factors that make measurements difficult, which subsequently result in a high user measurement uncertainty, some techniques to reduce the measurement uncertainty are described. These were implemented in the Metrology Unit of the Civil Engineering Regional Laboratory of the Azores archipelago. Topics to be covered include the following.

- General characteristics and calliper types;
- Metrological characteristics;
- Common operator errors;
- Calliper calibration structure and force control device;
- Operator uncertainty calculation method and examples resulting from the applied techniques, using the ANOVA method;
- Global uncertainty analysis using the Monte Carlo method.

2. General characteristics and calliper types

Generally, a calliper has the following components (see figure 1):

- 2 jaws for external measurements;
- 2 jaws for internal measurements;

[§] To whom correspondence should be addressed

Content from this work may be used under the terms of the Creative Commons Attribution 3.0 licence. Any further distribution of this work must maintain attribution to the author(s) and the title of the work, journal citation and DOI.
Published under licence by IOP Publishing Ltd

- 1 depth measuring blade.

It can be either analogue or digital.

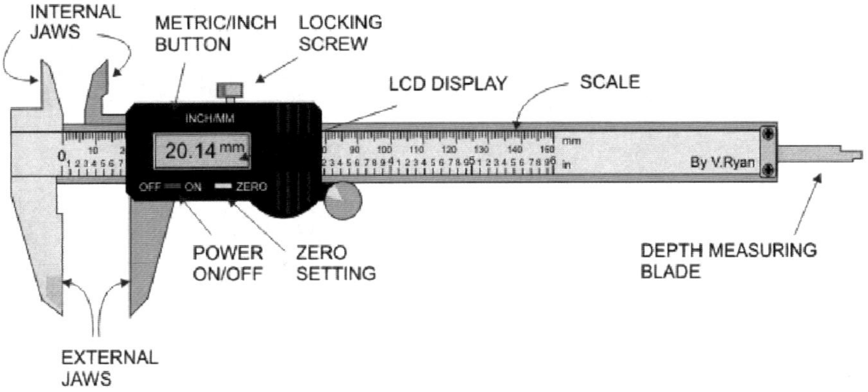

Figure 1. Digital calliper example

3. Metrological characteristics

According to ISO [2], the main metrological characteristics applicable to the calliper's external measurement jaws are the following:

- Partial surface contact error;
- Repeatability of partial surface contact error.

The partial surface contact error is the "error of indication when a partial measuring face contact is employed".

The repeatability of partial surface contact error may be obtained by the standard deviation of the partial measurements made along the length of the external jaws. The determination of these metrological characteristics includes deviations caused by the:

- Measuring position error (within the measuring range);
- Clearance between the slider and the beam;
- Parallelism and form deviation of the measuring surface of the external jaws.

4. Common operator errors

The most common operation errors caused by an inexperienced user can be divided into the following categories:

- Calliper alignment errors;
- Sliding (measuring) jaw force application error;

The alignment errors can be either horizontal or vertical, as represented in figures 2 and 3, respectively.

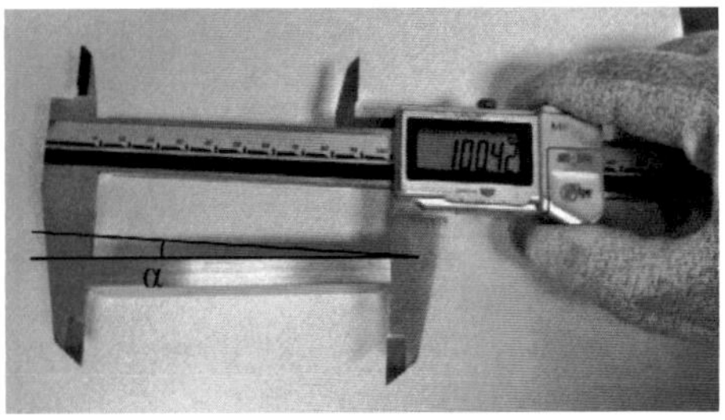

Figure 2. Horizontal alignment error

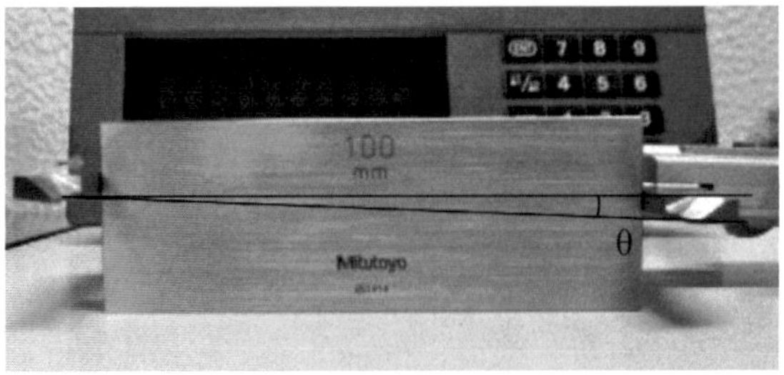

Figure 3. Vertical alignment error

The sliding jaw force application error is characterized by the deformation of the sliding jaw, caused by the strength exerted by the operator while doing several measurements on the same object. The deformation of the sliding jaw will, consequently, produce inaccurate readings, especially at the tip of the jaws for external measurements. This error occurs because the calliper does not fulfill the Abbe's principle of alignment, due to the existing clearance between the sliding jaw and the calliper's beam. This clearance is inevitable; otherwise it would be impossible to move the sliding jaw, although some callipers allow some mechanical adjustment for the minimization of this clearance.

5. Calliper calibration structure and force control device

A possible solution for the reduction or elimination of calliper alignment errors is the implementation of a structure (see Figure 4), to support the calliper while it is being calibrated, and a force control device to fit on the calliper's beam (see Figure 5).

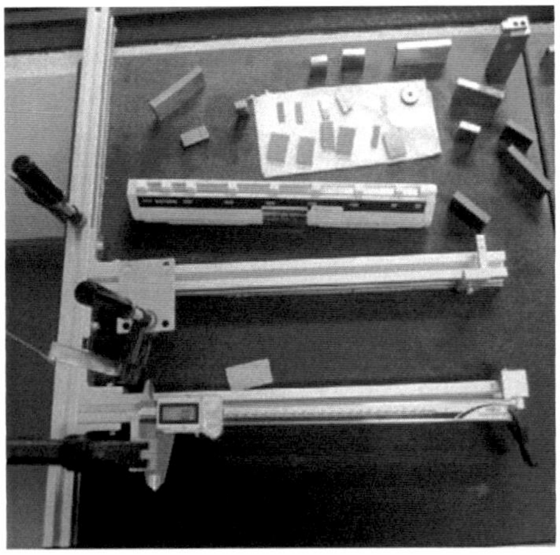

Figure 4. Structure, made in aluminum profiles, to support a calliper – made in LREC (Azores)

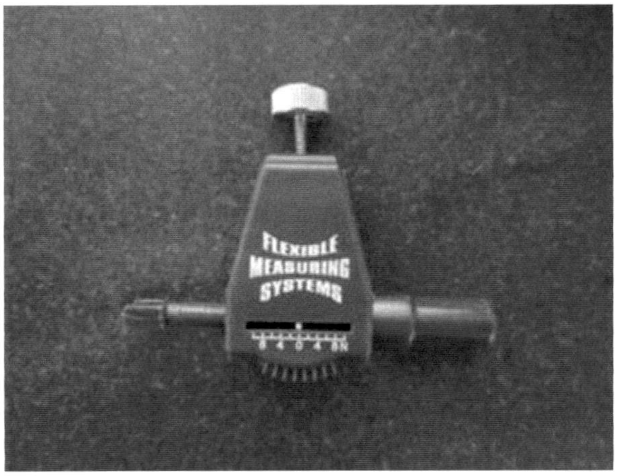

Figure 5. Force control device – made in Flexible Measuring Systems (Denmark)

6. Operator uncertainty calculation method

Operator's uncertainty is determined from studies of repeatability and reproducibility. The ANOVA is a method for the analysis of variance, which allows for the combined study of repeatability, reproducibility and the interaction between different operators or between the same operator and time. The two-way ANOVA presented in this paper is based on the method used by Engineered Software [4]. The examples presented in the following sections will characterize the variation of measurements made, by a single operator, in different time periods.

The total measurement system variation is given by equation (1).

$$V_T = \left(\sqrt{R_{ep}^2 + R_{epro}^2 + I^2} + V_O^2 \right)^{1/2} \quad (1)$$

Where R_{ep}, R_{epro}, I and V_O represent, respectively, the repeatability and reproducibility of the measuring system, the interaction between factors A (time) and B (operator) and the operator variation.

The operator's uncertainty will be determined as a rectangular distribution of this total system variation, according to equation (2).

$$u_O = \frac{2V_T}{\sqrt{12}} \quad (2)$$

In the following examples, the null hypothesis for the ANOVA method can be interpreted as:

1. The operator is not a significant source of measurement variability during different time periods;
2. The operator is not a significant source of measurement variability on measurements made on the same time period;
3. There is no interaction between measurements made by the operator through different time periods.

6.1. Operator uncertainty calculation - Example 1

The objective is to evaluate the reduction in the operator's uncertainty after applying a structure to support the calliper (see Figure 4). The procedure consisted on making 2 series of measurements, one in the morning and the other in the afternoon, on a 100 mm gauge block and with a digital calliper (resolution of 0,01 mm). Test data for 3 days is presented on Table 1.

Table 1. Measurements made using the structure shown in Figure 4 and determination of parameters

Measurements (mm)						Parameter	Value (mm)
Day 1		*Day 2*		*Day 3*			
1st Series	2nd Series	1st Series	2nd Series	1st Series	2nd Series	Repeatability	0,030
99,98	100,00	100,01	100,00	100,01	100,01	Reproducibility	0,039
100,00	100,00	100,01	100,00	100,01	100,01		
100,00	99,99	100,00	100,00	100,01	100,00	Interaction (time, operator)	0,000
99,99	99,99	100,00	100,00	100,00	100,01		
99,99	100,00	100,00	100,00	100,01	100,01	Measurement variation (on same day)	0,000
99,98	100,00	100,01	100,00	100,01	100,01		
99,98	99,99	100,00	100,00	100,01	100,01	Total system variation	0,049
99,98	99,99	100,01	100,00	100,01	100,00		
100,00	100,00	100,00	100,00	100,00	100,00	Operator uncertainty	0,028
99,99	100,00	100,00	100,01	100,01	100,01		

Table 2. F-Statistic - Significance and critical points for an operator uncertainty of 0,028 mm

Source of Variation	F Statistic	Degrees of Freedom	Significance	Critical Point (α=0,05)
Time	33,3(3)	2	2,39E-8	3,32
Operator	0,87	9	5,59E-1	2,21
Interaction (Time, Operator)	0,95	18	5,31E-1	1,96

Observing the operator's uncertainty value on table 1, it can be concluded that it is about 3 times the calliper's resolution value, which is significant. This means that the structure used to minimize the alignment errors during the measurement is not enough. The next example shows that the force component weighs heavier on the total system variation and consequently on the operator's uncertainty. Observing the calculated values on Table 2, it can be verified that all *F-Statistic* values, with the exception of the *Time* source of variation value, are inferior to the *Critical Point* values and consequently, hypotheses 2 and 3 are accepted and hypothesis 1 is rejected. This means that the operator is a significant source of variability on measurements made on different days.

6.2. Operator uncertainty calculation - Example 2

In the sequence of example 1, the objective is to apply a force control device (see Figure 5) on the calliper and test it on the support structure, just as before. Test data is presented on Table 3.

Table 3. Measurements made using the structure plus force device and determination of parameters

Measurements (mm)						Parameter	Value (mm)
Day 1		*Day 2*		*Day 3*			
1st Series	2nd Series	1st Series	2nd Series	1st Series	2nd Series	Repeatability	0,007
99,99	99,99	99,99	99,99	99,99	99,99	Reproducibility	0,000
99,99	99,99	99,99	99,99	99,99	99,99		
99,99	99,99	99,99	99,99	99,99	99,99	Interaction (time, operator)	0,000
100,00	99,99	99,99	99,99	99,99	99,99		
99,99	99,99	99,99	99,99	99,99	99,99	Measurement variation (on same day)	0,000
99,99	99,99	99,99	99,99	99,99	99,99		
99,99	99,99	99,99	99,99	99,99	99,99	Total system variation	0,007
99,99	99,99	99,99	99,99	99,99	99,99		
99,99	99,99	99,99	99,99	99,99	99,99	Operator uncertainty	0,004
99,99	99,99	99,99	99,99	99,99	99,99		

Table 4. F-Statistic - Significance and critical points for an operator uncertainty of 0,004 mm

Source of Variation	F Statistic	Degrees of Freedom	Significance	Critical Point (α=0,05)
Time	1,00	2	0,38	3,32
Operator	0,9(9)	9	0,46	2,21
Interaction (Time, Operator)	1,00	18	0,49	1,96

From Table 3, it can be seen that repeatability, reproducibility and uncertainty have decreased in relation to the previous example, which means that operator performance has significantly improved. From Table 4, it can be verified that all *F-Statistic* values are inferior to the *Critical Point* values and, consequently, all null hypotheses are accepted. This means that the operator is no longer a significant source of variability on measurements made on different days.

7. Global uncertainty analysis

Evaluating the global measurement uncertainty associated with the calibration of this type of equipment, might give useful indications on the importance of the operator's uncertainty on the global measuring system. Assuming fixed values for the other sources of uncertainty, namely, (a) working standards uncertainty (± 0,9 μm, Gaussian); (b) working standards drift (± 0,3 μm, rectangular); (c) temperature influence (± 0,7 μm, rectangular); (d) finite resolution (± 2,9 μm, rectangular), we can analyse the influence of considering (i) operator uncertainty = ± 28 µm, rectangular, as from Table 1 (0,028 mm), or (ii) operator uncertainty = ± 4 µm, rectangular, as from Table 3 (0,004 mm), using a Monte Carlo method. In the first situation the measurand output is clearly rectangular with an expanded factor k = 1,66 for a 95 % level of confidence, as in Figure 6, whereas the smaller value of operator uncertainty will produce a much more Gaussian shape curve (Figure 7), indicating that the conditions of the Central Limit Theorem (CLT) are fulfilled to a much higher degree.

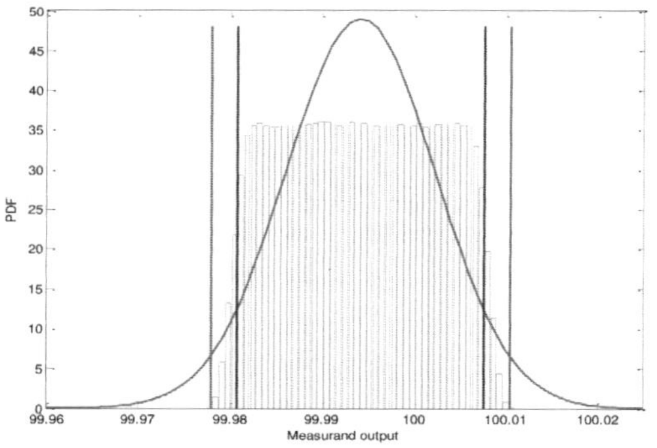

Figure 6. Measurand output for an operator uncertainty of 0,028 mm

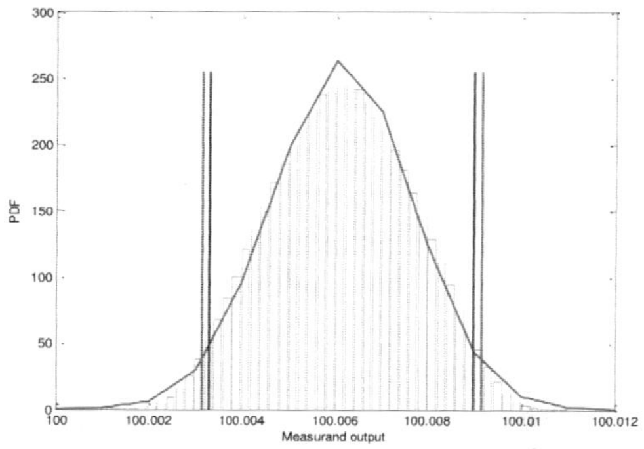

Figure 7. Measurand output for an operator uncertainty of 0,004 mm

These findings are in line with those stated in the EA-4/02 document [5], of mandatory nature for accredited calibration laboratories. The case represented in Figure 6 is typical of situations where there is one dominant source of uncertainty of rectangular shape (k = 1,65 according to EA [5], page 49, section S8.2). The case represented in Figure 7, on the other hand, shows that even when the conditions of the CLT are followed (to a degree), the use of the law of propagation of uncertainty might entail an approximation, in cases where not all input quantities are Gaussian. The degree of that approximation is hard to predict *a priori*, and in this particular case a k = 1,90 was found (instead of k = 1,96) to obtain the same 95 % confidence interval.

8. Conclusions

In the sequence of the techniques here presented, it is possible to conclude that the application of both a structure for supporting the calliper and a device for controlling the force on the calliper's sliding jaw, has significantly reduced the operator uncertainty to about 40 % of the calliper's resolution value. It has been shown by both ANOVA and Monte Carlo methods that the operator's uncertainty is significant and has a dominant effect on the global calibration uncertainty, when there is no force control on the measuring process.

The advantages in the implementation of these techniques are the following:

- Easier, quicker and stress free measurement operations;
- Better accuracy and repeatability on measurements;
- Lower training and test execution costs.

Moreover, it is important to have a full knowledge on the consequences, in terms of uncertainty evaluation, of having a large source of uncertainty associated with the operator handling skills, compared to the rest of the sources considered for the uncertainty budget, or having a much smaller value and therefore much more evenly distributed values for all the uncertainty sources.

Journal of Physics: Conference Series **588** (2015) 012005 doi:10.1088/1742-6596/588/1/012005

References

[1] Prins J, McCormack D, Michelson D and Horrel K 2012 *e-Handbook of Statistical Methods*. NIST/SEMATECH Albany, NY chapter 7.4.3.7; [acessed december 2013]. Available from: http://www.itl.nist.gov/div898/handbook/

[2] ISO 2011 *ISO 13385-1 Geometrical product specifications (GPS) – Dimensional measuring equipment – Part 1: Callipers; Design and metrological characteristics*. ISO copyright office Geneva

[3] JCGM 2012 *International vocabulary of metrology – Basic and general concepts and associated terms (VIM)*. JCGM

[4] Engineered Software Inc. *Repeatability and Reproducibility*. Engineered Software, Inc. Belleville; [acessed december 2013]. Available from: http://www.engineeredsoftware.com/papers/msa_rr.pdf

[5] EA 2013 *EA-4/02 M:2013 Evaluation of the Uncertainty of Measurement in Calibration*, European Accreditation. Available from: www.-european-accreditation.org.

IMEKO IOP Publishing
Journal of Physics: Conference Series **588** (2015) 012006 doi:10.1088/1742-6596/588/1/012006

Lumped Capacitance Model in Thermal Analysis of Solid Materials

P Koštial, I Špička, Z Jančíková, J David, J Valíček, M Harničárová and V Rusnák

VŠB-Technical University of Ostrava, Czech Republic

E-mail: pavel.kostial@vsb.cz

Abstract. The paper is devoted to the presentation of a method for measurement of thermal conductivity k, specific heat capacity c_p and thermal diffusivity α applying the lumped capacitance model (LCM) as a special case of Newton's model of cooling. At the specific experimental conditions resulting from the theoretical analysis of the used model, we present relatively very precise method for experimental determination of all three above mentioned thermal parameters for materials with different thermal transport properties. The input experimental data provide a cooling curve of the tested material obtained in special experimental arrangement. The evaluation of experimental data is realized by software the fundamental features of which are presented here. The statistical analysis of experimental data was performed *(99% confidence interval P_{99}).*

1. Introduction

The cooling of objects is often described by a law, attributed to Newton, which states that the temperature difference of a cooling body with respect to the surroundings decreases exponentially with time. Such behaviour has been observed during many laboratory experiments, which led to a wide acceptance of this approach. However, the heat transfer from any object to its surrounding is not only due to conduction and convection but also due to radiation. The latter does not vary linearly with temperature difference, which leads to deviations from Newton's law. A discussion of this topic is provided in work [1, 2, 3].

Specific properties of materials demand a large spectrum of measuring methods suitable for thermal parameters measurements. Some of them are based on a photoacoustic effect and they are described in [4].

Another group of experiments is based on the application of thermal impulse on the sample surface and its propagation through a sample. These impulse applications either have contact of the heat source with the sample or they are in a contactless (flash) mode. The fundamental theoretical work in this sense is [5]. The so-called flash method has a large spectrum of realizations [6].

Different contact impulse methods with different position of the heat source were described in [7]. The heat sources in this case usually have a meander shape. Heat generation in the measured sample can also be realized by mechanical stress. The sample is pulled in a tensile test machine and the generated heat is caused by deformation of its structure. The temperature decay of the pre-pulled sample is measured by a thermovision camera. This measuring system allows measuring specific heat capacity,

Content from this work may be used under the terms of the Creative Commons Attribution 3.0 licence. Any further distribution of this work must maintain attribution to the author(s) and the title of the work, journal citation and DOI.
Published under licence by IOP Publishing Ltd

thermal diffusivity and thermal conductivity as well as the mechanical parameters usually obtained from a tensile test [8].
In this paper, we present a method for measurement of thermal conductivity k (W·m^{-1}·K^{-1}), specific heat capacity c_p (J·kg^{-1}·K^{-1}) and thermal diffusivity α (m^2·s^{-1}) applying the lumped capacitance model as a special form of Newton's model of cooling. Used method offer relatively reliable results and has a very good repeatability. Obtained results are statistically treated and they are also compared with known table values or results obtained on independent measurements.

2. Theory

For a heat flow from a sample to an environment with temperature T_∞ Newton´s cooling law can be written in the form [9,10]

$$\frac{dQ}{dt} = -h_t.S.(T - T_\infty) \quad , \qquad (1)$$

where Q is the heat given to a sample, h_t is the total heat transfer coefficient of a sample, S is the total heat flow area, T_∞ is the environment temperature.
For the lumped capacitance method solution of surface temperature T can be found from differential equation

$$m.c_p \frac{dT}{dt} = -h_t.S.(T - T_\infty) \quad , \qquad (2)$$

further

$$m=\rho.V= \rho.S..L \ , \qquad (3)$$

where ρ is density of the sample, L is sample thickness and S is the effectively cooled surface.
In our case the sample is cooled on both side and so we can write

$$S^*=S_{top} + S_{lower}=2S. \qquad (4)$$

After integration of equation 2 we subsequently obtain

$$\left[\rho SLc_p \, ln(T - T_\infty)\right]_0^t = \left[-2hSt\right]_0^t . \qquad (5)$$

If we set

$$\tau = \frac{\rho.c_p.L}{2.h_t} \quad , \qquad (6)$$

then

$$\left[ln(T - T_\infty)\right]_0^t = -t/\tau \qquad (7)$$

and finally for master function one can write

$$T = T_\infty + (T_0 - T_\infty) exp(-t/\tau). \qquad (8)$$

Relation among k, c_p and α has the form [10]

$$k = \alpha \, . \, \rho \, . \, c_p \qquad (9)$$

where k is sample thermal conductivity, α is diffusivity, c_p, is specific heat capacity and ρ is the sample density.

Validity of the model is verified by the Biot number Bi in the form

$$Bi = \frac{h_t L}{k} << 1. \qquad (10)$$

where

$$h_t = h_c + h_r \qquad (11)$$

represents so called combined heat transfer coefficient through convection with the coefficient of h_c and radiation with the coefficient of h_r.
The analysis of simulation results has shown that the radiation heat transfer coefficient observed at the temperature range of 25-26 °C is practically a linear function of temperature with increasing trend. However, the value of hr coefficient within the monitored temperature interval does not change by more than 0.5%, which is why, taking into account the approximate 5% model accuracy, it can be approximated by its arithmetic average or median, which is less sensitive to deviation and extreme values .
The comparison shows that even at room temperatures and relatively very little temperature differences between the cooling body and the surroundings, the value of radiation heat flow density is considerably high. Radiation heat flow in the given case has an average of more than 23.9% share on the total heat discharge from the surface of the body. That is why the description of the cooling process must usually take into account both convective and radiation heat transfer mechanism, and it must take into account the combined nature of the thermal interactions of the body with the surroundings, which more accurately corresponds to the macroscopic description of the heat exchange processes than during the application of a simple exponential model of the first order which neglects radiation.
With increasing temperature differences between the cooling body and its surroundings, the possibility of approximation of the radiation heat transfer coefficient using a constant value is definitively lost. At higher temperatures and temperature differences, the temperature dependence of the convective heat transfer coefficient begins to show [9].
Higher temperatures can also show thermal dependence of other physical parameters concentrated in the relaxation time, thus losing the character of a constant value.

3. Experimental procedure and results

The measured samples must have an area of about (10x10x2) mm^3 and must be finely ground. Matt black spray-paint is applied on all sides of the samples in order to ensure they have the same emissivity. Every sample has been measured ten times; standard deviation as well as the values of 99 percent confidence intervals (P99) were calculated [11].
The equipment consists of a thermally insulated measuring chamber [12]. The cover of the chamber has an opening for a pyro-electric sensor Raytek THERMALERT MID 02. The sample is heated above room temperature in a thermostat (~40°C). After removal from the thermostat, the sample is quickly placed into the measuring chamber which will be closed.
It is clear that at the beginning of the sample cooling the transient process take place. In the process of the sample cooling the relaxation time is not constant in whole range. To judge this process we have to choose proper interval where τ is approximately constant. From equation 9 we obtain relation

$$\tau = -t / \ln\varphi \qquad (12)$$

where

$$\ln\varphi = \ln\left|(T - T_\infty) - (T_0 - T_\infty)\right| \qquad (13)$$

Relation 13 allows to set proper, nearly constant interval of τ from its time dependence. This interval is used for next calculation of desired thermal parameters. Typical transient process for Cu sample is in Figure 2

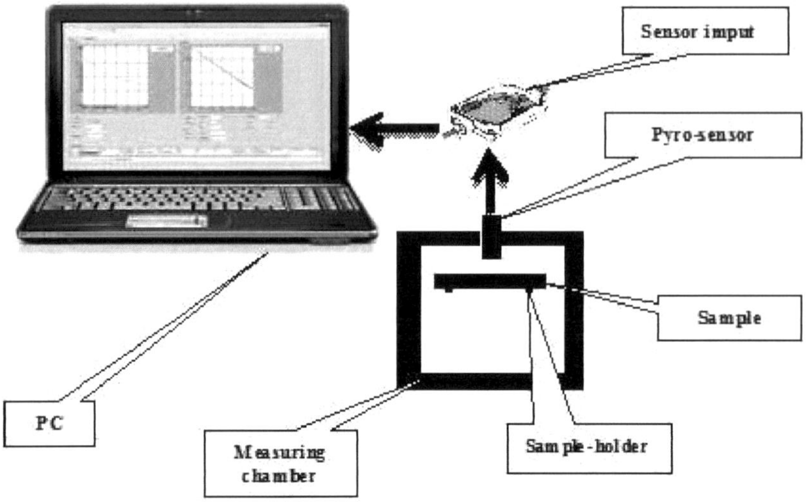

Figure 1. Block schema of the apparatus

The heart of the device is software that provides data processing. The time constant is the basis for determining other heat-transport parameters of the material. It is a reason, why the program uses in the first step search for τ. Equation 8 which represents the master equation of the problem is linearized by logarithmic function and we obtain relation $\ln(T-T_\infty) = \ln(T_0-T_\infty) - t/\tau$ which is fitted by using the method of the smallest squares and the term $1/\tau$ indicates the slope of line. The initial temperature T_0 is known from the measurement. The relaxation time τ has been determined and this value is used to determine of c_p.

In the second step we start to find further unknown physical value c_p. Because the Equation 6 expressing the relation of c_p. versus h_t has infinite number of solutions, which means that only fitting procedures can be used to determine the required thermal and physical quantities. The basic prerequisite is the knowledge or at least an estimate of the total heat transfer coefficient h_t from the material to the surrounding environment. According to experimental analysis in this step, we can set $h_t \approx (15\text{-}25)$. For further unknown c_p we set qualified estimate according to the type of material – metals, plastics, laminates, etc. (table values).

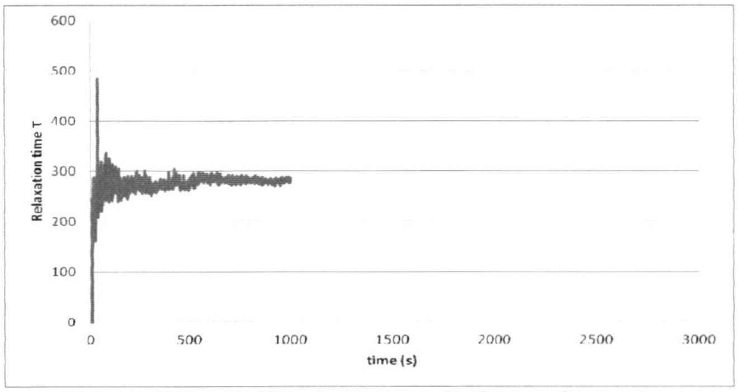

Figure 2. Transient process in *Cu* sample

The algorithm in the initial step is based on the intersection of the given and possible intervals of both quantities. Once the common intervals of meaningful values have been determined, the algorithm tries to find a single value using the interval division method by gradually decreasing the interval of both required values. After this procedure c_p and experimental h_t are known and they are used for evaluation of thermal conductivity and thermal diffusivity. Similar procedure is used for the further unknown pair of thermal parameters α and k, where the equation 9 is applied and we can rewrite it as $\alpha \rho c_p = k$, $B c_p = k$, $B = \alpha \rho$ and where for both quantities (α,k) will use qualified estimates according to the type of material at the beginning of fitting (see Table 1).

The size of the sample and its weight is also used to determine the density of the sample material.

Finally, if we have determined all k, c_p and α it is necessary to verify validity of the condition given by relation 4, which is satisfying for model validity (Biot number have to be smaller than 0.1).

After this analysis, the experiment proceeds with the presentation of data obtained from selected materials with relatively high or low value of k, c_p and α. Firstly, we present the measurements of thermal parameters obtained for Cu (Table 1).

To test the validity of results obtained for Cu by used LCM model, we have measured the same sample of copper also by Netzsch laser equipment LFA 427. Obtained result for diffusivity is $1{,}20\cdot10^{-4}$ ($m^2\cdot s^{-1}$) at ambient temperature which is very encouraging result supporting the validity of presented results together with statistical analysis.

In the next experiment we tested samples of lead with lower transport parameters α and k then in previous case. Measured and table values together with statistical parameters are in Table 2.

As we can see from the presentation of the measured data, the conformity of the measured, tabular or referential data is very good, both for low-conductive material, as well as for materials with relatively good thermal transport properties. The width of P99 confidence interval is smaller than 10 percent in all samples under investigation. Biot number is smaller than 0.1 in all cases and so used LCM is valid.

Table 1. Thermal parameters, tabular values, *Bi* number and statistical parameters for copper

Material Cu	Measured	P_{99}	Table value [13]
c_p ($J\cdot kg^{-1}\cdot K^{-1}$)	387.4±4.69	(372.17 - 402.63)	383
α ($m^2\cdot s^{-1}$)	$(1.24\pm0.01).10^{-4}$	$(1.19 - 1.28).10^{-4}$	$1.15\ 10^{-4}$
k ($W\cdot m^{-1}.K^{-1}$)	393.30±4.82	(377.65- 408.95)	400
Bi	**0.021**		

Table 2. Thermal parameters, tabular values, *Bi* number and statistical parameters for lead

Material Pb	Measured	P_{99}	Table value [13]
c_p ($J\cdot kg^{-1}\cdot K^{-1}$)	126,80±0,95	(123,71-129,89)	129
α ($m^2\cdot s^{-1}$)	$(2{,}42\pm0{,}01)10^{-5}$	$(2{,}38\text{-}2{,}46).\ 10^{-5}$	$2{,}44.10^{-5}$
k ($W\cdot m^{-1}.K^{-1}$)	34,70±0,30	(33,73-35,68)	35
Bi	**0,016**		

The third sample under investigation was PVC (polyvinyl chloride). This sample has the smallest values of k and α. The results are presented in Table 3.

Table 3. Thermal parameters, tabular values, *Bi* number and statistical parameters for *PVC*

Material PVC	Measured	P_{99}	Table value [13]
c_p ($J\cdot kg^{-1}\cdot K^{-1}$)	1399±5,04	(1382,61-1415,39)	1400
α ($m^2\cdot s^{-1}$)	(6,88±0,05).10^{-8}	(6,71-7,05) .10^{-8}	6,90.10^{-8}
k ($W\cdot m^{-1}.K^{-1}$)	0,14±0,01	(0,14-0,15)	0,14
Bi	**0,08**		

Now we compare presented method with those from other authors which are based on the exponential body cooling.

In the work [14] authors tested polyurethane matrix by pulse transient method (Thermophysical Transient Tester 1.02 [15]). They tested the application of first order exponential model. Process of cooling of specimen after its thermal stimulation was monitored during the following approximately 6000 seconds. In these experiments relaxation time τ and c_p was determined only.

Finally we can compare our method with this described in [16]. Experimental method used in this case may be summarized as follows. A thin slice of the material to be tested is held between a hot copper block and a cold copper base, which remains essentially isothermal during the measurement. After an initial transient the temperature difference between the block and the base decays exponentially, and the thermal conductivity of the sample of the material may be calculated from the exponent. Heat losses from the block to the surroundings and the influence of thermal contact resistance at the sample-block interface and the sample-based interface are established by calibration.

In comparison with previous works our method offers:

-Complete set of thermal constants obtained from one cooling curve, not only α or k.

-Method is fully contactless and demands on the sample preparation are small.

-Our apparatus is also able to measure materials with large scale of thermal parameters (see presented results).

-As we can see from the presentation of the measured data, the conformity of the measured, tabular or referential data is relatively very good.

-The width of P_{99} confidence interval is smaller than *10* percent in all samples under investigation.

- Biot number is smaller than *0.1* in all cases and so used *LCM* is valid.

4. Conclusion

The presented method used to measure k, c_p and α is based on Newton's model of cooling body, respectively on its modification in the form of lumped capacitance model. Software processing, based on qualified estimates of k, c_p and α and h_t, provides approximate solution of the characteristic equations, which allow relatively highly accurate determination of k, c_p and α. The advantage of the method is that it is contactless; it requires only small amounts of tested sample, whose surface treatment is not very demanding.

5. References

[1] Sedlaček J, Dolejší J 2010 *Trends in Agricultural Engineering.*
[2] Cheng K C 2009 Applied Mechanics Rewiews, **62** 6
[3] Vollmer M 2009 *European Journal of Physics,* **30** 5
[4] Mandelis A 1991 Elsevier, New York, 207–284
[5] Parker W J, Jenkins R J, Butler C P, Abbott G L 1961 *Journal of Applied Physics* **32** (9) 1679
[6] Akoshima M, Baba T 2006 *Int. J. of Thermophysics*, **27** (4) 1189

[7] Malinarič S, Dieška P, Krupa P 2012 *Int. J. of Thermophysics* **33** (3) 528
[8] Koštial P, Ružiak I, Jonšta Z, Kopal I, Hrehuš R, Kršková J 2010 *Int .J. of Thermophysics* **31** (3) 630
[9] Kopal I, Koštial P 2010 Investigation of thermal parameters of materials I, VŠB-TU Ostrava
[10] Lienhard J H 2004 *A Heat Transfer Textbook*, Phlogiston Press, Cambridge
[11] Meloun, M, Militký J 2002 *Compendium of Statistical Data Treatment* (in Czech), Academia
[12] Koštial P, Ružiak I, Jančíková Z 2012 *Patent* NO 303269 CS
[13] Brož J, Roskovec V, Valouch M 1990 Physical and mathematical tables (in Czech) SNTL Praha
[14] Kováč K, Kopal I, 2012. Proc. of Machine Modeling and Simulations, *The 17th International Polish-Slovak Conference*, Poznaň
[15] *Transientms*. [Online] http://www.transientms.com
[16] Lisker I S, Solovyev S V, Axcell B P, Varlow B R, Donnelly, K 2001, *Experimental Thermal and Fluid Science* **25** 377

IMEKO IOP Publishing
Journal of Physics: Conference Series **588** (2015) 012007 doi:10.1088/1742-6596/588/1/012007

Measurement data preprocessing in a radar-based system for monitoring of human movements

Roman Z. Morawski[1], **Andrzej Miękina and Paweł R. Bajurko**
Warsaw University of Technology, Faculty of Electronics and Information Technology, Institute of Radioelectronics, Warsaw, Poland

E-mail: *r.morawski@ire.pw.edu.pl*

Abstract. The importance of research on new technologies that could be employed in care services for elderly people is highlighted. The need to examine the applicability of various sensor systems for non-invasive monitoring of the movements and vital bodily functions, such as heart beat or breathing rhythm, of elderly persons in their home environment is justified. An extensive overview of the literature concerning existing monitoring techniques is provided. A technological potential behind radar sensors is indicated. A new class of algorithms for preprocessing of measurement data from impulse radar sensors, when applied for elderly people monitoring, is proposed. Preliminary results of numerical experiments performed on those algorithms are demonstrated.

1. Introduction

It is expected that the proportion of European people reaching the age of 65 years or more will increase by at least 50% during the next 30 years. The problem of organised care over elderly people, especially those suffering dementia is, therefore, of increasing importance. Hence the demand for research on new technologies that could be employed in care services for such people. Its primary objective is to examine the applicability of various sensor systems for non-invasive monitoring of the movements and vital bodily functions, such as heart beat or breathing rhythm, of elderly persons in their home environment. The capability of such systems to detect dangerous events, such as person's fall, is of principal importance. A fall is defined as *"unintentionally coming to the ground or some lower level and other than – as a consequence of sustaining a violent blow, loss of consciousness, sudden onset of paralysis as in stroke or an epileptic seizure"* [1]. A fall can occur not only when a person is standing, but also while sitting on a chair or lying on a bed during sleep. Falls among elderly people are the main cause of their admission and long-term stay in hospitals: it is the sixth cause of death for people over the age of 65, the second for people between 65 and 75, and the first for people over 75 [2]. The fall risk factors are of various nature:

- intrinsic: age, low mobility and bone fragility, poor balance, chronic disease, cognitive and dementia problems, Parkinson disease, sight problems, use of drugs that affect the mind, incorrect lifestyle (inactivity, use of alcohol, obesity), previous falls;
- extrinsic: individual (incorrect use of shoes and clothes), drugs cocktail;
- environmental: internal (slipping floors or stairs, need to reach high-located objects) and external (damaged roads, crowded places, dangerous steps, poor lighting).

[1] To whom any correspondence should be addressed.

Content from this work may be used under the terms of the Creative Commons Attribution 3.0 licence. Any further distribution of this work must maintain attribution to the author(s) and the title of the work, journal citation and DOI.
Published under licence by IOP Publishing Ltd

2. Existing monitoring techniques

There are three main categories of monitoring techniques already applied in care practice: vision-based, environmental and wearable [2–4].

The vision-based techniques are based on fixed cameras that continuously record the movement of a patient; the acquired data are processed by means of algorithms of pattern recognition that trigger an alarm in case of fall. They usually contain three types of operations: detection of inactivity, based on the idea that after a fall, the patient lies on the floor without moving; analysis of the body-shape change, based on the change of posture after the fall; and analysis of the head motion, based on the monitoring of the position and velocity of the head. The main limitations of the vision-based techniques are: the time and cost of installation, the limited space of application (within the range cameras) and privacy violation.

The environmental techniques are based on the installation of sensors in the places to be monitored – *e.g.* pressure sensors on chairs, cameras, and RFID tags embedded throughout the home of the elderly people, as well as in their furniture and clothing.

The wearable – unlike visual-based and environmental techniques which require a pre-built infrastructure – may be used outdoor. The signals form movement sensors (mainly accelerometers, and gyroscopes), worn by a patient, are transmitted via radio to a computer and analysed. This solution makes also possible the acquisition of physiological data (blood pressure, ECG, EEG, *etc.*).

3. Formulation of the research problem

Since several years numerous attempts have been made to apply radar technology for monitoring of elderly persons. They are mainly motivated by the conviction that this technology may be less intrusive than vision-based solutions, less cumbersome than the wearable solutions and less invasive with respect to the home environment than the environmental solutions. The research directions related to this topic may be broadly classified according to the spectrum of the radar signals applied. So, some researchers prefer to focus on narrow-band solutions, especially those using the Doppler principle [5–36], others opt for broad-band solutions, especially those using pulse-type signals [37–54]. The research reported in this paper belongs to the latter category: it is devoted to the analysis of measurement data acquired by means of an impulse radar sensor. A typical sequence of such data, after removing the static background, is shown in Figure 1.

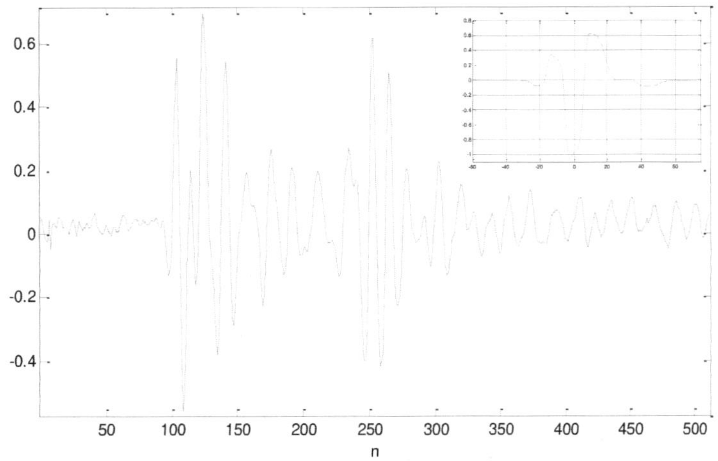

Figure 1. A typical sequence of measurement data acquired by means of a radar sensor (blue line) and the shape of the emitted pulse (read line).

The problem addressed here consists in identification of sub-sequences of data containing information on echoes. Each echo should be characterised by an estimate of its location on the time axis and an estimate of its magnitude. This is called the preprocessing of radar data, because the estimates of echo parameters, obtained for a series of emitted pulses, are to be next processed by decision-making algorithms detecting various kinds of person's abnormal behaviours, falls in particular.

4. Proposed methods for data preprocessing

4.1. Sectioning of data sequences

The first stage of data preprocessing is aimed at partitioning of the sequence $\{\tilde{y}_n | n = 1, ..., N\}$ into sub-sequences – each carrying information on a single echo only. It is, therefore, assumed here that such partitioning is possible without deconvolution-type preprocessing of the data sequence $\{\tilde{y}_n\}$. Three methods for data preprocessing, proposed in this subsection, are based on the use of an operator $\mathcal{M}[\circ]$ generating the so-called *max-envelope* of a data sequence being its argument. The max-envelope of the data sequence $\{x_n\}$ is a sequence whose elements are determined by the piecewise-linear interpolation of the non-strict maxima of $\{x_n\}$, *i.e.* the maxima satisfying the inequalities: $x_n \ge x_{n-1}$ and $x_n \ge x_{n+1}$. For example, the max-envelope of the data from Figure 1 is shown in Figure 2.

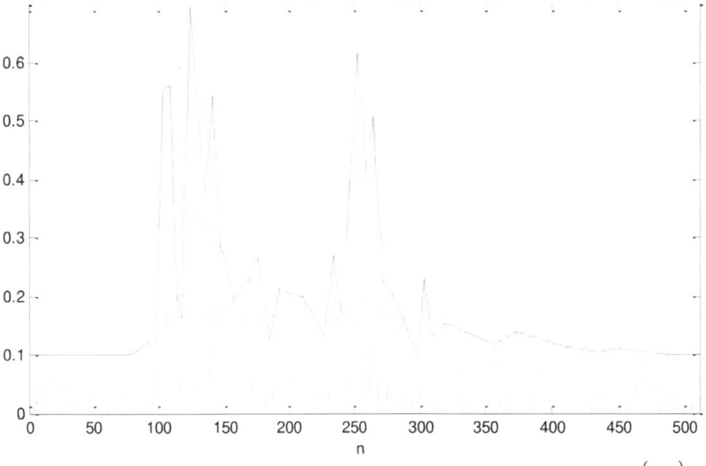

Figure 2. The sequence of the absolute values of the data from Figure 1 $\{|\tilde{y}_n|\}$ (blue line) and its max-envelope $\mathcal{M}[\{|\tilde{y}_n|\}]$ (black line).

Three, structurally similar, algorithms of sectioning are compared here; each of them is composed of three steps:

- transformation of the original sequence of radar data $\{\tilde{y}_n\}$ into an auxiliary sequence of non-negative numbers $\{\tilde{y}_n^+\}$;
- determination of the envelope $\{y_n^E\}$ of the sequence $\{\tilde{y}_n^+\}$ according to the formula:
$$\{y_n^E\} = \mathcal{M}\left[\left\{\max\left(\tilde{y}_n^+ - y_{th}^{(0)}, 0\right)\right\}\right] \tag{1}$$
where $y_{th}^{(0)}$ is a discrimination level fit to the level of errors in the data $\{\tilde{y}_n\}$;
- determination of the borders between sections according to the equation:
$$n_k = \tfrac{1}{2}n'_k + \tfrac{1}{2}n'_{k+1} \quad \text{for} \quad k = 1, ..., K-1 \tag{2}$$
where n'_k are ordered abscissas of all the maxima of the envelope, *i.e.*:

$$n'_k = \arg_n \max\left\{\max\left(y_n^E - y_{th}^{(1)}, 0\right) \middle| n = 1, \ldots, N\right\} \text{ for } k = 1, \ldots, K \tag{3}$$

with $y_{th}^{(1)}$ being a discrimination level fit to the level of errors in the data $\{\tilde{y}_n^+\}$.

The compared algorithms are labelled with two-letter acronyms (ME, HE and FE) associated with the corresponding definitions of the sequence $\{\tilde{y}_n^+\}$:

- $\{\tilde{y}_n^+\} \equiv \{|\tilde{y}_n|\}$ in the ME algorithm;
- $\{\tilde{y}_n^+\} \equiv \left|\{\tilde{y}_n\} + j\mathscr{H}[\{\tilde{y}_n\}]\right|$ in the HE algorithm;
- $\{\tilde{y}_n^+\} \equiv \left|\mathscr{F}[\{\tilde{y}_n\}; n_{coi}]\right|$ in the FE algorithm;

where $\mathscr{H}[\circ]$ is the operator of the discrete Hilbert transformation, and $\mathscr{F}[\circ; n_{coi}]$ is the operator of the ideal low-pass filtering with the cut-off index n_{coi}.

The comparative study has been based on the semi-synthetic data generated according to the formula:

$$\tilde{y}_n = \sum_{k=1}^{K} r_k \cdot x_{n-n_k} + \eta_n \text{ for } n = 1, \ldots, N \tag{4}$$

with $K = 4$, $N = 1000$; $r_1 = 1$, $r_2 = 2$, $r_3 = 0.3$, $r_4 = 3$; $n_1 = 100$, $n_2 = 400$, $n_3 = 450$, $n_4 = 800$; and η_n being pseudorandom numbers following the uniform distribution $\mathscr{U}(-d\eta, +d\eta)$. The sequence $\{x_n\}$ is representative of the impulse emitted by the impulse radar sensor NVA6100, manufactured by Novelda AS (*https://www.novelda.no /content/radar-ics*), whose shape is shown in the right upper corner of Figure 1. The study has been carried out under the following assumptions:

- $d\eta \in \{0.1, 0.08, 0.06, 0.01, 0.001\}$;
- $\frac{y_{th}^{(0)}}{d\eta}, \frac{y_{th}^{(1)}}{d\eta} \in \{0.5, 1.0, 1.5, 2.0, 4.0. 6.0, 8.0, 10.0, 20.0\}$;
- $n_{coi} \in \{100, 120, 140, 160, 180, 200, 220, 240\}$.

The statistical evaluation of the results of study has been based on $R = 1000$ repetitions of each experiment defined by a fixed combination of the above mentioned parameters – the repetitions differing in the realisations of the sequence of random errors $\{\eta_n\}$. For each experiment the number of repetitions resulting in the correct sectioning, R_{cor}, has been determined, as well as the rate of failure:

$$r_f \equiv \frac{R - R_{cor}}{R} 100\% \tag{5}$$

Selected results od comparison – obtained for the combinations of parameters (the thresholds $y_{th}^{(0)}$ and $y_{th}^{(1)}$, plus n_{coi} in case of the FE algorithm) optimised for each algorithm and for each level of errors $d\eta$ – are shown in Table 1. An example of sectioning is provided in Figure 3.

Table 1. The values of the rate of failure (r_f) characterising the compared algorithms of data sectioning, and the optimised values of their parameters: $y_{th}^{(0)}$, $y_{th}^{(1)}$, and – in case of FE – n_{coi}.

$d\eta$	Algorithm of sectioning		
	ME	HE	FE
0.001	100.0%	0.0% (0.01, 0.008)	0.0% (160, 0.01, 0.015)
0,010	100.0%	0.0% (0.06, 0.060)	0.0% (160, 0.04, 0.040)
0.060	100.0%	9.0% (0.09, 0.090)	0.0% (160, 0.06, 0.060)
0.080	100.0%	14.9% (0.12, 0.080)	2.2% (180, 0.08, 0.040)
0.100	100.0%	25.2% (0.05, 0.150)	13.7% (180, 0.10, 0.050)

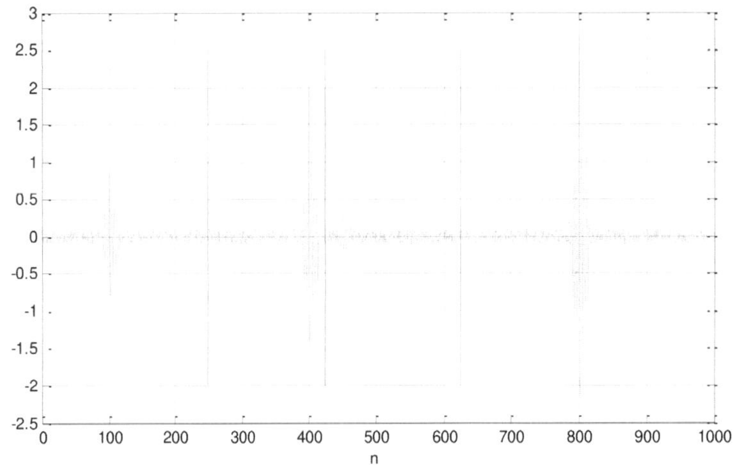

Figure 3. The results of sectioning of the data sequence defined by Eq.(4).

The higher the discrimination levels $y_{th}^{(0)}$ and $y_{th}^{(1)}$, the greater is the probability that smaller echoes may be overlooked. Two complementary strategies, aimed at diminishing the possible loss of information, seem worth being considered, *viz.*: re-sectioning of each section of data and partitioning of the discrimination process. Due to the space limitation, they will not be presented in this paper.

4.2. Estimation of echo parameters

The second stage of data preprocessing consists in estimation of the position and magnitude of an echo in each sub-sequence of the data $\{\tilde{y}_n\}$. For the sake of simplicity, a data sub-sequence will be characterised here using the same symbols as those used for the whole sequence: $\{\tilde{y}_n\}$, N, *etc.* First, the time limits of the echo, n_L and n_R, will be identified using piecewise-linear least-squares approximation of the cumulant sequence defined by the formula:

$$\tilde{y}_\nu^C \equiv \sum_{\nu=1}^{n} \max\left(y_\nu^E - y_{th}^{(1)}, 0\right) \text{ for } n = 1, \ldots, N \tag{6}$$

Next, the estimates of the echo position, $\hat{n}$, and of its magnitude, $\hat{m}$, will be computed, *viz.*:

$$\hat{n} = \tfrac{1}{2}(n_L + n_R) \text{ and } \hat{m}^2 = \sum_{\nu=n_L}^{n_R} \tilde{y}_\nu^2 \tag{7}$$

The piecewise-linear approximation is based on the use of three "independent" linear functions, *i.e.*:

$$\hat{y}_n^C = \begin{cases} a_1 n + b_1 & \text{for } n \in [1, n_L - 1] \\ a_2 n + b_2 & \text{for } n \in [n_L, n_R] \\ a_3 n + b_3 & \text{for } n \in [n_R + 1, N_R] \end{cases} \tag{8}$$

For a fixed pair $\langle n_L, n_R \rangle$, the least-squares estimates $\hat{a}_k$ and $\hat{b}_k$ of a_k and b_k $(k = 1, 2, 3)$ are expressed by the formulae:

$$\begin{bmatrix} \hat{a}_1 \\ \hat{b}_1 \end{bmatrix} = \begin{bmatrix} 1 & 1 \\ \vdots & \vdots \\ n_L - 1 & 1 \end{bmatrix}^+ \cdot \begin{bmatrix} \tilde{y}_1^C \\ \vdots \\ \tilde{y}_{n_L-1}^C \end{bmatrix}, \begin{bmatrix} \hat{a}_2 \\ \hat{b}_2 \end{bmatrix} = \begin{bmatrix} n_L & 1 \\ \vdots & \vdots \\ n_R & 1 \end{bmatrix}^+ \cdot \begin{bmatrix} \tilde{y}_{n_L}^C \\ \vdots \\ \tilde{y}_{n_R}^C \end{bmatrix} \text{ and } \begin{bmatrix} \hat{a}_3 \\ \hat{b}_3 \end{bmatrix} = \begin{bmatrix} n_R + 1 & 1 \\ \vdots & \vdots \\ N & 1 \end{bmatrix}^+ \cdot \begin{bmatrix} \tilde{y}_{n_R+1}^C \\ \vdots \\ \tilde{y}_N^C \end{bmatrix} \tag{9}$$

where $[\circ]^+$ is the Moore-Penrose pseudo-inverse of the matrix $[\circ]$. The improved estimates of n_L and n_R may be now obtained by solving two linear algebraic equations:

$$\hat{a}_1 n_L + \hat{b}_1 = \hat{a}_2 n_L + \hat{b}_2 \text{ and } \hat{a}_2 n_R + \hat{b}_2 = \hat{a}_3 n_R + \hat{b}_3 \tag{10}$$

The iterative estimation of the vectors $\mathbf{a} \equiv [a_1\ a_2\ a_3]^T$ and $\mathbf{b} \equiv [b_1\ b_2\ b_3]^T$, followed by correction of n_L and n_R, has turned out to be convergent and quite efficient algorithm for determination of n_L and n_R.

Example 1: The result of piecewise-linear approximation – obtained for a realisation of the data generated according to Eq.(4) with $K=1$, $N=400$, $r_1=1$, $n_1=200$ and $d\eta=0.1$ – is shown in Figure 4. In this case, the worst-case error of position estimation, assessed for $R=100$ realisations, of data is 1.5, and the relative worst-case error of magnitude estimation is 10.8%.♣

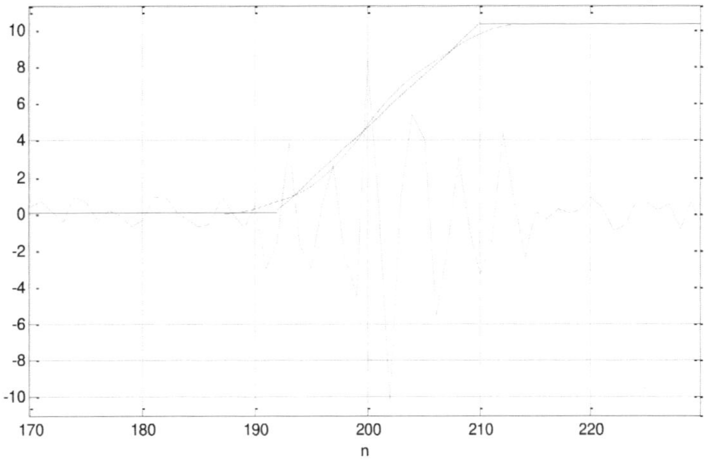

Figure 4. The piecewise-linear approximation (black line) of the sequence $\{\tilde{y}_n^C\}$ (red line) on the background of the original data $\{10\tilde{y}_n\}$ (blue line).

The uncertainty of estimation of the echo parameters depends on the echo position: it increases quickly when the echo position is approaching 1 or N. The extrapolation of the approximated sequence of data is a partial remedy for that negative effect. According to the experience acquired during numerical experiments described here, the prolongation of the sequence by $N/5$ elements to the left and by $N/5$ elements to the right seems to be sufficient.

Example 2: Some exemplary results of estimation of echo parameters – obtained on the basis of the data generated according to Eq. (4) with $K=1$, $N=400$, $r_1=1$, varying n_1, and $d\eta=0.1$ – are collected in Table 2. The indicators of estimation uncertainty, presented there, have been computed for $R=100$ realisations of each data sequence. ♣

Table 2. Results of numerical experiment described in Example 2.

Estimated parameter	Indicator of estimation uncertainty	Exact echo position						
		10	20	50	200	350	380	390
position	Bias	1.015	0.69	0.61	0.715	0.63	0.515	−0.13
	Standard deviation	0.49	0.56	0.58	0.52	0.48	0.61	0.79
	Worst-case error	2	2	2	1.5	2	2	3
magnitude	Relative bias	−4.13%	−3.54%	−4.10%	−4.01%	−3.93%	−3.93%	−7.32%
	Relative standard deviation	2.92%	3.09%	3.22%	3.00%	3.21%	2.65%	3.55%
	Relative worst-case error	11.02%	10.84%	11.00%	10.80%	11.92%	12.04%	15.92%

5. Conclusion

A set of algorithms for preprocessing of data from an impulse radar sensor has been proposed. It consists of three alternative algorithms for sectioning of those data (labelled ME, HE and FE), and an algorithm for estimation of echo parameters in each section. If the sectioning algorithms are concerned:

- The ME algorithm is the simplest (in terms of the numerical complexity), but least reliable. It may be used provided the estimation of echo parameters is preceded by an additional operation aimed at elimination of false echoes.
- The HE and FE algorithms, if correctly implemented, are comparable in terms of numerical complexity, but the second of them is more reliable. The HE algorithm is performing perfect sectioning till the error level $d\eta = 0.01$, while the FE algorithm – till the error level $d\eta = 0.06$.
- The rate of failure for the FE algorithm is relatively low (*ca.* 14%) even for $d\eta = 0.1$; therefore, it may be applied for data subject to such considerable errors, provided a mechanism for elimination of outliers is included into a set of algorithms for data preprocessing.
- The FE algorithm is also most tolerant with respect to the values of its parameters ($y_{th}^{(0)}$, $y_{th}^{(1)}$, and n_{coi}).

If the algorithm for estimation of echo parameters is concerned:

- The estimation of the echo position is very robust with respect to the errors in the data: even for $d\eta = 0.1$, the random errors of estimation do not exceed 2 sampling units. This is an important advantage of the proposed algorithm since the estimates of the position are to be used for evaluation of movements of the target; so, their sequences will be subject to explicit or implicit differentiation, as a rule amplifying random errors.
- The estimates of the echo magnitude are subject to considerable systematic errors due to the piecewise linear approximation of the data resulting in a systematic limitation of the support of their subsequence identified as an echo.
- At the same time, the random error of those estimates is *ca.* 3 times smaller than the errors in the data. This is again an important advantage of the proposed algorithm since the estimates of the magnitude are to be used for evaluation of movements of the target; so, their sequences will be also subject to differentiation eliminating the systematic errors.

Acknowledgement

This work has been accomplished within the project PL12-0001 financially supported by *EEA Grants – Norway Grants* (*http://eeagrants.org/project-portal/project/PL12-0001*).

References

[1] Gibson M Andres R Isaacs B Radebaugh T and Worm-Petersen J 1987 *Danish Medical Bulletin* **34**(4) pp 1–24
[2] Abbate S Avvenuti M Corsini P Light J and Vecchio A 2010 [in] *Wireless Sensor Networks: Application – Centric Design*, Eds Geoff V Merrett and Yen Kheng Tan (Intech Pub) Chapter 9
[3] Alemdar H and Ersoy C *Computer Networks* 2010 **54** pp 2688–2710
[4] Noury N Fleury A Rumeau P Bourke A K Laighin G Ó Rialle V and Lundy J E 2007 *Proc 2007 IEEE-EMBS International Conference* pp 1663–1666
[5] Bilik I and Tabrikian J *IEEE Transactions on Aerospace and Electronic Systems* 2007 **43**(4) pp 1510–1522
[6] Birsan N Munteanu D Iubu G and Niculescu T 2011 *Proc 2011 E-Health and Bioengineering Conference* pp 1–4
[7] Changzhi-Li and Jenshan-Lin 2010 *Proc 2010 Asia-Pacific Microwave Conference* pp 283–290
[8] Droitcour A D Boric-Lubecke O and Kovacs G T A 2009 *IEEE Transactions on Microwave Theory and Techniques* **57**(10) pp 2498–2507

[9] Fairchild D P and Narayanan R M 2013 *Proc SPIE 'Active and Passive Signatures IV'* **8734** pp 0701–0711

[10] Fletcher R and Jing-Han 2009 *Proc 2009 IEEE MTT-S International Microwave Symposium* pp 1325–1328

[11] Fok-Hing-Chi-Tivive Son-Lam-Phung and Bouzerdoum A 2013 *Proc SPIE 'Active and Passive Signatures IV'* **8734** pp 0601–0612

[12] Fortuny-Guasch J Sammartino P F and Petit J L 2009 *Proc 2009 International Carnahan Conference on Security Technology* pp 221–226

[13] He-Tan Dengyu-Qiao and Ye-Li 2012 *Proc 2012 International Conference on Systems and Informatics* pp 1711–1714

[14] Jingli-Li Son-Lam-Phung Fok-Hing-Chi-Tivive and Bouzerdoum A 2012 *Proc 2012 International Joint Conference on Neural Networks* pp 1–6

[15] Li-Fei Huang-Binke Zhang-Hang and Du-Hao 2012 *Proc 2012 Global Symposium on Millimeter Waves* pp 326–329

[16] Liang-Liu Popescu M Ho K C Skubic M and Rantz M 2012 *Proc IEEE-EMBS International Conference on Biomedical and Health Informatics* pp 256–259

[17] Liang-Liu Popescu M Rantz M and Skubic M 2012 *Proc IEEE-EMBS International Conference on Biomedical and Health Informatics* pp 180–183

[18] Liang-Liu Popescu M Skubic M Rantz Yardibi T and Cuddihy P 2011 *Proc International Conference on Pervasive Computing Technologies for Healthcare* pp 222–225

[19] Massagram W Lubecke V M and Boric-Lubecke O 2011 *Proc 2011 International Conference of the IEEE Engineering in Medicine and Biology Society* pp 1544–1547

[20] Moulton M C Bischoff M L Benton C and Petkie D T 2010 *Proc SPIE 'Millimetre Wave and Terahertz Sensors and Technology III'* **7837** pp 0L1–0L7

[21] Obeid D Sadek S Zaharia G and Zein G E 2011 *Proc 2011 E-Health and Bioengineering Conference* pp 1–4

[22] Othman M Sinnappa M Azman H Aziz M Ismail M Hussein M Sulaiman H Misran M Said M Ramlee R Soh-Ping J and Ahmad B 2013 *Proc 2013 Conference 'Radioelektronika'* pp 367–370

[23] Petrochilos N Rezk M Host-Madsen A Lubecke V and Boric-Lubecke O 2007 *Proc IEEE International Conference on Acoustics Speech and Signal Processing* pp I333– I336

[24] Pfanner F Allmendinger T Flohr T and Kachelrieß M 2013 *Proc SPIE 'Medical Imaging 2013: Physics of Medical Imaging'* 2013 8668 pp 3731–3712

[25] Phillips C E 2012 *Walk Detection Using Pulse-Doppler Radar* MS Thesis Faculty of the Graduate School at the University of Missouri-Columbia

[26] Phillips C E Keller J Popescu M Skubic M Rantz M Cuddihy P E and Yardibi T 2012 *Proc IEEE-EMBS International Conference on Biomedical and Health Informatics* pp 5863–5866

[27] Postolache O Girao P Pinheiro E Madeira R Pereira J M D Mendes J Postolache G and Moura C 2012 *Proc 2011 IEEE Instrumentation and Measurement Technology Conference* pp 1–5

[28] Rahman M S Haque M M Byung-Jun-Jang and Ki-Doo-Kim 2012 *Proc 2012 International Conference on Electrical & Computer Engineering* pp 264–267

[29] Sekine M and Maeno K 2011 *Proc 2011 IEEE Sensors Applications Symposium* pp 318–322

[30] Sekine M Maeno and K Kamakura T 2012 *Proc 2012 Health Informatics Workshop* pp 1–6

[31] Shobha-Sundar-Ram Yang-Li Adrian-Lin and Hao-Ling 2008 *Journal of the Franklin Institute* **345** pp 679–699

[32] Tahmoush D 2013 *Proc SPIE 'Active and Passive Signatures IV'* **8734** pp 0301–0307

[33] Tariq A and Shiraz H G 2010 *Proc 2010 Loughborough Antennas and Propagation Conference* pp 293–296

[34] Yazhou-Wang and Fathy A E 2013 *Proc SPIE 'Active and Passive Signatures IV'* **8734** pp 0401–0409

[35] Yazhou-Wang Quanhua-Liu and Fathy A E 2013 *IEEE Transactions on Geoscience and Remote Sensing* **51** No 5 pp 3097–3107

[36] Zakrzewski M Raittinen H and Vanhala J 2012 *IEEE Sensors Journal* **12**(3) pp 627–634
[37] Baboli M Boric-Lubecke O Lubecke V 2012 *Proc IEEE-EMBS International Conference on Biomedical and Health Informatics* pp 3947–3950
[38] Baldi M Appignani F Zanaj B and Chiaraluce F 2012 *Proc 2012 IEEE First AESS European Conference on Satellite Telecommunications* pp 1–6
[39] Baldi M Chiaraluce F Zanaj and B Moretti M 2011 *Proc. 2011 IEEE International Conference on Ultra-Wideband* pp 341–345
[40] Bernardi P Cicchetti R Pisa S Pittella E Piuzzi E Testa O 2012 *Proc 2012 International Symposium on Electromagnetic Compatibility* pp 1–6
[41] Chi-Hsuan-Hsieh Yi-Hsiang-Shen Yu-Fang-Chiu Ta-Shun-Chu and Yuan-Hao-Huang 2013 *Proc 2013 IEEE International Symposium on Circuits and Systems* pp 1079–1082
[42] Immoreev I and Ivashov S 2008 *Proc 2008 International Conference on Ultrawideband and Ultrashort Impulse Signals* pp 34–38
[43] Jing-Li Lanbo-Liu Zhaofa-Zeng and Fengshan-Liu 2013 *IEEE Journal of Selected Topics in Applied Earth Observations and Remote Sensing* **PP**(99) pp 1–9
[44] Lazaro A Girbau D Villarino R and Ramos A 2011 *Proc 2011 European Microwave Conference* pp 135–138
[45] Leib M Menzel W Schleicher B and Schumacher H 2010 *Proc 2010 European Conference on Antennas and Propagation* pp 1–5
[46] Sakamoto T Sato T Yuan-He Aubry P J and Yarovoy A G 2013 *Proc. 2013 URSI International Symposium on Electromagnetic Theory* pp 119–122
[47] Sang-Hyun-Chang Mitsumoto N and Burdick J W 2009 *Proc 2009 IEEE Radar Conference*, pp 1–6
[48] Sang-Hyun-Chang Ta-Shun-Chu Roderick J Chenliang-Du Mercer T Burdick J W and Hashemi H *Proc 2011 IEEE International Conference on Ultra-Wideband* pp 355–359
[49] Sang-Hyun-Chang Wolf M and Burdick J W 2010 *Proc 2010 IEEE International Conference on Robotics and Automation* pp 452–457
[50] Sharafi A Baboli M and Eshghi M 2010 *Proc 2010 International Conference on Bioinformatics and Biomedical Engineering* pp 1–4
[51] Sharifahmadian E and Ahmadian A 2009 *Proc IEEE-EMBS International Conference on Biomedical and Health Informatics* pp 5717–5720
[52] Yarovoy A G Ligthart L P Matuzas J and Levitas B 2006 *IEEE Aerospace and Electronic Systems Magazine* **21**(3) pp 10–14
[53] Yarovoy A G Zhuge X Savelyev T G and Ligthart L P 2007 *Proc. 2007 European Microwave Conference* pp 1574–1577
[54] Yifan-Chen and Rapajic P 2008 *International Journal of Automation and Computing* **4**(5) pp 325–333

IMEKO IOP Publishing
Journal of Physics: Conference Series **588** (2015) 012008 doi:10.1088/1742-6596/588/1/012008

A property type invariant model of measurement applied to nominal evaluations

Luca Mari[1,#] and Arianna Mencattini[2]

[1] Università Cattaneo – LIUC, School of Industrial Engineering, Italy
[2] University of Rome Tor Vergata, Dept. of Electronic Engineering, Italy

lmari@liuc.it

Abstract. The scientific community and the standardization world are exploring the concepts of nominal evaluation and nominal property in the metrological context. In this paper we show that measurement can be framed in a structural context that is independent of the algebraic structure assumed for the measurand and the measured values. In such a framework the basic metrological concepts are consistently applied also to the nominal case.

1. Introduction: the problem of measurability of nominal properties

Measurement is loaded of stereotypes, and some of them are related to the concept itself of measurability: *what can be measured?* Indeed, expressions such as "counting and measuring" and "weights and measures" suppose that counting and weighing are not ways to measure. It is the plausible trace of the Euclidean standpoint, in which only geometric quantities were considered object of mensuration, intended as a mathematical activity aimed at comparing quantities by their ratio. The theory underlying Lebesgue measure is the outcome, where not confusing "measure" and "measurement" here is then crucial [1]. Referring to this tradition to justify that only ratio (or even only additive) quantities are measurable is then a weak strategy. It is true that the variables appearing in physical laws generally denote ratio quantities, and that the powerful tools of quantity calculus / dimensional analysis apply only to ratio quantities, that are structurally more informative than, e.g., purely ordinal properties. But since values of ratio quantities can be obtained by means other than measurement, e.g., subjective estimation and guess, being a ratio quantity is not sufficient for being object of measurement: is it necessary? [2] Not so surprisingly, a negative answer comes from the *International Vocabulary of Metrology* (VIM) [3], that in its latest edition extends measurement to ordinal entities and makes the move lexically explicit by terming them "ordinal quantities", an expression that would be considered contradictory in the Euclidean framework. The VIM also defines 'nominal property', at the same time stating that "measurement does not apply to nominal properties" (and in fact the chosen term is not "nominal quantity"), because they "have no magnitude", thus implicitly implying that "having magnitude" is a necessary condition for a property to be measurable (note that 'magnitude' is not defined by the VIM; is it, in particular, identical to 'rank', as in "shape cannot be measured because objects cannot be ranked with respect to shape" [4] p.99 ?). Since measurement is a designed-on-purpose, instead of natural, process, one is surely free to characterize it as she considers the best, and in fact multiple strategies of definition have been followed [5]. What

[#] One of the authors is a member of the Joint Committee on Guides in Metrology (JCGM) Working Group 2 (VIM). The opinion expressed in this paper does not necessarily represent the view of this Working Group.

Content from this work may be used under the terms of the Creative Commons Attribution 3.0 licence. Any further distribution of this work must maintain attribution to the author(s) and the title of the work, journal citation and DOI.
Published under licence by IOP Publishing Ltd

does not seem so conventional is the good quality assumed of the measurement results (intended as accuracy, reliability, ...), that is not generally expected from, e.g., subjective estimations and guesses. On the other hand, this criterion is in principle unrelated to the condition of having magnitude, so that activities are in progress (both in the scientific community, e.g., [6], and in the standardization world, e.g., [7]) to better explore the concepts of nominal property and its evaluation, in a metrological context and with a metrological approach.

Our contribution here is to show that *the invariant features of a measurement process are independent of the algebraic structure of the measurand*, and therefore that from the point of view of the structure of the process the concept of measurement could be safely generalized so to encompass nominal properties. The analysis will lead us to show, in particular, that the usual contraposition measurement vs classification is misleading, as classification is just a component of the process of "reasonable attribution of one or more property values to a nominal property" (a paraphrase of the VIM definition of 'measurement', where "quantity" is used instead of "nominal property"). Whether this generalization will be lexically accepted is mainly a matter of social conventions.

2. Nominal properties and nominal evaluations

In the context of measurement science nominal properties have been often treated in confused way. Consider for example: "The usual method of assigning code numbers to entities or unordered classes is, in effect, to select integers randomly (without replacement) from the sequence 1, 2, ..., k. It is because numbers so selected bear no relationship to the 'relative amount or degree of a property possessed by the object' or class that the construction of such 'nominal scales' is not measurement." [8] p.127. Together with the peculiar hypothesis that a scale construction is a measurement, the implicit message is conveyed here that nominal properties are not empirical entities, and therefore assigning values to them is a matter of conventions (as in the unfortunate example given by the VIM: "ISO two-letter country code") or even randomness. In what follows we show that this is just wrong.

Two quantities, and more generally two properties of the same kind are by definition mutually comparable. The simplest condition of comparability is indistinguishability: if a property P_a of an object a and a property P_b of an object b are of the same kind P then the two properties are either indistinguishable, $P_a=P_b$, or $P_a \neq P_b$ ($P_a=P_b$ can be considered conveying the same information as $a \approx_P b$, i.e., the objects a and b are indistinguishable with respect to the general property / kind P, "P-indistinguishable" for short). While for some properties this information can be refined, for example by splitting the case $P_a \neq P_b$ as either $P_a < P_b$ or $P_a > P_b$, for some other properties the only comparison that conveys meaningful information is the one of P-indistinguishability. The entities for which this happens are called *nominal properties*.

By now recalling the VIM definition of 'measurement', "process of experimentally obtaining one or more quantity values that can reasonably be attributed to a quantity", the basic issue arises: is it possible, and in the case how, to design, setup and perform *a process of experimentally obtaining one or more property values that can reasonably be attributed to a nominal property*, in such a way that the same "reasonability" criterion applies in both cases?

We believe that the process of "reasonable attribution" of a value to the property P_x, for an object x under consideration, can be intended in measurement science in a purely structural way, as follows.

S1. *Define a set of reference properties*: a set P^* of properties P_i^* of the same kind P is defined, such that if $i \neq j$ then $P_i^* \neq P_j^*$ for all $P_i^*, P_j^* \in P^*$.

S2. *Establish a criterion of P-indistinguishability with reference properties*: a criterion is defined to operatively assess that for any candidate property P_x there is one and only one reference property $P_i^* \in P^*$ such that $P_x=P_i^*$. This may require producing a set S of objects s_i, each of them realizing a reference property P_i^*, so that the assessment is performed by comparing objects, $x \approx_P s_i$, or it could be structured as an algorithm that recognizes the patterns P_i^*.

S3. *Attribute a value to P_x*: the defined criterion is applied, for example by comparing the candidate object x with the objects in S with respect to the general property P, so that once the object s_i is found such that $P_x=P_i^*$, P_i^* as value of P_x is attributed.

Some notes on this process.

First and most important. The steps S1-S3 are independent of the type of the involved properties, and thus apply indifferently to both, e.g., nominal properties and ratio quantities. The fact that in the latter case some specific techniques can be adopted (see the next note) does not eliminate this commonality.

Second. The step S1 is sometimes considered of "scale construction". In the case of ratio quantities, for which there is a "natural zero" property, it can be performed by choosing a "unit" property, P_1^*, and then obtaining P_i^* by concatenating P_1^* i times with itself. A nominal property does not have a "natural zero" nor a concatenation, and therefore this construction procedure cannot be exploited and the reference properties in P^* must be explicitly listed (note that the same applies to ordinal quantities).

Third. The step S2 usually includes the process of "standards dissemination": the objects with which the candidate object is possibly compared are not those originally produced (the "primary standards") but some suitably calibrated replicas of them, so to guarantee the (metrological) traceability to P^*. This also shows that a critical requirement is the stability of the reference properties in P^*.

Fourth. While the generic linguistic form of property values is "P_i^* in P^*" (as in "3.5 in Richter scale"), in the case of ratio quantities the name of the unit, "P_1^*", can be considered as the name of the whole reference set, and property values can be denoted as "$i\ P_1^*$" (as in "3.5 m"), actually meaning "i times P_1^*".

Fifth. Let us call *nominal evaluation* the process S1-S3 when applied to a nominal property. Then nominal evaluations and classifications (intended as clustering of objects according to property) are distinct processes, and while classification is induced by nominal evaluation (two objects belong to the same class if there are evaluated as P-indistinguishable) the converse does not hold (consider, e.g., that a classification might be obtained without the previous choice of reference properties, and therefore with no constraints on traceability).

Sixth and as a synthesis. The good quality assumed of the measurement results is obtained by suitably designing, setting up and performing the process S1-S3, in particular:
- by a properly refined criterion of P-indistinguishability, so that the set P^* of reference properties includes many elements (and quantity values have a high number of significant digits in the case of ratio quantities);
- by selecting stable and universal reference properties (such as the speed of light in vacuum or the electron charge);
- by producing stable primary and working standards, thus building an appropriate traceability chain; by experimentally performing the comparison in S3 by means of high quality instrumentation, thus guaranteeing the objectivity of the outcome.

Interestingly, all these conditions are independent of the type of the involved properties.

3. A note on multivariate nominal evaluation

Exactly as quantities, also nominal properties can be multivariate, but the lack of algebraic structure allows us to avoid the peculiar consideration of the VIM: "A quantity as defined here is a scalar. However, a vector or a tensor, the components of which are quantities, is also considered to be a quantity." [3] 1.1 Note 5. The simpler characterization is here that a nominal property can be multivariate, and nevertheless the structure of the process S1-S3 is maintained.

This can be easily shown. Let $P_{(k)i}^*$ be the i-th reference property for the k-th component of the multivariate general property $\langle P_{(k)} \rangle$ under consideration. Then:

S1. *Define a set of reference properties for each $P_{(k)}$ in $\langle P_{(k)} \rangle$*: a set $P_{(k)}^*$ of properties $P_{(k)i}^*$ of the same kind $P_{(k)}$ is defined, such that if $i \neq j$ then $P_{(k)i}^* \neq P_{(k)j}^*$ for all $P_{(k)i}^*, P_{(k)j}^* \in P_{(k)}^*$.

S2. *Establish a criterion of $P_{(k)}$-indistinguishability with reference properties for each $P_{(k)}$ in $\langle P_{(k)} \rangle$*: a criterion is defined to operatively assess that for any multivariate candidate property $\langle P_{(k)x} \rangle$ there is one and only one reference property $P_{(k)i}^* \in P_{(k)}^*$ such that $P_{(k)x} = P_{(k)i}^*$.

S3. *Attribute a value to* $\langle P_{(k)x}\rangle$: the defined criterion is applied for each $P_{(k)}$ in $\langle P_{(k)}\rangle$ so to find $P_{(k)x} = P_{(k)i}^*$, and on this basis $\langle P_{(k)i}^*\rangle$ as value of $\langle P_{(k)x}\rangle$ is attributed.

This formulation frames nominal evaluations in a very general context, in which the multivariate properties $\langle P_{(k)x}\rangle$ are classified by their evaluation, and marginal (and less refined) classifications are obtained by clustering according to a partial criterion of $P_{(k)}$-indistinguishability that applies only to some components of $\langle P_{(k)}\rangle$.

On this basis, in the following section we provide a practical example of nominal evaluations.

4. An example of nominal evaluations in clinical practice

The perception of breast tumors may be difficult, especially for small lesions. To assist radiologists in standardly reporting the detected breast lesions, the American College of Radiology (ACR) occasionally publishes or updates the Breast Imaging-Reporting and Data System (BI-RADS) [9], used by medical professionals to communicate a patient's risk of developing breast cancer and the characteristics of the abnormal signs found in the mammogram. According to the BI-RADS, a breast lesion should be described in terms of its margin, shape, and density. Hence, using the evaluation process proposed above, it is assumed that a lesion is characterized by a bivariate property $\langle P_{(1)}$ = margin, $P_{(2)}$ = shape$\rangle$, whose components are respectively the margin of the lesion and the shape of the region delimited by the margin. Steps S1-S3 can be then applied to perform an evaluation.

S1. *Define a set of reference properties for* $P_{(1)}$ and $P_{(2)}$: in this case the sets of reference properties are described as follows (the meaning of the adjectives by which the properties are described is not explained here for the sake of brevity):

$P_{(1)1}^*$ = Circumscribed, $P_{(1)2}^*$ = Microlobulated, $P_{(1)3}^*$ = Indistinct, $P_{(1)4}^*$ = Obscured, $P_{(1)5}^*$ = Spiculated;

$P_{(2)1}^*$ = Round, $P_{(2)2}^*$ = Oval, $P_{(2)3}^*$ = Lobular, $P_{(2)4}^*$ = Irregular.

From these adjectives the reader can easily, and correctly, infer that each component is nominal, as in turn the bivariate property is.

S2. *Establish a criterion of* $P_{(k)}$*-indistinguishability with reference properties for* $P_{(1)}$ and $P_{(2)}$: in this case this requires establishing a criterion that, given a lesion x, is able to recognize the i-th reference property $P_{(1)i}^*$ from which its margin $P_{(1)x}$ is indistinguishable and the j-th reference property $P_{(2)i}^*$ from which its shape $P_{(2)x}$ is indistinguishable. To this goal a computer-assisted analysis system is usually exploited, that performs the following steps:

i) automatically detect the lesion margin and the region delimited by the margin [10];
ii) compute a set of numerical features $\{f_M\}$ on the margin such as fractal features [11];
iii) compute a set of numerical features $\{f_S\}$ on the shape of the region delimited by the margin, such as eccentricity, compactness, circularity, elongation [12];
iv) apply a supervised machine learning algorithm to the extracted features $\{f_M, f_S\}$ in order to assign margin and shape values among those identified in step S1.

S3: *Attribute a value to* $\langle P_{(1)x}, P_{(2)x}\rangle$: in this case this requires applying the computer-assisted analysis system setup in step S2 to a lesion x identified in a mammogram and assign a pair $\langle P_{(1)i}^*, P_{(2)j}^*\rangle$ to the bivariate property margin-and-shape, MS for short.

On this basis, what follows aims at showing two facts. First, that systems based on nominal evaluations can convey high quality information and lead to high performance decisions. In particular, margin and shape may allow to correctly recognize the histological characteristic of a large amount of lesions. Second, that results of nominal evaluations may be further elaborated by means of "indirect evaluation" procedures to compute further properties, mass *malignancy* in this case.

In order to provide a demonstrative example of these facts, using statistically well-founded data, let us consider the largest public dataset of reported mammographic images available for the community: the Digital Database for Screening Mammography (DDSM) [13]. We collected the margin and shape assessed by expert radiologists from different hospitals for the 1714 lesions in the DDSM (828 malignant tumors and 886 benign masses), and we implemented, as an indirect evaluation, a decision tree classifier to automatically assign malignancy/benignity to each lesion, using a leave-one-patient-out cross-validation strategy. To show the first fact, we implemented a tree-based classifier trained on a single component, i.e., one for margin and one for shape. Performance is evaluated in terms of the proportion of actual results identified by each algorithm (malignant tumors and benign masses correctly classified by the algorithm). To prove the second fact, we implemented a tree-based classifier trained on the bivariate property, margin and shape, and computed the same performance metrics as before.

The table below lists the values for the bivariate classifier and those obtained for each of the univariate classifier. It can be noticed, considering the bolded values highlighted in the table, that nominal evaluation of margin alone or of shape alone can lead to good results in terms of the recognition of true malignant tumors and true benign masses respectively. However, when the two targets should be simultaneously achieved (i.e., recognition of malignant tumors as well as of benign masses), then nominal evaluation based on the bivariate property (margin and shape) may increase the overall proportion of correctly identified results. This solution simultaneously reduces the recall rate (a patient with a lesion erroneously recognized as malignant is recalled for biopsy) or the missing rate (a patient with a tumor erroneously recognized as benign is not recalled as it should be with a consequent cancer spread out in the breast).

	Margin-Shape classification tree	Margin classification tree	Shape classification tree
Proportion of correctly recognized results	0.78	0.75	0.73
Proportion of malignant tumors recognized	0.74	0.84	**0.59**
Proportion of benign masses recognized	0.81	**0.67**	0.86

5. Conclusions

In other papers, e.g., [14], we proposed that measurement should be characterized in pragmatic terms, as a property evaluation process structured in such a way to be *able to guarantee, in principle, the objectivity and intersubjectivity of its results*, where then the provided information should be:
- specific to the measurand and independent of any other property of the object or the surrounding environment: hence objective;
- interpretable in the same way by different users in different places and times: hence intersubjective.

In order to comply with the expectations of objectivity and intersubjectivity, measurement science and technology refined their means in the course of time, but their strategy remained basically the same, being based in particular on:
- measuring systems that are stable and selective, i.e., that maintain their metrological features in the inter-calibration time intervals and minimize the effects of influence quantities on produced results;
- traceability chain that allow refer measurement results to universally accessible units.

In the perspective explored in this paper, the steps S1-S3 can be thought as the structural low-level implementation of these features / requirements. The fact that these steps are independent of the property types shows that nothing prevents to design, setup, and operate processes of nominal evaluation that, in the sense mentioned above, lead to objective and intersubjective results. Of course, the differences in reference set / scale construction and invariant transformations (and therefore, critically, methods for uncertainty evaluation) between, e.g., nominal properties, ordinal quantities, and ratio quantities cannot be neglected [15], but whether nominal evaluation should be included in the context of measurement or maintained in a parallel realm seems to be just unrelated to criteria of experimental quality of results: the public trust that distinctively characterizes measurement can be given also to measurement-like (i.e., structured according to S1-S3) nominal evaluations.

References

[1] Bunge M 1973 On confusing 'Measure' with 'Measurement' in the methodology of behavioral science *The methodological unity of science* ed Bunge M (Dordrecht-Holland: D. Reidel)
[2] Mari L, Maul A, Torres Irribarra D and Wilson M 2013 Quantification is neither necessary nor sufficient for measurement *IOP J Phys Conf Ser* **459 012007**
[3] JGCM 200:2012 *International Vocabulary of Metrology – Basic and general concepts and associated terms* (VIM, 2008 edition with minor corrections)
[4] Savage CW 1970 *The measurement of sensation: a critique of perceptual psychophysics* (Berkeley: University of California Press)
[5] Mari L 2013 A quest for the definition of measurement *Measurement* **46** 2889-2895
[6] Watanabe H 2005 Coarse-grained information in formal theory of measurement *Measurement* **38** 295–302
[7] Nordin G, Dybkaer R, Forsum U, Fuentes-Arderiu X, Schadow G and Pontet F 2010 An outline for a vocabulary of nominal properties and examinations – Basic and general concepts and associated terms *Clin Chem Lab Med* **48(11)** 1553–1566
[8] Duncan OD 1984 *Notes on social measurement. Historical and critical* (New York: Russell Sage Foundation)
[9] American College of Radiology (ACR) 2003 *Breast Imaging Reporting and Data System Atlas (BI-RADS® Atlas)* (Reston, Va: American College of Radiology)
[10] Mencattini A, Salmeri M, Casti P, Raguso G, L'Abbate S, Chieppa L, Ancona A, Mangieri F and Pepe ML 2011 Automatic breast masses boundary extraction in digital mammography using spatial fuzzy c-means clustering and active contour models *IEEE Int. Workshop on Medical Measurements and Applications (MeMeA)*
[11] Rangayyan RM and Nguyen TM 2007 Fractal analysis of contours of breast masses in mammograms *J. Digital Imaging* **20(3)** 223–237
[12] Mencattini A, Salmeri M, Rabottino G and Salicone S 2010 Metrological characterization of a CADx system for the classification of breast masses in mammograms *IEEE Trans Instr Meas* **59(11)** 2792–2799
[13] Heath M, Bowyer KW, Kopans D, Moore R, Kegelmeyer P 1998 Digital Mammography, chapter Current status of the Digital Database for Screening Mammography, 457–460, *Kluwer*
[14] Mari L, Carbone P and Petri D 2012 Measurement fundamentals: a pragmatic view *IEEE Trans Instr Meas* **61(8)** 2107–2115
[15] Stevens SS 1946 On the theory of scales of measurement *Science* **103(2684)** 677–680

IMEKO IOP Publishing
Journal of Physics: Conference Series **588** (2015) 012009 doi:10.1088/1742-6596/588/1/012009

An innovative method of generating current and thermoelectric equipment for its realization

M Kušnerová, J Valíček, M Harničárová, P Koštial, Z Jančíková

VŠB-Technical University of Ostrava, Czech Republic

E-mail: pavel.kostial@vsb.cz

Abstract. The paper assesses an innovative thermoelectric device used to generate a current by the conversion of thermal energy into electrical energy. The device has been created and verified. Until now, the efficiency of conventional thermoelectric batteries is for technical practice not sufficiently appreciated, because it does not exceed 3 %. A necessary condition needed for the implementation of the Seebeck thermoelectric effect is a sufficient and stable source of heat that provides an optimum temperature difference, but there are two other sufficient conditions for a significant increase in efficiency of thermoelectric batteries. These are concerned with the development of new materials for thermoelectric batteries and with the development of new construction of thermoelectric batteries.

For nearly 200 years, four thermoelectric effects, which coexist in reality, have been known: Seebeck, Thomson, Peltier and Benedick. By means of technical adjustments of the device, one of the thermoelectric phenomena may be preferred over the others. In the present case of thermoelectric current generation, it is primarily related to the application of Seebeck effect, and secondarily with the closely related but hardly measurable Thomson effect.
The Seebeck thermoelectric voltage occurs in a closed electrical circuit composed of two conductors, which are made up of two different materials and whose ends have a perfect conductive contact ensured by soldering. When a temperature difference between these two connections has been established, the two voltages working against each other arise in the circuit and give rise to an electromotive voltage. As a result of this thermo-voltage it is created the thermoelectric current, which generates Joule heat. However, the conductors are not heated uniformly along their entire length, but they have different temperatures at different locations, therefore the Thomson effect is applied as well. As a consequence of this phenomenon, the thermoelectric voltage occurs also in a single homogeneous conductor and on the assumption that the conductor is heated to different temperatures in different locations.

Thermoelectric couples [1, 2, 3, 4] are connected in series and in a cascade connection in order to increase the thermoelectric voltage, forming a thermoelectric battery (Figure 1).

Content from this work may be used under the terms of the Creative Commons Attribution 3.0 licence. Any further distribution of this work must maintain attribution to the author(s) and the title of the work, journal citation and DOI.
Published under licence by IOP Publishing Ltd

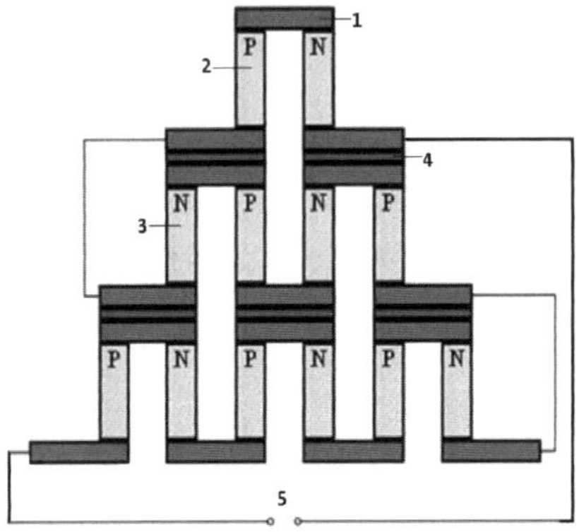

Figure 1. Classic cascade thermoelectric battery scheme
1 - connecting bridge, 2 - column with P-type conductivity, 3 - column with N-type conductivity, 4 - insulating layer, 5 - DC voltage reading due to heating of the bottom part of the thermoelectric battery

In terms of technical applications, however, the generated thermoelectric voltage is relatively very small (when using conventional thermoelectric materials, it is necessary to actuate a temperature difference as high as 100 °C to create a thermal voltage in the amount of millivolts), while, on the contrary, thermoelectric current is relatively high and the internal resistance, compared to the external load resistance, is relatively very small.

Because there are relatively little changes in the open circuit voltage of the load, thermoelectric battery can be considered as a source of current, precisely, a source of hard voltage. However, this kind of energy source is not very usable either in weak current or in heavy current electrical engineering.

Its efficiency tends to be very low, i.e. lower than 3 %. The development of advanced alloys and semiconductor materials has achieved up to 100 times higher values of thermoelectric voltage than when metal materials were used [5, 6, 7, 8]. But even the use of advanced thermoelectric materials does not lead to the anticipated results in technical practice.

Semiconductor materials [9, 10, 11] (in this paper we deal with ternary alloys) have good thermoelectric properties (relatively low specific electric resistance and low thermal conductivity), but their efficiency is undermined by an important construction element, which is a copper bridge connecting the semiconductor pillars.

Although copper has a relatively high specific electric conductivity and allows good soldering, however, at the same time, it also diffuses into the semiconductor material and impairs its thermoelectric properties. In the point of contact between the bridge and the semi-conductor, there is an undesirable contact resistance, which may adversely affect the achievable thermocouple efficiency, as well as the maximum achievable temperature difference between the two contact points.

The development of new thermoelectric materials will be useless for applications in technical practice, unless a technological process of production capable of ensuring low contact resistance of the thermocouple is discovered. Thermal amortization represents another factor significantly influencing the efficiency of thermoelectric battery in technical practice, as the semi-conductor materials are more efficient than the metal ones, but their heat load is limited. Finding relatively substantial waste heats as suitable sources of conversion would be useless as well if the thermoelectric battery „disintegrates" due to thermal amortization. Soldering has a significant technological impact on the thermoelectric battery elements, which shows itself only during long-term and more demanding operation in technical practice or in a laboratory.

Thermoelectric battery composed of thermoelectric coils (Figure 2) is a proposed innovative design solution that requires the incorporation of a problematic construction element - the copper bridge connecting the pillars.

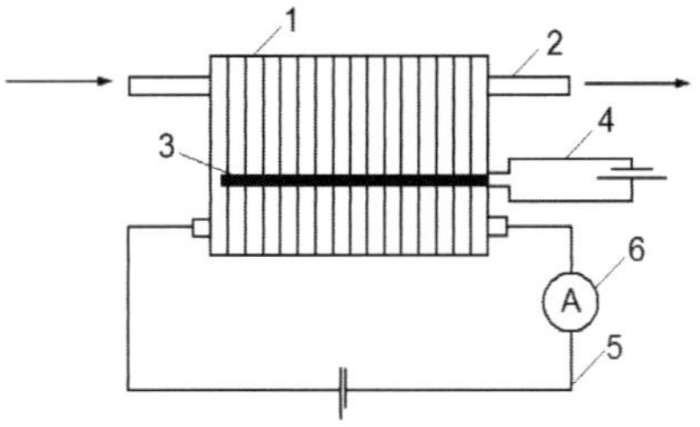

Figure 2. Equipment for generating thermoelectric current using temperature difference

1 - coil with special metal-coated threads from another metal material, 2 - cold water tube, 3 - metal heating element from a resistance conductor, 4 - power supply of a metal heating element from a resistance conductor (3), 5 - electrical circuit with a source of direct current, 6 - sensitive ampere-meter

The proposed solution to generate thermoelectric current (Figure 2) consists of a special metal-coated coil forming threads (1), with an implemented inlet cold water tube (2) and also a rodlike shaped heating element (3) powered by an electric power source, (4), while the coil output conductors are powered by a DC source (5) and the electric current is controlled by measurements using a sensitive ampere-meter (6).

Seebeck thermoelectric voltage is practically difficult to measure, which is why it is more suitable to theoretically evaluate and measure the thermoelectric current.

Thermoelectric voltage generating thermoelectric current can be determined as a maximum Seebeck thermoelectric voltage U_{Smax} at the so called neutral temperature T_n, by empirically verified relation [12, 13, 14, 15]

$$U_S = (a_1 - a_2)(T_1 - T_2) + \frac{1}{2}(b_1 - b_2)(T_1 - T_2)^2 \tag{1}$$

The peak V of the parabolic course of the thermoelectric voltage $V = [U_{Smax}, T_n]$ is given as a local extreme of function $U_S = U_S(\Delta T)$, while the thermoelectric material coefficients can be considered approximately constant, and temperature T_2 of the colder location can be considered constant as well. The maximum value of Seebeck voltage and the neutral temperature can be derived as follows [4]:

$$U_{S\,max} = -\frac{(a_1 - a_2)^2}{2(b_1 - b_2)};\; T_n = -\frac{a_1 - a_2}{b_1 - b_2} + T_2. \tag{2}$$

It doesn't make sense to further increase the temperature of the heating point above the neutral temperature, because the value of the maximum Seebeck thermoelectric voltage would decrease. The optimal thermoelectric current value I can be evaluated in the following way, using Ohm's law for closed circuit.

The numerator is the Seebeck thermoelectric voltage, the denominator is the sum of the partial resistances, i.e. the resistance of R load (appliance) and the resistances of both conductors R_1, R_2, which are heated in the joints. The conductor resistance values are evaluated according to their geometrical dimensions, their material and their electrical resistance temperature change.

$$I = \frac{(a_{V1} - a_{V2})(T_1 - T_2) + \frac{(b_{V1} - b_{V2})(T_1 - T_2)^2}{2}}{R + \rho_{V1} \cdot \frac{l_{V1}}{S_{V1}} \cdot \left[1 + \alpha_{V1}\left(\frac{T_1 + T_2}{2}\right)\right] + \rho_{V2} \cdot \frac{l_{V2}}{S_{V2}} \cdot \left[1 + \alpha_{V2}\left(\frac{T_1 + T_2}{2}\right)\right]},$$

$$a_{V1} - a_{V2} = \frac{k}{e} \cdot ln \frac{n_{0V1}}{n_{0V2}}. \tag{3}$$

Description (3). I thermoelectric current; R load resistance (appliances), a_{V1},b_{V1} Seebeck thermoelectric coefficients of the first conductor *V1*, a_{V2},b_{V2} Seebeck thermoelectric coefficients of the second conductor *V2*, T_1 connection heating temperature (wall), T_2 colder connection temperature (wall); ρ_{V1}, ρ_{V2} electric resistivity of conductors *V1*,*V2*; l_{V1},l_{V2} *V1*, *V2* conductor lengths, S_{V1},S_{V2} *V1*, *V2* conductor cross-sections; α_{V1}, α_{V2} thermal coefficient of linear expandability of conductors *V1*, *V2*; k the Boltzmann constant $1.38 \cdot 10^{-23}$ J·K^{-1}; e the elementary electric charge $1.6 \cdot 10^{-19}$ C; $n_{0\,V1}$ concentrations of free electrons in conductor V1; $n_{0\,V2}$ concentrations of free electrons in conductor *V2*.

The choice of optimal parameters in an electrical circuit must be related to the knowledge of the maximum achievable parameter values. The highest possible value of temperature difference is not at the same time the optimum value. The question of how much we expose electrochemical copper to heat load is irrelevant (it is impossible to exceed the melting temperature that is high in case of metals, in case of semiconductor alloys, the temperature load limit is relatively low). It is necessary to calculate the neutral temperature for the combination of the selected joined materials (with constant temperature of the colder connection) for the maximum achievable thermoelectric voltage. A heat source, which would heat the joint of thermoelectric materials to a temperature higher than the neutral temperature, would not be desirable.

The present advanced and suitable materials include various semiconductor alloys, such as TAGS (i.e. tellurium, antimony, germanium, silver), or ternary alloys of bismuth-tellurides (Bi-Te-Se for N-type conductivity, Bi-Sb-Te for P-type conductivity). A possible development of other advanced materials seems to be promising in order to achieve higher efficiency of thermoelectric batteries by adding rare earth metals (e.g. 1 % of cerium or ytterbium) into the materials of traditionally used TAGS metals. The advanced thermoelectric materials include Heusler alloys, i.e. inter-metallic compounds (such as manganese and aluminium, tin and antimony) and materials with skutterudite structure, especially ternary skutterudite compounds, which are characterized by a complex structure and a possibility of applying even larger and heavier atoms into free cavities of this structure.

The thermoelectric coil was implemented within the scope of the functional model only for the purpose of verification of its functionality, without any optimization of its parameters.

Material. The coil is made from the material, which does not show good thermoelectric properties. An air core coil was wound from a commonly available metal conductor (i.e. conductor made of electrochemical copper, with Seebeck coefficient of $\alpha_{Cu} = 3.2$ μV·K^{-1}in reference to platinum). This conductor was evenly coated with a thin metal layer of another material (tin, Seebeck coefficient of $\alpha_{Sn} = 4.2$ μV·K^{-1} in reference to platinum). Contact voltage is due to low value of the resulting Seebeck coefficient extremely low. The electrical and thermal conductivity of copper is excellent and therefore, the particles of copper diffuse within a conductive bridge into a semiconductor material and degrade its excellent thermoelectric properties. It can be estimated that when chossing this type of material, tens of degrees Celsius from a possible source of waste heat and hundreds of optimized coil threads would increase the thermoelectric voltage only in the range of millivolts.

Geometric dimensions of the coil. The coil threads should ideally have the highest possible radius because, in reality, the distance between the cold and heated locations should be significant in order to prevent the temperatures affecting each other. In addition to that, the thermal conductivity of copper is excellent and it impairs the temperature difference. The thermal coil had 20 threads, the threads had a radius of 4.5 cm, and they were wound using a conductor of radius 1 mm. The thickness of the coated layer has not been measured.

Temperature difference. The coil was simultaneously heated and cooled in opposite locations (Figure 2), using a metal tube with cold water and a metal rod shaped heating element. The metal tube was inserted into the coil cavity just below the coil thread inner surface and it was connected to the cold water inlet from a water conduit with constant temperature of 12 °C. Rod shaped heating element was also inserted into the coil cavity just below the coil thread inner surface opposite the cooling tube, while powered by a 12 V DC power source and it generated heat with a contact temperature of 300 °C. The temperature difference between the heating and cooling locations was measured using thermoelectric thermometers, by means of contact measuring from the temperatures scanned on the surface of the coil threads. As a result of the temperature difference and the contact voltage, thermoelectric voltage measurable by a micro-voltmeter was induced and, thermal voltage measurable by a sensitive miliampere-meter was a result of that. All the thermal phenomena are characterized by considerable inertia, which is why all the partial measurements (currents and temperatures) were performed in equidistant half-minute intervals.

Efficiency. A comparative device with the same geometric dimensions and from the same material, but without implementation of proposed innovations, has been created in order to measure the thermal current value depending on the temperature difference. Effectiveness of innovation was evaluated experimentally as a ratio of the measured currents generated at the same temperature difference by using both devices. The innovated thermoelectric battery has been evaluated as a functional device, but the mean efficiency was insufficient (0.65 %) at a maximum temperature difference of 50 °C. Low equipment efficiency was a result of the selection of readily available, but very inefficient thermoelectric materials (electrochemical copper and tin).

The paper assesses an innovative thermoelectric device generating thermoelectric current, which was realized as a functional model. Because there was no material with suitable thermoelectric properties available for production of the thermoelectric battery, the thermoelectric battery has been verified as functional, but with poor efficiency. A classic cascade construction of the thermoelectric couple was replaced by an innovative thermo-coil construction, the threads of which (made of electrochemical copper) were coated with a metal of another material (tin). The temperature difference required for the formation of thermoelectric voltage and thermoelectric current was simulated artificially in a laboratory, but in technical practice, comparable temperature differences may arise, for example, due to heating, preferably due to heating taking advantage of waste heat. The thermoelectric device will serve for two purposes: disposal of unwanted waste heat and clean conversion into electrical energy.

References

[1] Goldsmid H J 2010 Introduction to Thermoelectricity. Springer, Berlin

[2] Harman T C, Honig J M 1967 Thermoelectric and thermomagnetic effects and applications. Mc Graw Hill, New York

[3] Nolas G S, Sharp J, Goldsmid H J 2001 Thermoelectrics - Basic Principles and New Material Developments, Springer, Berlin, Heidelberg, New York

[4] Cobble M H, 1995 Handbook of Thermoelectrics - Calculations of Generator Performance, CRC Press, Inc.

[5] Burns G W 1993 Temperature-Elektromotive Force Reference Functions and Tables for the Letter-Designated Thermocouple Types Based on the ITS-90, NIST Monograph 175, U.S. Gouvernment Printing Office, Washington

[6] Decher R 1997 Direct Energy Conversion - Fundamentals of electric power production, Oxford University press, New York
[7] Bejan A 1997 Advanced Engineering Thermodynamics, John Wiley & Sons
[8] Rowe D M, Min G 1998 Evaluation of thermoelectric modules for power generation, Journal of Power Sources, Vol. 73
[9] Ioffe A F 1957 Semiconductor thermoelements and thermoelectric cooling. Infosearch, London
[10] Wu C 1996 Analysis of waste-heat thermoelectric power generators - Applied Thermal Engineering, Vol. 16
[11] Snyder G J, Ursell T S, 2003 Thermoelectric efficiency and compatibility, Physical Review Letters, Vol. 91
[12] Heikes R R, Ure R W 1967 Thermoelectricity. Interscience Publ., New York
[13] Esarte J, Min G, Rowe D M 2001 Modelling heat exchangers for thermoelectric generators - Journal of Power Sources, Vol. 93
[14] Pramanick A K, Das, P K 2006 Constructual design of a thermoelectric device, Int. Journal of Heat and Mass Transfer, Vol. 49
[15] Stevens J.W 2001 Optimal design of small ΔT thermoelectric generation systems, Energy conversion and Management, Vol. 42

Changes in a basic course of electrical measurement caused by development in microelectronics

V Haasz
Czech Technical University in Prague, Fac. of Electrical Eng., Dept. of Measurement, Technicka 2, 166 27 Prague, Czech Republic

E-mail: haasz@fel.cvut.cz

Abstract. Changes in electrical measurement and instrumentation caused by a development in microelectronics should be reflected also in education. This trend runs more than 50 years. At present it relates especially to substitution of analogue signal processing by digital processing of sampled signal in a number of measuring instruments. The actual essential changes in the basic course of electrical measurement at the Faculty of Electrical Engineering of the Czech Technical University (CTU) in Prague corresponding to this progress in last years are presented.

1. Introduction

The first changes in electrical measurement caused due to development in microelectronics appeared with coming of monolithic operational amplifiers and first analogue-digital converters in seventies of the last century. Their further development gradually leads to a replacement of analogue instruments for current and voltage measurement (ammeters, voltmeters etc.) and Wheatstone bridges for resistance measurement by digital multimeters, replacement of AC bridges for capacitance and inductance measurement by digital LCR meters etc. A higher sampling rate of AD converters made it possible to develop digitizing scopes and special methods of AD conversion improve accuracy of digital voltmeters so that they could replace potentiometers for a precise DC measurement [1], [2]. In addition, development of microcontrollers made possible to implement methods of digital signal processing directly in measuring instruments in the last years. Nevertheless the "classical" (out of date) methods remained partially in some courses or books till the first years of the 21st century [3], [4].

The progress mentioned above has influenced education of measurement and instrumentation and also evidently curricula of basic courses of electrical measurement and instrumentation before now. The changes made more than 20 years ago reflected this development, but analogue instruments (which were still used) remained as an indispensable part there. At present, when signal digitalization and digital signal processing are leading principles of a number of instruments [5] and industrial partners ask for this knowledge by graduates, curriculum of basic courses in measurement and instrumentation should be radically changed again. The actualised structure of the basic course, which it took into account, was implemented at the faculty of electrical engineering meanwhile in one study program three years ago and in the further two programs it is prepared for the academic year 2014-15.

The basic course *Electrical measurement and instrumentation* for students of the bachelor study stage at the Faculty of Electrical Engineering of CTU in Prague is offered in the study programs

Content from this work may be used under the terms of the Creative Commons Attribution 3.0 licence. Any further distribution of this work must maintain attribution to the author(s) and the title of the work, journal citation and DOI.
Published under licence by IOP Publishing Ltd

Communications, Multimedia and Electronics (in the 3rd semester) and *Electrical Engineering, Power Engineering and Management* (in the 2nd semester) in the amount 2 hours of lectures + 2 hours of labs per week. The course *Sensors and measurement* (in the 4th semester) intended for study program *Cybernetics and Robotics* in amount of 3 hours of lectures + 2 hours of labs per week covers in addition a significant part concerning sensors. This course arose after last re-building of study programs by combination of two previous (separated) courses *Electrical measurement* and *Sensors and transducers* [8] 3 years ago.

2. New approach to a structure of basic course in electrical measurement

In connection with the radically rebuilding of the study program *Cybernetics and Robotics* and creation of the new course *Sensors and measurement*, the new approach was implemented. The changes respected the requirement to give students the knowledge that they will need after finishing their master study, thus at 4 years or later after ending this course. Especially applications of digital signal processing in instrumentation were taken into account. Analogue instruments are mentioned only very briefly (from a historical point of view). Digital signal processing relates especially to methods and instruments for measuring of active electrical quantities. The chapters relating methods and instruments for measuring of resistance, capacitance and inductance were updated only less now (the main changes were done already before).

Compared to the classical approach from the end of last century (see e.g. [6], [7]), when the first part of the course was devoted to the instruments and methods for voltage, current and power measurement including principles of analogue instruments, the new approach started with signal digitalization and corresponding methods and devices. The principles of AD conversion are explained including their properties. It enables to apply the suitable kinds of ADC in different applications in measuring instruments and systems. Since the course is user-oriented, the teaching process remains based on the way "the knowledge of principle" → "properties of method" → "the area of their application".

Principles of sampling and quantization are briefly explained initially including the Shannon theorem and a problem of an aliasing. Description of methods, which enables to convert analogue voltage signal to sequence of numbers follows. Three basic sorts of AD converters are explained:

- integrating (dual slope, voltage-frequency and $\Sigma-\Delta$);
- successive approximation;
- parallel comparison (flash).

Their properties (sampling rate, resolution, disturbance immunity) and typical utilizations corresponding to them are mentioned. This knowledge is used in following lectures concerning electrical quantities measurement.

3. Changes in selected parts of the course

In the part concerning DC voltage or current measurement is accented a possibility of serial disturbance suppression using integrating principle that is used in a number of modifications in the most of multimeters as far as the integrating time is equal to period (or integer number of period) of a disturbing signal (usually from a power net). A typical application of the integrating ADC, which uses a voltage → frequency conversion is presented in the part concerning electrical energy measurement. It concerns of watt-hour meters with analogue multiplication of voltage and current, where output voltage of multiplier is proportional to the power while the number of output pulses is proportional to integral of it, thus the energy. Easy possibility of a change of the ratio resolution/sampling rate is underlined for $\Sigma-\Delta$ ADC in selected applications.

Since not only the analogue circuits for determination of parameters of AC or AC+DC signals (active rectifiers, RMS to DC converters, analogue multipliers) but a number of instruments utilize digital signal processing for measurement of RMS value, rectified mean value and DC component of voltage or current, active power etc., also typical algorithms for their calculation from the sampled signal are presented in the part concerning AC voltage, current and power measurement. The describer simple algorithms are based on changes from integration to summation, accordingly

$$U_{\mathrm{RMS}} = \sqrt{\frac{1}{T}\int_T u^2(t)\,\mathrm{d}t} \quad \Rightarrow \quad U_{\mathrm{RMS}} \cong \sqrt{\frac{1}{N}\sum_{i=1}^{N} u_i^2} \tag{1}$$

(similarly for I_{RMS}) and using the same way, the resultant relation for the rectified mean value can be derived to be

$$U_{\mathrm{AVG}} = \frac{1}{N}\sum_{k=0}^{N-1} \left| u_k \right| \tag{2}$$

for the DC component (mean value) it is

$$U_{\mathrm{DC}} = \frac{1}{N}\sum_{k=0}^{N-1} u_k \tag{3}$$

and for active power is valid

$$P = \sqrt{\frac{1}{T}\int_T u(t)\,i(t)\,\mathrm{d}t} \quad \Rightarrow \quad P \cong \sqrt{\frac{1}{N}\sum_{i=1}^{N} u_i\, i_i} \tag{4}$$

where N is the number of samples per period or per integer number of periods.

Also the parts concerning frequency and phase shift measurement and voltage vector measurement were completed by corresponding algorithms for determination of measured values from a sequence of samples. In this case only the simplest methods are presented. At the same time it is underlined that while in the previous cases the mentioned methods can be used also for non-sinusoidal signals, the presented simplest methods of frequency and phase shift measurement are applicable only for limited types of signals. The period T and thus also the frequency $f = 1/T$ can be calculated from a sequence of discrete values (samples) of u_k (see figure 1). The values T can be approximately determined as

$$T = (k_2 - k_1)\, T_{\mathrm{S}} \tag{5}$$

where k_1 is the sample number after the 1st zero crossing of u, k_2 is the sample number after the 2nd zero crossing of u (with the same sign of derivation) and T_{S} is sampling interval.

In the case of two sinusoidal signals the phase difference φ can be found as

$$\varphi = \frac{2\pi t_0}{T} \ (\mathrm{rad}), \quad \text{or} \quad \varphi = \frac{360\, t_0}{T} \ (^0) \tag{6}$$

where t_0 is the time delay (time shift) of the zero crossing (with the positive derivative) of the signals

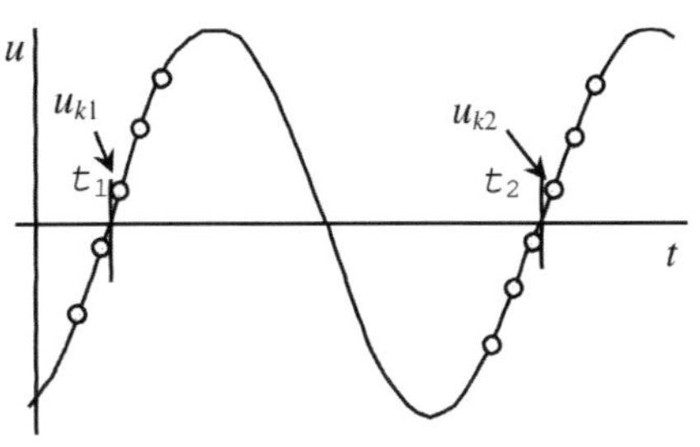

Figure 1. Determining the frequency from a sampled waveform

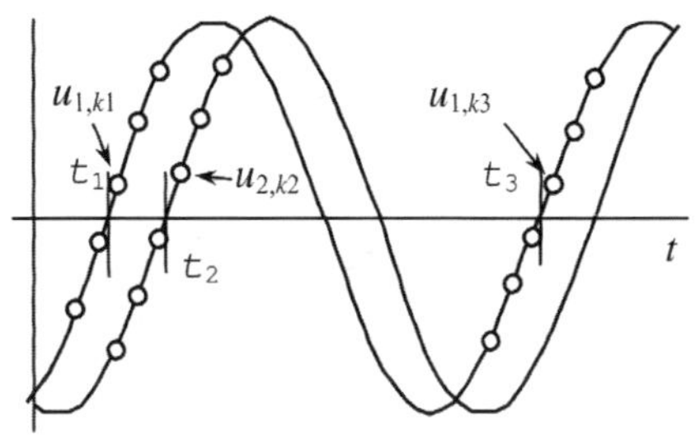

Figure 2. Determining the phase shift from sampled waveforms

$x_1(t)$ and $x_2(t)$ and T is the period time. The ratio of t_0/T can be calculated from a sequence of discrete values (samples) of $u_{1,k}$ and $u_{2,k}$. (see figure 2). The values of t_0 and T can be approximately determined as

$$t_0 = (k_2 - k_1)\,T_S \quad \text{and} \quad T = (k_3 - k_1)\,T_S \tag{7}$$

where k_1 is the sample number after the 1st zero crossing of u_1, k_2 is the sample number after the 1st zero crossing of u_2 (with the same sign of derivation), k_3 is the sample number after the 2nd zero crossing of u_1 (with the same sign of derivation) and T_S is sampling interval. Then

$$\varphi = 2\pi\frac{t_0}{T} = 2\pi\frac{k_2 - k_1}{k_3 - k_1}\ . \tag{8}$$

The Fourier transform which enables to determined the frequency spectrum of non-sinusoudal signals is mentioned only in brief.

Contemporary to the explanation of the methods mentioned above the students are got acquainted with the fact that they are implemented not only in professional up-to-date stand-alone measuring instruments (in digitizing oscilloscopes, power analyzers etc.) but they can be used also in small measuring systems with embedded microcontrolers designed for special applications (some examples of them developed for industrial partners are presented).

A separate chapter concerning applications of microcontrolers in instrumentation was also added to the curriculum. Principles of autocalibration and software calibration and a difference between these two methods are explained for example.

The new approach is used not only in lectures, but also in labs. E.g. a commercial digital clamp-on wattmeter that measures the active and apparent power, the RMS value of voltage and current, the power factor and the frequency based on signal sampling and following digital processing is used in comparison to classic analogue instruments. It shows possibility how to replace several classical instruments that use analogue signal processing by one instrument which use signal sampling and following digital processing.

An application of a digitizing scope in magnetic measurement is presented in another task. Permeability and the iron loss of a ferromagnetic material (ring specimen) are measured by the sinusoidal voltage magnetizing (thus by τηε non-sinusoidal magnetizing current). The peak value of the magnetizing current (necessary for H_{max} calculation) is determined from the peak-peak value of a voltage drop on a serial resistor, which is measured automatically by a digitizing scope. Similarly is determined the value of induced voltage, which is necessary for B_{max} determination. It is currently used application, of course. However, the embedded microcontroller in the digitizing scope with a corresponding firmware enables also the iron loss measurement because a number of digitizing scopes enable multiplication of signals of two input channels and following determination of the mean value of the result, thus the active power.

4. Conclusion

Knowledge of methods used for digital signal processing in up-to-date commercial measuring instruments is important. It enables not only to understand their principles, advantages and disadvantages, but also their usage in non-traditional applications, e.g. active power measurement using a digitizing scope mentioned above etc. These methods, which are now presented already in the basic course of electrical measurement, can be applied also in simple PC measuring systems based on a multifunction AD module or in embedded systems using a microcontroller with one or more analogue inputs, too.

The positive response of this last actualization of curricula of basic course of electrical measurement and instrumentation in the study program *Cybernetics and Robotics* both of industrial partners and of students confirm eligibility of this way. Therefore these changing are implementing now also in the further two programs and will be applied already in the academic year 2014-15.

5. References

[1] Coombs C F 1995 *Electronic Instrument Handbook* (New York: McGraw-Hill, Inc)
[2] Witte R A 1993 *Electronic Test Instruments* (Englewood Cliffs: Prentice Hall PTR)
[3] Biesen L V 2009 Metrology Education in the Curriculum of the Accredited Bachalor in Engineering Programme of the "Verie Universiteit Brussel" *Fundamental and Applied Metrology: Proc. XIX IMEKO World Congress (Lisbon Portugal, 6-11 September 2009)* pp 10-14
[4] Northrop R B 2005 *Introduction to Instrumentation and Measurements (2nd edition)* (New York: Taylor & Francis)
[5] Placko D *Fundamentals of Instrumentation and Measurement* (London: ISTE Ldt)
[6] Schrüfer E *Grundpraktikum Elektrische Messtechnik* (Munich: Technische Universität München)
[7] Zwicky R, Weiler J 1987*Messtechnik I, II* (Zürich: AMV Verlag)
[8] Haasz V 2002 Changes in Education of Measurement and Instrumentation at CTU in Prague Caused by Implementation of Bologna Declaration *Proc 12th IMEKO TC4 International Symposium Electrical Measurements and Instrumentation (Zagreb, Croatia 25–27 September, 2002)* pp 535-538

Impact of the codec and various QoS methods on the final quality of the transferred voice in an IP network

Oldřich Slavata[1,2], **Jan Holub**[1]

[1] Czech Technical University, Faculty of Electrical Engineering, Dept. of Measurement K13138, Technicka 2, 166 27 PRAGUE 6, Czech Republic

[2] Czech Technical University in Prague, University Center for Energy Efficient Buildings, Nam. Sitna 3105, Kladno, Czech Republic

slavao1@fel.cvut.cz

Abstract. This paper deals with an analysis of the relation between the codec that is used, the QoS method, and the final voice transmission quality. The Cisco 2811 router is used for adjusting QoS. VoIP client Linphone is used for adjusting the codec. The criterion for transmission quality is the MOS parameter investigated with the ITU-T P.862 PESQ and P.863 POLQA algorithms.

1. Introduction

At the beginning of the 21st century, increasing transmission capacity of the network and improved methods for digital processing of video and acoustic signals allowed the Internet to be used for real-time voice and video communication. VoIP (Voice over Internet Protocol) allows the transmission of voice in digital form in UDP / TCP / IP packets. Networks that are used for simultaneous data, voice or video transfer are known as Converged Networks. Before networks converged, protocols were developed to occupy most of the available network capacity and to deliver data as fast as possible. This can lead to packet loss, or to receiving the packet in the wrong order. Lost packets are then resent. In the end point, all packets are re-grouped in the correct order.

For voice transmission, it is important that voice packets are delivered with minimum delay, with little variability of delay, and in the correct order. To some extent, we can accept loss or dropping of some packets in return for low delay and increased fluency. Different transmission requirements have led to the need to handle simultaneous transmission of voice and data (or video) over a single network. The simplest way to solve these problems is to increase the network bandwidth so that all protocols have sufficient capacity. However, this method is expensive and is not always realizable (because of some technological limits). Another option is to introduce QoS classes and to prioritize network traffic according to its importance. Typical classes are:

Voice - Class for voice transmission.

Content from this work may be used under the terms of the Creative Commons Attribution 3.0 licence. Any further distribution of this work must maintain attribution to the author(s) and the title of the work, journal citation and DOI.
Published under licence by IOP Publishing Ltd

Business-Critical - Class for important data (business applications, access to databases, etc.)
Best-Effort - Class for normal traffic (e-mail, web access, etc.)
Scavenger - Class for unwanted traffic that can also be inhibited

Voice and business-critical traffic should have sufficient bandwidth, and voice packets should additionally be sent as preferred. The following mechanisms are used to provide this capacity:

Classification - manually configured or automatic classification of packets / frames
Marking - specific values are written in the header of the packet / frame
Congestion Management - header tags are used in deciding on the class / queue to which the packet will be assigned
Congestion Avoidance - preventive dropping of packets is used to prevent congestion of the line
Policing and shaping - packets are delayed or dropped in traffic that has reached a defined limit
Link Efficiency Mechanisms - the efficiency of the line is increased increasing by using header compression and packet fragmentation

The MOS scale (mean opinion score) is used to assess the quality of voice transmission (Table 1). The term MOS is defined in Recommendation ITU-T P.800.

Table 1 MOS scale

MOS	Quality	Impairment
5	Excellent	Imperceptible
4	Good	Perceptible but not annoying
3	Fair	Slightly annoying
2	Poor	Annoying
1	Bad	Very annoying

There are several methods for obtaining MOS values. The most accurate is a subjective test, where the MOS value is obtained directly from users. It is time-consuming and expensive to carry out these subjective tests. Objective methods based on computer algorithms are therefore used instead.

Intrusive methods provide results that are near to the results of the subjective tests. They are based on comparing the original sample with the transferred sample. These algorithms use psychoacoustic models of human perception, seeking to provide a mathematical description of human sound perception and to find variables that have a direct impact on the perceived quality of the voice signal. Intrusive methods include PAMS (Perceptual Analysis Measurement System), developed by British Telecommunications, PSQM (Perceptual Speech Quality Measurement), described in Recommendation ITU-T P.861, PESQ (Perceptual Speech Quality Evaluation of), according to ITU-T P.862 (P.862.1) [1], [2], and newly ITU-T P.863 - POLQA (Perceptual Objective Listening Quality Analysis) [3], [4].

Non-intrusive methods are a different type of quality measurement. These methods do not use the reference signal, and the final MOS is calculated using the parameters of the transferred sample only. The disadvantage of these methods is their lower accuracy and reliability. An example of a non-intrusive method is 3SQM, which is defined in recommendation ITU-T P.563.

2. Description of the experiment

2.1. Test-bed

The test bed (Figure 1) consisted of two small LAN networks connected with a serial line simulating the WAN connection with bandwidth limited to 384 kbit/s. We used the following network elements: Cisco 7942G IP phones, Switches: Cisco Catalyst 3560, Routers: Cisco 2811, Cisco 1841, PC and servers: Dell.

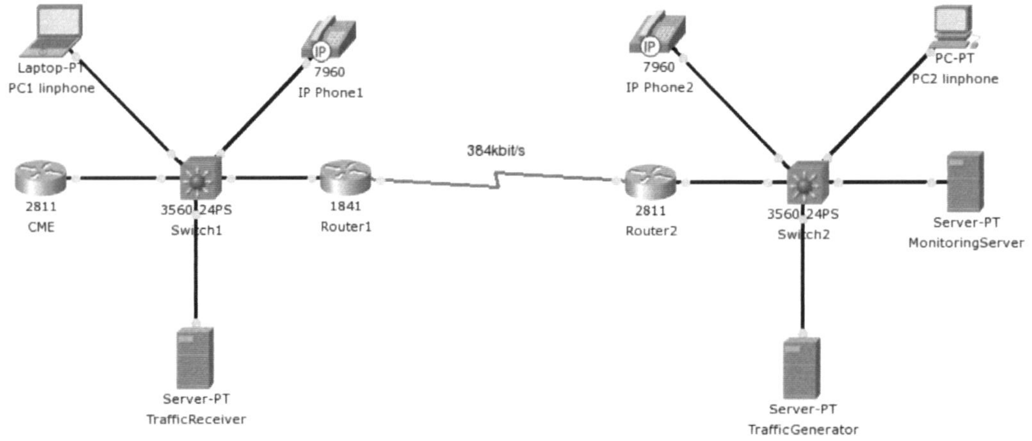

Figure 1. The test-bed

A concatenated speech file in WAV format (8kSa/S, 16bit), 16.75 s in length, was used (Figure 2). The file contains 4 short sentences spoken by 4 different speakers (two men, two women) and adequately covers the entire human speech spectra. The concatenated file that is used therefore effectively replaces tests using multiple speech samples.

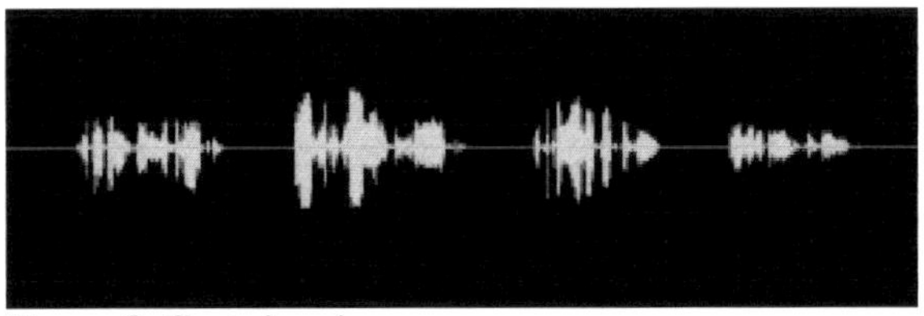

Figure 2. Test signal

A VoIP telephone call was made between PC1 and PC2 using Linphone. On the side of the caller, the sample was directly played using Linphone. On the receiver side, it was recorded using Audacity. Various QoS methods were adjusted on Router2. One of the purposes of the experiment was to verify the reliability of the MOS estimates that are provided in Cisco IP phones. Test calls for each QoS setting were also made using these phones. To accelerate the experiment, the dialing and recording was controlled by a bash script. Softphone Linphone was chosen because it allows playback of a voice sample and script control. In previous experiments, it was evaluated as the best in terms of quality of transmission [5]. The following codecs were used in Linphone: G.711, G.722, G729, iLBC. The iperf traffic generator was used to simulate the real traffic in the network.

The PESQ and POLQA algorithms were used to evaluate the voice transmission quality. The output of the PESQ algorithm was recalculated to the value of MOS-LQO (Listening Quality Objective) according to a mathematical prescription defined in ITU-T P.862.1. As follows from the official wording of P.862, it can be used for testing the effect of packet loss on CELP coded transmission. It has not been tested for PCM transmissions affected by packet loss. However, the recommendation itself does not prevent any user from doing this, and it was successfully used in previous experiments [6,7,8,9]. The results can also be affected by Packet Loss Concealment (PLC), implemented by Linphone. Comparing the results of our previous experiment [6,7] with [10], it is obvious that our version of Linphone had no PLC implemented.

2.2. *Methods for controlling and preventing congestion*

FIFO – First In First Out

Handles packets in the same order as they come. For larger loads, the queue fills up quickly, and this causes a delay and packet losses. This is the simplest type of queue.

WFQ – Weighted Fair Queuing
Forms 16-256 queues according to the bandwidth. Used for classifying packets to queues using automatic classification.

CBWFQ – Class-based Weighted Fair Queuing
Extends WFQ to support traffic classes. For each class, a queue is formed in which all the traffic of this class is routed. Bandwidth is allocated to classes by priority.

LLQ – Low-latency Queuing
Adds a priority queue for real-time traffic (voice, video) to CBWFQ. Real time data is cleared first.

CB-WRED – Class-based Weighted Random Early Detection
Provides a RED mechanism that randomly drops TCP packets before a queue is full, TCP traffic slows down and there is no unsteady flooding of lines. Each class can be assigned a different RED profile.

ECN – Explicit Congestion Notification
Extends WRED by adding information about the network congestion to the packet header.

LFI – Link Fragmentation and Interleaving
Fragments long data packets in order to reduce the delay of voice packets.

3. Results

Ten samples were measured and processed for each codec setting and QoS method. Ten repetitions are enough to achieve satisfactorily low dispersion results and uncertainties. The confidence intervals (CI95) were computed using the following formulae (1):

$$u_{Ax} = \tilde{\sigma}(\bar{x}) = \sqrt{\frac{1}{n \times (n-1)} \sum_{i=1}^{n} (x_i - \bar{x})^2} \qquad (1)$$

The graphs show a significant correlation between the QoS method and the final voice quality. Generally speaking, the more advanced the method is, the better the transmission quality will be. Some methods have the same results in transmission quality, but they differ in the length of the delay, represented by the ping response (figure 3.). The length of the delay is important for the quality of the conversation.

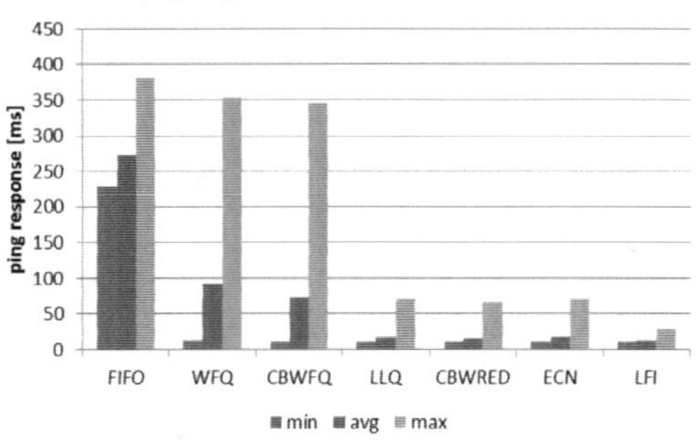

Figure 3. Ping response as a function of the QoS method

The following set of graphs (Figure 4 – 7.) shows the results for various codecs. The PESQ and POLQA algorithms produce very similar results. An estimate of MOS using an IP phone frequently differs from the results of specialized algorithms, but it is adequate for showing whether or not the call is possible.

Journal of Physics: Conference Series **588** (2015) 012011 doi:10.1088/1742-6596/588/1/012011

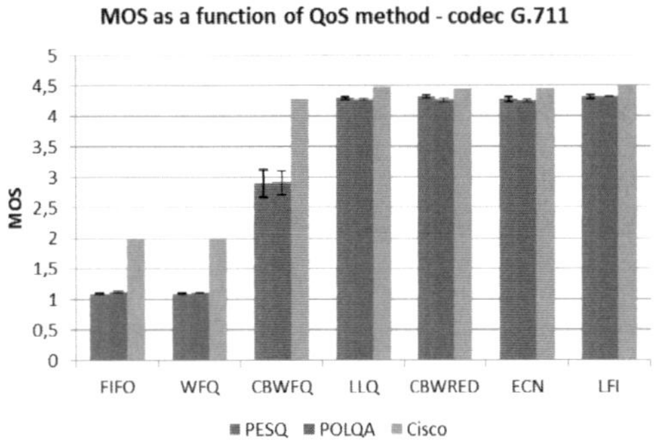

Figure 4. MOS as a function of the QoS method - codec G.711

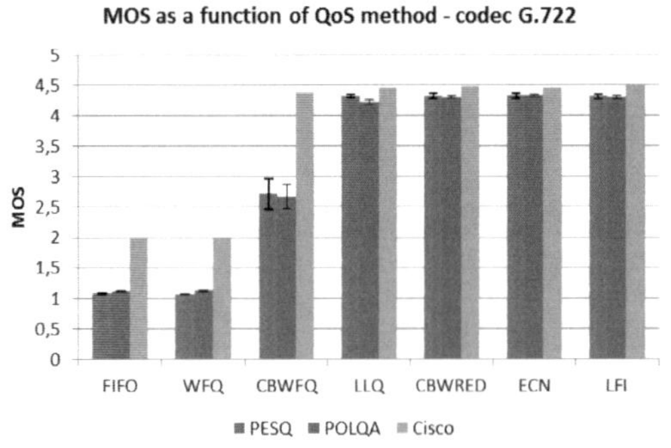

Figure 5. MOS as a function of the QoS method - codec G.722

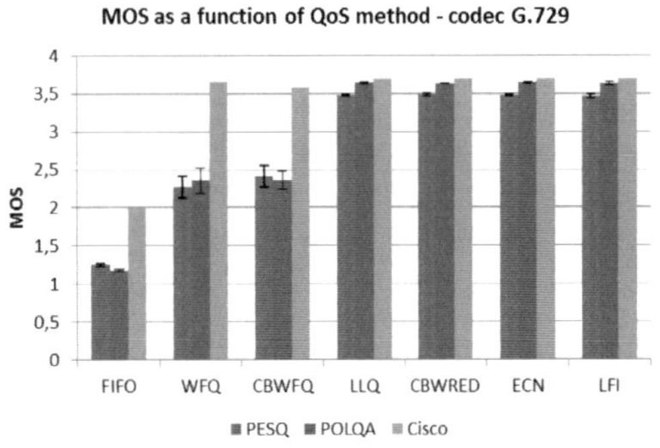

Figure 6. MOS as a function of the QoS method - codec G.729

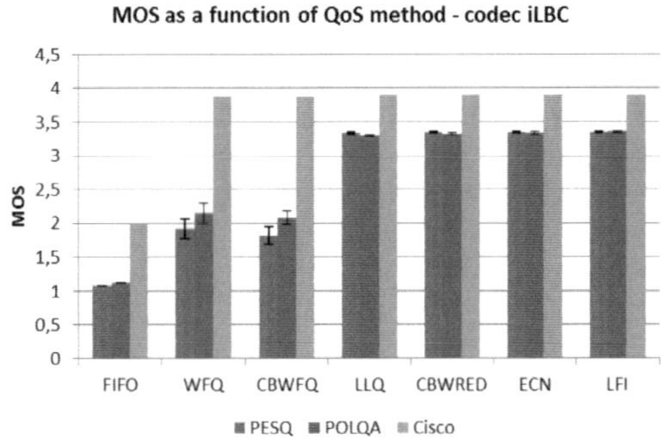

Figure 7. MOS as a function of the QoS method - codec iLBC

The next set of graphs (Figure 8 – 9.) shows the results sorted according to the algorithm that is used. The results show that high bitrate codecs G.711 and G.722 usually provide better transmission quality than low bitrate G.729 and iLBC. However, a low bitrate codec can provide better results when there is low bandwidth or a worse QoS method.

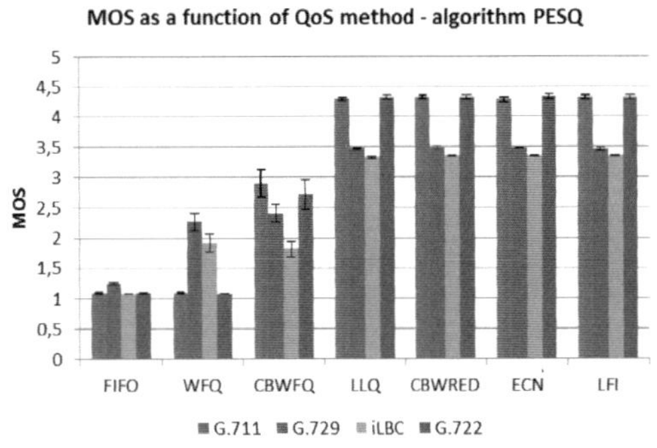

Figure 8. MOS as a function of the QoS method - PESQ algorithm

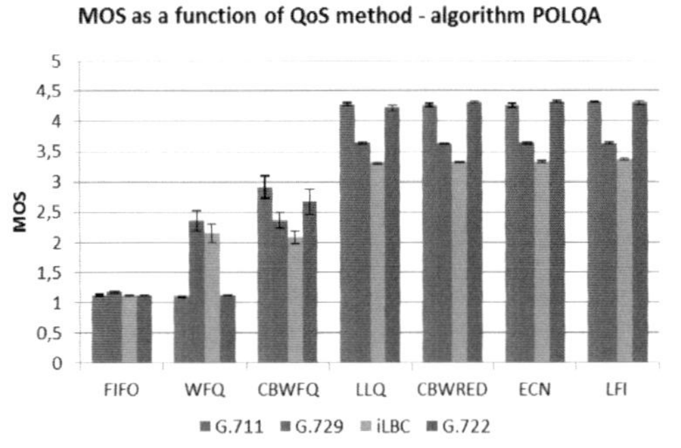

Figure 9. MOS as a function of the QoS method - POLQA algorithm

The results confirm the assumption that the more advanced the QoS method is, the better the transmission quality will be. A simple FIFO queue is not suitable for the purpose of traffic management in converged networks. The WFQ method offers a slight improvement for low bitrate codecs, but the resulting quality is still very bad. In the CBWFQ method, there is a noticeable improvement even in high bitrate codecs. The LLQ method and others already provide transmission quality at the maximum limit for the current codec. However, they differ in the length of the response, and therefore in the delay and the conversation quality.

4. Conclusion

This paper has checked the impact of the codec that is used and the QoS method on the quality of voice transmission in IP networks.

The main objective was to identify the relation between the QoS method and the MOS values as delivered by different objective algorithms. Generally speaking, the more advanced the method is, the better the quality of the transmission will be. Some advanced methods differ only in the delay.

The second objective was to explore differences between widely-used codecs. Low bitrate codecs provide worse results than high bitrate codecs, but they may be useful for low bandwidth.

The results of this experiment can be used in planning and implementing other measurements.

5. Acknowledgment

We would like to thank Bc. Tomas Kral for his participation in this experiment

References

[1] ITU-T P.862: Perceptual evaluation of speech quality (PESQ): An objective method for end-to-end speech quality assessment of narrow-band telephone networks and speech codecs, February 2001

[2] ITU-T P.862.1 Mapping function for transforming P.862 raw result scores to MOS-LQO, November 2003

[3] ITU-T P.863 Perceptual objective listening quality assessment, January 2011

[4] POLQA Technical White Paper, July 2010, Opticom

[5] Slavata, O.:Měření kvality přenosu hlasu pro sítě typu VoIP, Bachelor project, FEE CTU, August 2007

[6] Slavata, O.: Neintruzivní měření kvality přenosu hlasu v telekomunikačních sítích, Diplomová práce, Master's thesis, CTU FEE, January 2010

[7] Slavata, O., Holub, J.: Evaluation of objective speech transmission quality measurements in packet-based networks, Computer Standards & Interfaces 36 (2014)

[8] Počta, P., Holub, J.: Predicting the Quality of Synthesized and Natural Speech Impaired by Packet Loss and Coding Using PESQ and P.563 Models

[9] Stein, Y., Druker I.: The Effect of Packet Loss on Voice Quality for TDM over Pseudowires, RAD Data Communications, October 2003

[10] Gündüzhan E., Momtahan K.: A Linear Prediction Based Packet Loss Concealment Algorithm for PCM Coded Speech, IEEE TRANSACTIONS ON SPEECH AND AUDIO PROCESSING, November 2001

[11] Ditech networks: Limitations of PESQ for Measuring Voice Quality in Mobile and VoIP Networks, December 2007

[12] http://www.voip-info.org

[13] http://www.itu.ch

Conceptualising computerized adaptive testing for measurement of latent variables associated with physical objects

F R Camargo[1] and B Henson[2]
[1,2]School of Mechanical Engineering, University of Leeds
Woodhouse Lane, Leeds, UK, LS2 9JT

[1]E-mail: mnfrc@leeds.ac.uk

Abstract. The notion of that more or less of a physical feature affects in different degrees the users' impression with regard to an underlying attribute of a product has frequently been applied in affective engineering. However, those attributes exist only as a premise that cannot directly be measured and, therefore, inferences based on their assessment are error-prone. To establish and improve measurement of latent attributes it is presented in this paper the concept of a stochastic framework using the Rasch model for a wide range of independent variables referred to as an item bank. Based on an item bank, computerized adaptive testing (CAT) can be developed. A CAT system can converge into a sequence of items bracketing to convey information at a user's particular endorsement level. It is through item banking and CAT that the financial benefits of using the Rasch model in affective engineering can be realised.

1. Introduction

The value of affective responses with regard to underlying attributes of products can be defined for practical purposes as the users' evaluation of physical characteristics that expresses some degree of positive or negative response with certain consistency. In affective engineering (AE) the information obtained from users' responses is converted by mathematical models into an improved design of products. However, measuring those impressions is not straightforward as it is typically to measure the properties of the physical elements that aggregate the product. One reason is that the underlying attribute solely exist as an element of a hypothesis or a premise, such as pleasantness or comfort, which are referred to as latent variables. Those attributes cannot directly be measured and inferences based on their assessment can be error-prone.

An approach in AE for eliciting affective responses is to establish a pool of variables, collect data from people's responses to those variables and apply statistical methods of analysis [1]. Typically, analysts have employed self-report questionnaires where persons give a rating of agreement against a set of words or statements, which will be termed ***items*** hereafter. Statistical treatments are given to the scores, resulting in a numerical representation of the observations subject to the analyst's interpretation. In many cases, the items emerge empirically from a multivariate analysis rather than being prescribed. This implies in difficulties to identify items that provide useful information in terms of measurement. Furthermore, scores from two different pools of items for measuring the same attribute cannot usually be compared directly. One of the reasons is that the treatments of scores can be sample dependent, limiting a quantitative interpretation [2].

One solution to overcome those difficulties is to construct through a stochastic framework a wide range of well designed items referred to as an item bank, which covers a variety of situations. Item

Content from this work may be used under the terms of the Creative Commons Attribution 3.0 licence. Any further distribution of this work must maintain attribution to the author(s) and the title of the work, journal citation and DOI.
Published under licence by IOP Publishing Ltd

bank is, therefore, a repository of variables that can be used to measure a number of contextual applications. The approach follows theoretical propositions and successful applications in the fields of education and health sciences [3][4].

As a result of the item bank approach, the development of computerized adaptive testing (CAT) is possible. The concept of the CAT is concerned with establishing a sequence of items that seem most appropriate for a particular person. Although it is possible to draw upon the techniques for the development of CAT from educational testing and clinical assessment, applications in AE require particular characteristics in a CAT system such as the inclusion of stimulus objects.

The aim of this paper is to present a concept of a CAT system for the measurement of latent variables associated with physical objects. An anticipated benefit of this approach is that analysts can develop an ad hoc structure with additional items without losing the properties of the core of the original, off-the-shelf calibrated scale to make whatever general comparisons they require.

Item calibration is a set of procedures attached to the development of an item bank and CAT. That is, the procedures test whether independent items work well all together for measuring the latent attribute as a unidimensional structure. The concept of item bank and CAT proposed in this paper is underpinned by the Rasch measurement model (RM). The RM refers to a family of probabilistic models that provide mechanisms to test the hypothesis that the observations meet the necessary assumptions for a structure with quantitative properties. The Rasch model, named after the Danish mathematician Georg Rasch who developed it in the 1950s [5], holds the property of separation of the parameters for persons and items, which allows the design of a range of variables used as a yardstick in scales of measurement.

2. Rasch-calibrated metrics for underlying attributes of physical objects

Measurement is fostered in this paper as a way of making meaningful inferences on latent variables based on the numbers obtained from observed events. The main assumption is that the numbers represent a property of the relevant attribute. That is, a metric must show valid evidence for a one-to-one relationship between the structure of mathematical operations on real numbers and the properties of the attribute being measured [6].

The RM's procedures test the observations against necessary measurement principles for quantifying the numerical validity of the data [7]. Such procedures are denoted calibration, a term coined by Wright and Panchapakesan [8], referring to measurement scales that are independent of the sample of persons used to estimate parameters of items and independent of the set of items used to obtain scale scores. The resultant probabilities allow the analysis of the expected values and what is actually observed because some results will certainly not follow the expected pattern. The relationship between person locations (β) on the continuum and the probability of a positive response is represented in the RM as item characteristic curve (Figure 1), where δ is the item parameter.

A number of RMs have been used for different applications. Camargo and Henson [9] have adapted for applications in AE the many-facet Rasch model (MFRM), a derivation of the RM developed by Linacre [10]. Thus, assuming that the data fit the model, it is possible to transform a categorical scale into an interval level with a constant unit of measurement.

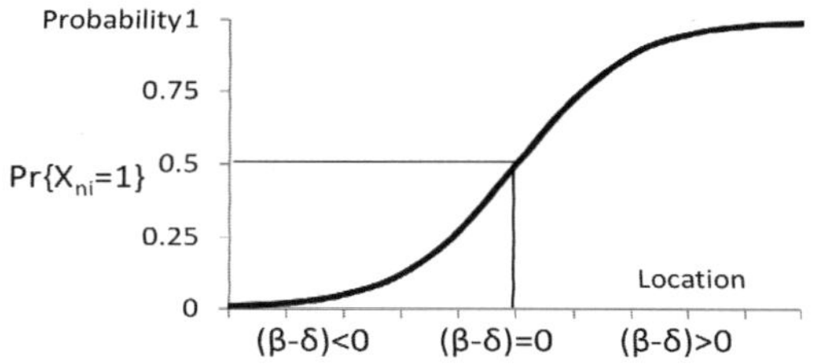

Figure 1– Probability of a positive response associated with persons' locations on the continuum.

3. How a Rasch-calibrated metric works in AE

A metric attains interpretation when the difference between persons as well as between items and between stimulus objects is established by the distance between different locations on the linear continuum [9]. A generic representation of a metric for latent variables associated with physical objects is shown in Figure 2.

Person, item and stimulus locations are based on estimation of parameters. Most of the estimation procedures are based on the method of maximum likelihood [11], calculating the standard error for each estimate through a second derivative of a likelihood function [12]. Parameters are preliminarily obtained and compared with the observations. Estimates are then revised and new estimates are computed. This process of iteration is carried out until the changes of the estimates are smaller than a stopping rule controlled by a convergence criterion. Subsequently, Rasch analysis is carried out to evaluate the extent to which the data fit the model [7][13].

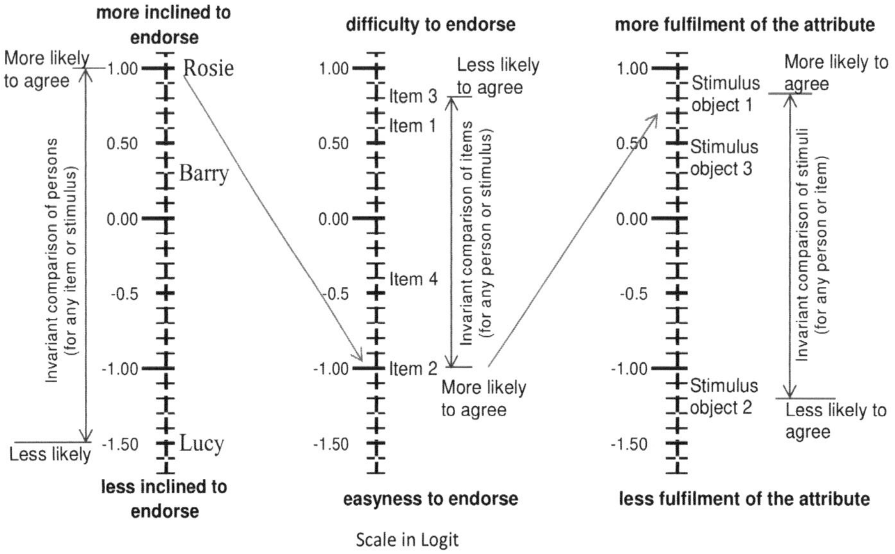

Figure 2 – Generic representation of a metric operation.

The interpretation of metric is based on the model's property of invariant comparisons. In Figure 2, for example, it is possible to interpret that Rosie has a higher probability of endorsement than Lucy independently on any item and stimulus object. Those differences are interpreted through the adapted MFRM, such that the probability of a response X is obtained by a function f of the parameters given by $\Pr(X) = f(\beta - \delta + \zeta)$, where β represents the person parameter, δ is the item parameter and ζ represents the stimulus object [9]. Using a symmetric argument, the comparison of difficulty of endorsement between any pair of items and the fulfilment of latent attribute between any pair of stimulus objects can also be made.

It is noteworthy that the unit in Rasch modelling is the log-odds unit or logit. The unit logit represents the distance on the continuum that indicates changes of the odds of observing the relevant event computed through natural logarithm [14]. The unit in logit denotes the same interval with regard to changes on the continuum.

4. A concept of computerized adaptive testing in AE

4.1. Construction of item banks for underlying attributes of products

The development of an item bank [15] entails a repository of variables represented by words or statements that hold different levels of readiness of endorsement when people are using a scale. Each item is chosen as an independent element of a strand, which is defined to represent the relevant underlying attribute of a product. After calibration of the scale, responses for each item will indicate the degree of endorsement to a physical characteristic of the product at a specific point on the linear continuum.

The initial step toward an item bank is to obtain the pool of words or statements that can provide sufficient information in relation to a particular purpose. Those words or statements, referred to as items, are obtained from observations of the users' interaction with a product, online reviews, interviews, search in relevant literature and expert opinion [1]. Those targeted items will be calibrated and established as a yardstick for measurement and therefore, they should carefully be designed through an essential specification within the context of the scale, ensuring consistency and replication validated by the RM.

4.2. Representation of the CAT system embodying physical characteristics

The calibrated items of an original structure can be used as the core of a preliminary item bank. The original, calibrated items determine anchoring points that yield the same predictions. Parameters may be anchored for either persons or items. Anchoring the person's parameter emphasises score distribution. In contrast, anchoring the item's parameter gives emphasis to the meaningfulness because items are more directly interpreted [16].

In the CAT system conceptualised in Figure 3, items are selected through a computer program such that if a respondent endorses an item, a slightly more challenging item for endorsement is automatically presented in the sequence, and contrariwise if the item is too difficult. This technique usually converges into a sequence of items bracketing and convey information at the respondent's effective endorsement level. Consequently, each respondent responds to neither all items of the item bank nor all stimulus objects, but only a subset bracketing the threshold of endorsement.

The system can allow different stopping rules. In Figure 3 three rules have been used. The test can, for example, be ceased if a determined number of items are responded. A second rule can determine that a certain standardised measurement error be met. A third rule can establish a limited number of stimulus objects being assessed.

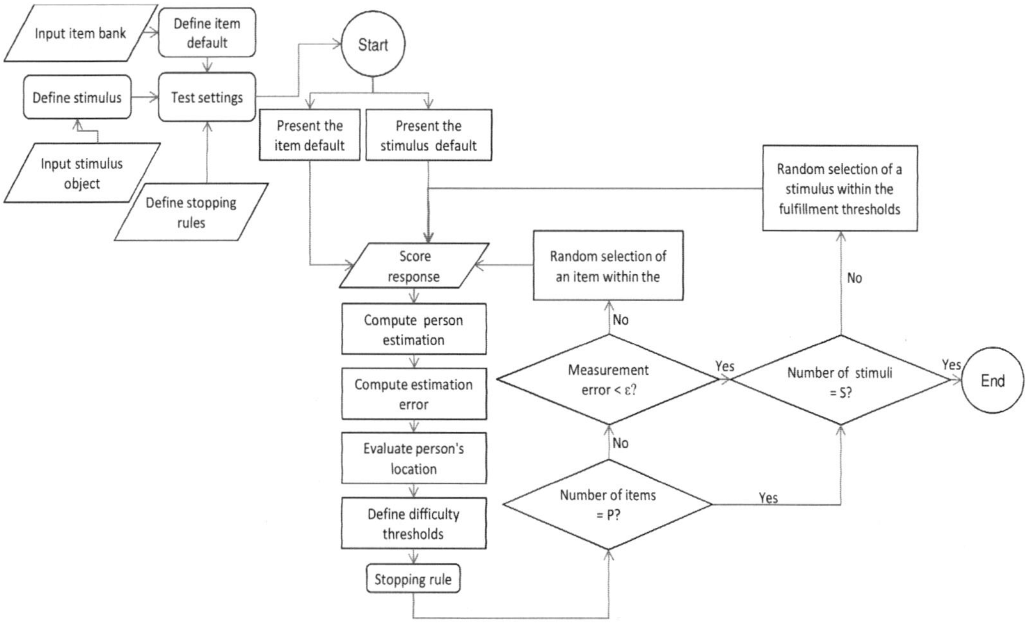

Figure 3 – Flowchart of a conceptual CAT system for latent variables associated with physical objects.

4.3. Operation and quality of measurement

Preliminarily, all items in an item bank are calibrated, i.e., they are tested for meeting the requirements of the RM. The locations of items on the linear continuum are anchored and the metric can then be used with a far smaller sample of individuals.

Figure 4 represents a generic example of the system operation based on the calibrated scale shown in Figure 2. The computer program presents the first item and the first stimulus by default. Items are subsequently selected according to two objectives. One objective is to minimise the test length. Another objective is to maximise the information function at all person locations. Information function is a technical term firstly defined by R.A. Fisher as the maximum precision which a person parameter can be estimated by the model. The precision is statistically obtained by the variability of the estimates around the value of the parameter [17], in which the maximum value at person level is found when $\beta - \delta = 0$ (see Figure 1).

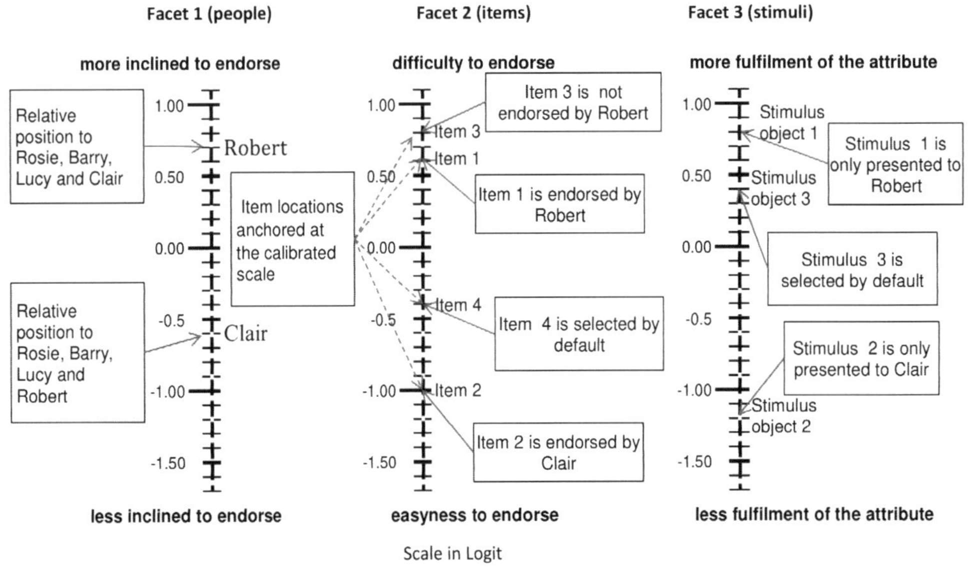

Figure 4 – Example of the CAT system operation in AE.

Empirical evidence to support the quality of measurement using the CAT system conceptualised in this paper can be given by tests of repeatability and reproducibility of measurement [18].The first replicates measurements on the same stimulus objects over a short period of time under the same conditions used to calibrate the scale. The latter involves setting conditions using the same or similar stimulus objects, different locations and operator to replicate the test, including a reasonably targeted sample of respondents with dissimilar characteristics.

5. Discussion and Implications

5.1. Potential difficulties

The concept of CAT presented in this paper for the measurement of latent variables associated with products has included in the system the assessment of physical features, distinguishing it from other domains. However, the system itself cannot ensure that the relevant measure of an underlying attribute of the product is accurate. Accuracy requires the comparison against external measurement systems or agreed standards, which hold robust evidence of validity. In the domain, such systems have been absent and precise definitions of affective attributes have hitherto been limited or inexistent. As a consequence, external systems of affective responses, to date, cannot be used as a criterion of acceptance entirely adequate to assess the quality of the measures.

Another potential difficulty of implementation of the CAT system in AE is to resolve outstanding issues with regard to dependence of words or statements when constructing an item bank. Response dependence can be found in redundant items and when a positive rating of a respondent depends on the responses to the preceding items and where that rating will interfere in the way that the responses on the following items are rated.

5.2. Anticipated benefits

A practical value from a Rasch-calibrated metric is the possibility of construction of an item bank with elements that vary in a range of difficulty for different endorsement levels. The immediate benefit is that an analyst can tailor another structure without losing the properties of the core of the original calibrated structure to make comparisons between persons and between any pair of physical characteristics of a product. Furthermore, it is possible to make comparisons between two different studies using parallel scales.

Another benefit of item banking is to know the characteristics of the information for each item, allowing thus the specification of a subset of items with better discrimination. This feature supports the incorporation of further items to elicit affective user experiences as well as additional stimulus objects with different or improved physical characteristics.

It is through item banking and CAT that the financial benefits of using the RM in AE can be realised. The CAT system using Rasch-calibrated measurement structures is potentially useful to reduce cost in consumer research. Fewer items are necessary for estimation along with higher measurement precision. The system, after calibration, will allow the use of smaller samples and offer the advantages of convenience to respondents concerning flexible scheduling, improved security and data collection. Nevertheless, research on CAT in AE is still limited and, therefore, its application will require further investigation.

6. References

[1] Barnes C and Lillford S 2009 *J. of Eng. Design* **20** 477 – 492.
[2] Camargo F R and Henson B 2012 *Int. J. of Hum. Factors and Ergon.* **1** 204 – 219.
[3] Choppin B H 1978 *Item banking and the monitoring of achievement* (Slough: National Foundation for Educational Research).
[4] Eckes T 2011 *Psychol. Test and Assessm. Model.* **53** 414 – 439.
[5] Rasch G 1960, 1980 *Probabilistic models for some intelligence and attainment tests* (Copenhagen: Danish Institute for Educational Research), expanded edition (1980) (Chicago: The University of Chicago Press).
[6] Krantz D H, Luce R D, Suppes P and Tversky A 1971 *Foundations of measurement* **1** (New York: Academic Press).
[7] Andrich D 1988 *Rasch models for measurement* **68** (London: Sage Publications)
[8] Wright B and Panchapakesan N 1969 *Educ. and Psychol. Meas.* **29** 23 – 48.
[9] Camargo F R and Henson B 2013 *J. Phys.: Conf. Ser.* **4** 59.
[10] Linacre J M 1989 *Many-facet Rasch measurement* (Chicago: MESA Press)
[11] Fisher R A 1922 *Philos. T. R. Soc.* A **222** 309 – 368.
[12] Linacre J M 1999 *J. Outcome Meas.* **3** 382 – 485.
[13] Tennant A and Conaghan P G 2007 *Arthritis Rheum.* **57** 1358 – 62.
[14] Wright B D 1993 *Rasch Measurement Transactions* **7** 288.
[15] Hahn E A, Cella D, Bode R K, Gershon R and Lay J 2006 *Med. Care* **44** S189 – S197.
[16] Embretson S E and Reise S P 2000 *Item response theory for psychologists* (Mahwah: Lawrence Erlbaum).
[17] Timminga E and Adema J J 1995 *Rasch models: foundations, recent developments and applications* ed G H Fisher and I W Molenaar (New York: Spring-Verlag) 111 – 127
[18] VIM - International Vocabulary of Metrology 2012 *Basic and general concepts and associated terms*, 3rd ed (The Joint Committee for Guides in Metrology – JCGM)

Importance of education and competence maintenance in metrology field (measurement science)

J. Dobiliene[1], **A. Meskuotiene**

Institute of Metrology, Kaunas University of Technology, Studentu str. 50-454, Kaunas, Lithuania
E-mail: justina.dobiliene@ktu.lt

Abstract. For certain tasks in metrology field trained employers might be necessary to fulfill specific requirements. It is important to pay attention that metrologists are responsible for fluent work of devices that belong to huge variety of vide spectrum of measurements. People who perform measurements (that are related to our safety, security or everyday life) with reliable measuring instruments must be sure for trueness of their results or conclusions. So with the purpose to reach the harmony between the ordinary man and his used means it is very important to ensure competence of specialists that are responsible for mentioned harmony implementation. Usually these specialists have a university degree and perform highly specified tasks in science, industry or laboratories. Their task is quite narrow. For example, type approval of measuring instrument or calibration and verification. Due to the fact that the number of such employers and their tasks is relatively small in the field of legal metrology, this paper focuses on the significance of training and qualification of legal metrology officers.

Introduction

To ensure the competence and qualification of legal metrology officers who carry out verification, inspection and supervision in the field of legal metrology there is a certain need for a specific training program. International Document OIML D 14 Training and qualification of legal metrology personnel is the legal basis for training and qualification [1].

This document also is the main basis for assurance of national metrological system in Lithuania creating and implementing complex program for qualification and training of metrology officers. The aim of the program is to harmonize competence of metrology officers with the practical appliance. Regarding to measurement areas as use of balances where each gram determines the price, the water or heat meters where the used amount is directly related to our expenses or such sensitive area as medicine the need of qualified and competent metrology personnel is evidence. Also our security involves all possible ways to reduce fraud or corrupt practices.

The review of metrological situation in Lithuania and other Baltic States (especially Latvia) seems to be strange because of the latest actions of the great officers of state lead our countries to destruction of metrological system. Still our country has a certified training system that is provided by Metrology institute. But resent practice showed that Latvia destroyed this system some years ago and now they confront with the difficulties when new personnel do not have primary basis and knowledge of metrology terms, uncertainty evaluation even basic tasks as verification performance.

[1] Corresponding author: E-mail: justina.dobiliene@ktu.lt (Justina Dobiliene), phone: +370 37 300776, fax: +370 37 300 488.

Content from this work may be used under the terms of the Creative Commons Attribution 3.0 licence. Any further distribution of this work must maintain attribution to the author(s) and the title of the work, journal citation and DOI.
Published under licence by IOP Publishing Ltd

1. Model of metrology personnel education

As it was mentioned metrology personnel education is organised by the OIML D14 with adjustment to our country needs. We have 5 programs that allow ensuring fluent work of metrology system:

1. Training courses of metrology inspectors and metrologists (This program is applied to individuals who want to gain the metrologist and/or metrology inspector qualification);
2. Training course for tachograph inspector (Attendees have to be trained by tachograph manufacturers and here they gain qualification of tachograph inspector with metrologist competence and basic knowledge);
3. Refresher course-seminar for metrology personnel – annual courses (This program is applied to continued training of metrologists (and tachograph inspectors);
4. Seminar for metrologists (Here can be gained deeper knowledge of one specific field, for example, mass measurements, metrology in chemical laboratories, advanced metrology, etc.);
5. Reassessment of metrology inspectors and tachograph inspectors (Program is applied to improve the qualification. If inspectors participated in refresher courses annually (for 5 years) then their qualification is improved automatically. Metrology inspectors, who did not participate in such courses or participated partly, have adequate examination.)

Educational program for metrology qualification has changed for some last years. Earlier these courses were divided into 4 specialized fields: non-electrical measurements, electrical measurements, physical-chemical measurements, and radiation measurements. Analysis of needs showed that extensive specialization is much more useful for fast growing and changing technologies century. So now non-electrical and electrical measurements specializations are joined into metrology officer specialization and other various specializations can be improved in specially organised seminar. Also tachograph inspectors training is provided as separate program.

A system of metrology personnel education and qualification development is presented in Figure 1.

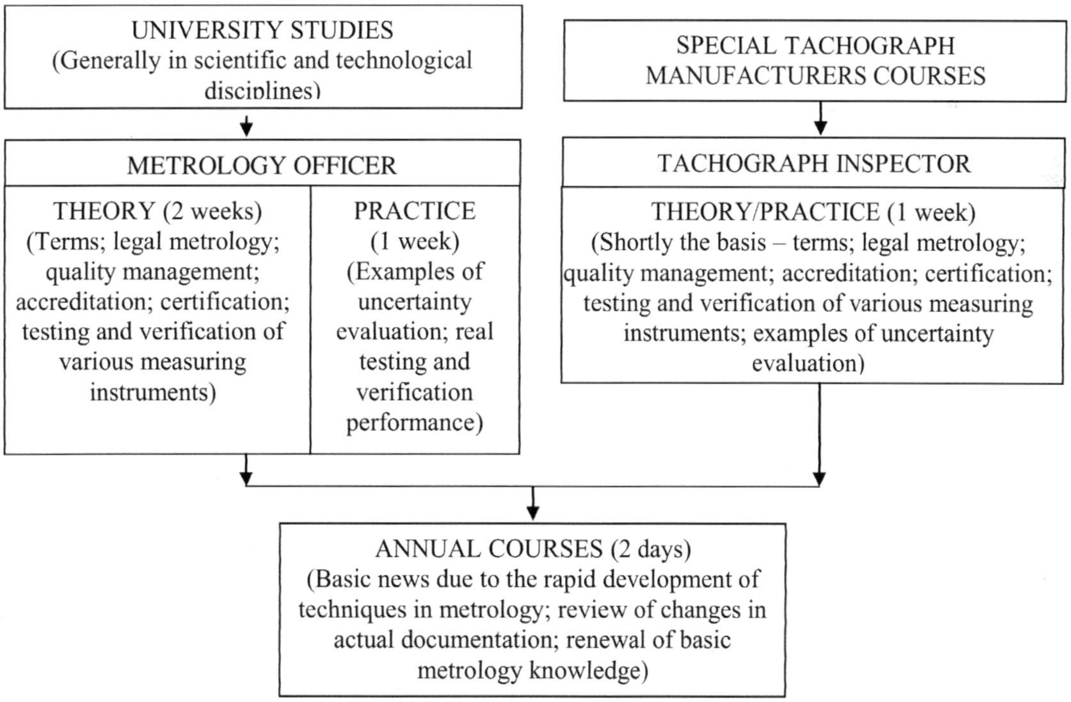

Figure 1. Model of metrology personnel education

2. The relationship between education and safety and security

Measurements are met in almost all human activities ranging from various production controls, measurements and status evaluation of environmental quality, health and safety assessment, conformity assessment of products to consumer protection and fair trade assurance [3].

Because of metrology science includes control of all above mentioned fields and also units of measurement and their standards, measuring instruments and their field of application, theoretical and practical problems, here appears direct relationship between human safety (security) and qualified / competent personnel that is responsible for the primary step of whole measurements (Fig. 2).

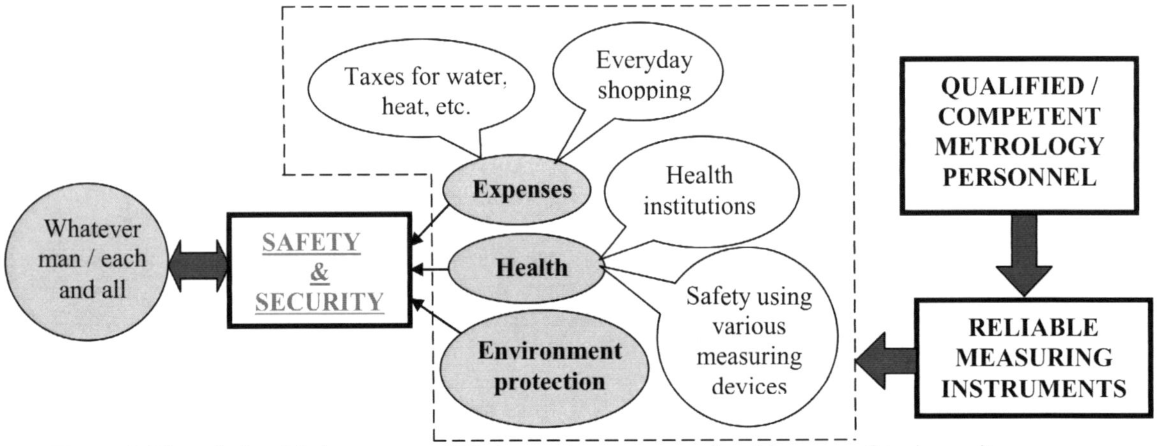

Figure 2. The relationship between qualified metrology personnel and human safety / security

Trying to reach better understanding of relationship between qualified metrology personnel and human safety / security it is purposeful to fathom the measurements. What is that? Where they are met? Who perform them? And where is that relation? Firstly, measurements enter into practically all commercial transactions from the trading of goods to the retail sale of goods in the market place. Here legal metrology must ensure that during the each sale the actual amount of every good must be not less or higher than the quantity contracted and paid for. The same rule is applied to prepackaged goods; various facilities (for example, consumed water, heat, time in the internet, time of phone conversations, etc.). Net content (it is not very important the particular object) carried out by the legal metrology protects every of us when we cannot verify the net quantity of contents.

Secondly, the health is especially sensitive field. The measurements here are absolutely different from other metrological areas. The measurement object is a human. Every "measurand" is different. Measurements in this field are variable due to unrepeatability of measurements. So the need of more reliable, comparable and efficient diagnostic and therapeutic methods remains open as the development of new technique for health care. Responsible control on the performed measurements and used measuring instruments in all health institutions is equally important from the consumer protection viewpoint. Now we can talk about medical instruments and about all other measuring instruments that can cause any trouble. For example, in the medicine thermometer or a blood pressure instrument that is not properly verified may lead to wrong diagnosis and incorrect medication [3]. And these are so small side of possible examples.

The field of environmental protection and pollution monitoring is very wide and very close to humanity. Our planet extremely needs of more accurate measurements. For example, measurement of energy consumption helps to better understand and control the use of energy, measurement of various properties of our soils helps to control contamination of used food or ground waters, measurement of pollutants in the atmosphere let us evaluate impact of human activities and etc. New stable and reproducible measurement standards for environmental changes monitoring have to be created. Just perfect knowledge, competence and understanding can help in research of new innovative systems and technologies that could help stop or prevent negative environmental impacts. Also additional measurements must be made in many areas in order to accurately and correctly determine the sources of these changes.

3. Metrology education and situation in Europe

Metrology is quite unusual profession but it is met very wide: from measurement systems or instruments design to calibration works, scientific research, and everyday life. The new products and technologies, their legislation depends on metrology [4]. Considering the large variety of people in this field we can realise how broad the metrology matter is. Metrology comprises concepts from physics, chemistry, biology and medicine, mathematics, all kinds of engineering. Person who needs to gain a metrology speciality should have a strong grounding in physics or other technological science. During metrology courses the specialized knowledge in metrology is gained but it is preferred that the learning person should have the most common knowledge in mechanical and electrical measurements, also have understanding about the main measurands, for example time, mass, temperature, pressure and etc. In many cases the understanding of more than one field is needed. With an increasing demand for higher accuracy, traceable measurements play a crucial role in almost every aspects of modern life. The measured ranges become wider from very small quantities to extra-large and this tendency has growing character in year's viewpoint. And relevant qualification plays very important role for fast changing and growing measurement world (Fig. 3).

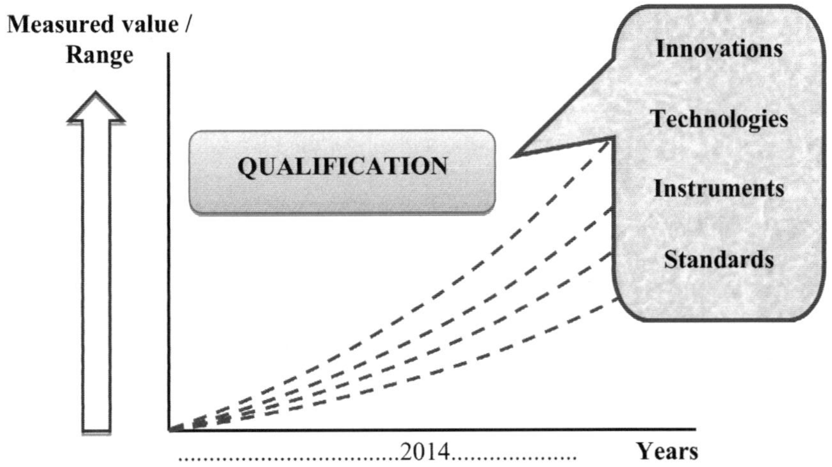

Figure 3. Qualification demand

Good metrology system and qualified specialists can help in increasing productivity and efficiency, improving repeatability by minimizing human errors, increasing customer satisfaction and saving money applying new or suitable technologies for various tasks. Many Europe and also World countries with established metrology system have metrology training centres, specialized educational institutions and formal metrology education at technical schools and/or university level [4]. Mostly metrology education is accomplished through short special courses where theoretical and practical skills are gained.

4. Conclusions

A key challenge is the need to ensure human safety and security. Continuous and sustainable growth whilst reducing negative impacts in many fields leads to more qualitative understanding of importance of education and competence maintenance of responsible personnel.

Journal of Physics: Conference Series **588** (2015) 012013 doi:10.1088/1742-6596/588/1/012013

The nature of the challenges faced in protecting the environment and in developing new sustainable technologies dictates a multi-disciplinary approach that will drive the requirement for research that brings together different disciplines within measurement science. One of the most important tasks is to pay more attention to expedient and continuous training.

Metrology education and competence maintenance following innovations, standardization should reach organizations, industry and be spread between metrology institutes inside and outside Europe.

References

[1] Training and qualification of legal metrology personnel. OIML D 14, Edition 2004(E).
[2] Meskuotiene A, Zilinskas R P, Zabolotnas V. Education and competency evaluation system of metrology practitioners in Lithuania, 2008. ISBN 2951645384. p. 349-355.
[3] K. Ramful. Legal metrology and international trade, 2004. International trade centre
[4] Zaimovic-Uzunovic N, Lemeš S. Importance Of Metrology Education - B&H Case, 2008. 20th International Metrology Symposium. p. 262-265. Croatia.

Implementation of Multiple Host Nodes in Wireless Sensing Node Network System for Landslide Monitoring

Faizulsalihin bin Abas[1] and Shigeru Takayama[2]

Department of Electric and Electronics Engineering, Ritsumeikan University BKC, 1-1-1 Noji –higashi, Kusatsu, Shiga, 525-8577 Japan

Email: [1] re0034sf@ed.ritsumei.ac.jp , [2]s-tkym@se.ritsumei.ac.jp

Abstract. This paper proposes multiple host nodes in Wireless Sensing Node Network System (WSNNS) for landslide monitoring. As landslide disasters damage monitoring system easily, one major demand in landslide monitoring is the flexibility and robustness of the system to evaluate the current situation in the monitored area. For various reasons WSNNS can provide an important contribution to reach that aim. In this system, acceleration sensors and GPS are deployed in sensing nodes. Location information by GPS, enable the system to estimate network topology and enable the system to perceive the location in emergency by monitoring the node mode. Acceleration sensors deployment, capacitate this system to detect slow mass movement that can lead to landslide occurrence. Once deployed, sensing nodes self-organize into an autonomous wireless ad hoc network. The measurement parameter data from sensing nodes is transmitted to Host System via host node and "Cloud" System. The implementation of multiple host nodes in Local Sensing Node Network System (LSNNS), improve risk-management of the WSNNS for real-time monitoring of landslide disaster.

1. Introduction

Landslides are one of the most dangerous natural hazards which give a great impact upon occurrence, cause huge damage to infrastructures as well as loss of human life. They can strike in minutes when stability of a slope changed from stable to unstable condition. They usually triggered by heavy rainfall, but can also result from earthquakes, volcanic activity, and erosion. Although they cannot be prevented, they can be forecast. By reason of dangerousness, man power only cannot be used to observe this pattern of disaster. Therefore, in order to abate the damage and loss, a system that can monitor this disaster is necessary.

Wireless Sensing Node Network System (WSNNS) is an effective way to monitor landslide disaster in a wide area for long time. The system provides dual-way asynchronous communication between Host System (HS) and Local Sensing Node Network System (LSNNS). Host System is placed outside of the monitored area. Measuring person can access Host System; send commands to local sensor network (LSNNS) to get information about the monitored area without any risk. LSNNS then will transmit the measurement data required to the Host System. The mesh-type network formation in LSNNS enabled WSNNS to continue the landslide measurement even if some sensing nodes have been damaged during landslide occurrence.

2. Wireless Sensing Node Network System (WSNNS) Configuration

Wireless Sensing Node Network System (WSNNS) is structured into Host System (HS), Local Sensing Node Network System (LSNNS) and "Cloud" System (CS) as shown in Figure 1.

Content from this work may be used under the terms of the Creative Commons Attribution 3.0 licence. Any further distribution of this work must maintain attribution to the author(s) and the title of the work, journal citation and DOI.
Published under licence by IOP Publishing Ltd

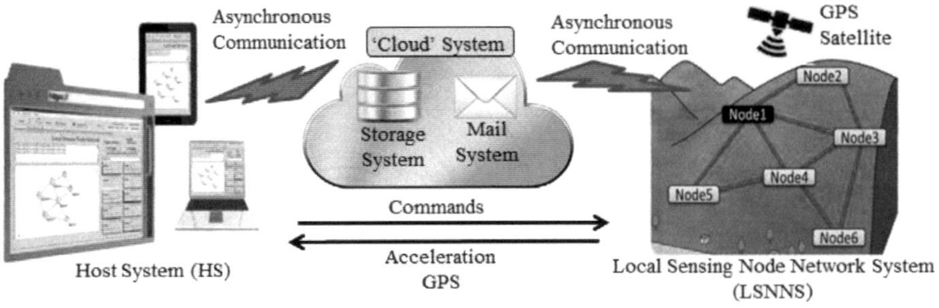

Figure 1. System construction of Wireless Sensing Node Network System (WSNNS).

2.1. Local Sensing Node Network System (LSNNS)

LSNNS is divided into host node and plural distributed sensing nodes.

Sensing node is a node in LSNNS which capable of gathering measurement data needed in landslide detecting and it can perform communication with other connected nodes in LSNNS. A sensing node consists of 3D acceleration sensor, GPS, controller board, liquid crystal display (LCD), and wireless communication unit MU-1. 3D acceleration sensor is used to monitor ground acceleration. By analyzing 3D acceleration angles and power spectrum, types of landslides (sliding down, rolling down) can be recognized. Besides that, GPS provides ground position information for each sensing node. By analyzing the position information, Host System enables to determine network topology of LSNNS which is necessary in constructing routing table and analyzing data during landslide occurrence. The position precision of GPS in sensing node is within 5m. Sensing nodes are located at the area which has high land-sliding probability to sense the occurrence of landslides. Figure 2 shows a sensing node prototype and Figure 3 shows mesh-type network connection between sensing nodes.

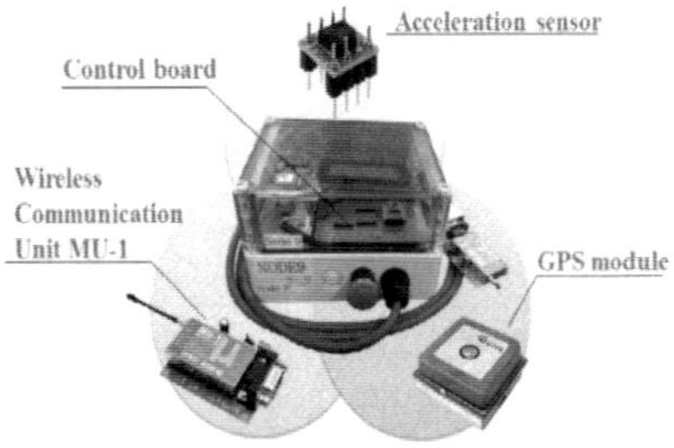

Figure 2. Sensing node prototype

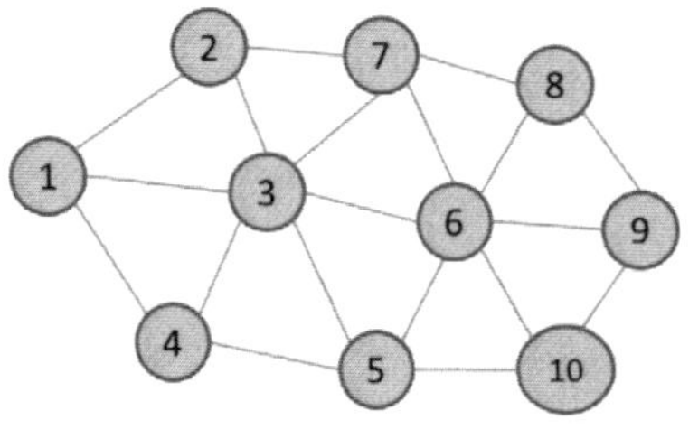

Figure 3. Mesh-type network

Host node is a gateway of LSNNS for receiving commands from Host System and transmitting measurement data from sensing nodes in LSNNS to Host System. The communication between host node and sensing nodes in LSNNS is by using long range wireless unit while the communication between host node and Host System is performed by using internet connection via Android. Host node is an important node which cannot be damaged during landslide disaster. Therefore, host node is set up at a place which has low probability of slope failure and has stable communication with Host System.

2.2. "Cloud" System (CS)

"Cloud" System contains storage system and mail system. Storage system and mail system are very useful as they transmit commands from Host System to LSNNS and measurement data from LSNNS to Host System. Host node in LSNNS can access the "Cloud" System by 13 messages per minute. Host node is connected to Android via Bluetooth in order to get internet access. Figure 4 shows series of data transmitted by mail messages in Android while Figure 5 shows the growth of LSNNS based on the measurement data from deployed sensing nodes.

Figure 4. Series of data measurement

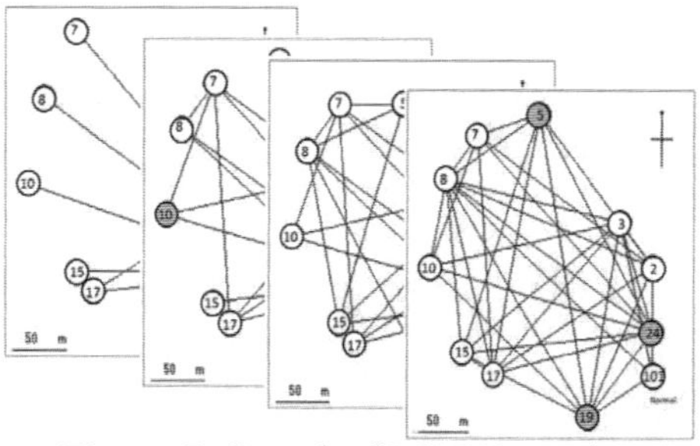

Figure 5. Growth of LSNNS topology

2.3. Host System (HS)
Host System sends operation commands to manage LSNNS. It also analyzes measurement parameter condition, construct routing table from network topology of LSNNS and alarm the occurrence of landslide. Figure 6 shows command operation window in Host System. Measurement data such as GPS and acceleration sensor data, network topology, routing table and node condition can be monitored from this command operation window.

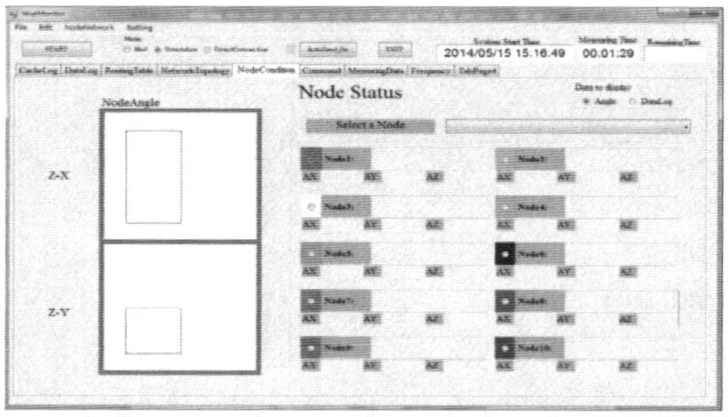

Figure 6. Command operation window in Host System

3. Implementation of Multiple Host Nodes in WSNNS
During landslide disaster, although a host node is set up at a place which has low probability of slope failure, there is still a possibility for the host node to be damaged. When the one host node is damaged, no command can be transmitted from Host System to sensing nodes in LSNNS and no measurement data can be received by Host System from deployed sensing nodes. Therefore, to avoid this kind of communication failure, it is a good solution to have more than one host node in LSNNS so that if one host node is damaged, other host node can still perform the data communication. Therefore, we propose the implementation of multiple host nodes in WSNNS for landslide monitoring. Figure 7 shows application of four host nodes in WSNNS.

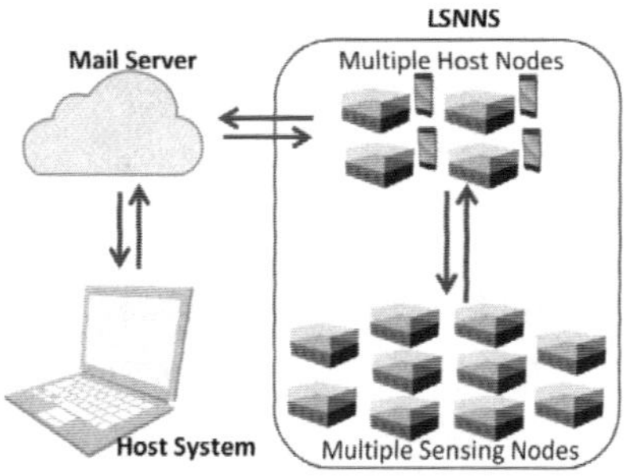

Figure 7. Application of four host nodes in WSNNS

3.1. Robustness of WSNNS by Using Multiple Host Nodes

In order to conduct an evaluation on WSNNS by using multiple host nodes, four host nodes (number 101,102,103,104) and six sensing nodes (number 2, 3, 7, 10, 15, 19) are deployed as shown in Figure 8. Four host nodes are located at different places so that in case of landslide occurs at one host node area and the host node is damaged, other host nodes can still survive. It is necessary to write four email addresses that commands will be sent to as shown in Figure 9.

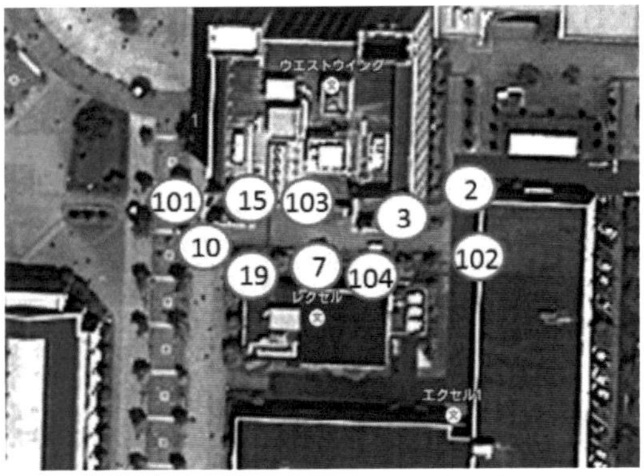

Figure 8. Arrangement of nodes

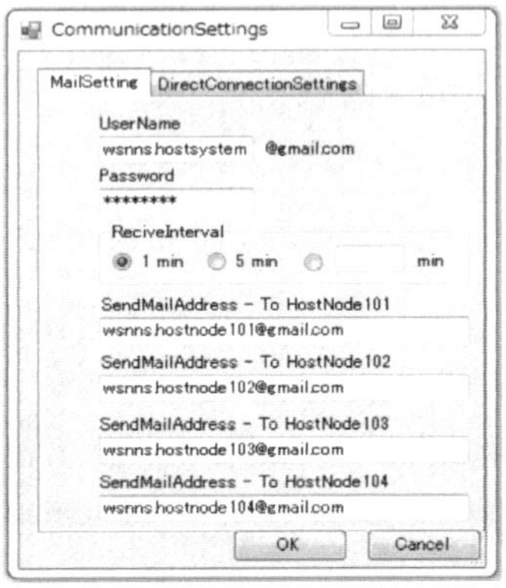

Figure 9. Mail Communication Setting

The evaluation started by sending a command to each host node to gather all sensing nodes measurement data information including GPS, acceleration sensor, and nodes adjacent information. Common data received by each host nodes are analysed at Host System. After each 3 minutes, one host node is artificially damaged starting from Host Node 101, 102 and 103. Measurement data received until the end of experiment is monitored at the Host System as shown in Figure 10, 11, 12 and 13.

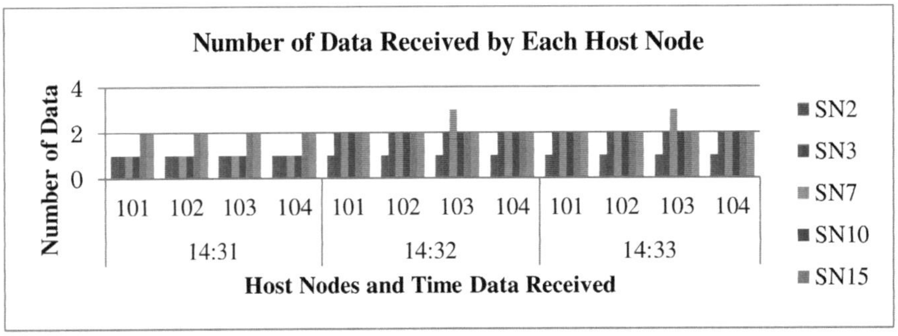

Figure 10. Number of data received by each host node

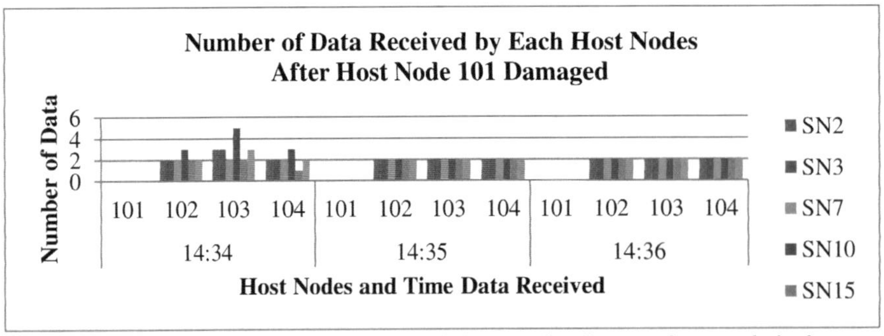

Figure 11. Number of data received by each host node after one host node is damaged

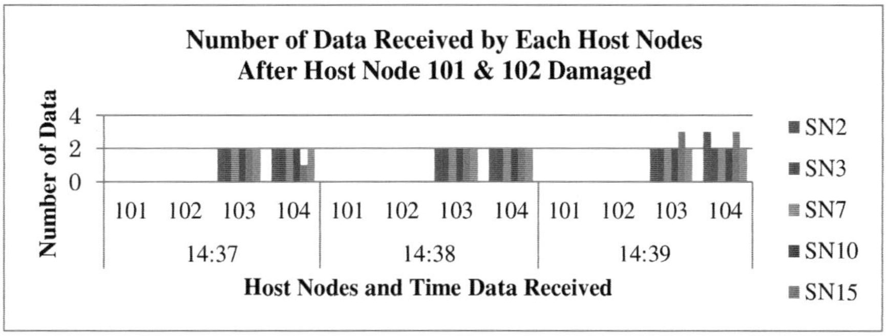

Figure 12. Number of data received by each host node after two host nodes are damaged

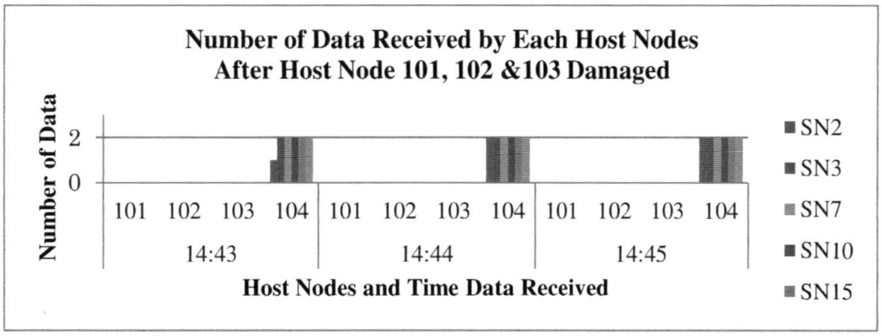

Figure 13. Number of data received by each host node after three host nodes are damaged

From Figure 10, we can see that each host node received measurement data from all sensing nodes. All host nodes have the same characteristics and load balance data collection and they can connect to the Host System at the same time. Figure 11, 12 and 13 shows that after one host node is damaged, other host nodes can still perform the data communication. There must be a backup host node at one time. If we use only two host nodes at the first place, after one host node is damaged, the other one host node will not have any backup. Therefore, we propose the use of not only two host nodes in LSNNS but three or four host nodes are necessary to improve the robustness of WSNNS in landslide monitoring.

3.2. Less Number of Hop for Transmitting Warning Information

In emergency mode (landslide occurrence), it is important to determine route taken by sensing nodes in sending the emergency data to Host System. Three host nodes (number 101,102, 103) and six sensing nodes (number 2, 3, 5, 7, 10, 24) are deployed as shown in Figure 14 while Figure 15 shows the number of data received by each host nodes during normal mode. All data sent to each host node contains different routing information that later will be used to transmit warning information.

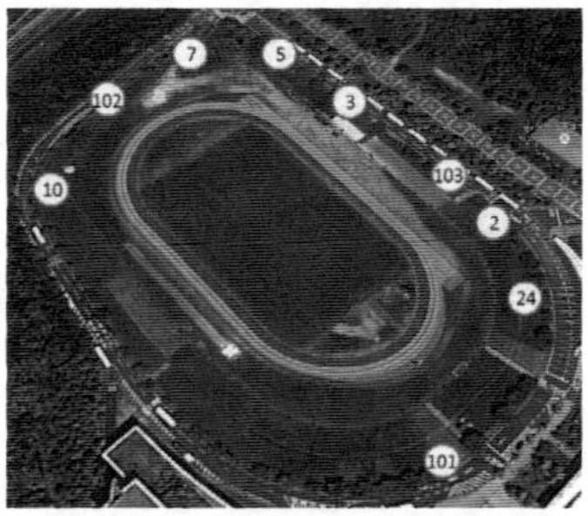

Figure 14. Arrangement of nodes

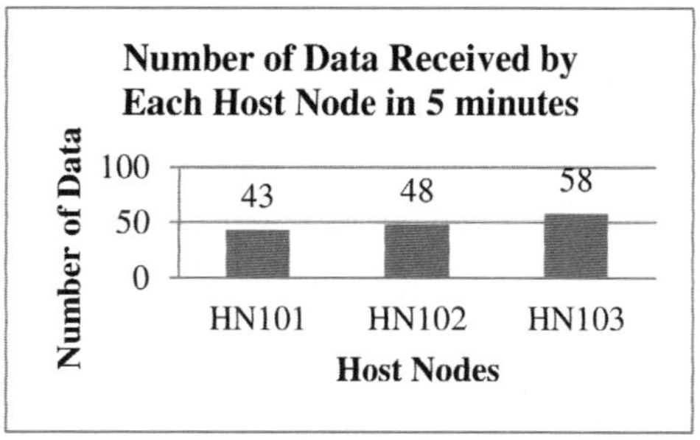

Figure 15. Number of data received by each host node

In the middle of the experiment, one sensing node (number 24) is artificially damaged and the route taken for emergency data transfer is analysed. Figure 16 shows the emergency data received at the Host System. "[024-9]" means that the emergency data is transmitted directly from sensing node number 24 to host node. "Host103" means that Host System received the emergency data via Host Node 103 which is the nearest host node for sensing node number 24.

```
[2014-05-13/181546]#DR=20#83[024-9]N024T091544A22:Host103
[2014-05-13/181546]#DR=20#83[024-9]N024T091544A22:Host103
[2014-05-13/181546]#DR=20#83[024-9]N024T091546A22:Host103
```

Figure 16. Number of data received by each host node

The presences of more host nodes enable sensing nodes to be closer to a host node. Therefore, sensing nodes can transmit the emergency data via the shortest route to the Host System. This can help minimize the time difference between time of landslide occurrence and time of Host System receives the emergency data. The result shows that Host System received the emergency data at 18:15:46 which is 46 seconds after the artificial landslide. This is because it takes time for sending and receiving email between host node, "Cloud" system and Host system.

4. Conclusion

The combination of Host System, "Cloud" System, and Local Sensing Node Network System in WSNNS enables this system to be a reliable system in monitoring landslide disaster. The implementation of multiple host nodes in LSNNS improves the robustness of WSNNS and enables warning information to be transmitted to Host System with less number of hops. Thus, emergency warning can be informed faster to take earlier safety precautions. Besides that, evaluation of multiple host nodes implementation in WSNNS at larger space area with more number of sensing nodes will be a good method to determine its merits and demerits.

References

[1] S Takayama, J Akiyama, T Fujiki and N A B Mokhtar *Wireless Sensing Network Management for Landslide Disaster Monitoring* J. Phys. Conf. Ser. 459 012045

[2] Nur Atiqah binti Mokhtar and S Takayama *Wireless Sensing Node Network Management for Landslide Disaster* SICE Annual Conference 2013, pp 67

[3] S Taenaka, Y Watanabe and S Takayama *Construction of Wireless Sensing Network System for Landslide Disaster Monitoring*

[4] S Taenaka, J Akiyama, T Fujiki and S Takayama *Design of Flexible Wireless Sensing Network system for Landslide Disaster Monitoring* Proc. Of SICE International Conference 2012, pp 1443-1459.

[5] Aoki Hiroshi, Steven Glaser, Ikegawa Youjiro, Boda Shingoand Miki Kazuhide *Development and Application for The Wireless Sensor Networks for The Ground Monitoring* October 2006 Vol 46 No. 2 pp 1-5

[6] K Taneda, S Niimi, S Takayama *Measurement of Dynamic State of Landslide Disaster* ACTA IMEKO September 2012 1, 1931-1936

[7] R Ohbayashi, Y Nakajima, H Nishikado and S Takayama *Monitoring System for Landslide Disaster by Wireless Sensing Node Network* SICE Annual Conference 2008 p 1704-1710

[8] F Rahman, A Kumar, G Nagendra and G S Gupta *Network Approach for Physiological Parameters Measurement* IEEE Transactions on Instrumental and Measurement, 2005 Vol. 54, pp 322-329

Physical Monitoring in Daily Life by Remote Body Area Network System

Keisuke Kamada[1], Nakanishi Yuichi[2], Hideaki Shimizu[3], Hiroto Takahashi[4], Nur Alia Athirah Binti Hj Mohtadzar[5], Shigeru Takayama[6]

Department of Electronics System, Faculty of Science and Engineering, Ritsumeikan Univ. BKC, 1-1-1,Nojihigashi,Kusatu city,Shiga,525-8577, Japan

E-mail: [1]re0003iv@ed.ritsumei.ac.jp, [2]re007081@ed.ritsumei.ac.jp, [3]re0005hh@ed.ritsumei.ac.jp, [4]re0005sf@ed.ritsumei.ac.jp, [5]re0023hi@ed.ritsumei.ac.jp and [6]s-tkym@se.ritsumei.ac.jp

Abstract. To spend daily life in high QOL, it is important to keep our health condition. Physical diseases are caused by various body parameters. People must get body parameter in daily life. Therefore people need wearable body area network system for getting body parameter in daily life. Authors made wearable body area network system which can get heart rate, SpO_2, body temperature, skin temperature, air temperature, impact and acceleration of waist, shoulder, both ankles and wrist. Moreover authors made some applications by using these parameters. This paper describes the wearable sensing network system, host system to monitor dynamic physical conditions of user at remote location and applications.

1. Introduction

People should spend in high QOL (Quality Of Life), however we sometime get physical diseases such as slipped disc and over tiredness. To prevent them we have to monitor own body parameter in daily life before getting them, and we have to care by ourselves. To share information with family and home doctor at remote location is important for week person. They can monitor the state of physical weak person by the information. To realize them, we develop wearable body area network system, high frequently asynchronous measurable system, and remote wireless communication system. The BAN and Remote Interactive Care system (RICS) can monitor and feedback for user's life activities. This paper describes the BAN system, RICS system and Applications.

2. System Construction

2.1 Body Area Network system (BAN)

The figure 1 shows the construction of BAN. This system is constructed by 5 wearable modules (waist, shoulder, wrist, right ankle and left ankle). In this system, waist module works as host module, and other modules work as local nodes. They use 2 kind of wireless communications; blue lines show Bluetooth and green lines show Zigbee. Each module has PIC micro controller made by Microchip to control sensors and communication units. The shoulder module has earphone and sound module to inform physical and conditions methods of improvement to user.

Content from this work may be used under the terms of the Creative Commons Attribution 3.0 licence. Any further distribution of this work must maintain attribution to the author(s) and the title of the work, journal citation and DOI.
Published under licence by IOP Publishing Ltd

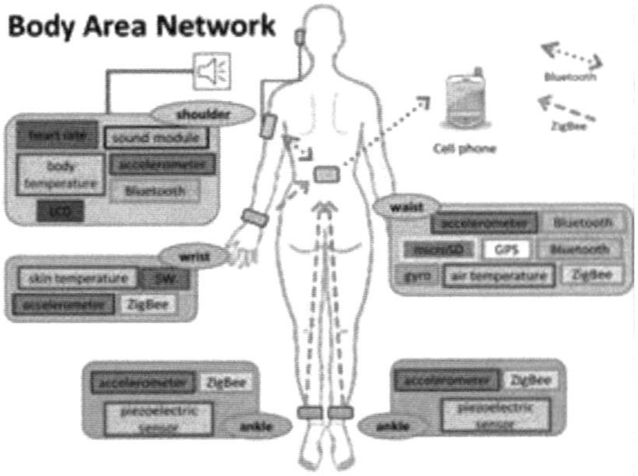

Figure 1. Construction of Body Area Network

2.2 Remote Interactive Care System (RICS)

Figure 2 shows the construction of RICS. This system is both direction data communication system. This system can report error to external when BAN detection body error. And this system can send advice to BAN user at remote location. This system is constructed with waist module, smart phone, cloud system and host PC. This system is connected to waist module and smart phone by Bluetooth of waist module and smart phone. This system sends and receives the data between waist module and smart phone by Bluetooth. Smart phone communicates with cloud system by 3G communication. This system sends and receives data between cloud system and smart phone by mail. Host system sends and receives data by accessing to cloud system.

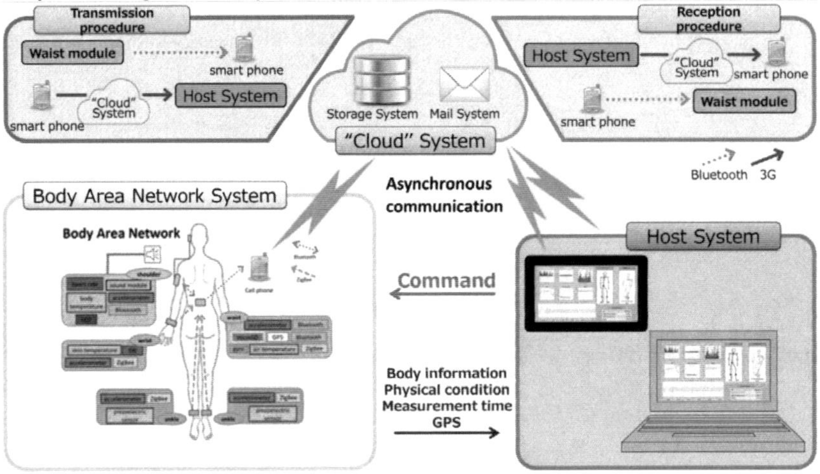

Figure 2. Remote Interactive Care System (RICS)

3. Applications

3.1 Training Assistance by Monitor HR

This application leads user to exercise intensity by using the heart rate. User can keep adequate exercise by this application.

This application use Karvonen Formula, threshold and trend as index of making advice. Relationship of heart rate and exercise intensity is defined by Karvonen Formula. Figure 3 shows Karvonen Formula. HRmax mean 220-age. HRmin mean heart rate of normal time. HR mean max value or minimum value of threshold.

Journal of Physics: Conference Series **588** (2015) 012015 doi:10.1088/1742-6596/588/1/012015

This application use Karvonen Formula for deciding threshold of heart rate. In the Karvonen Formula, 40-50 [%] mean Light exercise (walking), 50-60 [%] mean Medium exercise (slow running), 60-85 [%] mean Stressed exercise and 85-100 [%] mean Heavy exercise (athlete level).

$$EI[\%]=(HR-HR_{min})/(HR_{max}-HR_{min})\times 100 \quad \text{[Karvonen Formula]}$$

HRmin : heart rate in quiet HRmax : heart rate in the most heavy exercise

Figure3. Karvonen Formula

Thresholds value are calculated by Karvonen Formula stated above. It is obtained by splitting the value between max value (max) and min value (min) into three. High splitting value called thr1. Low splitting value called thr2. For example, if max value is 145 and min value is 120, splitting values are 136(thr1) and 128(thr2). Figure 4 shows this example.
Index of the trend are [HR < min], [min < HR <Thr1], [Thr1 < HR <Thr2], [Thr2 < HR < Thr max], [Thr max < HR].

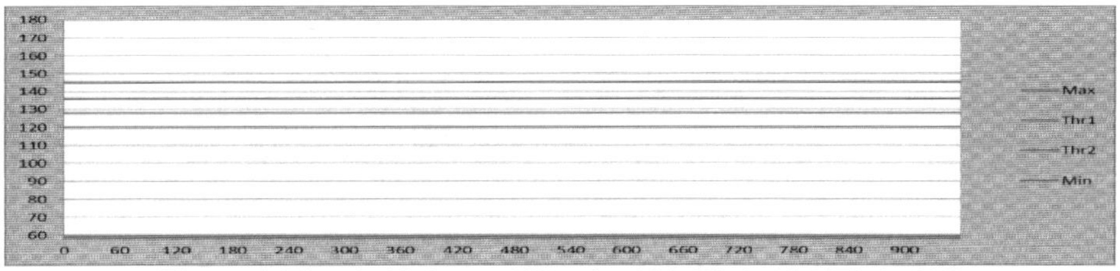

Figure 4. Thresholds(max=145, min=120, Thr1=136, Thr2=128)

Trend mean change of heart rate. Trend and threshold judge kinds of advice. Thresholds are classified three parts. They are as falling trend or flat, as mild trend and as steep trend. The maximum trend calculation of data acquisition for heart rate is three seconds. Index of the trend are [trend< =0],[0< trend<2] and [2< =trend]. Figure 5 (a) shows advice generation by heart rate of existing range and trend.
Figure 5 (b) shows kinds of sound advice. This experiment feeds back to user by six kinds of advice. They lead to proper exercise subjects.

Trend / Existing Of HR	Trend<=0	0<Trend<2	2<=Trend
HR < Thr min	Advice1		Advice3
Thr min< HR <Thr1	Advice1	Advice2	
Thr1 < HR <Thr2	Advice2		
Thr2< HR <Thr max	Advice2	Advice4	Advice5
Thr max< HR	Advice6		

Advice 1	Pace up slowly
Advice 2	Keep pace
Advice 3	Pace down slowly
Advice 4	Pace down
Advice 5	Pace down slowly
Advice 6	Stop exercise now, have a short rest

(a) Judge Kinds of Advice (b) Sound Advice

Figure 5. Kinds of Advice

3.1.1 Experimental Method

Experiment is conducted on treadmill. The experiment take 2 patterns for proving effectiveness of the system. 2 patterns are training without advice and training with advice. This paragraph prove validity by comparing experimental results. Experiment time schedule is as follows. Training without advice; Firstly, a subject waits for 1 min. Secondly, a subject runs for 10 min by 7km/h (no advice). Finally, a subject waits for 5min. Training with advice; Firstly, Waiting for 1 min. Secondly, user runs for 10 min by advised running. Finally, user waits for 5min.

3.1.2 Impact on The Movement by Sound Advice

Table 1 shows body parameter of subjects. All subjects are men.

Table 1. Physical characters of subjects

Subject	Age	Weight	Normal Heart Rate	Threshold max	Threshold min
	24	67kg	70bps	145	120
	22	57kg	64bps	144	117

Colours on figure 6 (b) and figure 7 (b) are kinds of advice. Proper advices can lead heart rate in threshold. Percentage of HR on threshold of training without advice is 5.4% in case of subject A. Percentage of HR on threshold of training with advice is 84.1% in case of subject A. Percentage of HR on threshold of training without advice is 4.2% in case of subject B. Percentage of HR on threshold of training with advice is 80.3% in case of subject B. Both subjects are better. This result proves that advices lead user to aim heart rate zone.

This paragraph proves effectiveness of the system by experimental results.

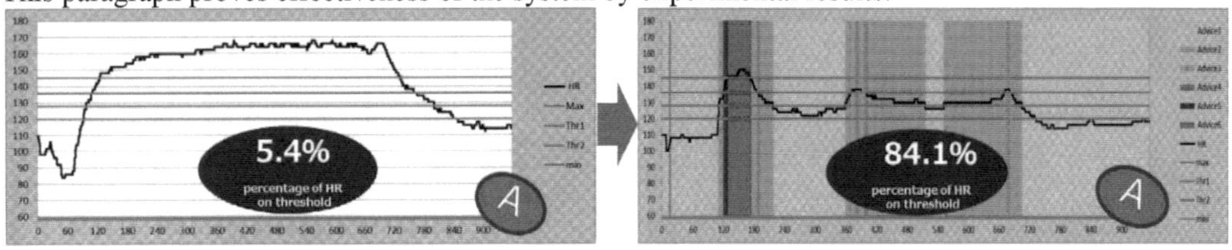

(a) Training without advice (**b**) Training with advice

Figure 6. Comparison of no advice and advising to subject A

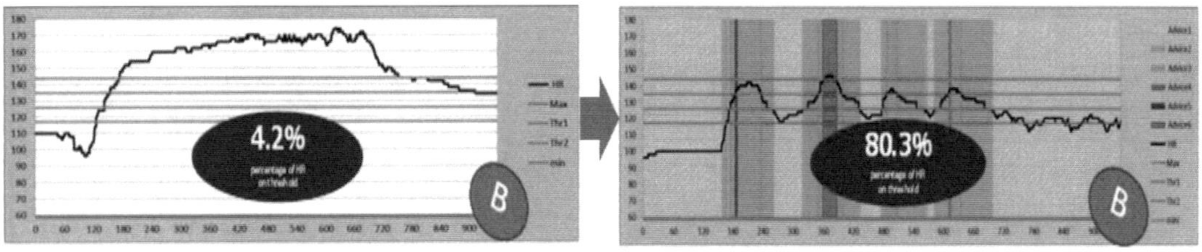

(a) Training without advice (b) Training with advice

Figure 7. Comparison of no advice and advising to subject B

3.2 Estimation of Body Inclination in Walking

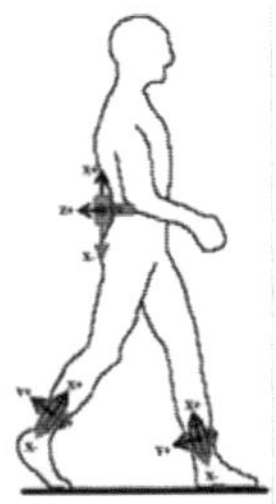

Figure 8. Arrangement of Acceleration axis on body

This application shows body inclination to user. Acceleration sensor of waist module gets body inclination and body swing. Acceleration sensor of ankle modules gets gap of landing impact of both ankles. Figure 8 shows the structure of the acceleration axis of each module.

The tendency and degree of inclination of the body are judged from a measurement result. Table 2 and Figure 9 show the types of the tendency of inclination and how to determine it.

In this system, the tendency of inclination is judged from Table 2 The case that the body inclines to the same side as the foot of a greater landing impact is set to "Trunk Distortion

Type". This type is the type that inclination of the trunk causes a greater landing impact to the foot. Then the case that the body inclines to the opposite side as the foot of a greater landing impact is set to "Tread Reaction Type". This type is the type that the gap of landing impact causes the inclination of the body.

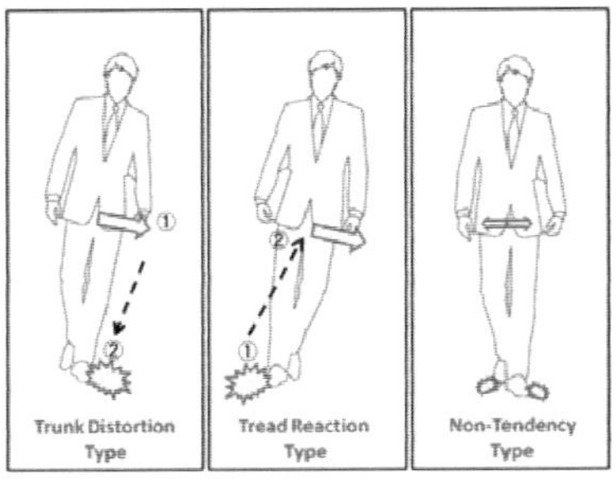

Figure 9. 3-Tendency Type

Table 2 Tendency of Inclination

Landing Impact \ Body Inclination	Right	Left
Right	Trunk Distortion Type	Trend Reaction Type
Left	Trend Reaction Type	Trunk Distortion Type

3.2.1 Experimental method

Experiment time schedule is as follows. Firstly, a subject stood on treadmill for 30 seconds. Secondly, a subject walked for 20 minutes at 4km/h. Finally, a subject stopped walking and stood for 5 minutes.

The experiment was carried out three times. Subject A was once, subject B was one times, and subject C was once. Table 3 shows body parameter of subjects. All subjects are men.

Table 3 Physical characters of subjects

Sbuject	Age	Height(cm)	Weight(kg)	Smoling	Sport experience
A	22	169	56	No	Baseball plyer for 8 years
B	22	174	57	No	None
C	22	168	59	Yes	Football plyer for 8 years

3.2.2 Judgment Result

The processing result is shown below. Judgment of tendency is expressed by bar chart in each section. A red stick in two upper graphs expresses the inclination of right side, and a blue stick in them expresses the inclination of left side. Further, a red stick in the lower left graph expresses Trunk Distortion Type, and a blue stick in it expresses Tread Reaction Type.

Figure 10 shows the first experimental result of subject A. The burden placed on his left ankle is heavier than his right ankle. However, the inclination of his body is small, there is no trend. Moreover, the swing of his body is small as a whole. To summarize these results, this system determines that the walking posture of the subject A is good balance.

Figure 11 shows the experimental result of subject B. The burden placed on his left ankle is heavier than his right ankle. His body leans to the left. Moreover, the swing of his body is large. To summarize these results, this system determines that the walking posture of the subject B is Trunk Distortion Type at the rate of 87%. In addition, the burden of his body becomes bigger because the swing of his body is large.

Figure 12 shows the experimental result of subject C. The burden placed on his right ankle is heavier than his left ankle. His body leans to the left. However, the swing of his body is small. To summarize these results, this system determines that the walking posture of the subject C is Tread Reaction Type at the rate of 77%.

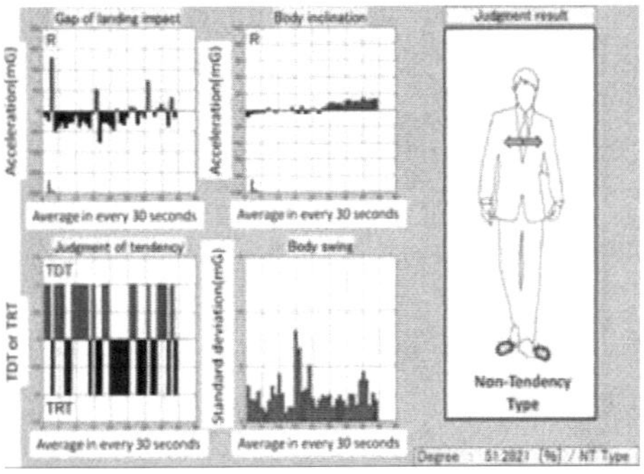

Figure 10. Judgment result of subject A

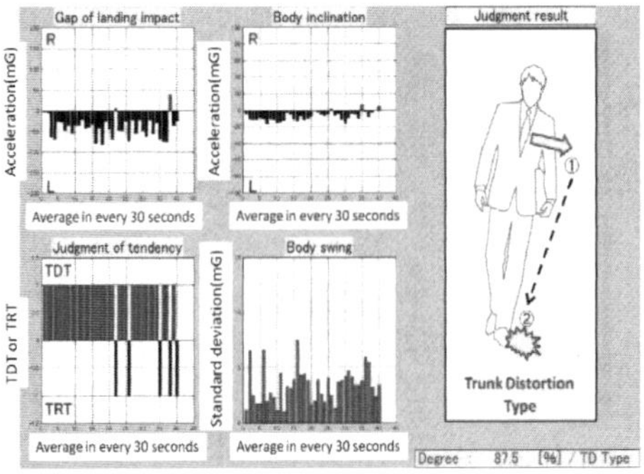

Figure 11. Judgment result of subject B

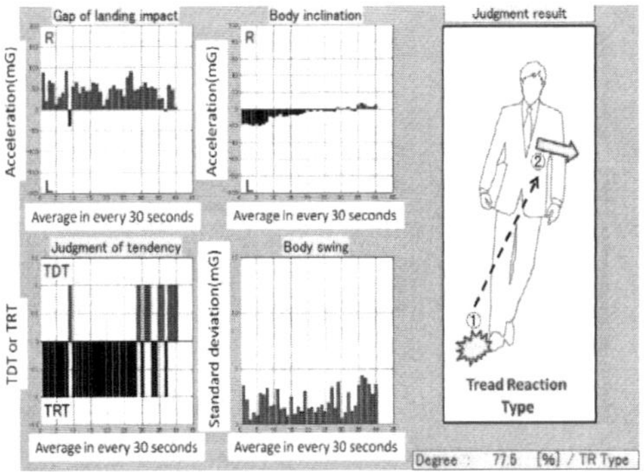

Figure 12. Judgment result of subject C

4. CONCLUSION

BAN can get body information in daily life. RICS share information getting by BAN to Family, doctor and trainer. Applications can do training assistance by monitor and estimation of body inclination in walking. They system can lead people to better movement. People can spend in high QOL by using this system.

References

[1] Mienkovic M., Jovanov E., Chapman J., Raskovic D. and Price J.: "An Accelerometer-Based Physical Rehabilitation System", ECE Dept., University of Huntsvile, Alabama Department of Electrical Engineering & Comp. Eng. University of Alabama in Huntsvile, IEEE, pp.57 – 60, Huntsvile, AL 35899 USA (2002).

[2] Y.Nakanishi, Y.Matsuda, K.Toriyama, R.Watabe, K.Sentoku, S.Takayama, "Design of Wearable Sensing Network System for Monitoring Dynamic Physical Condition", TUA09-02, SICE, Akita Univ. , 2012More references

[3] T.Fujiki, S.Takayama, "Composition of Email Assisted Data Communication System in Sensing Network for Monitoring Landslide Disaster", B-20-1, IEICE, Gifu Univ. , 2013

[4] Gafurov D., Helkala K. and Søndrol T.: "Gait Recognition Using Acceleration from MEMS", IEEE First International Conference on Availability, Reliability and Security (ARES '06). 2006

[5] Youngsung Kim, Il-yeon Cho, "Wearable ECG Monitor: Evaluation and Experimental Analysis", Information Science and Applications (ICISA), 2011 International Conference, pp.1-5, Jeju Island, 2011

[6] Stefano Melzi, Luca Borsani, Matteo Cesana, "The Virtual Trainer: Supervising Movements Through a Wearable Wireless Sensor Network", Sensor, Mesh and Ad Hoc Communications and Networks Workshops, 2009. SECON Workshops '09. 6th Annual IEEE Communications Society Conference, pp.1-3, Rome, 2009.

Journal of Physics: Conference Series **588** (2015) 012016 doi:10.1088/1742-6596/588/1/012016

Base metal thermocouples drift rate dependence from thermoelement diameter.

P Pavlasek[1] , S Duris[1] and R Palencar[1]
[1] Slovak University of Technology, Faculty of Mechanical Engineering, Bratislava, Nam. Slobody 17, Slovakia

E-mail: peterpavlasek@gmail.com

Abstract. Temperature measurements are one of the key factors in many industrial applications that directly affect the quality, effectiveness and safety of manufacturing processes. In many industrial applications these temperature measurements are realized by thermocouples. Accuracy of thermocouples directly affects the quality of the final product of manufacturing and their durability determines the safety margins required. One of the significant effects that affect the precision of the thermocouples is short and long term stability of their voltage output. This stability issue occurs in every type of thermocouples and is caused by multiple factors. In general these factors affect the Seebeck coefficient which is a material constant, which determines the level of generated voltage when exposed to a temperature gradient. Changes of this constant result in the change of the thermocouples voltage output thus indicated temperature which can result in production quality issues, safety and health hazards. These alternations can be caused by physical and chemical changes within the thermocouple lead material. Modification of this material constant can be of temporary nature or permanent. This paper concentrates on the permanent , or irreversible changes of the Seebeck coefficient that occur in commonly used swaged MIMS Type N thermocouples. These permanent changes can be seen as systematic change of the EMF of the thermocouple when it is exposed to a high temperature over a period of time. This change of EMF by time is commonly known as the drift of the thermocouple. This work deals with the time instability of thermocouples EMF at temperatures above 1200 °C. Instability of the output voltage was taken into relation with the lead diameter of the tested thermocouples. This paper concentrates in detail on the change of voltage output of thermocouples of different diameters which were tested at high temperatures for the overall period of more than 210 hours. The gather data from this testing was used to establish the relation between the level of EMF drift and the lead diameter of the thermocouple thermoelements. Furthermore this data was also used to create a drift function which mathematically expresses the dependency between the drift rate and the diameter of the thermocouple leads.

[1] To whom any correspondence should be addressed.

Content from this work may be used under the terms of the Creative Commons Attribution 3.0 licence. Any further distribution of this work must maintain attribution to the author(s) and the title of the work, journal citation and DOI.
Published under licence by IOP Publishing Ltd

1. Introduction

Measurements of temperature have a great importance in wide range of industrial applications. As temperature affects the quality, safety and effectives of many of these applications, a great effort has been made to enhance the precision and reliability of temperature measuring sensors. One of the main types of temperature sensors that are used in industry are thermoelectric sensors, more commonly known as thermocouples. These sensors play a irreplaceable role in high temperature industrial measurements. Their robust construction, ability to withstand high temperatures and harsh conditions had made them popular among many users. These active temperature sensors are used for monitoring and control of many industrial processes. As a widely used type of temperature sensor, there are still effects which occur in thermocouples under various conditions that need further investigation. This process of testing and investigation is necessary to understand the behaviour under boundary condition. Although base metal thermocouples have been used in a wide variety of industrial applications for a long period of time there are still cases in which these sensors accuracy is affected. Accuracy of thermocouples directly affects the quality of the final product of manufacturing and their durability determines the safety margins required. To efficiently reduce the effects, that result in lower levels of precision it is necessary to explore various influences that can cause measurement errors and investigate possibilities of their reduction or removal.

2. Long term stability of thermocouples

Short and long term stability of thermocouples voltage output is one of the substantial effects that affect the precision and reliability of these sensors. This stability change is caused in general by the change of the Seebeck coefficient which is a material constant. This coefficient determines the level of generated voltage when exposed to a temperature gradient. It is unique for each type of material and combination of materials. Changes of this constant result in the change of the thermocouples voltage output thus indicated temperature which can result in production quality issues, safety and health hazards. These alternations can be caused by physical and chemical changes within the thermocouple lead material and they occur in every thermocouple type. Modification of this material constant can be of temporary or permanent nature. The reversible changes in Mineral Insulated Metal Sheathed thermocouples (MIMS) happen most commonly by the annealing process which happens at temperatures above 600°C and by hysteresis effect which occurs at temperatures to 1000 °C [1, 4, 5]. Irreversible changes happen at temperatures above 600 °C for the bare wire thermocouple configuration and at 1000 °C for the MIMS configuration [1, 4, 5]. These changes happen due to chemical contamination and cause a degradation of electro motive force (EMF). In this paper we are going to concentrate on the permanent , or irreversible changes of the Seebeck coefficient that occur in commercially used MIMS Type N thermocouples. Type N thermocouples leads are made with metal alloy consisting of 14.4% chromium, 1.4% silicon, 0.1% magnesium which is used for one branch of the thermocouple and another alloy consisting of 95,6% nickel, 4.4% silicon is used for the other branch. These permanent changes can be seen as systematic change of the EMF of the thermocouple when it is exposed to a high temperature over a period of time. This change of EMF by time is commonly known as the drift of the thermocouple. The presented work deals with the drift occurring in base metal Type N thermocouples in MIMS configurations which were exposed to temperatures over 1200 °C over a period of time exceeding 210 hours. Thermocouples with different diameters were used to be able to determine the relationship between the level of drift and the thermoelement diameter.

2.1. Cause of instability of nickel based thermocouples

The stability of voltage output of thermocouples as mentioned earlier is one of the significant issues that occur in each type of thermocouples. This study is going to deal with the chemical induced changes thus the permanent change in Seebeck coefficient. These permanent changes arise in nickel based thermocouples (type N and K) at temperatures above 600°C. An increasing voltage output can be seen when thermocouples without a protective sheathing are exposed to these temperatures. This positive drift can be seen in nickel based thermocouples with metal sheathing and mineral insulation (MIMS) but only at temperatures between 600°C to 900°C. By higher temperatures a significant and constant voltage drop occurs. This behaviour was described in various publication [1, 2, 3] with the same results for type N and type K thermocouples. The mentioned publications describe the process of drift by the migration of particles between the thermocouple thermoelements and the mineral isolation material and the metal sheath. The main source of contamination according to publications [1, 2] is considered manganese (Mn) as the main element that causes drift. This elements can be found in the sheathing material of the thermocouple and at temperatures over 1100°C it contaminates the thermocouple thermoelements affecting the Seebeck coefficient. The publications [1, 2] also points out that the concentration of manganese in the sheathing material also determine the level of the drift. For instance when a Inconel 600 which has a 1% concentration of manganese was used as sheathing material the voltage output drop wasn't so high as when a AISI 310 material with 2% manganese concentration was used. Publications [4, 5] show a decrees of indicated temperature of type N and K MIMS thermocouples with a 3mm outer diameter. At temperature of 1100°C the measured temperature difference form the initial state was 10°C and at temperature of 1200°C the drop was 24°C. These results were obtained after a 1000 hour testing cycle.

Several publications deals with the thermocouple drift and time stability in which they point out that a considerable degree of long term stability of nickel based thermocouples is an issue that needs further investigation. In this presented paper we are going to deal with this drift problematic but in a relation to the diameter of the thermocouple thermoelements wires.

3. Measuring setup

The drift of the voltage output and diameter relationship was measured on type N thermocouples in MIMS configuration. The sheath material for the tested thermocouples was made of Inconel 600 and with mineral insulation inside the sheath. Eight thermocouples of the highest precision class for the mentioned type were tested. The outer diameters together with the corresponding wire diameter are presented in Table 1. One pair of the same thermocouple diameter and type from the same manufacturer was tested to avoid possible error caused by the manufacturing process. Furthermore two runs of the drift testing were planned to proof the repeatable behaviour of drift for individual diameter of sensors.

Table 1. Outer and lead diameters of tested type N thermocouples.

Outer diemeter of Type N thermocouples (mm)	Thermo element wire diameter (mm)
0.5	0.085
1.0	0.140
1.5	0.280
2.5	0.340

The temperature stability and homogeneity of the testing furnace was determined by initial furnace homogeneity scans with a calibrated noble metal type R thermocouple. After establishing the temperature profile the ideal depth for the thermocouples was determined and was set for 550mm (position of the tip of sensor from the opening of the furnace). Temperature time stability was also determined with the same type R thermocouple. The resulting stability over a 5 hour test was not more than ± 0.06 °C which was considered as sufficient for our study. Furthermore to be confident about the furnaces temperature stability a calibrated type R thermocouple was used to monitor the temperature inside the furnace. All the initial tests were done at work temperature of 1200°C which was the later used testing temperature as well. The temperature stability of the reference point is also of great importance because it determines the voltage output of the thermocouple and its temperature instability would result in the voltage output instability of the thermocouple which would make the drift detection difficult. This reference point temperature stability was ensured by a dry block cell with a high long term stability of ± 0.010 °C. The uncertainty budget of the experiment taking into account uncertainties of each element used in the experiment can be found in the Table 2

Table 2. Uncertainty budget of the experiment taking into account uncertainties of each element used in the experiment. The uncertainty values of each source are presented in °C.

Source of uncertainty	Values (°C), where k = 2
Furnace stability	0.03
Type N thermocouple	1.1
Type R thermocouple	0.3
Reference point stability	0.005
Multimeter	0.0045

To be able to determine the level of drift from the thermoelements diameter the thermocouples had to be exposed to an identical temperature conditions. This was ensured by putting all the tested sensors to a narrow ceramic tube which ensured that the temperature conditions would be sufficient for our measurement.

After setting up the measuring equipment and the initial testing phase the first batch of thermocouples was exposed to a temperature of 1200°C for a time period of more than 80 hours. The second run of tests was done on the second batch of thermocouples at temperatures of 1250 °C for a time period of more than 210 hours. The thermocouples were tested from their original state without applying any annealing process. The results of this continuous testing are presented in the following part of the paper.

4. Measuring results

The data presented was recorded using an automatic recording system which consisted of a switch system, multimeter device and PC with recording software. The recording interval of all the sensors voltage output was set for one minute. By this quick and automatic recording we ensured the comparability of the data from each sensor due to virtually identical data record time. The recorded data has a certain level of noise which was compensated using mathematical filtration methods. After this initial filtration an average value vas calculated for each five hour sections to make the interpretation of the drift level clearer. The results of the measurements that were obtained at temperature of 1200 °C are presented in Figure 1 and the results measured at 1250 °C are presented in Figure 2. The figures shows the voltage output difference of type N thermocouples of different diameters from the initial state of the voltage value.

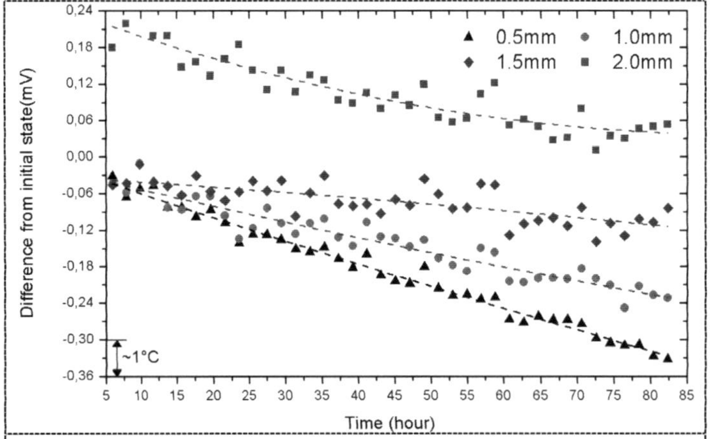

Figure 1. Voltage output difference form initial state for type N

thermocouples of different diameters (measured at 1200°C).

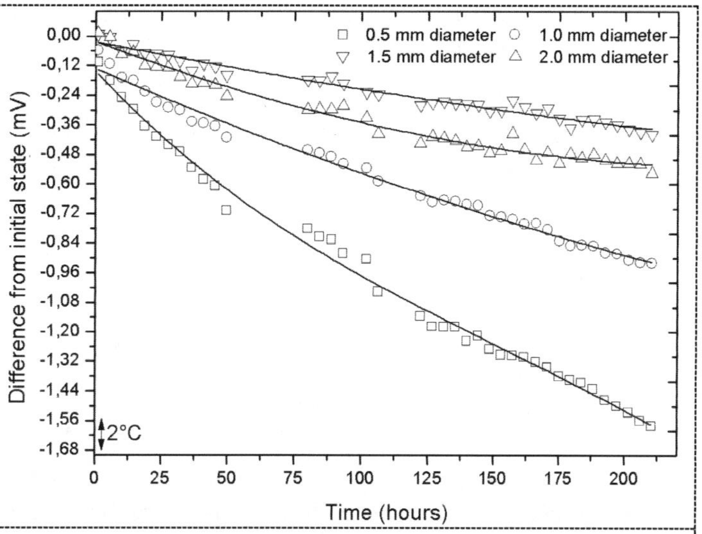

Figure 2. Voltage output difference form initial state for type N thermocouples of different diameters (measured at 1250°C).

As we can see from the Figure 1 a decrees of the voltage output is visible. The highest level of voltage decrees was noticeable on the smaller thermoelement diameters. This highest level of decrees can be seen on the thermocouple with the smallest outer diameter of 0.5mm. These result were the same for the measurements performed by the temperatures of 1200 °C and 1250 °C. The level of decrees was 5 °C from the initial state after 84 hour (at 1200 °C) and for the same diameter at 1250 °C the difference from the initial state after 214 hours was 25 °C. Other diameters of thermocouples show a different maximum level of decrees and their values are presented in Table 3. As can also be seen there is a difference between drift rates at different temperatures. The higher temperature resulted in a more rapid drift as can also be seen in Figures 1 and 2. The results have confirmed that the exposure of nickel based thermocouples to high temperatures causes a drop of voltage with time. These results have also proven that the levels of voltage output decrease and thermocouple thermoelements diameter size are related.

Table 3. Maximum temperature difference from the initial state for various thermoelement diameters at 1200°C and 1250°C.

Thermo element wire diameter (mm)	Temperature difference from the initial state after 84 hours at 1200 °C (°C)	Temperature difference from the initial state after 214 hours at 1250 °C (°C)
0.085	- 5	- 25
0.140	- 3	- 14
0.280	- 2	- 7
0.340	- 4	- 8

Values of temperature decrees and the corresponding wire diameters have been used to establish a drift function for the type N thermocouples in MIMS configuration. These functions can be seen in Figure 3 and Figure 4 and they express the dependence of average temperature decrees by one hour and the thermoelement diameter when thermocouples are exposed to a temperature of 1200 °C and 1250 °C. As can be seen from the there figures the smaller thermoelement diameters show a higher level of average °C/hour decrease than the larger diameters. This is not the case for the 0.340mm thermoelement where an anomaly occurs that need to be further examined and analysed.

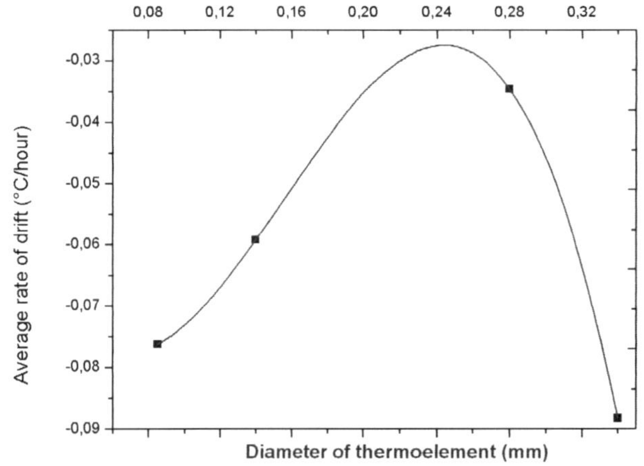

Figure 3. Function of type N thermocouple thermoelements diameter and an average temperature decrees in temperature from the initial state after one hour (measured at 1200°C).

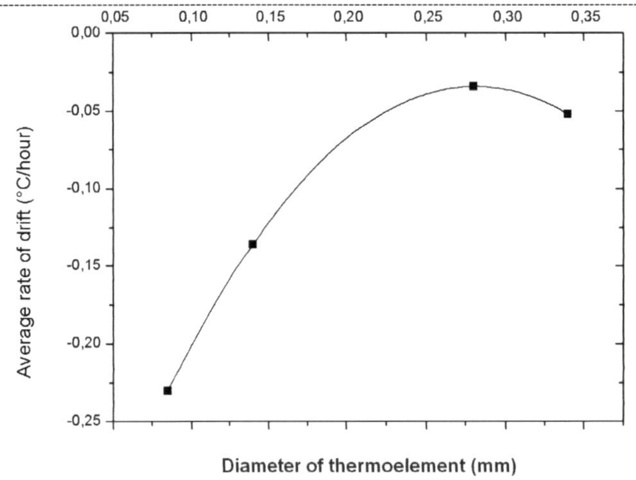

Figure 4. Function of type N thermocouple thermoelements diameter and an average temperature decrees in temperature from the initial state after one hour (measured at 1250°C).

5. Conclusion

The measured data show a clear relationship between the level of voltage output decrease and the thermocouple thermoelements diameter. This drift behaviour was observed at temperatures above 1200 °C. The sources form which this conclusion was determined can be seen in Figure 3 and 4. These figures show the different trends of voltage decrease and average °C/hour drop for different thermoelements diameters. This behaviour of nickel based thermocouples in NIMS configuration agrees with the publications [1, 2, 3] in terms of general drift behavior. The publications describe the cause of the drift as a chemical contamination of the thermoelement material by manganese (Mn) which can be found in the sheath material of the thermocouple. According to the publications the contamination process starts to occur at temperatures above 1000°C. By analysing the previously made studies in this field we have come to the conclusion that the levels of drift for different thermoelement diameters rely from the amount of material that is contaminated. Smaller diameters are therefore naturally doped faster by the Mn than the larger diameter thermoelements.

Acknowledgment

The authors would like to thank NPL (National Physical Laboratories) and the Slovak University of Technology for their support. Furthermore the authors would like to thank the grant agency VEGA – grant number 1/0120/12, APVV – grant number 0090-10 and program KEGA grant number 005STU-4/2012.

References

[1] Robin E. Bentley: Theory and practice of Thermoelectric Thermometri, Springer, Csiro, Volume 3, ISBN 981 – 4021 – 11 – 3, 1998

[2] University of Cambridge - Department of Materials Science and Metallurgy: Thermoelectric Materials for Thermocouples – Type K Thermocouples: MIMS Configuration, 2009

[3] University of Cambridge - Department of Materials Science and Metallurgy: Thermoelectric Materials for Thermocouples – Type K Thermocouples: Bare wire Configuration, 2009

[4] Robin E. Bentley and T. L. Morgan: Ni-Based thermocouples in the mineral insulated metal-sheathed format: thermoelectric instabilities to 1100°C, J. Phys. E., 19:262-68, 1986

[5] R.L. Anderson, J.D.Lyons, T.G. Kollie, W.H. Christie, R.Eby, "Decalibration of sheathed thermocouples". Temperature: its measurement and control in science and industry.,Volume5,1982

The estimation of the measurement results with using statistical methods

O Velychko[1], T Gordiyenko[2]

[1]State Enterprise "Ukrmetrteststandard", 4, Metrologichna Str., 03680, Kyiv, Ukraine
[2]State Scientific Institution "UkrNDIspirtbioprod", 3, Babushkina Lane, 03190, Kyiv, Ukraine

E-mail: velychko@ukrcsm.kiev.ua

Abstract. The row of international standards and guides describe various statistical methods that apply for a management, control and improvement of processes with the purpose of realization of analysis of the technical measurement results. The analysis of international standards and guides on statistical methods estimation of the measurement results recommendations for those applications in laboratories is described. For realization of analysis of standards and guides the cause-and-effect Ishikawa diagrams concerting to application of statistical methods for estimation of the measurement results are constructed.

Introduction

Technical measurement is the set of measurement activities used to provide the supplier and/or acquirer insight into progress in the definition and development of the technical solution. Technical measurement is planned early in the life cycle and then performed with increasing levels of fidelity as the technical solution is developed [1].

The important source of increase of production efficiency is a permanent increase of technical level and quality of the made products. The most complete and all-round quality evaluation is provided at taking into account of all properties of analyzable object that show up on all stages it life cycle: making, transporting, storage, application, repair and technical service.

Control of process envisages the exposure of causal factors that influence on the trouble-free functioning of productive process. Therefore the modern statistical methods of quality control of products and process now acquire all greater confession and distribution in industry. These methods can to use, first of all, for estimation of the technical measurement results in productive process.

For successful application of statistical methods for estimation of the technical measurement results a large role has corresponding normative documents and standards [2, 3].

Row of international standards describes various statistical methods that apply with the aim of realization of the analysis of measurement data. Having regarded to mark sufficiently actual is an analysis of those normative documents and standards for application of statistical methods for estimation of the measurement results. On the basis of the conducted analysis it is necessary to work out recommendations in concerning application of statistical methods in laboratories. Those recommendations and recommendations concerning application of general international guides and standards in the field of metrology [4] will be very profit for laboratories.

Content from this work may be used under the terms of the Creative Commons Attribution 3.0 licence. Any further distribution of this work must maintain attribution to the author(s) and the title of the work, journal citation and DOI.
Published under licence by IOP Publishing Ltd

2. Application of statistical methods for estimation of measurement results

The term "statistics" is commonly associated with an idea of lists of numbers. It is thus advisable to make clear at the outset what in fact this statistical method is that may gainfully be applied in the field of quality, standardization, and metrology.

Statistical methods have numerous practical applications in the industries, research, laboratories and other spheres. Their effectiveness depends firstly on the suitability of the chosen method for the intended purpose and secondly on the application, the way it is used. Incorrect choice or poor application can lead to improper deductions and therefore to crucial errors.

International standard ISO/TR 18532 [2] describes the row of statistical methods, that apply for a management, control and improvement of processes, and international standard ISO 13528 [3] describes the detailed description of statistical methods and recommendations for estimation of the measurement results.

Untechnical resumes, as a rule, are short; the number of technical terms in this type resume is erected to the minimum. Technical resumes more long than untechnical resumes, for its use more technical terms. For many documents two resumes complement each other.

On Figure 1 a cause and effect Ishikawa diagram for application of statistical methods for estimation of the measurement results is represented. It applies where it is required to show, pictorially, cause and effect relationships. There are several types, based on the formation of the main branches (categories), including technical consideration. The primary factors on a diagram are included: terminology; describe; compare; explain; accept; follow in time; measurement control. Secondary and tertiary factors of diagram of Figure 1 are represented in the Table 1, standards and guides – in Table 2.

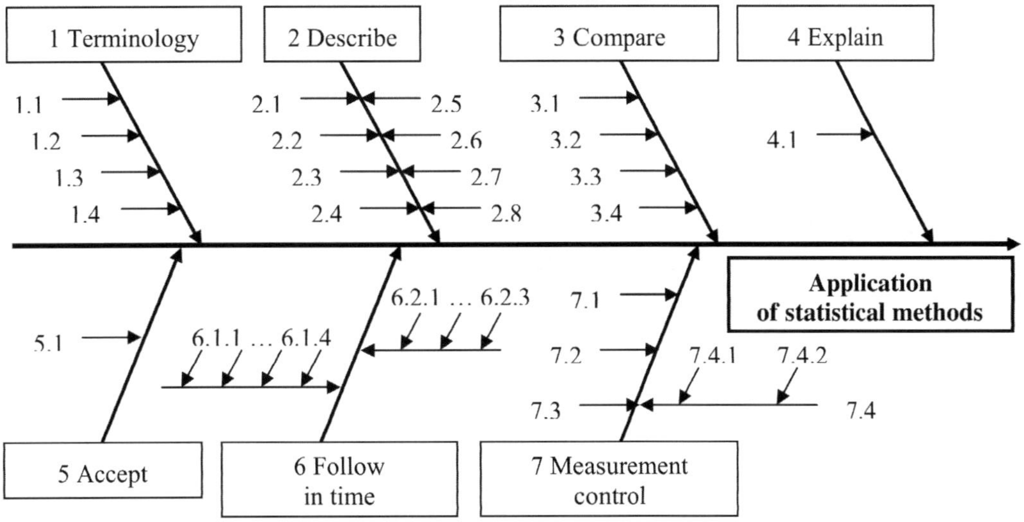

Figure 1. Cause and effect Ishikawa diagram for application of statistical methods

Practically all international standards (81) on a code 03.120.30 "Applications of statistical methods" International Classification for Standards (ICS) are developed by ISO/TC 69 "Applications of statistical methods". Standards for conformity assessment are developed by Committee ISO on conformity assessment (CASCO); guidance for reference materials – Committee ISO on reference materials (REMCO). Other international standards (10) are developed by IEC/TC 56 "Reliability" and subcommittee of IEC/SC 86B "Fibre optic interconnecting devices and passive components".

Guides ISO/IEC Guide 99 "International vocabulary of metrology" (VIM) and ISO/IEC Guide 98-3 "Guide to the expression of uncertainty in measurement (GUM)" [4] developed by eight international organizations in fields of metrology, standardization, accreditation, physics and chemistry.

Table 1. Factors of cause and effect Ishikawa diagram for application of statistical methods

Secondary factor	Tertiary and other factors
1 – terminology	1.1 – statistics probability; 1.2 – metrology; 1.3 – design of experiments; 1.4 – applied statistics
2 – describe	2.1 – prediction interval; 2.2 – median; 2.3 – difference of 2 median; 2.4 –mean; 2.5 – variance; 2.6 – proportion; 2.7 – tolerance interval; 2.8 – normal distribution
3 – compare	3.1 – variances; 3.2 – means; 3.3 – proportion; 3.4 – power of test
4 – explain	4.1 – design of experiments
5 – accept	5.1 – sampling (5.1.1 – guide); 5.2 – process performance
6 – follow in time	6.1 – reliability (6.1.1 – maintainability; 6.1.2 – adjustment; 6.1.3 – estimation; 6.1.4 – sampling plan); 6.2 – evolution (6.2.1 – statistical process control, SPC); 6.2.2 – control charts; 6.2.3 – process capability)
7 – measurement control	7.1 – accuracy; 7.2 – laboratory capability; 7.3 – measurement uncertainty; 7.4 – reference materials

Table 2. Factors for application of statistical methods and international standards or guides

Tertiary and other factors	International standards or guides
1.1 – statistics probability	ISO 3534-1, 2
1.2 – metrology	ISO 5725-1; ISO 11843-1
1.3 – design of experiments	ISO 3534-3
1.4 – applied statistics	ISO 3534-2
2.1 – prediction interval	ISO 16269-8
2.2 – median	ISO 16269-7
2.3 – difference of 2 median	ISO 2854
2.4 – mean	ISO 2854; ISO 2602
2.5 – variance	ISO 2854
2.6 – proportion	ISO 11453
2.7 – tolerance interval	ISO 16269-6, ISO 3207
2.8 – normal distribution	ISO 5479
3.1 – variances	ISO 2854
3.2 – means	ISO 2854; ISO 3301
3.3 – proportion	ISO 11453
3.4 – power of test	ISO 3301; ISO 3494
4.1 – design of experiments	ISO 3534-3
5.1 – sampling	ISO 2859-0…5, 10; ISO 3951-1…5; ISO 8422; ISO 8423; ISO/TR 8550-1…3; ISO 10725; ISO 11648-1, 2; ISO 13448-1, 2; ISO 14560; ISO 18414; ISO 21247
5.2 – process performance	ISO 21747
6.1.1 – maintainability	IEC 60706-2, 3, 5
6.1.2 – adjustment	IEC 60605-6; IEC 61649
6.1.3 – estimation	IEC 61073-1; IEC 60605-4
6.1.4 – sampling plan	IEC 60410; IEC 61123; IEC 61070
6.2.1 – statistical process control	ISO 11462-1, 2
6.2.2 – control charts	ISO 7870-1, 3, 4; ISO/FDIS 7870-2, 5; ISO 8258; ISO 7873
6.2.3 – process capability	ISO 21747; ISO 22514-1…4, 6…8
7.1 – accuracy	ISO 5725-1…6; ISO 15725-1, 2; ISO/TR 22971
7.2 – laboratory capability	ISO/IEC 17043; ISO 13528
7.3 – measurement uncertainty	ISO/IEC Guide 98-1, 3, 4; ISO/IEC Guide 99; ISO TS 17503; ISO/TR 21748; ISO/TS 21749
7.4 – reference materials	ISO 11095; ISO Guide 31, 33, 35, ISO 10576-1, ISO 11843-1…5; ISO 11843-6

3. Standards for measurement control in laboratory

Among the presented standards most interest for laboratories is contained by the standards concerning of measurement control in laboratories (primary factor "measurement control" for diagram on Figure 1).

Features of introduction in the international guides and standards for evaluation of measurement uncertainty (secondary factor for diagram on Figure 1 "measurement uncertainty") and for reference materials ("reference materials") described in [5–10]. Having regard to it is expedient to be more in detail stopped for the feature of introduction in international guides and standards for evaluation of accuracy ("accuracy") and for laboratory capability ("laboratory capability").

The detail cause and effect Ishikawa diagram concerning primary factor of diagram for "measurement control" is resulted on Figure 2. The secondary factors for "measurement control" are: accuracy; laboratory capability; measurement uncertainty; reference materials. Tertiary and other factors for diagram on Figure 2 are driven in Table 3.

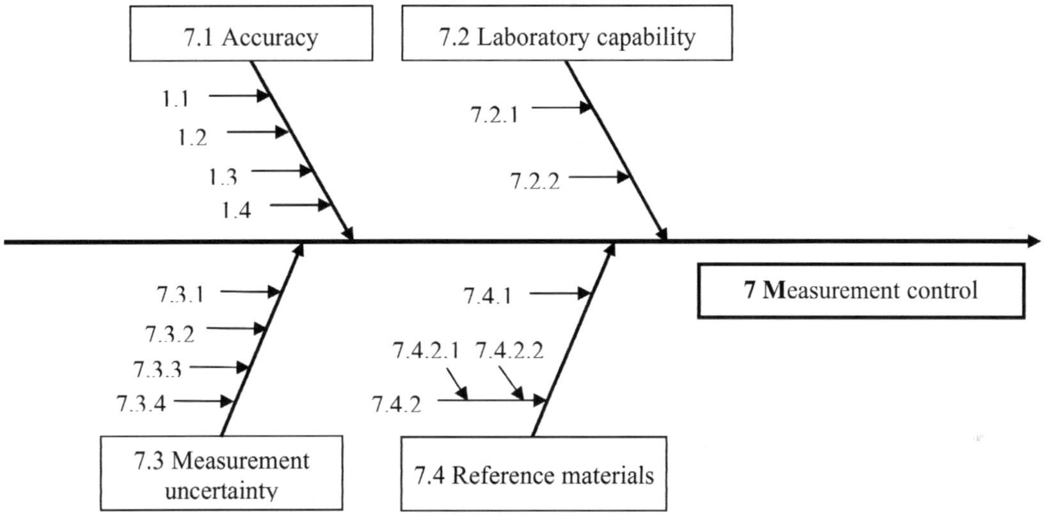

Figure 2. Cause and effect Ishikawa diagram for the secondary factors for "measurement control"

Table 3. Secondary and other factors of cause and effect Ishikawa diagram for measurement control

Secondary factor	Tertiary and other factors
7.1 – accuracy	7.1.1 – accuracy (trueness and precision) of measurement methods and results; 7.1.2 – repeatability and reproducibility standard measurement method; 7.1.3 – intermediate measures of the precision of a standard measurement method; 7.1.4 – trueness of a standard measurement method
7.2 – laboratory capability	7.2.1 – qualification of laboratory; 7.2.2 – participating in interlaboratory comparisons
7.3 – measurement uncertainty	7.3.1 – estimation method; 7.3.2 – expression; 7.3.3 – use of repeatability, reproducibility and trueness estimates; 7.3.4 – repeated measurements and nested experiments
7.4 – reference materials	7.4.1 – technical limit; 7.4.2 – capacity for exposure (7.4.2.1 – methodology of determination for the cases of linear and nonlinear calibration; 7.4.2.2 – methodology of determination in default of calibration data)

The most essential international standards for measurement control in laboratories are standards ISO 5725 with the group name "Accuracy (trueness and precision) of measurement methods and results" (6 parts) [5–10].
The international standard ISO/TR 22971 [11] provides users with practical guidance to the use of ISO 5725-2 [6] and presents simplified step-by-step procedures for the design, implementation, and statistical analysis of inter-laboratory studies for assessing the variability of a standard measurement method and on the determination of repeatability and reproducibility of data obtained in inter-laboratory testing.
The standard ISO/IEC 17043 [12] specifies general requirements for the competence of providers of proficiency testing schemes and for the development and operation of proficiency testing schemes. These requirements are intended to be general for all types of proficiency testing schemes, and they can be used as a basis for specific technical requirements for particular fields of application.
International standard ISO 13528 [3] contains the detailed descriptions of statistical methods, that use for data analysis, which got by means of charts of qualification verification, and gives to recommendation concerning using by the participants of such charts and authorized persons. It is appointed as adding to the standard ISO/IEC 17043 concerning verification of abilities by means of interlaboratory comparisons.
The International standard ISO 21748 [13] gives guidance for evaluation of measurement uncertainties using data obtained from studies conducted in accordance with standard ISO 5725-2 and comparison of collaborative study results with measurement uncertainty obtained using formal principles of uncertainty propagation. This standard is applicable in all measurement and test fields where an uncertainty associated with a result has to be determined. It does not describe the application of repeatability data in the absence of reproducibility data.
The standard ISO 21748 assumes that recognized, non-negligible systematic effects are corrected, either by applying a numerical correction as part of the method of measurement, or by investigation and removal of the cause of the effect. The recommendations in this standard are primarily for guidance. It is recognized that while the recommendations presented do form a valid approach to the evaluation of uncertainty for many purposes, it is also possible to adopt other suitable approaches. In general, references to measurement results, methods and processes in this standard are normally understood to apply also to testing results, methods and processes.
The International standard ISO 21749 [14] follows the approach taken in the GUM and establishes the basic structure for stating and combining components of uncertainty. To this basic structure, it adds a statistical framework using the analysis of variance (ANOVA) for estimating individual components, particularly those classified as Type A evaluations of uncertainty, i.e. based on the use of statistical methods. A short description of Type B evaluations of uncertainty (non-statistical) is included for completeness.
The standard ISO 21749 covers experimental situations where the components of uncertainty can be estimated from statistical analysis of repeated measurements, instruments, test items or check standards. It provides methods for obtaining uncertainties from single-, two- and three-level nested designs only. More complicated experimental situations where, for example, there is interaction between operator effects and instrument effects or a cross effect, are not covered.
When results from interlaboratory studies can be used, techniques are presented in the companion standard ISO/TS 21748. The main difference between standards ISO/TS 21748 and ISO 21749 is that the standard ISO/TS 21748 is concerned with reproducibility data (with the inevitable repeatability effects), whereas standard ISO 21749 concentrates on repeatability data and the use of the analysis of variance for its treatment.
Main approaches for uncertainty assessment in environmental and metrological guides are considered in [15–17]. The use of GUM in development of new and reconsideration of old international environmental guides is recommended.

Journal of Physics: Conference Series **588** (2015) 012017 doi:10.1088/1742-6596/588/1/012017

Summary

Plenty of international standards and guides regulate various statistical methods that apply for the analysis of technical measurements. Practically all these standards are developed by the specialized technical committee of ISO concerning application of statistical methods.

Having regard to plenty of standards that regulate application of statistical methods expedient and necessary development of the special recommendations concerning their application in laboratories. For realization of analysis of these standards the useful to use of cause and effect Ishikawa diagram.

References

[1] Roedler G. J., Martin L., Jones C. 2005. Technical Measurement. A Collaborative Project of PSM, INCOSE, and Industry. *Technical Report*. INCOSE-TP-2003-020-01.

[2] ISO/TR 18532:2009. Guidance on the application of statistical methods to quality and to industrial standardization

[3] ISO 13528:2005. Statistical methods for use in proficiency testing by interlaboratory comparisons.

[4] Velychko O., Gordiyenko T. 2010. The implementation of general guides and standards on regional level in the field of metrology. *J of Physics: Conf. Series*. **238**. Numb. 1. 012044. 6 p.

[5] ISO 5725-1:1994. Accuracy (trueness and precision) of measurement methods and results. – Part 1: General principles and definitions.

[6] ISO 5725-2:1994. Accuracy (trueness and precision) of measurement methods and results. – Part 2: Basic methods for the of repeatability and reproducibility of a standard measurement method.

[7] ISO 5725-3:1994. Accuracy (trueness and precision) of measurement methods and results. – Part 3: Intermediate measures of the precision of a standard measurement method.

[8] ISO 5725-4:1994. Accuracy (trueness and precision) of measurement methods and results. – Part 4: Basic methods for the determination of the trueness of a standard measurement method.

[9] ISO 5725-5:1994. Accuracy (trueness and precision) of measurement methods and results. – Part 5: Alternative methods for the determination of the precision of a standard measurement method.

[10] ISO 5725-6:1994. Accuracy (trueness and precision) of measurement methods and results. – Part 6: Use in practice of accuracy values.

[11] ISO/TR 22971:2005. Accuracy (trueness and precision) of measurement methods and results -- Practical guidance for the use of ISO 5725-2:1994 in designing, implementing and statistically analysing interlaboratory repeatability and reproducibility results.

[12] ISO/IEC 17043:2010. Conformity assessment. – General requirements for proficiency testing.

[13] ISO/TS 21748:2004. Guidance for the use of repeatability, reproducibility and trueness estimates in measurement uncertainty estimation.

[14] ISO/TS 21749:2005. Measurement uncertainty for metrological applications. – Repeated measurements and nested experiments.

[15] Velychko O., Gordiyenko T. 2012. The Uncertainty Estimation and Use of Measurement Units in National Inventories of Anthropogenic Emission of Greenhouse Gas. Greenhouse Gases – Emission, Measurement and Management. *InTech*: Croatia. 187–213.

[16] Velychko O., Gordiyenko T. 2009. The Use of Guide to the Expression of Uncertainty in Measurement for Uncertainty Management in National Greenhouse Gas Inventories. *International J of Greenhouse Gas Control*. **3**. Issue 4. 514–517.

[17] Velychko O., Gordiyenko T. 2007. The use of metrological terms and SI units in environmental guides and international standards. *Measurement*. **40**. Issue 2. 202–212.

Metrology in health: a pilot study

M Ferreira[1,2] and A Matos[2]
[1]Portuguese Institute for Quality, Rua António Gião, 2, 2829-513 Caparica, Portugal
[2]UNIDEMI, Faculdade de Ciencias e Tecnologia, 2829-516 Caparica, Portugal

E-mail: mcferreira@ipq.pt

Abstract. The purpose of this paper is to identify and analyze some relevant issues which arise when the concept of metrological traceability is applied to health care facilities. Discussion is structured around the results that were obtained through a characterization and comparative description of the practices applied in 45 different Portuguese health entities. Following a qualitative exploratory approach, the information collected was the support for the initial research hypotheses and the development of the questionnaire survey. It was also applied a quantitative methodology that included a descriptive and inferential statistical analysis of the experimental data set.

1. Introdution

Measurement, as a field of scientific and technical activity, encompass a set of activities aimed at determining the quantities and registers a remarkable evolution throughout the centuries The scientific context that serves these activities is called Metrology which includes a wide-range of activities with a key role in all sectors of society, being a fundamental basis for sustainable development [1], [2], [3] [4].

According to international recommendations, every system of measurement must use accurate and reliable results [5]. Metrological traceability [3] and uncertainty measurement [3] are the main elements that support this assurance. This means improving a chain of measurements from a single measurable result in a commonly accepted reference, as well as the uncertainty value associated to this measuring result [6]. Surely due to the inherent risk of life, these principles are strongly applied in health systems. Measurements and measuring instruments are present in day-to-day life and are fundamental processes in prevention, diagnosis and treatment of diseases. Measurement results are vital in the medical field, which requires they are accurate, real and comparable over time, i.e., they must exhibit repeatability and reproducibility [7]. Nevertheless, one must be aware that errors and uncertainties of measuring instruments could be the origin of unsuitable clinical procedures, such as a false diagnosis or inadequate treatments.

Clinical laboratories constitute a good example for the improvements achieved by metrology as regards traceability in medicine. The Joint Committee for Traceability in Laboratory Medicine (JCTLM) [8] [9], hosted by the International Bureau of Weights and Measures (BIPM), is the first international committee focused on clinical metrology and constitutes a platform envisaging the harmonization and dissemination of good practices within the framework of analytical chemistry [10].

Therefore, under the scope of the European Medical Devices Directives [11], new responsibilities and technical requirements for the measuring instruments were put in place, for preventing the existence of barriers as regards the free trade. trade. For one hand, although a wide range of instruments is covered by the medical devices directive, this sphere of regulation allows each Member State considering additional measures to protect public health and citizens. On the other hand, after

Content from this work may be used under the terms of the Creative Commons Attribution 3.0 licence. Any further distribution of this work must maintain attribution to the author(s) and the title of the work, journal citation and DOI.
Published under licence by IOP Publishing Ltd

placing them in the market and putting them into service, no further regulated control (according to EU policy) exists for those medical devices [12].

As the awareness of the benefits for having traceable measurement results increases, some researchers and technical experts (in metrology and medicine), have been increasing their efforts and interest for developing research within this scope. In recent years several pieces of research were published focusing some relevant measuring instruments used by health professionals. As regards specific diseases, some authors highlighted the key role of metrological traceability of sphygmomanometers as a tool for reducing health care costs and improving patient's care [13], [14], [15]. Likewise, other researchers pointed out the traceability of clinical temperature's measurement devices, such as thermometers [16], [17], [18]. In what concerns the evaluation of some ophthalmic diseases, the clinical metrology of tonometers' utilization has also been under metrological research [19], [20], [21]. Work has been developed regarding pulmonary and respiratory diseases, proving how important reproducibility is and its impact on diagnosis [22], [23].

Regardless the aforementioned work, much remains to be done. In fact, the various actors with jurisdiction in these subjects, and in particular in some areas of activity, didn't put in place enough actions -packed for assuring metrological traceability in health measurements. Furthermore, the growing concern with issues related to quality in health care made emerge new areas of activities in this important field of society. Moreover, it is also considered that the interoperability between these three disciplines is not contextualized within the current European regulatory framework. The lack of information regarding the perception of metrology by health professionals is a topic requiring evaluation.

2. Methodology

Taking into account the literature review, as well as the current framework of national health care facilities, the applied methodology aimed to identify the main indicators of health professionals' practices with their working instruments for measurement.

The Portuguese hospitals were the target of this work research, which was developed in 2011 and included private and public hospitals (national health system). A survey was applied to the hospitals included in the last database from Portuguese Directorate-General of Health (DGS) that was available at the time [24]. However, at the time, private hospitals were not available in the aforementioned national database. Hence, the universe of private hospitals was identified by regional random search.

According to that information and based on the real context of observational analysis, a multimethod design [25] approach was applied: i) inductive and deductive reasoning methods [26], ii) qualitative and quantitative research methodology [27].

Survey characterization

The issues in survey formulation were designed and developed based on the analysis of the National and European metrological requirements in health care facilities. Therefore, the research hypotheses and the development of the measurement instrument (survey) were performed through an exploratory qualitative approach without a previous similar validated survey.

From May to September 2011, 122 questionnaires were sent by e-mail to Portuguese hospitals. From those, 99 were addressed to public hospitals, including islands, and 35 responses were received. The remaining 23 were sent to private hospitals, and 9 responses were received. All surveys were sent by e-mail with a follow-up procedure and received through the same way. Thus, considering a stratified sample, a response rate of 35,4 % was obtained for public hospitals and 39,1 % for private hospitals, leading to an overall response rate of 36%, that were collected and analyzed.

A 'tree' questionnaire was developed, in which the choice of certain issues is determined by the answers in previous questions. The ten questions included in the developed survey were distributed between open and closed questions, including dichotomous (yes/no) and ordinal scales (Likert type). Envisaging the assurance of survey's validation, a pretesting was conducted [28]. Apart from slight changes, the questions were maintained after the pretesting.

Journal of Physics: Conference Series **588** (2015) 012018 doi:10.1088/1742-6596/588/1/012018

The questionnaire was designed to characterize information regarding the following issues:
a) Relationship between qualification systems implemented by health care facilities and associated metrological practices;
b) Metrological requirements that could influence the acquisition of medical instruments;
c) Budget for medical equipment;
d) Levels of literacy in health's metrology.

In a first stage a univariate analysis was adopted for characterizing the trend of responses in each scale item. Descriptive statistics were utilized for computing the mean, standard deviation, minimum, maximum and frequency response. The second stage focused the correlation between variables, including a bivariate statistical analysis from the dataset. Finally, in the third stage, it were utilized statistical inference, nonparametric tests, correspondence analysis and factor analysis.

3. Results

Data analysis was performed using the Statistical Package for Social Sciences (SPSS) version 19. A value for Cronbach alpha of 0,74 was achieved, thus making acceptable the internal consistency [29]. Once more, it is important to stress that the questionnaire constitutes a new survey model, entirely developed for this piece of research.

Quality systems of survey respondents

According to data analysis, 53,3 % of hospitals had a partial qualification, 25,0 % were accredited and 22,7 % of respondents referred that the hospitals were certified as a whole. In addition, the results' overview regarding the quality management system of hospitals shows that more than 50,0 % of public hospitals are qualified according to international standards. All responses from private hospitals revealed a certification by ISO 9001[30]. Both typologies of hospitals prefer national certifiers entities.

Procedure for acquiring equipment's vs metrology

The importance given to metrological requirements within the procedures for equipment's acquisition was assessed. Exploratory Factorial Analysis (EFA) was utilized with Varimax rotation. Table 1 shows the grouped variables by factor: i) the maintenance conditions and the historical of the equipment were named as *post-sales factor* (1st EFA factor); ii) the 2nd EFA factor was named as *legal requirements* and it includes the compliance of requirements by producers requisites as well as the hospitals' requirements for the notified bodies; iii) the price and supplier trust were grouped in the 3rd factor (*pre-sales factor*). The high value obtained for communalities shows that retained factors are appropriate for describing the latent correlation structure. The measure of sampling adequacy was validated by the result from KMO (Kaiser-Meyer-Olkin) and the Bartlett test [29].

Table nº 1: Factor Structure of Exploratory Factorial Analyses

	Factor			
Variable	1	2	3	**Communality**
Price	-0,211	0,004	**0,828**	0,730
Supplier trust	0,286	0,054	**0,752**	0,650
Requirements of ISO 13485	-0,074	**0,892**	0,136	0,819
Maintenance conditions	**0,801**	0,265	0,131	0,729
Historical of the equipment	**0,770**	-0,050	-0,048	0,597
Requirements of the Notified Body	0,440	**0,708**	-0,129	0,711
Eigenvalues	1,926	1,302	1,009	
Variance (%)	32,1	21,7	16,8	

Through the utilization of cross analyses, the maintenance requirement was considered the most relevant for private hospitals (Figure 1), whereas for the public hospitals (Figure 2), this issue is less important than the supplier's confidence (59,0 %). For these hospitals, price was the most important requirement (51,0 %) which constitutes a strong restriction for the enforcement of metrological traceability.

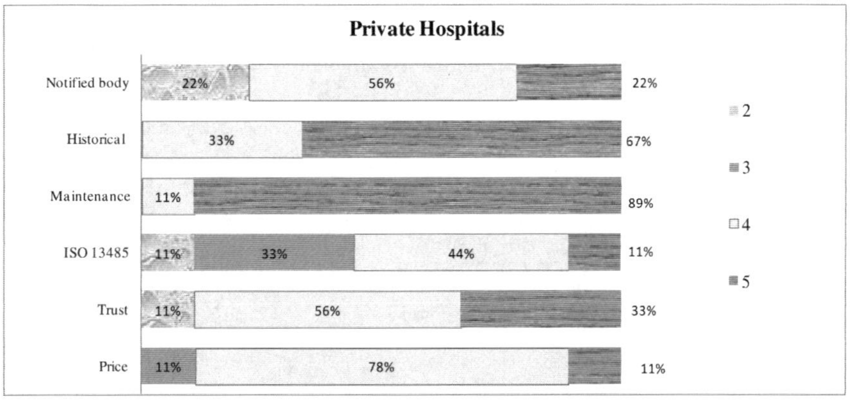

Figure nº 1: Indicators from private hospitals

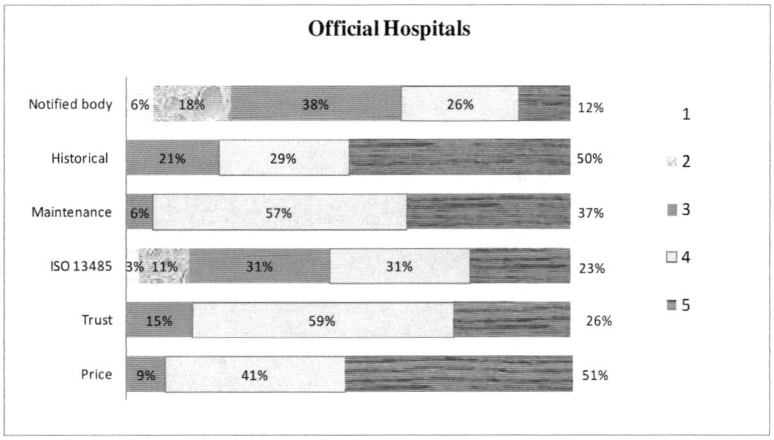

Figure nº 2: Indicators from official hospitals

Metrological documentation that is included in the terms of reference (contract document) was other aspect that was evaluated on the acquisition phase. Multiple Correspondence Analysis (MCA) was utilized [29], with engeinvalue □ 1 as a retention criteria for components. According to values shown in Table 2, orthogonal components explain 74,0 % of data variability. One component (first dimension) was represented by the verification certificate and the measures report. The calibration certificate and the conformity report have different scores, representing the 2nd and 3rd dimension.

The 1st dimension is influenced by respondents who have no preference for calibration certificates, which are replaced by verification certificates and test reports. Opposite reasoning is applied to the variables in the 2nd dimension. The analysis for the 3rd dimension indicates that respondents who receive evaluation reports are influenced by receiving test reports. The chi-square χ^2 ($\alpha = 0{,}05$) was utilized for analyzing the relation between those variables and the hospital's typology (public *vs* private). Since the null hypothesis H_0 is rejected, one concludes that the acquisition of medical measuring instruments with conformity report is dependent of hospital typology (Table nº 2). This situation was also assessed using the *odds ratio* θ evaluation, whose values show that the choice of the conformity report depends on the type of hospital, and this probability is more pronounced in private hospitals (θ_1 =0,650; θ_2 =8,000).

Table 2: Survey MCA

Variables	**Principal Components (dimensions)**				
	1	2	3	χ^2	**p-value**
calibration certificate	0,006	**0,706**	0,088	0,754	0,385
verification certificate	**0,523**	0,023	0,075	0,937	0,333
test report	**0,582**	0,000	0,090	0,075	0,784
conformity report (product)	0,287	0,026	0,**498**	**6,929**	**0,008**
no certificate	0,003	0,461	0,362	0,024	0,877
Eigenvalue	1,402	1,216	1,113		
Variance (%)	28,0	24,3	22,3		
Relevance of the traceability of medical instruments	0,315	0,339			0,000
Metrological Legislation	**0,644**	0,007			0,542
External calibrations	0,237	0,306			0,000
Internal calibrations	0,032	**0,497**			0,049
Metrological knowledge for measuring instruments	0,477	**0,642**			
Interpretation of certificates *vs* tests reports	**0,923**	-0,057			
Relation adverse event *vs* error of instrument	**0,924**	-0,068			
Relation error of instrument *vs* indication results	0,226	**-0,842**			
Eigenvalue	1,198	1,129			
Variânce (%)	49,6	28,3			
Preventive maintenance	0,544	**0,973**			
Curative maintenance	**0,847**	0,022			
Calibrations	**0,928**	-0,306			
Verifications	**0,938**	-,0282			
Eigenvalue	2,755	1,121			
α de Cronbach	0,849	0,144			

It was also analyzed the relation between hospitals that acquire equipments with calibration certificates and the related relevance as regards *i)* metrological knowledge and *ii)* technical competence for certificates' analyses. The opinion of relevant respondents pointed out that hospitals that purchase certified equipment (calibrated measuring instruments) consider relevant their certified interpretation (50,0 %). Furthermore, 46,0 % expressed their interest in metrological's knowledge applied on clinical practice.

Metrology vs risk management vs patient safety
Another issue that was evaluated focuses the key role of metrology in the risk management and patient safety. The MCA methodology (Table 2) was adopted as statistical approach . According to those values, it was possible to characterize the 1st dimension through the association between the needs to interpret the metrological certificates of equipment and the reflection of this interpretation on the errors provided by the measuring instruments. These errors can be widespread by the diagnostics and therapeutic procedures. Extrapolation of this interpretation for clinical practice, whose influence on adverse effects can be significant, is also present in this dimension.

Within this framework, the metrological knowledge that health professionals may acquire regarding their measuring instruments is very important and can influence the evaluation assigned to measurement's error as well as the outcomes. The high score value of 1st dimension for the interpretation of certificates *vs* tests reports and relation adverse event *vs* error of instrument show a significant concerns revolve around patient safety. Based upon these results, we sustain that the assessment of the risk associated to the utilization of medical equipment and patient safety might be considered exogenous latent variables, with special relevance in health metrology.

Budget for medical equipment's with measuring functions
It was also assessed the budget outcomes from hospitals that are applied to metrological operations and maintenance, notably preventive and curative maintenance, calibrations and verifications of measuring instruments. Having this in mind, the categorical variables were distributed by an interval range, from less than 10 % to more than 30 % of hospital's overall budget allocation. According with these results, curative maintenance is the operation selected by 13,6 % of hospitals with a budget more than 30,0 % whereas 75 % of hospitals allocated less than 10% of such overall budget.

It is worth mentioning that the number of medical equipments with measuring functions in the respondents' hospitals was above 1100 units. On the other hand, only 2,3 % of respondents reported more than fourty calibrated instruments by calibration period. Therefore, one can assume that the results obtained for the budget that is applied by hospitals represents a worrying indicator and a poor predictor for the metrology in the health services.

According the MCA analyses for such dataset (Table 2), the curative maintenance, the calibrations and verifications are included in the same principal component, being all of them close to saturation value. The preventive maintenance has the highest score for the second component. The value of Cronbach α shows a good internal consistence ($\alpha = 0{,}849$).

Metrological factors as a transversal tool in health care facilities
This issue is clearly cross-sectional, which led to its inclusion in several questions. For these purposes, the following variables were evaluated: *i)* Relevance of traceability for medical instruments, *ii)* Metrological Legislation, *iii)* External calibrations, *iiii)* Internal calibrations.

As the responses presented lack of harmonization, a binomial test was applied to the aforementioned variables. The null hypothesis was rejected as regards metrological legislation, which reveals that the proportion of hospitals that considers sufficient the present legislation in the field of traceability of medical devices, is statistically similar to the proportion of those who do not consider this factor as relevant. A MCA was applied for validating this conclusion (Table 2). Two dimensions were retained, whose values lead to designating dimension 1 as the component focused on the law (metrological traceability) and the second dimension as the traceability component. Among the respondents that classify the traceability as relevant, 93.2% make use of external calibration. Among the hospitals that perform internal calibrations, 63,6 % apply this procedure to evaluate the conditions of metrological instruments, without support of external calibrations (no evidence of a traceability reliably). The two approaches (external and internal calibrations) are practiced by only 34,1 % of respondents. The Odds Ratio (θ) was assessed for studying the relationship between internal calibration and hospital's typology. It was concluded that the probability of private hospitals to practice internal calibrations is 1,5 higher than the public hospitals and that there is no dependency relationship between both variables.

4. Discussion

This study showed that a large majority of respondents were representing organizations with recognized qualifications, which constitutes a valuable piece of information for establishing their profile.

Regarding the metrological issues, notably for public hospitals, the price of the product is the prevailing requirement. However, it was also verified that there is a significant correlation between trust in the supplier and the price of product to be delivered. This suggests that the score of the variable price may be undervalued when remaining confidence in the entity provider.. Nevertheless, as the economic downturn has raised major concerns for all sectors of activity, this component can represent an even greater concern in acquisition procedures. Thereby, the criterion of price is in fact the most important requirement to acquire equipment, which deserves a proper reflection.

Usually, the metrological documentation accompanying the equipment during its acquisition phase does is not dependent neither of the type of hospitals nor of systems of quality management, which reflects the reduced concern regarding the compliance to requirements of metrological equipment. Hence, despite the statistical significance for matters regarding the maintenance and taking into account the analysis performed on the metrological documentation mentioned on the previous items, it is suggested that the protocol established by hospitals to acquire instruments measurement does not reflect metrological concerns.

For both types of hospitals, the relationship between the adverse event and the instrument was fairly valued, being an indicator of the perception of health professionals regarding the dimension of risk management *vs* metrology.

In addition, entities that acquire equipment with calibration certificates are those who consider that its interpretation is important, which requires an improved metrological knowledge. Hence, the obtained results suggest that healthcare professionals have the perception that the metrological performance of equipment is important and that can influence the outcome of the measurements in their clinical practice. However, in most institutions, the metrological traceability is still concealed in maintenance operations, thus contributing for the low level of best practices of metrology in medicine.

The research results also stressed that a greater confidence in supplier is proportional to a lower overall budget for maintenance operations. The interpretation could be made under the framework of health and metrological literacy, encompassing a weak frontier between maintenance and metrology operations.

Therefore, considering the ratio of total number of equipments to the number of equipments subjected to calibration's operations (previous issue), one can assume that Portuguese hospitals has a reduced number of instruments that are submited to procedures able to assure the quality control of obtained results.

The hospitals tend to realize external calibrations, showing no tendency to perform internal calibrations. Those entities that perform it internally are predominantly private, and perform those operation for assuring the conditions of their metrological instruments. These entities have technicians that received specific training for this field of work.

The knowledge and opinions of respondents as regards metrological legislation proved to be incipient. It was also concluded that a trend for the application of metrological practices prevails in those entities that revealed more knowledge and awareness for metrology issues, emphasizing the implementation of procedures that promote the metrological traceability of measuring instruments.

5. Conclusion

Taking into account the obtained results, one concludes that metrological traceability is a subject that is absent from day-to-day life of healthcare professionals, although it is known by them.
Some factors contributing for this perception were identified, which might enrich the scientific knowledge regarding the fields of health and metrology.

Considering the quality in health care facilities, this piece of research aims to contribute for introducing the metrology subject in hospitals' framework.

Further work, focusing on the impact of measurements on Gross National Product (GDP) of globalized economies, particularly regarding health systems, is clearly necessary. Hence, further informations allowing updates and improvement developments on this important subject are obviously wellcome.

At the European level, a continuous international collaboration of experts is needed for promoting comprehensive Metrology within the framework of quality in health care. This is a major challenge.

References

[1] R White 2011 The meaning of measurement in metrology *Accreditation and Quality Assurance* **16**(1):31-41
[2] D Kind and H Lubbig 2003 Metrology–the present meaning of a historical term *Metrology* **40:**255-7
[3] JCGM 200. 2012 International Vocabulary of Metrology Basic and General Concepts and Associated Terms
[4] G Hratch, R Semerjian and J Watters 2000 Impact of measurement and standards infrastructure on the national economy and international trade *Measurement* **27**:179-96
[5] M Golze 2003 Why do we need traceability and uncertainty evaluation of measurement and test results? *Accreditation Quality Assurance* **8**:539-540
[6] E Desimoni and B Brunetti 2006 Considering uncertainty of measurement when assessing compliance or non-compliance with reference values given in compositional specifications and statutory limits: a proposal *Accreditation and Quality Assurance* **11**(7):363-6
[7] P Ashford 2010 Traceability. *Cell Tissue Bank* **11**:329–33
[8] W Hubert and L Vesper 2009 Traceability in laboratory medicine. *Clinical Chemistry* **55**(6): 1067-75
[9] D Armbruster and R Richard 2007 The joint committee for traceability in laboratory medicine (jctlm): a global approach to promote the standardisation of clinical laboratory test Results *Clinical Biochemestry Review* **28**(3): 105–14
[10] http://www.bipm.org/en/committees/jc/jctlm/
[11] http://ec.europa.eu/health/medical-devices/regulatory-framework/index_en.htm
[12] M Ferreira 2011The role of metrology in the field of medical devices *International Journal of Metrolology and Quality Engineering*. **2**:135–40
[13] C A'Court, R Stevens and R Sanders 2011 Type and accuracy of sphygmomanometers in primary care: a cross-sectional observational study *British Journal of General Practice* **61**(590):598-603
[14] E O'Brien, and A Neil 2007 Validation and reliability of blood pressure monitors *Clínical Hypertension and Vascular Diseases:Blood Pressure Monitoring in Cardiovascular Medicine and Therapeutics* pp 97-132. ISBN: 978-1-59259-978-3.
[15] M Turner, C Speechly and N Bignell 2007 Sphygmomanometer calibration why, how and how often? *Australian Family Physician* **36**(10):834-8
[16] I Juntaro 2008 Improving the realibility of temperature measurements taken with clínical infrared ear thermometers *Synthesiology-English Edition.* **I**: 47-58

[17] R Simpson, G Machin, H McEvoy and H., Rusby, H. (2006) Traceability and calibration in temperature measurement: a clinical necessity. *Journal of Medical Engineering & Technology* **30**(4):212-217
[18] W Chung and C Chen (2010 Evaluation of performance and uncertainty of infrared tympanic thermometers *Sensors (Basel)*. **10(4)**:3073-89
[19] S Sandhu, S Chattopadhyay, and G. Amariotakis 2009 The accuracy of continued clínical use of goldmann applanation tonometers with known calibration errors *Ophthalmology* **116**(1):9-13
[20] K Ogbuehi and T Almubrad 2008 Accuracy and reliability of the Keeler Pulsair EasyEye non-contact tonometer *Optometry and vision science* **85**(1):61-6
[21] S Choudhari, R George and M Baskaran 2009 Measurement of Goldmann applanation tonometer calibration error *Ophthalmology* **116**(1):3-8
[22] R Jensen, J Teeter, R England, H White and E Pickering 2007 Instrument accuracy and reprodutibility in measurements of pulmonary function *CHEST* **132**:388-95
[23] R Miller, J Hankinson and J Brusasco 2005 Standardisation of spirometry *European Respiratory Journal* **26**: 319–23
[24] DGS (2008 Elementos estatísticos: informação geral Direcção de Serviços de Epidemiologia e Estatísticas de Saúde. Divisão de Estatísticas de Saúde. Lisboa. ISSN 0872-1114.
[25] A Tashakkori and C Teddlie Handbook of Mixed Methods in Social & Behavioral Research Authors editors 2002 ISBN-13: 978-0761920731
[26] W Trochim and J Donnelly The Research Methods Knowledge Base 2006 Edition: 3ISBN-13: 978-1592602919
[27] J Brannen 2005 Mixing methods: the entry of qualitative and quantitative approaches into the research process *International Journal of Social Research Methodology* **8**(3):173-184
[28] R Ghiglione and B Matalon 1997 O Inquérito: Teoria e Prática (3ª ed.). Oeiras: Celta Editora
[29] J Maroco 2007 Statistical analysis using SPSS (3rd ed.). Walker Editions. ISBN: 978-972-618-452-2.
[30] EN ISO 9001:2000 Quality management systems-requirements

Innovative Mobile Smart Photonic Dimensional, Color and Spectral Measurement Engineering

Prof. Dr. Dietrich Hofmann, B.Eng. Paul-Gerald Dittrich, B.Eng. Dieter Höfner, Daniel Kraus

SpectroNet c/o Technologie- und Innovationspark Jena GmbH, Wildenbruchstraße 15, D-07745 Jena

E-mail: d.hofmann@spectronet.de

Abstract. Aim of the paper is the demonstration of a paradigm shift in dimensional, color and spectral measurements in industry, biology/medicine, farming/environmental protection and security, as well as in education and training: Measurement engineering and quality assurance become mobile, modular and smart. Smartpads, smartphones and smartwatches (smartcomps) in combination with innovative hardware apps (hwapps) and conventional software apps (swapps) are fundamental enablers for the transformation from conventional stationary working places towards innovative mobile working places with in-field measurements and point-of-care (POC) diagnostics. Furthermore mobile open online courses (MOOCs) are transforming the study habits. Practical examples for the application of innovative photonic micro dimensiometers, colorimeters and spectrometers will be given. The innovative approach opens so far untapped enormous markets for measurement science, engineering, applications, education and training. These innovative working conditions will be fast accepted due to their convenience, reliability and affordability. A highly visible advantage of smartcomps is the huge number of their real distribution, their worldwide connectivity via Internet and cloud services, the standardized interfaces like USB and HDMI and the experienced capabilities of their users for practical operations, obtained with their private smartcomps.

1. Mobilization of Computers

The computer market is changing. Miniaturization and mass production of smartcomps are in progress. The special advantages of smartcomps are their mobility and their multitouch user interfaces. Market leaders are Apple with iOS operation system and Google with Android operation system. Microsoft with Windows 8 operation system is a late but powerful and experienced follower (Figure 1a) [1]. Windows 8 smartcomps combine the typical Windows working style with the leisure look and feel of iOS and Android life style. The innovative computers are equipped with consumerized interfaces for wired and/or wireless communications (Figure 1b) [1]. Smartcomps are additional equipped with powerful hardware apps for image acquisition as for example light sources, cameras, camcorders and scanners and with software apps for image processings. Smartcomps are mobile because they are light weight and battery powered. In addition, the number, quality and diversity of external hardware apps for smartcomps are increasing (Figure 1c) [1]. They are applicable for industrial measurements, biological/medical diagnostics, farming/environmental protection and security as well as for scientific investigations or mobile education and training purposes.

Content from this work may be used under the terms of the Creative Commons Attribution 3.0 licence. Any further distribution of this work must maintain attribution to the author(s) and the title of the work, journal citation and DOI.
Published under licence by IOP Publishing Ltd

Figure 1a. Innovative Windows mobile computers

Figure 1b. Innovative mobile interfaces

Figure 1c. Consumerized measurement systems

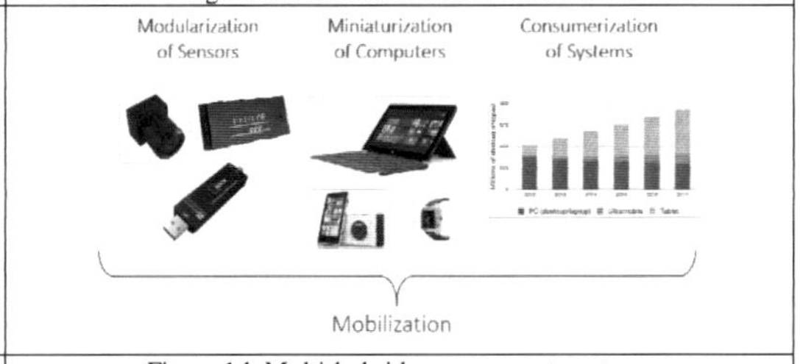

Figure 1d. Multi-hybrid measurement systems

Figure 1. Mobilization and standardization of computers and measurement systems

The Ozcan Group at the University of California, Los Angeles (UCLA) is working since years successfully on this topic. Their research results recently has been summarized in a very informative overview [2]. A fundamental purpose of the present paper is to show that many innovative mobile smart solutions can be designed with already existing functional modules according to the LEGO principle. Due to smartcomps with the quasi standardized industrial operation system Microsoft Windows 8 (see Figure 1a) and standardized interfaces (see Figure 1b) the existing consumerized hardware modules and software modules (see Figure 1c) must not be modified in their original structures and functionalities. A disadvantage of the shown multi-hybrid solutions might sometimes be that they are more complex in its structures and in its functionalities than specialized solutions. But the advantages of these multi-hybrid solutions are their fast availability because they are composed of commercially available components supplemented by mass-produced consumer goods (see Figure 1d) [1].

2. Mobile Smart Photonic Measurement Systems

Mobile smart photonic measurement systems are a combination of smartphones, smartpads and smart wearables (smartcomps) with additional task specific hardware apps (hwapps) and software apps (swapps). Dimensional measurements are the analysis of geometric shapes in 1, 2, 2.5, 3 and n dimensions (Figure 2a). Color measurements are the analysis of light and body colors in accordance with the human color perception (Figure 2b). Spectral measurements are the analysis of unknown spectra in comparison with evaluated terms of known spectra (Figure 2c).

Figure 2a. Dimensional measurement systems

Figure 2b. Color measurement systems

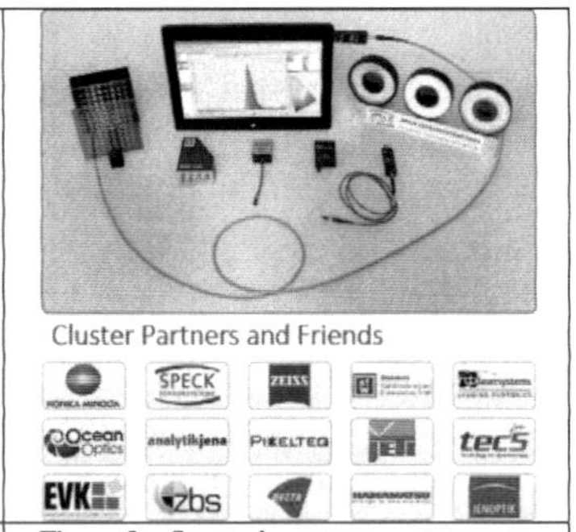

Figure 2c. Spectral measurement systems

Figure 2. Mobile smart photonic measurement systems

3. Mobile Smart Photonic Measurement Hardware Apps

Typical photonic dimensional measurement hardware apps are digital miniature cameras and microscopes with standardized CMOS sensors, USB interfaces and flexible S-Mount lenses (Figure 3a) [3] [4]. Typical photonic color measurement hardware apps are digital miniature colorimeters or photometers with standardized True Color sensors, USB interfaces and SMA-Connectors (Figure 3b) [5]. Typical photonic spectral measurement hardware apps are digital miniature spectrometers preferably with standardized linear CCD sensors, USB interfaces and SMA-Connectors (Figure 3c) [6].

CMOS sensors	Hardware modules	USB interfaces	S-Mount lenses
	Sensor: CMOS RGB Bayer Matrix, 5 MP 2592x1944 pixels, 1/2.5 Digital interface and power requirements: 5V USB 2.0, typ. 0.6W Dimensions and Weight WxHxD: 15 x 15 x 8 mm, 5 g		
Figure 3a. Dimensional hardware apps			

True Color sensors	Hardware modules	USB interfaces	SMA-Connectors
	Sensor: Tri-stimulus value function, DIN 5033, digital resolution (ADC) 16 bit Digital interface and power requirements: 5V USB 2.0 Dimensions and Weight WxHxD: 89 x 32 x 27 mm, 20 g		
Figure 3b. Color hardware apps			

Linear CCD sensors	Hardware modules	USB interfaces	SMA-Connectors
	Sensor: 2500 pixel linear CCD detector, 360 - 740 nm, 1.2 nm FWHM Digital interface and power requirements: 5V USB 2.0, 150 mA Dimensions and Weight WxHxD: 85.7 × 22.0 × 10.0 mm, 10 g		
Figure 3c. Spectral hardware apps			

Figure 3. Mobile smart photonic measurement hardware apps

4. Mobile Smart Photonic Measurement Software Apps

To design consistent measurement systems with smartcomp compatible specialized software apps must be available. Due to the longstanding successful implementation of specialized software tools for photonic measurements on computers with Microsoft Windows operation systems XP, Vista, 7 and 8 a big number of excellent software tools commercially are applied. Under the influence of modern speaking habits these software tools may be understood as software apps, if they are compatible with the Microsoft Windows 8 operation system. Normally Windows 8 is downwards compatible to Windows XP, Vista and 7. Therefore a remarkable variety of photonic measurement software apps for photonic dimensional (Figure 4a), color (Figure 4b) and spectral (Figure 4c) measurements are already traded.

Figure 4a. Dimensional software apps	Figure 4b. Color software apps	Figure 4c. Spectral software apps

Figure 4. Mobile smart photonic measurement software apps

5. Calibration tools for mobile smart photonic measurement systems

Calibration is an operation that, under specified conditions, in a first step, establishes a relation between the quantity values with measurement uncertainties provided by measurement standards and corresponding indications with associated measurement uncertainties and, in a second step, uses this information to establish a relation for obtaining a measurement result from an indication (VIM 2.39) (6.11) [7]. **Mobile** calibration tools are designed to enhance capabilities, performance, and ease-of-use of the mobile smart photonic measurement systems. There are various kinds of mobile calibration tools for photonic dimensional (Figure 5a) [8], color (Figure 5b) [9] [10] [11] [12] [13] and spectral (Figure 5c) [14] [15] measurements.

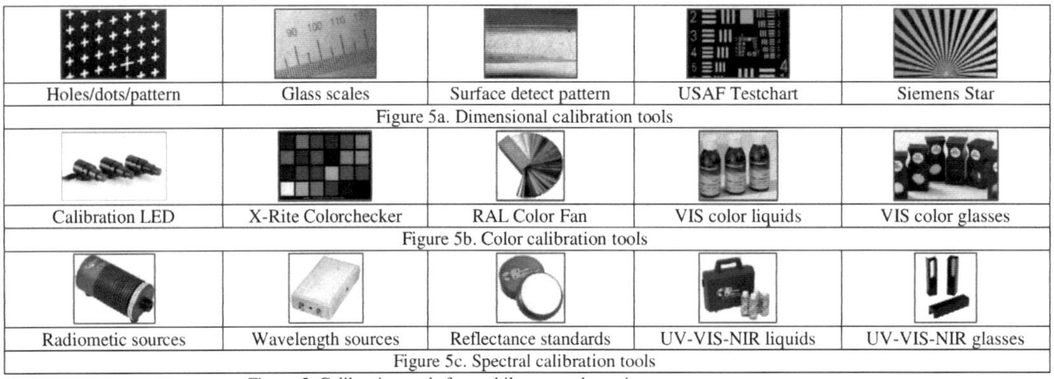

Figure 5. Calibration tools for mobile smart photonic measurement systems

6. Examples for Mobile Smart Further Educations and Shared Hands-on Trainings

Manufacturers and system integrators of hardware and software for photonic image acquisition and digital image processing are on the way to reduce the significant qualification deficits of potential users. Two efficient methods are promising:

1. Imaging specific multilingual digital text books (ebooks) and videos for end users on smartcomps,
2. Imaging specific hands-on trainings in industry and/or shared between universities and industry.

6.1. Mobile imaging specific ebooks and videos on smartcomps for end users

To increase the efficiency of individual education and training processes, about 2000 digital papers and 1000 videos, dealing with recent developments in mobile smart photonic dimensional, color and spectral measurements are open accessible on the cluster-platform spectronet.de (Figure 6.1). The search box in spectronet.de enables the users to identify experts, institutions and enterprises as well as professional contents, selectable by professional keywords.

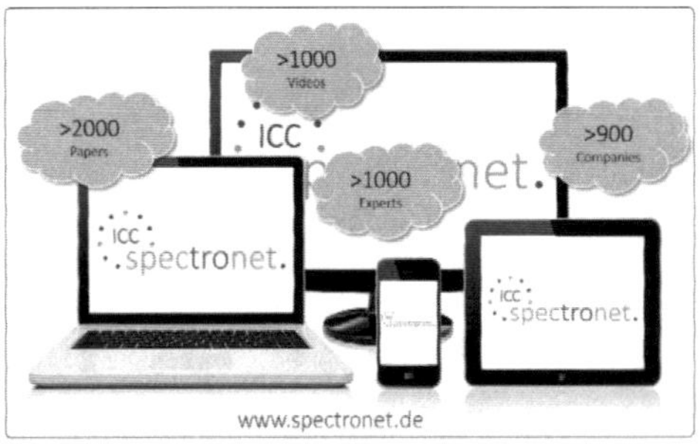

Figure 6.1. Digital papers, videos, experts, institutions and enterprises on spectronet.de

Highly visible and recognized contributions to the current education and training in image processing are provided by AIA (visiononline.org), SPIE (spie.org), EMVA (emva.org), VDMA-Industrielle

Bildverarbeitung (vdma.org) and Fraunhofer-Allianz Vision (vision.fraunhofer.de). Today any educational organization is affected by the transition of analogue paper books, paper pictures and film videos to their digital versions on smartcomps. This transition is irreversible due to their convenience, reliability and affordability for an efficient and flexible individual use at work and at home. A convenient, reliable and affordable conversion of .pdf-papers to ebooks is done for example by Yumpu (yumpu.com/de/browse/user/spectronet.de). Furthermore, clusterpartners of SpectroNet elaborated their own educational material (Figure 6.2).

analytikjena	ama	Baumer Passion for Sensors	INSPECT	KONICA MINOLTA
1	2	3	4	5
Mahr	MAZeT	Ocean Optics	OMRON	Pool-i.d.™
6	7	8	9	10
pyramid building IT	STEMMER® IMAGING	VC Vision Components® The Smart Camera People	ximea	ZEISS
11	12	13	14	15

1. analytik-jena.de/de/life-science/service-support/downloads.html
2. ama-weiterbildung.de
3. baumer.com/de-de/services/anwenderwissen
4. inspect-online.com/whitepaper
5. konicaminolta.eu/de/measuring-instruments/lernzentrum
6. mahr.de/de/Know-how/Know-how
7. mazet.de/de/downloads/produkt-und-kundeninformationen/white-paper#.U1YQKfl_s00
8. spectroscopytv.com
9. ns.europe.omron.com
10. pool-id.com/html/index.php?option=com_content&view=article&id=84&Itemid=4&lang=de
11. polytouch.de/de/pr-media.html#section-bilder
12. stemmer-imaging.co.uk/en/handbook
13. vision-components.com/service-support/wissensdatenbankfaq
14. ximea.com/support/wiki/allprod/Knowledge_Base
15. zeiss-campus.magnet.fsu.edu

Figure 6.2. Imaging specific ebooks for mobile smart photonic dimensional, color and spectral measurements

6.2. *Imaging specific hands-on trainings in universities and industry*

To streamline the enormous qualification deficits in imaging, several up-to-date manufacturers and system integrators increasingly are taking the qualification of their employees and customers into their own hands. Since many years an efficient collaboration between the Ernst-Abbe-University of Applied Sciences Jena and the Mahr Company Jena is performed. The Ernst-Abbe-University of Applied Sciences Jena is increasingly defined by interdisciplinary collaboration [16]. The Mahr Group is the world's third largest manufacturer of a complete range in dimensional measuring products from calipers to optical coordinate measuring systems [17]. Imaging specific hands-on trainings with students of the Ernst-Abbe-University of Applied Sciences Jena since many years are successfully shared with Mahr Company Jena to increase the efficiency of resources by the application of current hardware apps and software apps for qualification purposes (Figure 6.3.).

EUROPEAN IMAGING ACADEMY	Mahr	Ocean Optics	st w	ZEISS

1	2	3	4	5

1. stemmer-imaging.de/de/support/schulungen
2. spectronet.de/de?t=/contentManager/onStory&ParentID=1310139565449&StoryID=1294424405064&lang=1&active=no
3. oceanoptics.com/Products/education_new.asp
4. spectronet.de/de/vortraege_bilder/vortraege_2014/steinbeis-spectronet-collaboration-forum-2014-ilme_htfjcn4k.html
5. zeiss.de/microscopy/de_de/service-support/microscopy-labs.html

Figure 6.3. Imaging specific hands-on trainings

7. Conclusions

Smartcomps became global mass products with great convenience, high reliability and affordable prices. These market developments enable a paradigm shift in photonic measurement engineering, automation and quality assurance for industrial and non-industrial applications. With modular hardware apps and software apps new classes of hybrid instruments can be created. Modular hardware apps and software apps are reducing development time, cost and risk. Smartcomps are already revolutionary game changers in lifestyle. Now they are also penetrating into the workstyle. Miniaturization and consumerization inspire each other with increasing speed. Field instrumentations will be based on re-invented laboratory instrumentations. First steps on innovative routes are already gone successfully. An interesting convergence of measurements, diagnostics and quality control in industry, biology/medicine and environmental protection can be expected. In summary, it should be noted that the recently available mobile smartcomps and hardware apps in conjunction with proven software apps for Microsoft Windows operation systems are stimulating the re-invention of conventional solutions in measurement engineering, automation and quality assurance preferably for photonic image acquisition and digital image processing. This situation is also an enormous challenge for further education and shared hands-on trainings in these fields. Approaches to practical solutions have been explained in detail.

References

[1] http://spectronet.de/de/vortraege_bilder/vortraege_2014/steinbeis-spectronet-collaboration-forum-2014-ilme_htfjcn4k.html
[2] H. Zhu; S.O. Isikman; O. Mudanyali; A. Greenbaum; A. Ozcan: Optical imaging techniques for point-of-care diagnostics. Lab chip, 2013, 13, 51-67
[3] http://www.ximea.com/en/products/application-specific-cameras/subminiature-usb-cameras/mu9pc-mh
[4] http://www.lensation.de/en/shop/browse/6-s-mount-m12x05-lenses.html?sef=hc
[5] http://www.mazet.de/de/produkte/jencolor/jencolor-testsysteme/kit-mtcs-int-ab4/@@productdownloadsview#.Usmli9F3uUk
[6] http://www.rgb-laser.com/content_products/product_qstick.html
[7] http://www.bipm.org/en/publications/guides/vim.html
[8] http://www.pog.eu/en/ms/sms.html
[9] http://www.instrumentsystems.de/products/led-measurement-accessories/acs-calibration-leds/
[10] http://xritephoto.com/ph_product_overview.aspx?id=1192
[11] http://en.wikipedia.org/wiki/RAL_colour_standard
[12] http://www.lovibondcolour.com/liquid-standards
[13] http://www.lovibondcolour.com/glass-standards
[14] http://www.oceanoptics.com/Products/standards.asp
[15] http://www.precisioncells.com/news/article/39/UV-VIS-Spectrophotometer-Calibration
[16] http://www.fh-jena.de/fhj/fhjena/en/Seiten/default.aspx
[17] http://www.mahr.com/index.php?Language=enus

Journal of Physics: Conference Series **588** (2015) 012020 doi:10.1088/1742-6596/588/1/012020

New approach in subjective and objective speech transmission quality measurement in TCP/IP networks

Pavel Souček, Oldřich Slavata, Jan Holub

Dept. of Measurement, FEE CTU Prague, Technicka 2, CZ 166 27 Prague 6, Czech Republic

E-mail: soucepa3@fel.cvut.cz, slavao@fel.cvut.cz, holubjan@fel.cvut.cz

Abstract. This paper deals with problems of speech transmission quality measurement in modern telecommunication networks. It focuses on problems caused by specific types of distortions and errors caused present in transmissions using TCP/IP networks.

1. Introduction

Measurement methods used in telecommunications have evolved rapidly in last few years due to communication technologies developments, caused by the need of better utilization of existing transmission channel capacities or the need of minimization of new communication channels establishment, respectively. At the same time, the amount of transmitted data increases significantly. This trend requires many partial problems to be solved. Very actual topic is objective (algorithmic) measurement of transmitted speech, video or multimedia quality in (common) case of transmission channel errors including packet loss and jitter delay for packet-based transmissions. Moreover, the used algorithms must be continuously calibrated for given conditions like language and user portfolio as subjective quality assessment varies in time based on developing user experience (and thus expectation). Considering the fact that subjective service acceptability threshold approaches in some cases the offered service quality, even tiny details, frequently neglected in the past, must be taken into account, e.g. nativity or non-nativity of the language used for the communication. It was shown that non-native listeners have higher demands for telecommunication channel quality that can have significant (roaming in mobile networks) or even fatal (radio communication during multi-national military operations) impacts to the service users.

The main problem of present approaches in speech transmission quality measurement is that they are partially based on facts that are in some ways obsolete. Most problematic is their basis in analog telephony and they have difficulty in reflecting modern trends which lead to new type of distortions in signal and even more dramatic change in listeners perception.

Like technical part of communication, the listener's perception evolved over the years. This evolution is consequent of widespread availability of free communication tools like ICQ, Skype and many others, which are able to employ modern coders much more quickly than common telephone network or mobile operators. This is even more important with use of free communication software like Viber and Linphone, which are using cell phones data connection to make call instead of standard service. This allows much easier employment of modern coders and leads to shift in perceived speech transmission quality. One of the consequences of this widespread usage of such coders, which often

Content from this work may be used under the terms of the Creative Commons Attribution 3.0 licence. Any further distribution of this work must maintain attribution to the author(s) and the title of the work, journal citation and DOI.
Published under licence by IOP Publishing Ltd

deploy psycho-acoustic compression, is that respondents in research made in [1] often evaluated mp3 like higher quality record over lossless formats.

2. Conventional speech transmission quality measurement

Subjective speech transmission quality measurement is standardized in ITU Recommendation P.800 [2]. This recommendation defines listening conditions including listening chambers, its reverberation and length of testing session. It also defines the parameters of speech samples and scale used for evaluation of speech transmission quality. This scale is named Mean Opinion Score (MOS) and it is five point scale where 1 is the worst and 5 the best.

Subjective tests are divided into the conversational test and the listening test. Listening tests consist of a session with a group of subjects from whom an opinion on the speech transmission quality of the samples is gathered. Conversational tests are similar, but they are more demanding on time and organization than listening tests.

According to this recommendation the speech material should consist of simple, meaningful, short sentences, chosen at random as being easy to understand (from current non-technical literature or newspapers, for example). Every sample then should contain from two to five sentences, each from two to three seconds in length.

The specification of suitable sample is further narrowed by ITU-T Recommendation P.862.3[3]. For use with PESQ algorithm samples from eight to twelve seconds should be used.

Another part of P.800 defines number of votes needed to maintain high precision and to obtain statistically significant results. In case of listening tests it makes at least 100 votes per condition. It is usually difficult and impractical to have 100 respondents in one test, so in order to achieve this precision we usually use several samples for same condition.

The second way to obtain MOS values is by objective testing, which can be divided into two categories - intrusive and non-intrusive. These methods show good performance under most conditions, but it is also known that they fail under certain special circumstances.

Intrusive methods usually deliver results nearest to the results of subjective tests. They are based on a comparison between the original sample and the transferred sample. These tests are based on algorithms that use psychoacoustic models of human perception. Intrusive methods include several standardized algorithms, e.g. PESQ (Perceptual Evaluation of Speech Quality- ITU-T P.862 [4]) and POLQA (Perceptual Objective Listening Quality Assessment ITU-T P.863 [5]).

Non-intrusive methods are another type of objective measurement. These methods do not need to use a reference signal, and the final MOS is calculated only from the parameters of the transferred sample. A disadvantage of these methods is that they are less accurate and less reliable than intrusive methods. An example of a non-intrusive algorithm is 3SQM (Single Side Speech Quality Measurement - ITU-T P.563 [6]).

3. Specific problems with TCP/IP networks

Nowadays more and more of the calls is transferred using data and TCP/IP networks. Compared to original analog telephone network, these networks are relatively reliable and in normal conditions errors are rare. For example in case of packet loss the error rate should not exceed few percent.

If we take into consideration, that for most coders one TCP/IP packet carries from one to three voice packets, each containing 20 milliseconds of speech signal, we get 400 TCP/IP packets for whole sample. Error rate under 2 percent means that in worst case we will lose 8 packets. This together makes 480 milliseconds of speech signal, but not necessarily continuous.

With error rate this low, it is not difficult to imagine that lost packets will contain silent breaks between words or even a word that will not change the meaning of message. If only objective tests are used to evaluate speech transmission quality, the resulting MOS score will be below the actual quality, compared to subjective tests.

Another effect caused by use of these samples is, that the type A uncertainty will be very high and even though that objective tests should always give exactly same result. The way to eliminate this effect, used in current research, is to transfer every sample multiple times and as output speech transmission quality average of these acquired values is used.

In [7] researchers transfer every sample ten times and for every condition they use same speech material. However this approach cannot be used to gain subjective data for comparison, because subjects in such test will be affected by repetitiveness of the speech material and wouldn't be able to evaluate quality properly, if we played same sample to them multiple times.

4. New methodology using long samples

Based on our previous research we found out that samples long approx. 80 - 120 seconds would allow us to decrease time needed for objective tests and thus would lead to significant improvement. These samples would also make it possible to carry out subjective tests for the same set of samples and to compare the directly instead of comparing only similar samples.

In comparison with concatenated sample we can run only one test session and it would reflect real life conditions better. In this case sample is long enough for network related errors to affect quality in a way, listeners will be able to register it and distortion will be probably present in every sample.

The length of samples recommended by P.800 has its meaning. Samples should be short enough, so tested subjects can reliably evaluate their speech transmission quality. When we use longer samples, there is always a risk, that results will be influenced by so called recency effect. This means, that listeners are able to asses quality only of the end of sample, it leaves biggest impression. In order to eliminate this effect we are trying two approaches in order to gain reliable results:

- Continuous evaluation based on beep marks – samples are enhanced with beep marks placed equidistantly in sample every 10 seconds
- Continuous evaluation with custom measurement application – evaluation is realized with help of custom application and the time of change in score is fully in hands of subject

The second method is inspired by [8], where similar algorithm is used for video quality assessment. Application records speech transmission quality set by listener every second and computes average quality. There might be some improvements in the future, depending on gathered data, like weighting the values according to their position in sample. It is reasonable to assume that the beginning and the end of sample should have another importance than the rest of sample.

In order to evaluate, that these two new methods don't suffer same weakness as method currently used for speech transmission quality evaluation used with samples defined in P.800, we also plan to use reference test where original methodology is used, in order to confirm the recency effect with samples this long.

At this time we are carrying out initial tests in order to evaluate correctness of this methodology using speech material from four speakers (two males and two females). This speech material was processed using network simulator to emulate distortions.

5. Conclusion

Methodology currently used for speech transmission quality measurement is not entirely appropriate for use with conditions that include TCP/IP networks or any similar transfer method using packetization and buffers. Moreover speech material used for objective tests cannot be used for subjective testing in. Our new proposed methodology should eliminate this disadvantage and enable subjective tests to be carried out with same speech material as objective tests and without risk of results being affected by recency effect.

6. References

[1] Sterne, Jonathan,"MP3: The meaning of format", Duke University Press, 2012

[2] ITU-T Rec. P.800, "Methods for subjective determination of transmission quality," International Telecommunication Union, Geneva, 2001

[3] ITU-T Rec. P.862.3, "Application guide for objective quality measurement based on Recommendations P.862, P.862.1 and P.862.2", International Telecommunication Union,

Geneva, 2007
[4] ITU-T Rec. P.862, "Perceptual evaluation of speech quality (PESQ): An objective method for end-to-end speech quality assessment of narrow-band telephone networks and speech codecs", International Telecommunication Union, Geneva, 2001
[5] ITU-T Rec. P.863, "Perceptual Objective Listening Quality Assessment (POLQA)", International Telecommunication Union, Geneva, 2011
[6] ITU-T Rec. P.563, "Single-ended method for objective speech quality assessment in narrow-band telephony applications", International Telecommunication Union, Geneva, 2004
[7] Holub, J. - Slavata, O.: Impact of IP Channel Parameters on the Final Quality of the Transferred Voice In: Wireless Telecommunications Symposium 2012 Papers and Presentation [CD-ROM]. Pomona (CA): California State Polytechnic University, 2012
[8] ITU-T Rec. P.910, "Subjective video quality assessment methods for multimedia applications", International Telecommunication Union, Geneva, 2008

A new approach on restoration of dynamic measurement uncertainties in optical precision coordinate metrology

S Holder[1], E Reetz[1] G Linß[1]

[1]Ilmenau University of Technology, Faculty of Mechanical Engineering, Department of Quality Assurance and Industrial Image Processing, Gustav-Kirchhoff-Platz 2, 98693 Ilmenau, Germany

E-mail: `silvio.holder@tu-ilmenau.de`

Abstract. This paper presents a new approach to the restoration of dynamic influenced measurement uncertainties in optical precision coordinate metrology (OPCM) using image sensors to measure geometrical features. Dynamic measurements within the context of this paper are based upon relative motion between the imaging setup (CCD-camera and optical system) and the measuring object respectively the measuring scene. The dynamic image acquisition causes image motion blur effects, which downgrades the uncertainties of the measurand. The approach presented deals with a new technique to restore motion degraded images using different methods to analyze important image features by extending the famous state of the art Richardson-Lucy image restoration technique using a new convergence criteria based on the variation of the detectable sub-pixel edge position of each iteration.

Content from this work may be used under the terms of the Creative Commons Attribution 3.0 licence. Any further distribution of this work must maintain attribution to the author(s) and the title of the work, journal citation and DOI.
Published under licence by IOP Publishing Ltd

Journal of Physics: Conference Series **588** (2015) 012021 doi:10.1088/1742-6596/588/1/012021

1. Introduction

The primary objective of the new approach is to obtain accurate measurands on dynamic measurements in optical coordinate metrology (OPCM). Besides the measurement accuracy it becomes more and more important to provide measurement results at a reasonable time due to economic reasons. The computation time needed to process acquired images in a suitable way for metrological applications allocates a small amount of the complete measurement time needed to finalize a measurement task (often accelerated by multicore CPU-architectures and GPU-processors). Acquiring static images often consumes a higher amount of time due to repositioning the camera and freezing the drives. Therefore the image acquisition process actually is a bottleneck in the whole process chain. Within the context of this paper a static measurement describes the measurement object in rest relative to the imaging sensor respectively the camera and the lens. A dynamic measurement instead depicts the measuring object in motion during the measurement process [1]. The goal of the dynamic measurement is the decrease of the time consumption at the achievement of a constant measurement accuracy as on static measurement. High precise measurements in OPCM require high quality imaging of the measurement scene. Influences like distortion effects caused by the optical system, noise during the sensor integration time, analog-digital-conversion (ADC) and others reduce the measurement accuracy on static and dynamic measurements too. A dynamic measurement usually causes blur effects in the image due to the relative motion between measuring object and image sensor during image acquisition, so-called motion blur. For metrological purposes the blur effects at intensity transitions ("edges") are the major problem since they are used to interpolate subpixel-precise edge positions (SPEPs). Image blur effects increase the measurement uncertainty U [2] and therefore downgrade the measurement accuracy. The reachable uncertainty of the image measurement increases with the velocity of the relative motion. Figure 1 also depicts an dynamical acquired scene of a measuring object in horizontal motion and the overlay of a typical region-of-interest, which is used for a search line based detection of SPEPs.

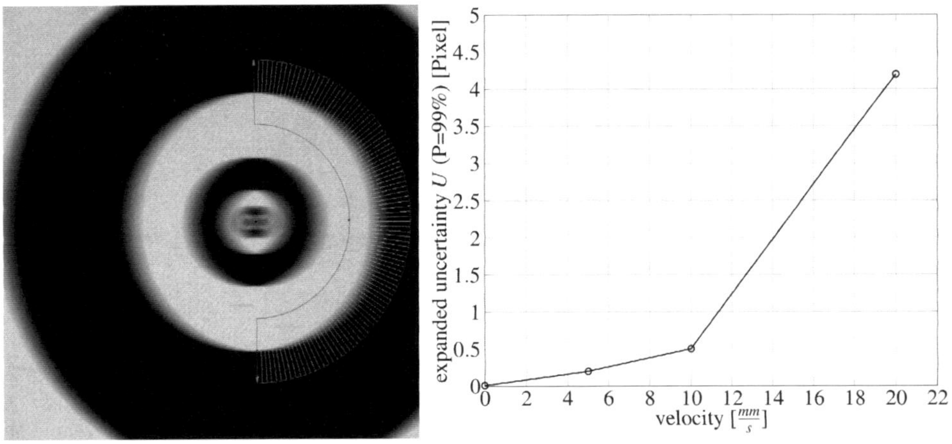

Figure 1. left: horizontal motion degraded image, right: measurement accuracy downgrade by increased uncertainties caused by motion-blur-effects [2]

2. State of the Art

2.1. Dynamic measurement in optical precision coordinate metrology

The state of the art approach to reduce image blur effects is to "freeze" the measuring scene for a short time. A common technique for this approach is to use strobed illumination during the shutter time of the CCD-camera, image capturing in a very short exposure time [3, 4]. These methods demand in both cases a powerful illumination of the measuring scene or a high electronic amplification, which yields higer image noise as the biggest disadvantage. The noise effects are degradational influences to the measurement of geometrical features and also to the reachable measurement uncertainties. Another simple method in combination with the other above mentioned approaches is the reduction of the motion velocity at the exposure time [3], which is counterproductive in regard to the advantage of saving of time of the dynamic image acquisition.

2.2. Image degradation process

When modeling the imaging process using linear system theory, the ideal image, described as $f(x, y)$, is transformed into the captured image $g(x, y)$ by convolving ($*$) the ideal image with a degradation function $h(x, y)$ which is superposed by noise effects $n(x, y)$ during the image acquisition process (equ. 1 [5]).

$$g(x, y) = h(x, y) * f(x, y) + n(x, y) \tag{1}$$

The image degradation function $h(x, y)$ is modeled as a superposition of different degradation kernels like the optical lens degradation kernel $h_{optic}(x, y)$ for the static measuring scene. The optical degradation kernel contains diffraction and optical distortion effects, depending on the analytic model used. For the dynamic measurement there is an additional image degradation function called motion blur kernel $h_{motion}(x, y)$ which depends on the CCD-sensors integration time, the relative motion velocity $\vec{v}$, the motion direction as angle α, the lens-magnification β and the constant pixel-distance on the image sensor d_{pixelX}, d_{pixelY} (see equ. 2).

$$h(x, y) = h_{optic}(x, y) + h_{motion}(x, y) \tag{2}$$

In case of linear uniform motion, a simple estimation of $h_{motion}(x, y)$ is used to restore the image $g(x, y)$ using iterative non-blind deconvolution methods like the Richardson-Lucy algorithm [6].

2.3. Iterative image restoration

Different non-blind deconvolution algorithms are applicable for the restoration of motion blurred images with knowledge of the blur kernel $h(x, y)$. Iterative deconvolution methods like the Richardson-Lucy algorithm (RLA) yield sufficient results on restoration of motion blurred images using a fixed number of iterations to calculate the unblurred estimation $\hat{f}(x, y)$. The image estimate of RLAs current iteration is calculated using the following equation 3 [7]:

$$\hat{f}_i = \hat{f}_{i-1} \cdot \left(h * \frac{g}{h * \hat{f}_{i-1}} \right) \tag{3}$$

The main issue on the restoration using the RLA is the estimation of the blur kernel function and the proper iteration stop criteria.

3. Extended Richardson-Lucy-Algorithm

The major problem when using the iterative RLA is to find an adequate number of iterations or any other useful convergence criteria. When using iterative deconvolution algorithms like RLA in OPCM, the image region to be restored especially the edge transitions are very important to yield precise SPEPs. In OPCM, an area in the image, so called region-of-interest (ROI), is selected to obtain measurements results. Such ROIs consist of search lines (see figure 1), which define the direction to interpolate the image-subpixeling. The new approach utilizes the variations of typical motion influenced properties in the image, which vary on each RLA iteration of the restoration. One of these properties is the width of the intensity transition (so-called edge-width [8]) x_W along the motion direction, another property is the detectable subpixel-precise edge positions $\vec{x}_{SPEP}$ [9]. Since common ROIs are not suitable for the restoration tasks, new types of ROIs (figure 2) are proposed, which compute the edge position $\vec{x}_{SPEP,j,i}$ and the edge-width $x_{W,j,i}$ for every search line j in every iteration i in the adjustable motion direction, which is requested from the drive controller of the coordinate measuring machine (CMM).

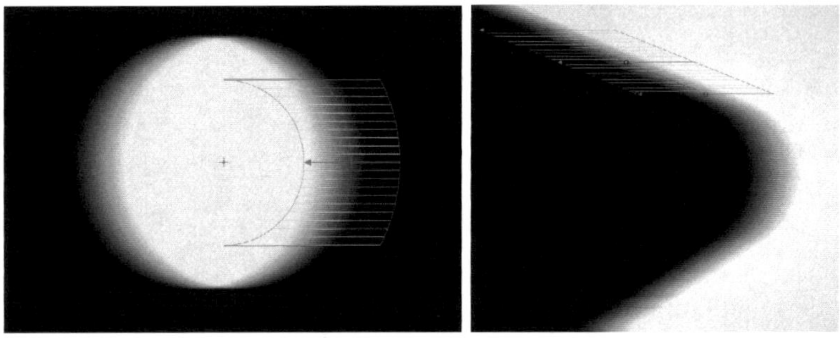

Figure 2. special regions-of-interest for the restoration task: left: *restoration-arc-ROI*, right: *restoration-line-ROI*

Figure 3 illustrates the full flow chart of the proposed extended RLA. The motion angle α (angle between horizontal image axis and motion vector) during exposure time of the sensor is used for the definition of the restoration-ROI in the image as illustrated in figure 2. The first step of the extended RLA starts with an initial analysis of the image based on the defined restoration-ROI, figure 3 - step 1. Every search line j of total $count_{SL}$ is used for the estimation of the subpixel-precise edge position $\vec{x}_{SPEP,j}$ and the edge-width $x_{W,j}$. The averaged edge-width $\overline{x}_W$ and the motion direction angle α are used for computing the motion point-spread-function $h(\overline{x}_W, \alpha)$. Under the pre-condition of an uniform linear motion, the point-spread-function in case of a pixel-grid-synchron motion is a vector of the size $[\overline{x}_W, 1]$ estimated by following equation:

$$h(\overline{x}_W, \alpha, x) = \frac{1}{\overline{x}_W} \qquad x = [0..\overline{x}_W] \tag{4}$$

Non pixel grid motion angles are produced by rotation of the motion vector and bilinear interpolation [6]. Afterwards the motion point-spread-function is extended to the image size by zero padding. The main iteration loop of RLA is processing the image estimate $\hat{f}_i(x, y)$ for every iteration i, figure 3 - step 2. The further step 3 is calculating the edge properties SPEP and edge-width again. The current iterations scalar distance value d of the SPEPs estimated from current and last RLA iteration is calculated component-by-component in the following way by the distance vectors absolute value:

$$d = |\vec{x}_{SPEP,j,i} - \vec{x}_{SPEP,j,i-1}| \tag{5}$$

$$d = |\begin{pmatrix} x_{j,i} \\ y_{j,i} \end{pmatrix} - \begin{pmatrix} x_{j,i-1} \\ y_{j,i-1} \end{pmatrix}| = |\begin{pmatrix} d_x \\ d_y \end{pmatrix}| = \sqrt{d_x^2 + d_y^2} \tag{6}$$

Afterwards the convergence criteria are checked by comparing the estimated values with pre-defined threshold values (figure 3 - step 4). This loop procedure is repeated until the conditions are true.

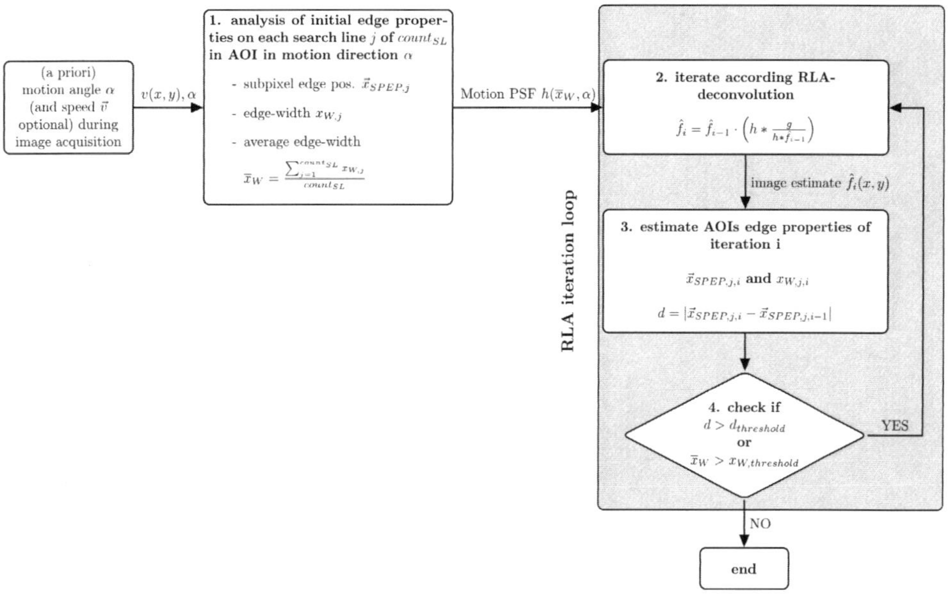

Figure 3. Extended RLA for motion blur restoration of images used for measurements in OPCM

4. Experimental results

The new algorithm was tested on different motion acquired images and delivered sufficient results on the restoration. Figure 4 depicts the restoration levels of an selected image restored by the proposed algorithm. The image edges were sharpened in a sufficient way to reduce the reachable uncertainty of the measurement. As seen in figure 5 a) the edge-width decreases on each iteration of the algorithm. The example image was sharpened from 58 pixel edge-width to a value of 10 pixel in only 38 RLA iterations. The distance of the SPEPs between two iterations are also decreasing, see figure 5 b). The charts of both values have an asymptotic characteristics, therefore pre-defined threshold values ($d_{threshold}$, $x_{W,threshold}$) as stopping criteria could be used. The distance value of the SPEPs $d_{threshold}$ converges to zero with a noise depending ripple.

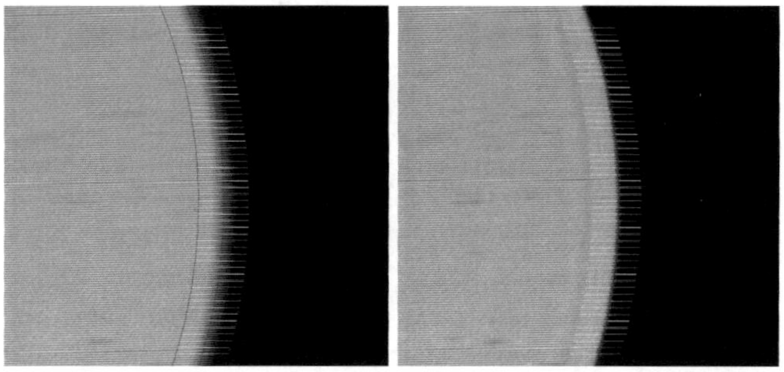

Figure 4. motion blurred image, left: before extended RLA restoration, right: after extended RLA restoration

For a subpixel-precise estimation of the edge position, an edge-width in the range of 5 to 10 pixel necessary for the interpolation. These values are estimated on real static images and mainly

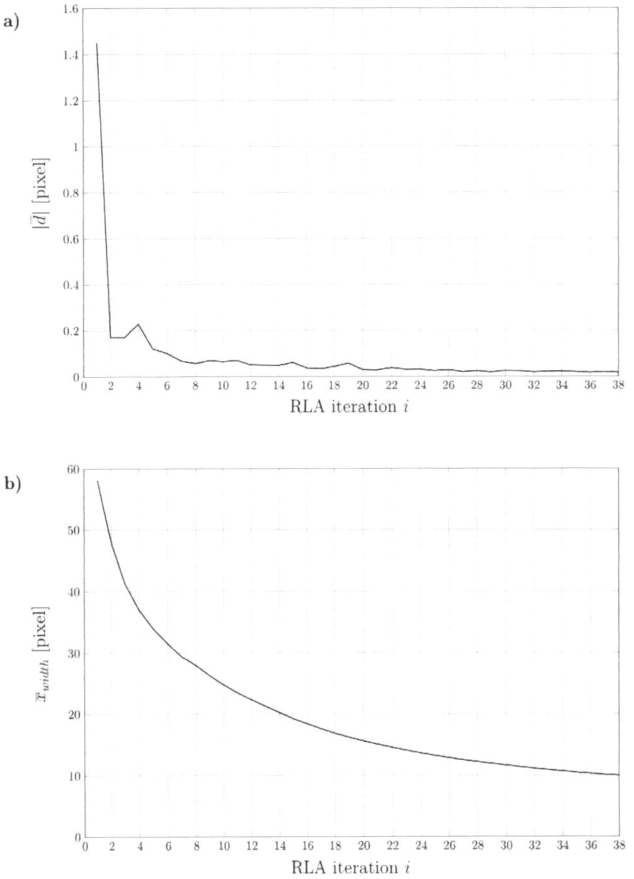

Figure 5. a) average edge-width, b) average SPEP distance vs. RLA iteration

depending on the diffraction effects of the lens system. Hence the treshold value $x_{W,threshold}$ is a-priori defined for the breaking condition to a value of 10 pixel. The SPEPs distance threshold $d_{threshold}$ is also a-priori defined to a value of 0.05 pixel and yields a good ratio of performance and accuracy of the algorithm.

5. Summary

We presented within this paper a new approach on the restoration of motion blurred images in the field of dynamic measurement using image sensors in the optical precision cooordinate metrology. The proposed algorithms are based on an extension of the Richardson-Lucy iteration with special remark to image properties regarding to the uncertainties for measurement applications. The deconvolution algorithm yields a good performance on motion blurred grayscale images and delivers sufficient results to the intensity edge transitions, which are important for the measurement application.

Aknowledgements

This work was supported by the Department of Quality Assurance and Industrial Image Processing at Ilmenau University of Technology and founded by the German Ministry of Education and Research (BMBF) within the research project "Mehrdimensionale Bildverarbeitung" 03WKP33A.

References

[1] Bantel M 2004 *Messgeräte-Praxis - Funktion und Einsatz moderner Messgeräte* (Leipzig: Fachbuchverlag Leipzig im Carl Hanser Verlag)
[2] Weissensee K, Holder S and Linß G 2011 *Uncertainty estimation of image-based measurements affected by motion blur* 56th Internatinal Scientific Colloquium, Ilmenau University of Technology (Ilmenau: TU Ilmenau)
[3] Christoph R and Neumann H J 2013 *Multisensor - Koordinatenmesstechnik* (München: Verlag Moderne Industrie)
[4] Mitutoyo Europe 2013 *Product brochure - Quick Vision Series PRE1320* (Neuss: Mitutoyo Europe)
[5] Gonzales R and Woods R 2008 *Digital image processing* (New Jersey: Pearson Prentice Hall)
[6] Bovik A 2008 *The essential guide to image processing - 2nd Ed.* (New York: Elsevier Academic Press)
[7] Molina R, Nunez J, Cortijo F and Mateos J 2001 *Signal Processing Magazine, IEEE* 18 pages 11–29
[8] Holder S et al 2011 *Method for robust search line based estimation of intensity edge-width in blurred gray scale images for the quantification of motion- and out-of-focus-blur* 56th Internatinal Scientific Colloquium, Ilmenau University of Technology (Ilmenau: TU Ilmenau)
[9] Kühn O 1997 *PhD Thesis: Ein Beitrag zur hochauflösenden zweidimensionalen Geometriemessung mit CCD-Zeilensensoren*, Ilmenau University of Technology (Ilmenau: TU Ilmenau)

A Novel Method of Zero Offset Reduction in Hall Effect Sensors with Applications to Magnetic Field Measurement

Cheng Liu, Ji-Gou Liu, Quan Zhang

ChenYang Technologies GmbH & Co. KG.
Markt Schwabener Str. 8, 85464 Finsing, Germany
http://www.chenyang.de

Tel. +49-8121-2574100, Fax: +49-8121-2574101
Email: cheng.liu@chenyang-ism.com

Abstract. The paper presents a novel method to reduce the zero offset in Hall-Effect based magnetic measurement with single power supply. This method consists of a coarse zero compensation and a fine zero adjustment afterwards. By using the proposed method the zero output offset of Hall Effect sensors under using single power supply can be controlled within 0.2%. This method can be applied to all Hall Effect sensors with analog output and other similar sensors, which are powered with a single voltage or current source.

Keywords: Magnetic Field Measurement, Magnetic Field Sensing, Magnetic Field Detection, Gaussmeter, Magnetometer, Hall-Effect Sensor, Zero Offset Adjustment, Measuring System.

1. Introduction

With the development and wide applications of permanent magnets and electromagnets, magnetic field sensing has become rapidly a very important section of measurement technology. Magnetic field measurement is widely used in many industrial branches such as motors, generators, wind turbines, automotive, loudspeakers, brain wave activities, magnetic sensors/transducers, controlling systems, medical instruments and other machines [1, 5].

There exist a lot of magnetic field measurement techniques. The most common methods are based on superconducting quantum interference device (SQUID), magnetometers (fiber optic, nuclear precession, optically pumped and fluxgate), magnetoresistive sensors (AMR, GMR), inductive sensors, Hall-effect sensors and magneto-optic sensors [1, 2]. Some of them are used for low-field measurements (< 1mT), like magnetometers and SQUID. Magnetoresistive sensors are more robust and cheaper to sense medium fields (< 100mT). Inductive sensors (< 2T) and Hall-effect sensors (< 8T) are good candidate for the high field measurement, and magneto-optic sensors is for very high field measurement (up to 10T).

The most useful magnetic field range is 1μT~5T for medical and industrial applications. Only Hall Effect sensor elements cover this magnetic field range. They have the benefits of easy handling, robustness, small package and cheap price in comparison with other sensors mentioned above. In addition to pure magnetic field measurement, Hall sensors can also be found in many position, proximity, angle, speed and current sensing applications [3]. Therefore Hall Effect sensors are preferred for magnetic field measurement.

In order to get a useful analog output signal for controlling systems and measuring instruments, the Hall voltage at the sensor output has to be further processed by a differential amplifier with high input impedance, gain and zero offset adjustments [2]. Under using double power supplies, adequate results are available. Nevertheless, double power supplies are exceptions in many applications. When the Hall Effect sensor is powered with a single constant voltage or current source, the sensor output will get a remarkable offset voltage even after a normal offset adjustment. For reducing the offset output

Content from this work may be used under the terms of the Creative Commons Attribution 3.0 licence. Any further distribution of this work must maintain attribution to the author(s) and the title of the work, journal citation and DOI.
Published under licence by IOP Publishing Ltd

voltage, various kinds of efforts are done, for instance, by using a DC-DC-voltage converter in order to generate an additional negative power supply internally. This causes a higher current consumption, more complicated circuit, and higher costs of the sensor systems.

This paper proposes a novel method of zero offset reduction in Hall-Effect based magnetic measurement with single power supply. This method consists of a coarse zero compensation and a fine zero adjustment afterwards. By using the proposed method the zero output offset of Hall Effect sensors under using single power supply can be controlled within 0.2%. This method can be applied to all Hall Effect sensors with analog output and other similar sensors, which are powered with a single voltage or current source.

2. Zero Offset Origin

Figure 1 shows a SIP package of a Hall sensor element, which has two supply pins (+Vcc, –Vcc/GND) and two output pins (+Vh, -Vh). The Hall element is powered with a constant voltage or current source through its supply pins, in order to get a reliable linear output voltage for measuring a magnetic field. In absence of magnetic field, a voltage difference exists between the two output pins (+Vh, -Vh), due to manufacturing tolerances and inhomogeneity in the semiconductor material [3]. A Hall element can be modeled as a resistive bridge (see figure 2). The output voltage difference is the result of mismatches between the resistances R1, R2, R3 and R4 [6].

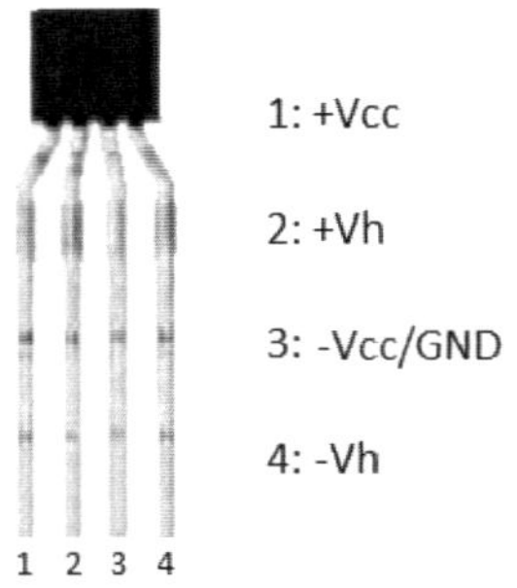

Figure 1. Pin assignment of a Hall element

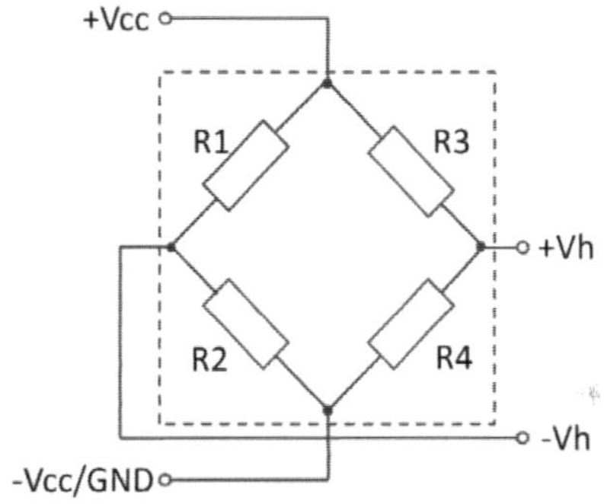

Figure 2. Equivalent circuit of a Hall element

When the output pins (+Vh, -Vh) are connected to a differential amplifier, a zero offset voltage at the amplifier output will be caused by the voltage difference mentioned above. Therefore zero offset adjustment circuit must be built in the signal processing unit of the measuring system, see figure 3.

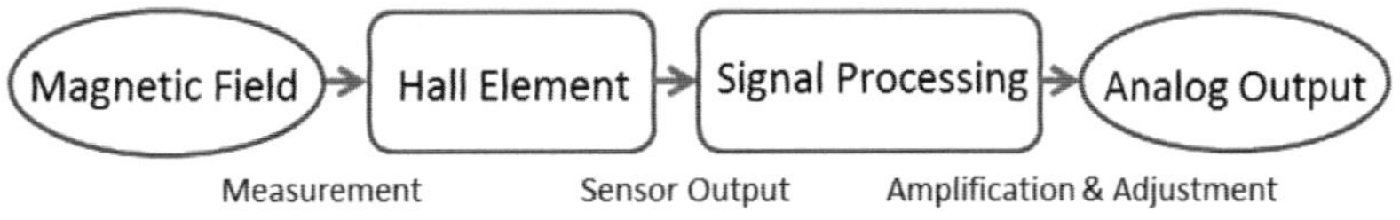

Figure 3. Diagram of Hall-effect based magnetic field measurement.

3. Conventional Zero Offset Adjustment

If the Hall element is powered with double voltage sources, the zero offset voltage can be easily eliminated by using an operational amplifier and a potentiometer (see figure 4).

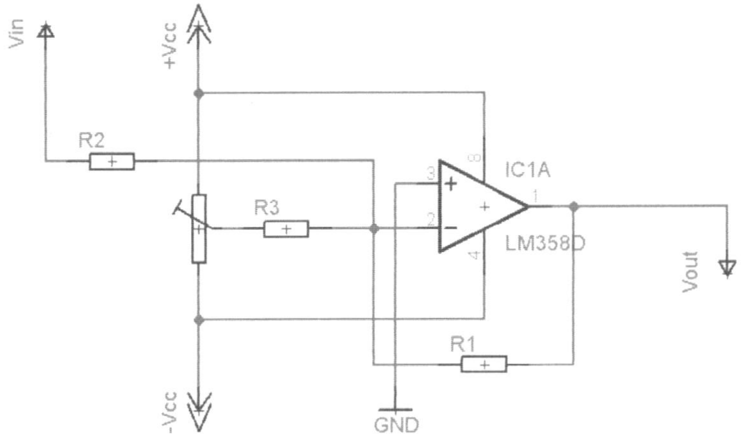

Figure 4. Zero Offset Adjustment with Double Voltage Supplies.

However, the zero offset voltage is difficult to reduce to zero under using a normal adjustment circuit (as like in figure 5) if the Hall element is powered with a single source, i.e. the pin 3 and 4 are connected to ground GND, see figure 1 and 2.

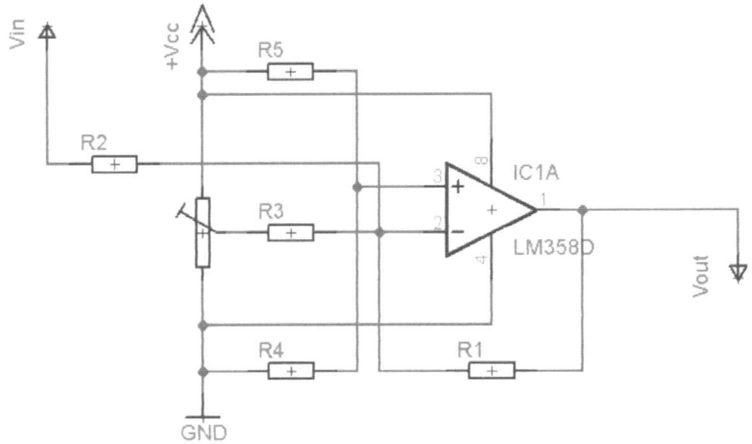

Figure 5. Zero Offset Adjustment with Single Voltage Supply.

In this case, -Vh=0 and the output voltage +Vh can be written by

$$+\boldsymbol{V_h} = \frac{\boldsymbol{R3}}{\boldsymbol{R3+R4}}\boldsymbol{V_{cc}} \qquad \textbf{(1)}$$

One gets +Vh=Vcc/2 when R3=R4. For offset compensation, a virtual ground of about +Vcc/2 is needed at the noninverting port of the operational amplifier for compensating the zero offset voltage. Therefore, a higher feedback loop current at R1 is available:

$$\boldsymbol{I_c} = \frac{\boldsymbol{V_{cc}}}{\boldsymbol{2 \times R1}} \qquad \textbf{(2)}$$

Due to the exist of inner resistance $\boldsymbol{R_{in}}$ at the output of an operational amplifier [4], this feedback loop current I_c causes an offset voltage:

$$\boldsymbol{V_{off} = R_{in} \times I_c = \frac{R_{in}}{2 \times R1} V_{cc}} \quad (3)$$

A noticeable amount of zero offset voltage still remains in the level of 50mV~100mV after the compensation for an output range of 0-5V. This means that the relative zero offset voltage is 1.0%~2.0%, which is not acceptable in many magnetic field measurement applications. Thus a new solution is needed for reduction of the zero offset in this case.

4. Novel Method of Zero Offset Reduction

The proposed novel method for zero offset reduction contains two stages of signal processing, see figure 6. In the first stage, a relative high zero offset voltage under single power supply can be reduced by a coarse zero offset compensation. There is still a rest offset voltage $\boldsymbol{V_{off1}}$, which is caused by the feedback loop current $\boldsymbol{I_{c1}}$ at the inner ouput resistance $\boldsymbol{R_{in}}$ inside the operational amplifier of the first stage and can be calculated by equation (3). The zero offset voltage for the second stage is now lower, so that the feedback loop current $\boldsymbol{I_{c2}}$ at R8 is much smaller than in the first stage or in conventional zero offset adjustment, see Figure 7. $\boldsymbol{I_{c2}}$ can be determined by:

$$\boldsymbol{I_{c2} = \frac{V_{off1}}{R8}} \quad (4)$$

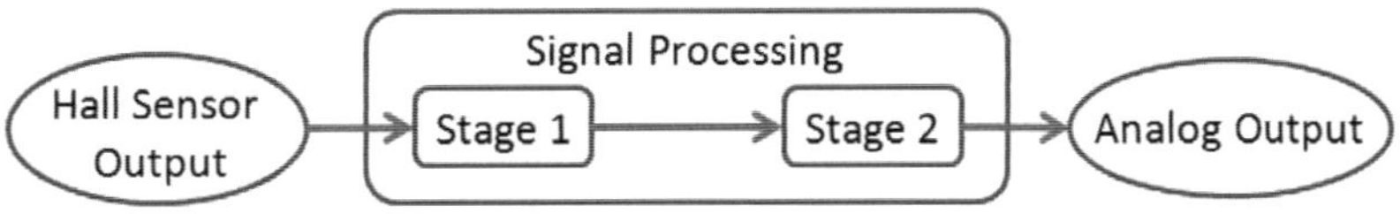

Figure 6. Description of a novel zero offset reduction method.

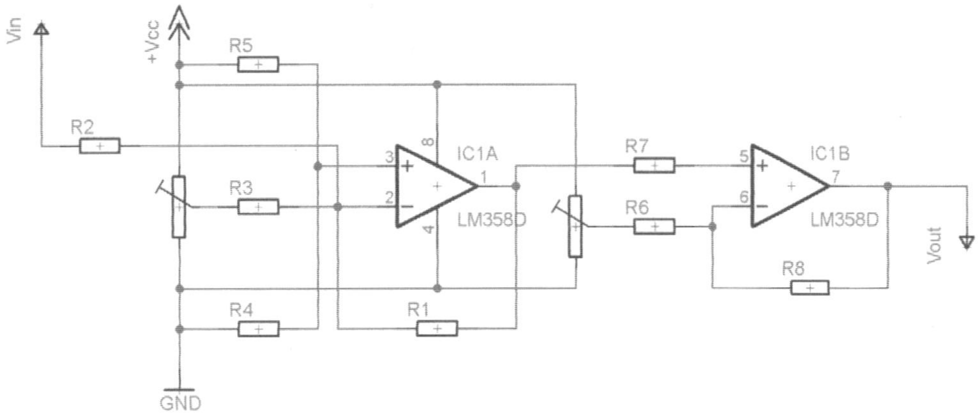

Figure 7. Possible Implementation of the novel Zero Offset Reduction method.

The zero offset voltage $\boldsymbol{V_{off2}}$ can be calculated by

$$\boldsymbol{V_{off2} = R_{in} \times I_{c2} = \frac{Rin}{R8} V_{off1} = \frac{R_{in}^2}{2 \times R1 \times R8} V_{cc}} \quad (5)$$

The zero offset voltage $\boldsymbol{V_{off2}}$ can be reduced within 10mV for voltage output range 0-5V by a fine zero offset adjustment in this second stage. This means a relative zero offset of 0.2%, which can be tolerated in the most magnetic field measurement. The advantages of this method are less power consumption and simplified circuit in comparison to other zero offset adjustment by using an internal negative power supply.

5. Results and Applications

This novel zero offset reduction method can be applied to Hall Effect sensors, magnetoresistive sensors and inductive sensors for magnetic field measurement etc. Figure 8 shows a Hall probe CYHP881 that uses the proposed method. This probe is designed for measuring 0-200mT (unipolar magnetic field) with an analog voltage output 0~4.5V. The probe is powered with a 5V source, and has a linearity of ±0.3%, a zero point offset of 0.2% and an accuracy of 1.0% [5]. The accuracy of the probe can be improved by using precise reference magnets with accuracy of better than 0.2%.

Figure 8. Product photo of Hall probe CYHP881.

Table 1 shows the calibration results of Hall probe CYHP881. By using an electromagnet, a magnetic field up to 198.8mT or even more can be generated. So a full-range calibration of the Hall probe is possible.

Table 1. Calibration results by using magnetic field generated by an electromagnet.

Magnetic field strength (mT)	Output voltage (V)	Relative error (%)	Linearity error (%)
0.0	0.0085	0.189	0.028
24.8	0.5650	0.156	-0.007
49.8	1.1235	0.067	-0.096
74.6	1.6816	0.069	-0.095
99.4	2.2403	0.084	-0.077
124.2	2.7987	0.093	-0.069
149.2	3.3582	0.027	-0.136
173.9	3.9147	0.042	-0.119
198.8	4.4719	-0.024	-0.188

A Hall probe with analog output can be used in many different ways for magnetic field measurement. It is suitable for quick measurements with a digital multimeter, and for integration in different measuring instruments and devices. By using an analog-to-digital converter, the analog output voltage of the Hall probe can be converted into digital signal so that it can be connected to microcontrollers and PC systems. Consumers can use their own software for data acquisition and processing etc. An overview of the Hall effect sensors for magnetic field measurement can be found in the literatures [1] - [3] and [6] - [13].

6. Conclusions

In this paper, a novel offset reduction method has been proposed, which has been used in Hall Effect sensors for magnetic field measurements. From the results one can draw the following conclusions:
For single power supply, the zero offset adjustment is much difficult, because a correction voltage close to one half of the supply voltage is needed. Nevertheless a noticeable amount of zero offset voltage still remains in the level of 50mV~100mV for an output range of 0-5V after zero offset adjustment. For eliminating this rest offset, a second zero offset adjustment stage is introduced. The offset can be reduced to under 10mV in the same output range. This novel zero offset reduction method can be applied to Hall Effect sensors, magnetoresistive sensors and inductive sensors for magnetic field measurement and current measurement and other similar sensors, which are powered with a single voltage or current source.

References

[1] M. J. Caruso, T. Bratland, Dr. C. H. Smith, R. Schneider: "A New Perspective on Magnetic Field Sensing", Honeywell Inc., 5/98.

[2] S. A. Macintyre: "Magnetic Field Measurement", CRC Press LLC, 1999.

[3] E. Ramsden: "Hall-Effect Sensors – Theory and Application", Elsevier Inc., 2006.

[4] E. Schrüfer: "Elektrische Messtechnik", 6th Edition, Hanser Verlag Munich, 1995.

[5] User's Manual of Hall Probe CYHP881. (http://www.cy-sensors.com/CYHP881-User's Manual.pdf)

[6] R. S. Popovic: "Hall Effect Devices", 2nd Edition, Institute of Physics Publishing, Bristol, 2004.

[7] S. C. Mukhopadhyay, R. Yueh-Min Huang: "Sensors: Advancements in Modeling, Design, Issue, Fabrication and Practical Applications", Springer, Berlin, 2008.

[8] S. Y. Yurish, M. T. S. R. Gomes: "Smart Sensors and MEMS", Kluwer, Dordrecht, 2004.

[9] Ch. S. Roumenin: "Microsensors for magnetic fields", in "MEMS: A Practical Guide to Design, Analysis, and Applications", Eds. J. G. Korvink and O. Paul, Andrew Publ., 2005.

[10] Honeywell, Hall Effect Sensing and Application, MICRO SWITCH Sensing and Control. (http://sensing.honeywell.com/honeywell-sensing-sensors-magnetoresistive-hall-effect-applications-005715-2-en.pdf)

[11] P. Ripka: "Magnetic Sensors and Magnetometers", Artech House, Boston, 2001.

[12] J. Fraden: "Handbook of Modern Sensors: Physics, Designs, and Applications", 3rd edition, Springer, New York, 2003.

[13] P. Ripka, A. Tipek, "Modern Sensors Handbook", ISTE, London, 2007.

Risk-aversive optimal planning of sensing

Mikko Lauri, Aino Ropponen, Risto Ritala

Tampere University of Technology, Department of Automation Science and Engineering, P.O. Box 692, FIN-33101 Tampere, FINLAND

risto.ritala@tut.fi

Abstract. Operable sensing is studied as a means for improving system performance. Both single and sequential planning are analysed, and the complexity due to risk-averseness in various formulations discussed. Three classes of sensing operation are defined and analysed.

1. Introduction

Autonomous machines and automated processes are increasingly equipped with sensing systems that are operable. Field of view in machine vision systems can be varied by zooming and/or turning the cameras, and the systems can switch between stereo vision mode and optical flow mode. In chemical process industries automated process analyzers are able to measure many chemical parameters – one at a time – and to handle samples from many locations of the process. Future autonomous mobile machines (AMMs) are expected to have separate measurement machines, such as unmanned Micro Aerial Vehicles (MAV) that will collect data about the working environment in order to improve the performance of the AMM. Similarly, in social or biological systems, before deciding of an operational action, there may be limited time available or cost associated to obtain new data about the system and it must be decided, which data source provided should be accessed by some observation action. Recently theories relevant to artificial systems have been proposed about human vision behavior [1-2].

Sensing actions add value only when the system state is not known completely. The incomplete information about the state is represented probabilistically. Sensing can be optimized either to maximize the quality of information, i.e. the prior expectation of posterior entropy, or to maximize the system performance for the current task. When deciding about the sensing actions, the data resulting from such actions is obviously not known. Thus even if the system were deterministic the performance resulting from sensing and action on the system will be subject to uncertainty. In decision about sensing the probability distribution of interest is the prior distribution of posterior optimal performance [3]. In such decisions under uncertainty the attitude towards risk may affect the choices but is rarely considered in control/sensing optimization.

For dynamic systems a sequence of decisions is to be planned instead of a single decision. Optimal planning takes into account that in all the future time instants spanning the sequence, there will be an opportunity to make further sensing actions. The receding horizon principle is applied: a sequence is planned, the first action is then implemented and at the next time instant the planned sensing action is carried out, and based on data obtained a new sequence is planned [4].

This paper discusses the formulation and solutions of the optimal sensing problem through problem classes of practical interest. In particular, the attitude towards uncertainty is discussed and problems arising in different formulations of the risk attitude in sequential decision tasks are analyzed. Section 2

Content from this work may be used under the terms of the Creative Commons Attribution 3.0 licence. Any further distribution of this work must maintain attribution to the author(s) and the title of the work, journal citation and DOI.
Published under licence by IOP Publishing Ltd

formulates optimal sensing. Section 3 presents three widely applicable problem classes of optimal sensing. Section 4 concludes the challenges when planning optimal sensing.

2. Objectives for operation of sensing: performance and attitude towards risk

Contrary to the design of experiments, in which the objective is to sense the system in order to minimize the uncertainty – in the most general form the entropy – about its state, here we consider sensing in the context of providing optimal support for a task given for the system. The sensing actions and the actions on the system to carry out the task are chosen to maximize a performance measure. The actions on the system are committed only after the data from sensing actions has been received.

Two typical performance measures are the reward (to be maximized) and the regulated end result (absolute difference form a desired end result to be minimized). These can be considered as linear, respective quadratic functions of the system state which is to be estimated based on sensing data.

According to utility theory optimal choices between distributions can be reduced to finding the distribution that maximizes the expectation of a utility function [5]. However, construction of a utility function is rather heuristic. There are also several other heuristic approaches for decisions between probability distributions. The uncertainty about a performance Q in sensing optimization can be taken into account, e.g. as follows:

- risk neutral: maximize the expected performance μ_Q without considering uncertainty
- risk constrained: maximize the expected performance μ_Q subject to that the probability of unsatisfactory performance Q_B is constrained, $P(Q < Q_B) < p_{constr}$
- risk aversive: concave utility function, e.g. $U_0(Q) = 1 - \exp(-\alpha(Q - Q_0))$ so that for Gaussian information about performance the expression $\mu_Q - \alpha\sigma_Q^2/2$ is maximized
- risk aversive: heuristic risk premium, $\mu_Q - \beta\sigma_Q$ is maximized
- risk conscious: maximize the probability of satisfactory performance, $P(Q > Q_S)$ [6]; this corresponds to a utility function which is constant above Q_S and negative infinite below.

The uncertainty may result from two main sources: either the predicted system state is uncertain, or the parameters of the performance function are uncertain. The latter case can be turned into state uncertainty by augmenting the uncertain parameters to the state. The decision about the sensing is based on the prior distribution of the posterior optimal performance. Let $s \in S$ be the system state, $a \in A$ an action on the system, $m \in M$ the observation action, and $y^{(m)}$ the data resulting from the observation action. The system develops from state s to state s' as a result of the action a with probability (density) $f(s'| s,a)$. Furthermore, let $Q(s',a,m)$ be the system performance, to be maximized, and $U(Q)$ be the utility of performance expressing the attitude towards risk. Then the posterior optimal action is

$$\begin{aligned} a_{opt}\left(m, y^{(m)}\right) &= \arg\max_{a\in A} \int_S ds' f\left(s'| y^{(m)}\right) U\left(Q(s',a,m)\right) \\ &= \arg\max_{a\in A} \int_S ds' ds f\left(s'| s,a\right) f\left(s \mid y^{(m)}\right) U\left(Q(s',a,m)\right) \end{aligned} \quad (1).$$

The posterior optimal performance is then

$$\tilde{Q}\left(m, y^{(m)}\right) = \int_S ds' ds f\left(s'| s,a\right) f\left(s \mid y^{(m)}\right) Q\left(s', a_{opt}\left(m, y^{(m)}\right), m\right) \quad (2),$$

This is a random variable before the measurement data has been received. Hence the measurement to be chosen is the one that maximizes the prior expectation of the utility of the posterior optimal performance:

$$m_{opt} = \arg\max_{m\in M} \int_{Y^{(m)}} dy^{(m)} f^{(ap)}\left(y^{(m)}\right) U\left(Q\left(m, y^{(m)}\right)\right) \qquad (3),$$

where the prior probility (density) of data is obtained from measurement description $f\left(y^{(m)} \mid s\right)$ and prior state information $f^{(ap)}(s)$ as

$$f^{(ap)}(y^{(m)}) = \int_S ds f\left(y^{(m)} \mid s\right) f^{(ap)}(s) \qquad (4).$$

A common class of optimal sensing problems is sequential sensing planning: the overall performance is a sum of instantaneous performances over a time horizon. As the expected performance is cumulative, the optimal sequence of decisions can be solved with dynamic programming, i.e. with the Bellman equation. However, if the decision considers the risk, e.g. through that the decision objective is a nonlinear utility of the performance or the risk premium is included, the objective function is in general not cumulative and Bellman's principle of optimality, the cornerstone of dynamic programming, is not valid. The utility function $U_0(Q)$ above is special in that if the performance is a sum of independent stage performances, the variances are additive. Hence the objective function is cumulative and Bellman's principle of optimality applies.

3. Case studies on optimal operation of sensing

This section discusses three classes of optimal sensing tasks, relevant in many real-life applications.

3.1. Discrete action options with Gaussian performances – path planning

Consider an AMM whose task is to optimize its route to a final destination. The map of routes is a directed acyclic graph. The information about the performance (e.g. time) of each leg is incomplete. AMM is equipped with an MAV as a measurement machine capable of improving the information about the cost of the next leg. When AMM enters a junction, it must decide if the MAV should be sent to scout the condition of the optional adjacent legs before choosing along which to proceed. Decision about measurements may be sequential or at one stage. Repeating the same measurement may improve information (regular measurement) or it may be non-informative (e.g. image-based measurement in which the uncertainty is in the interpretation of the image, not in capturing the image).

Let the prior information about the performances Q_i of the legs be Gaussian and independent on one another, $N\left(\mu_i, \sigma_i^2\right)$. The measurement of performance of the route is unbiased but uncertain, described by a Gaussian with measurement uncertainty $\sigma_{meas,i}^2$. There is a cost c_i of making the measurement. E.g. when repeated measurements improve information, the optimal value of utility function U_0 at a junction with two parallel legs satisfies the following functional equation:

$$V\left(\mu_1, \mu_2, \sigma_1^2, \sigma_2^2\right) = \max\begin{cases} \mu_1 - \alpha\sigma_1^2/2 \\ \mu_2 - \alpha\sigma_2^2/2 \\ -c_1 + \int_{-\infty}^{\infty} dy_1 N\left(y_1; \mu_1, \sigma_1^2 + \sigma_{meas,1}^2\right) V\left(\mu_1 + \frac{\sigma_1^2(\mu_1 - y_1)}{\sigma_1^2 + \sigma_{meas,1}^2}, \mu_2, \frac{\sigma_1^2\sigma_{meas,1}^2}{\sigma_1^2 + \sigma_{meas,1}^2}, \sigma_2^2\right) \\ -c_2 + \int_{-\infty}^{\infty} dy_2 N\left(y_2; \mu_2, \sigma_2^2 + \sigma_{meas,2}^2\right) V\left(\mu_1, \mu_2 + \frac{\sigma_2^2(\mu_2 - y_2)}{\sigma_2^2 + \sigma_{meas,2}^2}, \sigma_1^2, \frac{\sigma_2^2\sigma_{meas,2}^2}{\sigma_2^2 + \sigma_{meas,2}^2}\right) \end{cases} \qquad (5)$$

If repeated measurements are noninformative the decision tree is finite and junction problems can be solved straight forwardly from leaves to root. The decision tree at a node is shown in Figure 1.

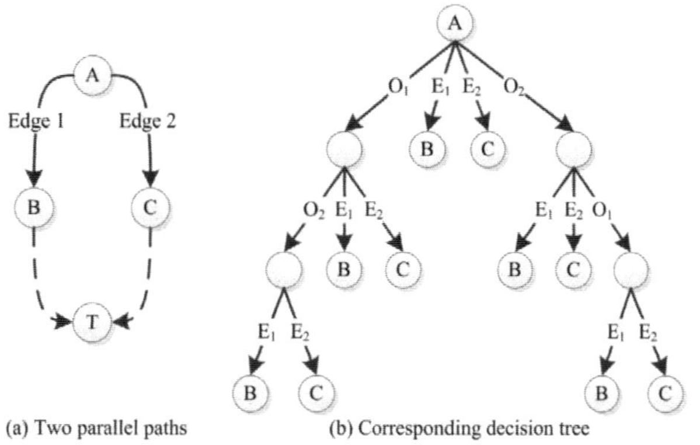

Figure 1. Sequential decisions at a junction of a directed acyclic graph when repeated measurements are noninformative. Prior optimal decisions from nodes B and C to target node T have been solved. In the case of additive objective function, decisions at A are made taking into account the measurement actions and choices of edges A-B and A-C and the optimal objectives B-T and C-T.

If the utility is cumulative – either the expected value of performance is maximized, or the utility U_0 is maximized and the performance terms are independent – a sequential route optimization on any directed acyclic graph can be solved by applying Bellman's principle of optimality both for informative and noninformative repeated measurements. This deterministic solution provides decision rules at each junction: to make measurement or not, and if measurement is made, how the next decision about measurement/advancing to next node depends on data obtained. However, the junction problems are computationally rather expensive as the prior expectations with respect to the measurement data of posterior optimal performances needs to be computed numerically. This limits practical solutions to maps where junctions have at most two exiting legs.

If the objective is to maximize the probability that the performance exceeds some given aspired value Q_S, $P(Q > Q_S)$, the problem can be solved backwards recursively as in the case of additive objective. However, the decisions at a junction obviously depend on how much performance has been accumulated when entering the junction. As this is a random variable not known when path is being planned, the optimal decisions must be solved for all values Q_S. Figure 2 shows an example of decisions at a node. First when entering the node there are four options in the root of Figure 1b. The probability that performance in these options exceeds Q_S (here: cost C is below a given aspired cost C_s) is given in Fig 2a. For example if $C_S = 33.6$, the optimal action is to measure the leg 2. Fig 2b shows $P(C < C_S)$ as a function of data $y^{(2)}$ obtained when measuring leg 2. If data obtained is $y^{(2)} = 33$, the optimal action is to measure also the leg 1. Correspondingly a decision rule between advancing along the route 1, respectively 2 can be given as a function of $y^{(1)}$. Fig 2c shows the entire system in which this decision takes place. the node considered is 5 after optimal decisions of its descendants 8 and 16 (trivial, being the target node) have been solved.

3.2. Optimal observation for occupancies in path planning

Consider an AMM traversing from a start to a goal location on a grid, where each traversed link between two adjacent locations results in a cost. Many real life planning problems can be formulated

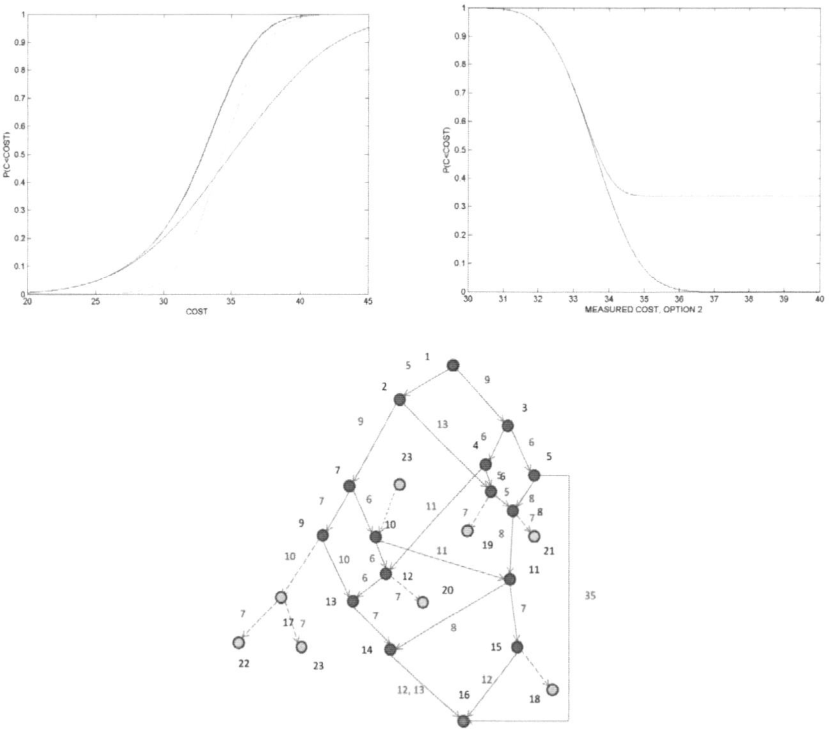

Figure 2. Maximizing the probability that the cost will not exceed a given aspired level. Top left: decision at a junction as a function of aspired cost; four options: measure leg 1 (blue), measure 2 (red), advance 1 (green), advance 2 (black). If aspired level of cost is 33.6, optimal is to measure 2. Top right: decision after measuring leg 2 as a function of measurement results; three options (colors as in top left). If measurement result is 33, next it is optimal to measure leg 1. A similar rule is established after obtaining measurement data from leg 1. Bottom: the entire path planning problem; initial site is 1 and target is 16. Red numbers indicate prior expected leg costs. The analysis presented in top figures is at node 5, after prvious optimal decisions have led to the route 1->3->5.

as path planning on an occupancy map with cost accumulation. We study the case in which the location rewards consist of known costs and unknown dynamic costs, dependent on Markovian binary occupancy states of the locations (high cost to penalize collisions). The location of the AMM at the grid is assumed known exactly. The AMM can choose between of moving to adjacent locations or observing the occupancies of some nearby locations. The npsotive and negative false observations are possible.

An approximate solution is generated as follows. First the paths from each location to goal location and the corresponding value functions are solved offline with the only the known rewards. Then the movement and observation actions are planned online over a short time horizon, typically 3-4 decision intervals, with the optimal value function from the first solution as the end reward. Typical result is shown in Fig. 3. When the agent is uncertain about obstacles not in the map, it chooses routes for which an observed obstacle can be by-passed from both sides. The computational complexity prevents longer on-line horizons, which in turn prevents the AMM to carry out complicated maneuvers to avoid observed obstacles.

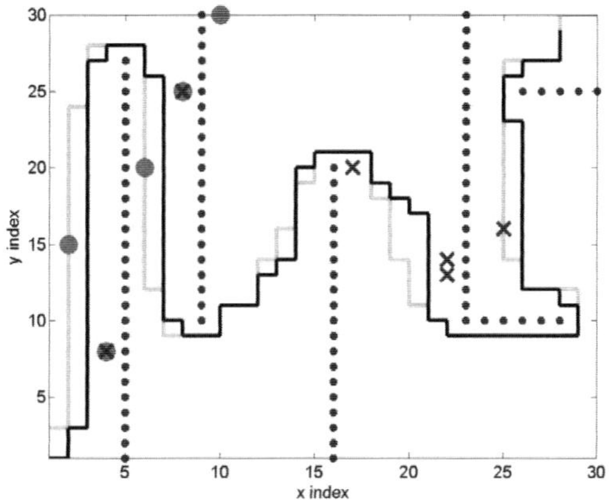

Figure 3. Path planning with a map in the presence of unmapped obstacles. Blue dots: obstacles in the map. Red dots: unmapped obstacles. Blue crosses: observations (true or false) about unmapped obstacles. Green line: optimal path according to known map. Black line: optimal route with optimal sensing for unmapped obstacles.

3.3. Optimal sensing of Linear-Quadratic-Gaussian systems

Quadratic objectives arise commonly in control problems. If the response of the system is linear and the initial state information is Gaussian, then the prior information about the end state is Gaussian. Furthermore, in risk neutral decisions, the optimal measurement action is independent of the estimate of the initial state and the optimal action on the system is independent on the initial state uncertainty [7]. Thus the optimal action and optimal sensing problems are decoupled and the optimal sensing can be solved as a policy offline. When the attitude is risk-aversive, the optimization of actions on the system and about sensing are coupled. In all risk-aversive forms the objective of sequential decisions is non-additive and Bellman's principle of optimality does not apply.

4. Conclusions

Operable sensing subsystem is a means of optimizing system performance. The formulation of risk-aversive planning of optimal sensing was discussed. Three classes of problems were given as examples. The non-neutral attitude towards risk was shown to lead radically increased complexity in sensing planning. In general the Bellman's principle of optimality in sequential decision making is no longer valid, except under exceptional circumstances. In LQG risk-aversive optimization sensing and optimal control are coupled. Some solutions to optimal sensing problems were outlined. This work is funded by Academy of Finland (project O3-SAM, 268152).

References

[1] Najemnik, J., & Geisler, W. S., Optimal eye movement strategies in visual search, Nature, 434(7031), 387-391 (2005).

[2] Johnson L., Sullivan B., Hayhoe M., Ballard D., Predicting human visuomotor behavior in driving task, Phil. Trans. R. Soc. B 369 20130044 (2014).

[3] Raiffa H., Schlaifer R., Applied Statistical Decision theory, Ch. 4.5, Wiley Classics (2000).

[4] see e.g. Thrun S., Burgard W., Fox D., Probabilistic Robotics, Ch. 17, MIT Press (2005).

[5] French S., Rios Insua D., Statistical Decision Theory, Kendall's Library of Statistics, Ch 2, Arnold Publishing (2000).

[6] Lim S., Sommer C., Nikolova E., Rus D., Practical Route Planning Under Delay Uncertainty: Stochastic Shortest Path Queries, Robotics, Science and Systems VIII (2012)

[7] Meier III L., Peschon J., Dressler R. M., Optimal Control of Measurement Subsystems, IEEE on Automatic Control, AC-12, 528-536 (1967).

IMEKO IOP Publishing
Journal of Physics: Conference Series **588** (2015) 012024 doi:10.1088/1742-6596/588/1/012024

On the input of a measurement process

Luca Mari[1,#], Alessandro Giordani[2]

[1] School of Engineering, Università Cattaneo – LIUC, Castellanza (VA), Italy
[2] Dept. of Philosophy, Università Cattolica, Milano, Italy

lmari@liuc.it

Abstract. It is assumed sometimes that the input of a measurement, and therefore the entity with which a measuring system interacts, is a quantity value, possibly the (true) measurand value, and from this hypothesis the model of ideal measurement as an identity process is formulated. In this paper we show that this position is based on an inappropriate superposition of quantities and quantity values, and therefore should be discarded.

1. Introduction

As any process, measurement can be firstly characterized as a black box that transforms inputs to outputs. Agreeing upon the nature of such entities is then the preliminary step of a top-down analysis or development process. Even at a so general level two incompatible positions are maintained on the input of a measurement process, a situation that might hinder a smooth, mutual understanding and therefore would be beneficially solved. Our aim in this paper is to argue about this duplicity and to show how the two positions can be reconciled.

A widespread standpoint on measurement is presented thus: «The input to the measurement system is the true value of the variable; the system output is the measured value of the variable. In an ideal measurement system, the measured value would be equal to the true value.» [1]. This position incorporates the customary intuition that measurement is generally affected by errors, which produce unwanted changes that should be minimized, so that in the ideal case in which all errors have been eliminated all changes are eliminated in turn. The immediate consequence is that an ideal measurement process is mathematically modeled as an identity function m, $m(x) = x$, where x is the (true) value of the measured quantity.

This characterization has been further and more analytically developed [2] by decomposing the mathematical model of measurement into two basic stages, called "observation" and "restitution", where «Observation is a mapping (a function) from the measurand value x to the instrument indication y, i.e., $y = f(x)$, where function f can be obtained by calibration. Restitution is a mapping from the instrument indication to the measurement value $\hat{x}$, which may be obtained by inverting f, i.e., $\hat{x} = f^{-1}(y)$. Measurement is the concatenation of the two, i.e., $\hat{x} = f^{-1}[f(x)]$, which, in the ideal uncertainty-free case, yields identity, $\hat{x} = x$.». Two interesting points emerge here. First, the term for the input entity is not "true value" but simply "value", perhaps in accordance with the *Guide to the expression of uncertainty in measurement* (GUM), where the term "true value" is avoided because the word "true" "is viewed as redundant" [3] (the argument of the GUM is not so clear, given that it seems to be

[#] One of the authors is a member of the Joint Committee on Guides in Metrology (JCGM) Working Group 2 (VIM). The opinion expressed in this paper does not necessarily represent the view of this Working Group.

Content from this work may be used under the terms of the Creative Commons Attribution 3.0 licence. Any further distribution of this work must maintain attribution to the author(s) and the title of the work, journal citation and DOI.
Published under licence by IOP Publishing Ltd

unable to distinguish between, e.g., measured and specified values; on the other hand the subject of "truth" of values is immaterial for what follows). Second, the entity that in the first quotation was generically mentioned as a variable is presented here a measurand, a crucial concept that the *International Vocabulary of Metrology* (VIM) defines "quantity intended to be measured" [4].

The model implied in [1] and [2] is then:

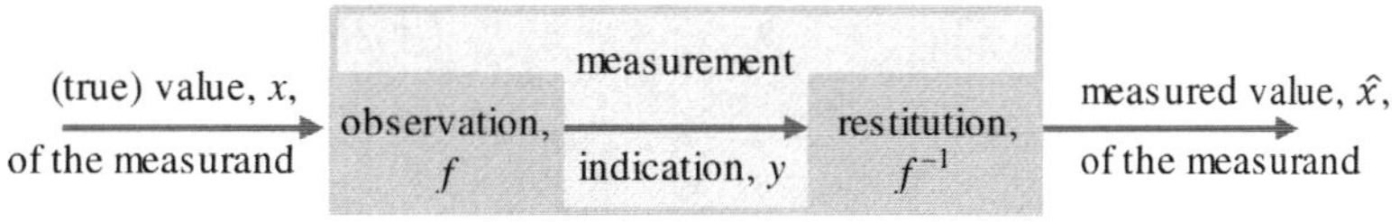

This is simple and elegant, but – we argue here – *too* simple. Indeed, the assumption that measuring systems interact with (true) values critically hides the basic feature of measurement of being a process mapping *properties* to property values or, more specifically, *quantities* to quantity values (the distinction between quantities and quantity values is developed in [5]). For example, according to the VIM measurement is "a process of experimentally obtaining one or more quantity values that can reasonably be attributed to a quantity" [4]. While the measured value, $\hat{x}$, is attributed to the measurand, the input quantity in the process cannot be generally expected to be the measurand itself, in the sense that what is experimentally under measurement can be different from what is intended to be measured. In the assumption of ideality, then:

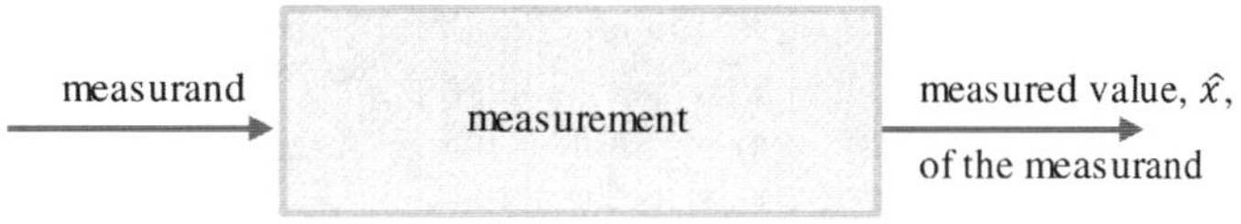

In order to highlight the difference between these two standpoints, we will develop the hypothesis that measurement, a process of information *acquisition*, is presented in the quotations above as structurally analogous to a process of information *transfer*, and measuring systems as structurally analogous to communication systems, where in fact the basic problem is to infer the entity x that was in input to the system from the channel output y, i.e., what was sent from what has been received. Clearly, sometimes similar can be said also about measurement, aimed at inferring a measured value x for the measurand from the obtained indication value y. Nevertheless the analogy fails in a critical point. But let us develop it for the moment.

2. Information transfer and information acquisition

An ideal transmission channel has null equivocation, because it does not lose information, and null noise, because it does not generate information: under the hypothesis that the channel is ideal, the knowledge of its behavior and the output is sufficient to infer the input with certainty. For example, let us suppose that x is the value of a physical quantity, say an electric tension, and that the channel attenuates the input signal at the receiver end uniformly by a factor of $k > 1$, so that its behavior is modeled as $y = f(x) = x/k$ (ideality is manifest here: infinite bandwidth and zero noise). By means of a suitable characterization of the channel (the functional analogous of calibration) the factor k is evaluated, so that the input can be "reconstructed" from the output as $\hat{x} = ky$, i.e., $\hat{x} = f^{-1}(y)$, exactly as proposed above for measurement. In fact, ideal transmission "yields identity", $\hat{x} = x$.

At the core of this interpretation there is the assumption that both the input and the output of the process (of measurement, of transmission) *are values*. This arises the question of the usefulness of an experimental process (ideally) implementing an identity function. In the case of transmission channels the answer is clear, and can be further highlighted by a second, parallel analogy, about information storage ("memory") systems.

The task of an *information transmission* channel is to transfer information at a distant location. It operates according to a two stage process. In the first stage its input and output (channel write and read) occur in different places, p_1 and p_2, so that the model of this stage is: $y_{p_2} = f(x_{p_1})$ If f is known, in the second stage f^{-1} is applied to y_{p_2} (then in p_2), and the channel behavior is maximally useful because the two equated values: $\hat{x}_{p_2} = x_{p_1}$ refer to different places.	The task of an *information storage* system is to maintain information through time. It operates according to a two stage process. In the first stage its input and output (memory write and read) occur in different times, t_1 and t_2, so that the model of this stage is: $y_{t_2} = f(x_{t_1})$ If f is known, in the second stage f^{-1} is applied to y_{t_2} (then in t_2), and the channel behavior is maximally useful because the two equated values: $\hat{x}_{t_2} = x_{t_1}$ refer to different times.

Hence identity is justified, and in fact desirable, for information transfer in space (i.e., transmission) and in time (i.e., memorization): the value that was in input is obtained identical *here* (transmission) or *now* (memorization).

(The behavior of information transfer systems is usually modeled even more abstractly, in terms of transfer of information entities such as "symbols" and sequences of them, i.e., "messages" (according to Shannon, «The fundamental problem of communication is that of reproducing at one point either exactly or approximately a message selected at another point.» [6]), so that quantity values act as carriers for such entities. This level, implying an invertible encoding of symbols to quantity values, is immaterial here and will not be further considered.)

On the other hand, this two stage model is too rudimentary, as it neglects a crucial difference between the first and the second stage, and can be refined by acknowledging that:
– the first stage (transmission, corresponding to observation above) is experimental: it is a *physical* process of transmitting a signal, that eventually will be interpreted;
– the passage from the first to the second stage is based on a mathematical function f that is just a model of the first stage, not the stage itself;
– the second stage (reconstruction, or "restitution" as it called above) is computational: it is a *numerical* process of inverting f and applying f^{-1}.

These points are effectively presented in terms of *the distinction between quantities*, hereafter denoted by uppercase characters, *and quantity values*, denoted by lowercase characters [5]. In fact, of the first stage of an information transfer system, intended as a part of the empirical world (W), two levels of modeling can be given:
(M1) *quantitative*: a device that causally changes its state in response to an external perturbation, where the perturbation is an input quantity X and the changing state is an output quantity Y (not necessarily of different kind), such that, under specified conditions, e.g., $Y = f(X) = X/K$, the units of such quantities and therefore the numerical value of K being still undetermined;
(M2) *numerical*: a device whose X-Y behavior has been identified, e.g., as $Y = f(X) = X/K$, for given units of such quantities, so that the numerical value of K in units of X/Y, k, is known; when the value x of the quantity X is in input, the value $y = f(x) = x/k$ of the quantity Y is obtained in output.

It should be noted that at each level the involved input-output entities *are of the same ontological kind*:
– in M1 they are quantities;
– in M2 they are quantity values.

The fact that the same mathematical equation formally characterizes both M1 and M2 ($Y = X/K$ in the example) might lead to suppose that there is no difference between them. On the other hand, the fact that M1 and M2 are distinct is evident because the law, as identified in M1, is independent of the conventional choice of the units: even though the decision about what to transfer regards a quantity value x, what is transmitted is a quantity X (a question such as "are you transmitting 1.234 V or 1234 mV?" – where 1.234 V and 1234 mV are indeed distinct values – is devoid of experimental content). The mathematical entity at the level M1 is in fact a quantity equation (a "mathematical relation between quantities in a given system of quantities, independent of measurement units" [4]), independent of the values of the involved quantities and with free parameters: *it is a model sufficient to describe the experimental process*.

This is just a different way to present the fact that physical laws connect quantities, not quantity values, and in fact they are independent of the choice of units. In other terms, whether a physical law apply (or: is true) does not depend on the calibration of the involved devices.

Hence, again in the mentioned example, were an additive composition available for X and Y, what could be done independently of quantity values is the validation of the linear dependence of the two quantities, i.e., of the structure of the quantity equation. A quantity X_1, possibly of unknown value, is applied and the obtained quantity Y_1 is recorded. Then a different quantity X_2, say $X_2 = 2X_1$ (generated through a quantity X' such that $X' \approx X_1$ and then combining $X' \circ X_1$) is applied and again the obtained quantity Y_2 is recorded. Through purely experimental means the ratio $Y_2/Y_1 = j$ can be obtained (e.g., in the simple case Y_2 is a multiple of Y_1, by generating j clones Y' of Y_1 and showing that $(Y' \circ [\dots j \text{ times}] \circ Y') \approx Y_1$). Then the linearity of the system behavior requires the constancy of the ratios X_2/X_1 and Y_2/Y_1, i.e., in this case such that $j = 2$: this shows that what is invariant at the experimental level, i.e., M1, in this example are ratios of quantities, not quantity values.

On the other hand, the second stage, i.e., the reconstruction f^{-1}, operates on quantity values: it is performed at the level M2, and therefore it is possible only when the entire stack W-M1-M2 is available, i.e.:

– the system has been actually (i.e., experimentally) operated (W);
– its causal X-Y behavior has been identified (M1);
– a numerical value of k has been identified through system characterization (M2),

and the value y of the output quantity Y has been obtained.

This framework can be graphically presented as follows, where continuous arrows denote actual process actions (*what we do*), and dotted arrows modeled actions (*what we know*):

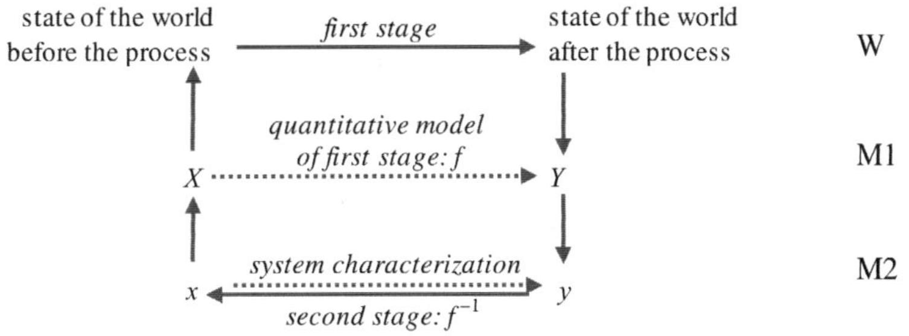

The vertical arrows from x to the state of the world before the process represent the *realization* of a quantity value by means of an empirical entity, and the vertical arrows from the state of the world after the process to y represent the *recognition* of a quantity value in an empirical entity.

This is an adequate description of an information transfer process. According to the quotations at the beginning of this paper this is also appropriate in the case of measurement: is it the case?

3. Information acquisition vs information transfer

Let us consider a basic difference between information transfer and information acquisition, measurement in particular. In the first case the value x is chosen by someone / something, i.e., the sender: hence it can be generally supposed to be perfectly defined, even though unknown to the receiver, before the process takes place. On the contrary, even those who endorse a metaphysics assuming that "numbers are in the world" plausibly admit that such numbers / values are *not* chosen (and the analysis above shows that at most "in the world" there are ratios of values, not values). In information transfer the value x is the *operative* starting point of the transfer process, from which the individual quantity X and more generally a state of the world are realized. On the contrary, the operative starting point of measurement is the experimental interaction between a state of the world, i.e., the state of the object under measurement in its environment, and the measuring instrument, e.g, between a body generating a force and a spring balance. While many measuring systems are based on transducers (the spring, in the example), their task is neither to perform a state transition (W, spring elongation), nor to transform the measurand X to an indication Y (M1, from a force to a length), nor to map the input value x to an output value y (M2), as instead it may be said of information transfer systems. In the case of measurement, transduction is just instrumental to the actual task of *acquiring information* on the measurand X and representing it as a measured quantity value $\hat{x}$ (neglecting uncertainties, as in the quotations above) [7]. This has nothing to do with an equality of values, $x = \hat{x}$, and in fact it is written:

$$X = \hat{x}$$

i.e., a quantity intended to be measured, X, is recognized equal to a quantity value, $\hat{x}$, e.g., force(applied by this object) = 1.234 N.

Understanding this equality is crucial ("equality gives rise to challenging questions which are not altogether easy to answer" [8]). It is not an identity, and it is not an equality of values (the left-hand term does not depend on a unit and is not a value). Rather, it is based on the, usual but subtle, observation that different expressions may have different meanings and nevertheless refer to the same entity. For example, $\sum_{i=0}^{\infty} 1/n!$ and exp(1) are clearly different concepts, but they refer to the same number, 2.718... Analogously, 'the force applied by this object' and 1.234 N are definitely different concepts, but might be recognized as different ways to refer to the same individual quantity, sometimes called their (common) magnitude.

This recognition is the outcome of a composite process such that:

(1) at the W level, the state of the object under measurement is transduced to the state of the measuring instrument;

(2) at the M1 level, the two states are modeled as quantities X and Y respectively, according to a mathematical model of transduction, $Y = f(X)$;

(3) in the passage between M1 and M2, the crucial move from an indication (as a quantity) to an indication value is accomplished, as modeled by a function $y = g(Y)$;

(4) at the M2 level, the values of the initial and the transduced quantities are introduced and the measured value is computed, according to $\hat{x} = f^{-1}(y)$.

By composing these functions we get $\hat{x} = f^{-1}(g(f(X)))$, where the generic concept of system characterization, $y = f(x)$, is here the result of instrument calibration and constitutes the pre-condition for the inversion of the numerical version of f (if the mapping g is left implicit and the notation $Y = y$ is used instead, then the previous formula becomes:

if $f(X) = y$ then $X = \hat{x}$, where $\hat{x} = f^{-1}(y)$

thus highlighting the remarkably symmetric structure of the measurement process).

The process is presented by the following modified version of the previous diagram:

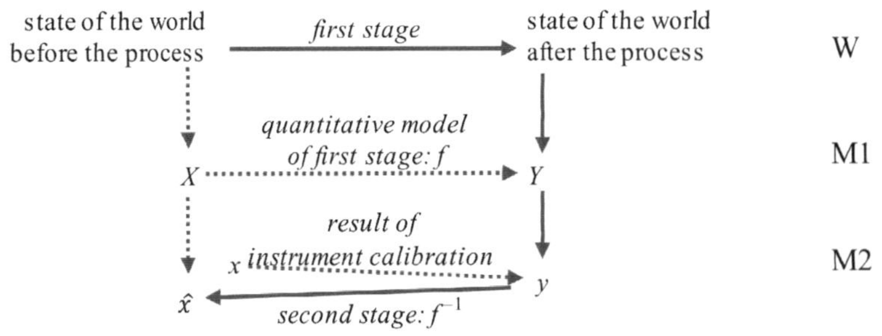

4. Conclusions

Measurement is loaded of stereotypes, and it is so a fundamental process that philosophical presuppositions seem to be hardly avoidable in understanding its basics. On the other hand, measurement is often perceived as arising only practical problems, so that these basics usually remain implicit. This could be the explanation of the peculiar fact that sometimes mutually incompatible positions emerge that are not reconciled and thus remain unsolved, under the hypothesis that this compresence does not actually affect the experimental operations of measurement. Whether this hypothesis is correct or not, a conceptually and terminologically consistent presentation of measurement and its results surely requires a consistent background. This applies, first of all, to a black box modeling of measurement, where then the nature of its input and output is unambiguously stated and agreed. Characterizing the nature of measurement input poses this kind of incompatibility problems, given that measurement input is alternatively intended as a (true) value of a quantity or a quantity of an object. We have argued here that the former assumes a model in which the processes of information transfer and information acquisition are mistakenly superposed: rather, measurement systems experimentally interact with quantities, not quantity values.

References

[1] Bentley JP 2005 (4th ed) *Principles of measurement systems* (Harlow: Pearson)
[2] Rossi GB 2012 Toward an interdisciplinary probabilistic theory of measurement *IEEE Trans. Instr. Meas* **61(8)** 2095–2106
[3] JCGM 100:2008 *Evaluation of measurement data – Guide to the expression of uncertainty in measurement* (GUM, originally published in 1993) Joint Committee for Guides in Metrology (http://www.bipm.org/en/publications/guides/gum.html)
[4] JCGM 200:2012 *International Vocabulary of Metrology – Basic and general concepts and associated terms* (VIM, 3rd Edition, 2008 version with minor corrections) Joint Committee for Guides in Metrology (http://www.bipm.org/en/publications/guides/vim.html)
[5] Mari L and Giordani A 2012 Quantity and quantity value *Metrologia* **49** 756–764
[6] Shannon C 1948 A mathematical theory of communication *The Bell System Technical Journal* **27** 379–423, 623–656
[7] Giordani A and Mari L 2012 Measurement, models, and uncertainty *IEEE Trans. Instr. Meas* **61(8)** 2144–2152
[8] Frege G 1892 (Engl transl Black M) *On sense and reference* (http://en.wikisource.org/wiki/On_Sense_and_Reference)

IMEKO IOP Publishing
Journal of Physics: Conference Series **588** (2015) 012025 doi:10.1088/1742-6596/588/1/012025

Using *MINITAB* software for teaching measurement uncertainty

Andrea Zanobini
Department of Information Engineering, University of Florence (Italy)
Via Santa Marta, 3 – 50139 Florence, Italy

Email:andrea.zanobini@unifi.it

Abstract. The concept of measurement uncertainty should be regarded not only related to the concept of doubt about the validity of the measurement result, but also to the quantization of this concept. In this sense the measurement uncertainty is that parameter associated with the result characterizing the dispersion of the values that could reasonably be assigned to the measurand (or more properly to its representation through a model). This parameter may be for example a multiple of the standard deviation but especially, and more importantly, the half width of an interval with a predetermined level of confidence or trust. In these terms in this paper I attempt, with the help of *MINITAB* software, to analyze this parameter; with simple and quick operations to evaluate the mean, the standard deviation and the confidence interval and by the use of several plotted graphs.

1. Introduction: the statistical procedures in educational research

The approach to learning statistical procedures presented in this paper will be useful through the behavioural sciences. Furthermore, and more importantly, educational research is an especially suitable source of examples for an introduction to statistical methods because of the multidisciplinary nature of research in education. An investigator studying students, teachers, and schools is sometimes a sociologist, frequently a psychologist, and at times anthropologist, psychiatrist, historian, economist, or architect.

The recent explosive expansion of activity in educational research, girded strongly by public and private financial support, is another justification for attention to this field. One of the chief concern is that there is now more money available to conduct educational research than there are people prepared to design, execute, report, and utilize such investigations. This need has not been met despite the fact that the new support has attracted to educational problems many competent scholars from many disciplines who previously had neither the money nor inclination to devote themselves to the study of educational problems.

Increased statistical sophistication will help reduce another difficulty in current educational research. The fact is that rather statistics model for guiding inquiry have dominated research activity in education. Statistical analysis plays a dominant role in two major types of educational research activities: ***surveys*** and ***experiments***. Although the student may not clearly see the distinction between them, he will see that there are many varieties of surveys and experiments. Some surveys, for example, involve simple counts on a single variable, while others comprise measures of many variables and their inter-relationships. However these data will serve as practice data so that the student can "play scientist" as he begins to master his new statistical tools.

Content from this work may be used under the terms of the Creative Commons Attribution 3.0 licence. Any further distribution of this work must maintain attribution to the author(s) and the title of the work, journal citation and DOI.
Published under licence by IOP Publishing Ltd

2. An educational experiment

An educational experiment occurs when a scientist assigns subjects at random to two or more treatment groups, imposes a different educational treatment on each group, and then measures the groups to see whether the different treatments have produced important differences among the groups on one or more outcome variables.

After the period in which the group of subjects have received different educational treatments, the scientist measure all subjects on one or more outcome variables (frequently these are achievement tests), and analyzes the scores to decide whether the groups differ in outcomes.

We can best define ***statistics*** by considering the problems the scientist faces in performing this analysis of outcome measurements. First, he has to characterize each group by some summary of the scores (***summary statistics***) for the subjects in the group. Computing an average score for the group is an example of such a summary. Statistics is in part the science of summarizing masses of measurements to obtain useful summary ***descriptions of trends in the data***, such as characterization of groups and of differences among groups. Second, the scientist has to decide whether the differences among the groups on the descriptive statistics could be due simply to the failure of his initial randomization procedure to equate the groups prior to the different treatments. He knows that if he had given precisely the same treatment to all the groups, they would nevertheless differ in outcomes by chance alone, as a result of accidents in his randomizing procedure. The other part of statistics is the science of making ***inferences about the significance of trends in the data***, such as observed differences among groups. In short, statistics is the science of making ***descriptions and inferences*** from measurements.

3. A brief introduction to the *MINITAB* software for a statistical analysis

The statistical analysis with *MINITAB* software, which often relies on a good background knowledge of the phenomenon to be analyzed, requires a series of steps and the main are:

- Data Analysis with graphs
- Perform statistical analysis and procedures
- Evaluation of the quality of a measure

Here I give a brief introduction to the use of this program by inviting the reader to download the trial software [6] and to carry out the .pdf file "*Meet Minitab*" a self-training mini-course that allows a basic use of all the main functions. When we open the application, the graphical user interface - GUI - Graphical User Interface - will present two windows.

The first is related to the current session and shows the results of the analysis in text format. It is also possible to enter commands in the form of instructions rather than using the menu bar.

The data window, very similar in appearance to a spreadsheet, contains the open worksheet. Looking at that sheet you can see the classic table structure of spreadsheets with the data arranged in columns, data that can be of three types: numeric, text, date. At any time you can change the properties of these data from the menu bar.

3.1 An example: the evaluation of the quality of a measure connected to its uncertainty

For quality of a measure we refer to the degree to which products or services meet the needs of customers. The ultimate goal of every manufacturer is to reduce the rate of defective products and, therefore, their variability; they produce manufacturing products that are compliant much as possible with the specifications [4].

MINITAB offers many methods to assess the quality of a measure qualitatively and quantitatively, definetly its uncertainty: control charts, statistical tools, quality planning, process capability and reliability.

Control charts are also useful for monitoring the stability of the process in time and to detect the presence of anomalies in the same process.

For example in **Figure 1,** easily obtained with *MINITAB* software, we can see:

- Centerline at the average value of the statistic
- Upper limit control (UCL) at 3σ above the center line
- Lower limit control (LCL) at 3σ below the center line

In this case, the student is asked to perform the exercises on "*Meet Minitab*" to understand the utility of control charts in a measurement process also to evaluate its measurement uncertainty.

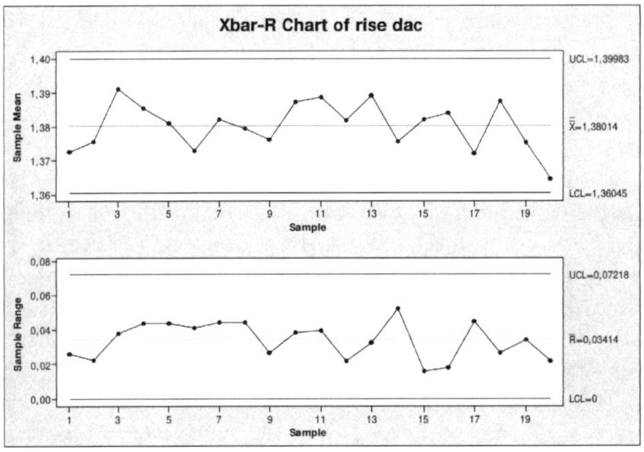

Figure 1. A *MINITAB* control chart plot for the readings of the rise times of the step response in a digital oscilloscope. How this plot is it connected to the measurement uncertainty?

4. Uncertainty evaluation with the help of statistical theory

The objective of a measurement is to determine the value of the measurand, that is the specific quantity subject to measurement. In general, no measurement or test is perfect and the imperfections give rise to errors in the result. Consequently, the result of a measurement is only an approximation to the value of the measurand and is complete only when accompanied by a statement of the uncertainty of the approximation [1]. The International Vocabulary of Basic and General Terms in Metrology defines uncertainty as "a parameter associated with the result of a measurement, that characterizes the dispersion of the values that could reasonably be attributed to the measurand" [2].

Taking into considerations all these sources of variability it is possible to elevate the quality level of the whole measurement process. It is important to reduce the uncertainty in the information and to enhance the detail of the measurand.

First of all it is important to point out that the measurement process cannot generate a rational number as result; in fact, some information connected to its validity and to the quality of the whole measurement process are necessary. Both these numbers should be referred to their measurement unit.

Now it should be clear that a measurement process should be able to transform an unknown measurand in a measurement result, with all the information necessary to state its quality level.

In this process the statistical theory plays a fundamental role but some questions have to be taken into account in the parameters estimation: what about expected value, variance and standard deviation by means of sample statistics?

4.1 Measurement, probability and random variables

Concerning the uncertainty it is possible to identify two main contributes: the first is originated by random effects *(GUM - type A)*; the other one depends on systematical effects *(GUM – type B)* [2]. In particular:

- The first contribute is essentially related to the stochastic variations of the influence quantities which will give different observation obtained in *repeatability* conditions. This means that the same person can collect these observations, in an independent way, with the same procedure, the same instrument, in the same conditions of use and in a restricted time interval.
- The second contribute is essentially related to those effects that are identical every time the measure is repeated and it depends, for example, on instrument accuracy and resolution, reference samples, and others.

If these two contributes are present, it is often possible to correct them and, eventually, to delete many of these effects. So the measurement can be treated like a *random variable, M,* distributed within the measurement interval. This random variable determines the measurement process and so the quality of the test results. In the paragraph 6, I will introduce the *confidence level* to attribute to every single event, associated to M in $S = \{ m_{\min} \le M \le m_{\max} \}$, the space domain of the results.

Obviously the maximum confidence level, equal to one, can be assigned when M belongs to S; vice versa the minimum confidence level, equal to zero, is assumed when the values of M do not belong to the space S of all the possible measurement results. Consequently, it is logical to assign a real positive number, in the range from zero to one, to the confidence level defined in a subinterval $\{ m_a \le M \le m_b \}$ of S; this number is named *probability* [5].

To each M is associated a probability distribution, or a function of random events which represents the probability that a measure belongs to one of the possible subintervals in which it is possible to divide the space S of all the possible measurement results.

5. Gauss/Laplace or Normal distribution

The probability distribution of the normal random variable M is expressed by:

$$f_M(m) = \frac{1}{\sigma\sqrt{2\pi}} e^{-\frac{(m-\mu)^2}{2\sigma^2}} \qquad -\infty < x < +\infty$$

Note that $f_M(m)$ contains two statistical parameters: μ is the expected value and σ denotes the standard deviation. Such distribution is denoted as *Gauss/Laplace or Normal distribution,* its plot, easily obtained with *MINITAB* software again, is shown in **Figure 2** and its notation is $M \sim N(\mu, \sigma^2)$ [3].
The graph is symmetrical with respect to μ, with higher concentration of measurement values near μ. It also appears that the density is the same for the two symmetrical points of μ; the points on the graph at which correspond μ-σ and μ+σ in the abscissa, represent the points of inflexion of the curve.
For different values of the mean μ and same σ, the plot changes in the position with respect to the abscissa. Instead, with different σ values and same μ, the graph changes in shape maintaining the symmetrical condition; in other words, the higher σ is, the higher is the dispersion of measures close to μ, and vice versa.

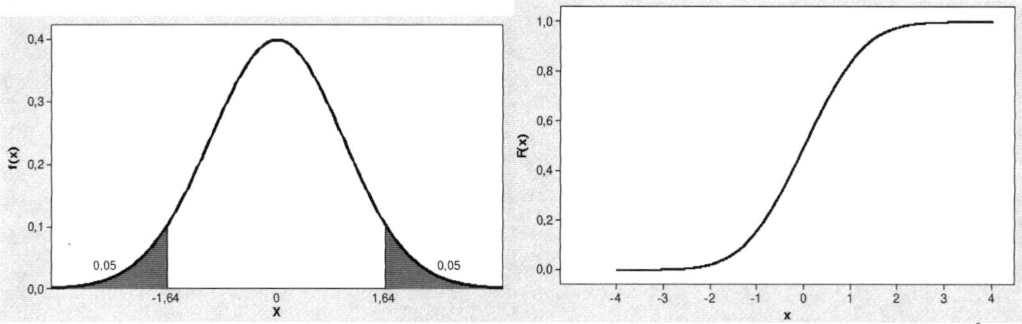

Figure 2 Probability density function (PDF) for Gauss/Laplace distribution $M \sim N(\mu, \sigma^2)$ and Cumulative distribution function for Standardized Gaussian $Z \sim N(0, 1)$.

The distribution function is:

$$P\{m_a \le M \le m_b\} = F_M(m_b) - F_M(m_a) =$$

$$= \frac{1}{\sigma\sqrt{2\pi}} \int_{m_a}^{m_b} e^{-\frac{(m-\mu)^2}{2\sigma^2}} dm = \frac{1}{\sqrt{2\pi}} \int_{m_a-\mu}^{m_b-\mu} e^{-\frac{m^2}{2}} dm = Z\left(\frac{m_b-\mu}{\sigma}\right) - Z\left(\frac{m_a-\mu}{\sigma}\right)$$

where $Z=(M-\mu)/\sigma$ denotes the standardized random variable [3].
Considering k the *coverage factor*, in **Table 1** the probability, in percent, that M is in the closed interval $\mu \pm k\sigma$ is shown in function of k.

$P\{\mu - k\sigma \le M \le \mu + k\sigma\}$	k
68,27	1
90	1,645
95	1,960
95,45	2
99	2,576
99,73	3

Table 1: Different probability values, in percent, in function of the coverage factor k.

6. Confidence interval of measurement results

The measure M, can be defined, as introduced in the previous paragraph, by the following expression:

$$P\{|M - E\{M\}| \le k\, u_M\} = P\{E\{M\} - k\, u_M \le M \le E\{M\} + k\, u_M\} = p$$

This is the probability that the measure M is comprised within an interval that is a function of its expected value and of the standard uncertainty u_M multiplied by the *coverage factor k*. Obviously the *confidence level p* should be greater than possible, better if near to one. The interval:

$$E\{M\} - k\, u_M \le M \le E\{M\} + k\, u_M$$

is called *confidence interval*; it represents the interval in which the probability to find a great number of possible values of M, within this interval, is near to one.
When the probability density function of M is known, it is possible to evaluate the confidence level p by the following expression:

$$p = \int_{E\{M\}-k\, u_M}^{E\{M\}+k\, u_M} f_M(m)\, dm$$

It is now possible to deduce the measurement result as the **uncertainty interval**, connected to the measurand with confidence level equal to p.
It is possible to write:

$$P\{m_\alpha \le M \le m_{p+\alpha}\} = \int_{m_\alpha}^{m_{p+\alpha}} f_M(m)\, dm = F(m_{p+\alpha}) - F(m_\alpha) = p$$

where α is a value, comprised between zero and one, defined by $0 \le p \le p + \alpha \le 1$. Finally:

$$F(m_\alpha) = P\{M \le m_\alpha\} = \alpha$$

The following **Figure 3**, easily obtained with *MINITAB* software again, represents an example of this interesting situation.

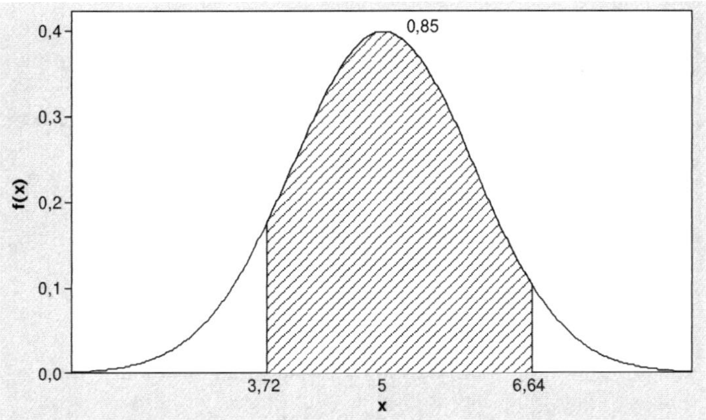

Figure 3 . PDF of a normal distribution with expected value equal to 5 and standard deviation equal to 1. The three regions under the curve, from the left, correspond respectively to α = 0,1; p = 0,85 and (1- p - α) = 0,05.

The choice of $\alpha=(1\text{-}p)/2$ generates, a **symmetric uncertainty interval**, with p as the uncertainty level, in the sense that the three sectors showed in Figure 3 seem identical.
If the probability distribution of M is symmetrical, in the sense that its PDF is also symmetrical, this means that the distribution is symmetrically centered around the expected value of M, and the borders of the narrower interval, with confidence level equal to p, are equidistant.
In this case, introducing again the standard uncertainty of M, $u_M = \sqrt{Var\{M\}}$, the uncertainty interval can be written as $E\{M\} \pm ku_M$ where k is the coverage factor opportunely chosen to realize that $E\{M\} - ku_M$ and $E\{M\} + ku_M$ are respectively the so called *[(1-p)/2]-quantile* and *[(1+p)/2]-quantile* of the cumulative distribution function *F(m)*.
In case of asymmetrical distribution, all things change but, conceptually, the problem remains the same and in this case the value of α in should be different by *(1-p)/2* in order to have the minimum amplitude of $(m_{p+\alpha} - m_\alpha)$ and the uncertainty interval is the narrowest interval with a given confidence level p.

7. Conclusions

In this paper the use of *MINITAB* software is presented as a very useful tool to introduce selected examples of behavioral sciences and education connected to "real live" data for any student practice. With the help of this software the students understand to turn the data analysis of statistics over to a computer in order to free his intelligence for the study of human problems of theory, design and interpretation in a challenging theory of knowledge.

8. References

[1] WEB BIPM: *http://www.bipm.org/*
[2] International Organization for Standardization, ISO/IEC Guide 98-3:2008- Uncertainty of measurement-Part 3: Guide to the expression of uncertainty in measurement, ISO, 2008. WEB GUM: *http://www.bipm.org/en/publications/guides/gum.html*
[3] William Navidi, "*Probability and Statistics for engineering and sciences*", Mc Graw Hill, 2006.
[4] Douglas C. Montgomery, "*Statistical Quality Control*", Mc Graw Hill, January 2006.
[5] Athanasios Papoulis, "*Probability, R.V. and Stochastic Processes*", Mc Graw Hill, 2002.
[6] WEB *MINITAB*: *http://www.minitab.com*

Journal of Physics: Conference Series **588** (2015) 012026 doi:10.1088/1742-6596/588/1/012026

Influence of short-term sampling parameters on the uncertainty of the L_{den} environmental noise indicator

M Mateus[1], **J Dias Carrilho and M Gameiro da Silva**

ADAI/LAETA – Department of Mechanical Engineering, Faculty of Science and Technology University of Coimbra – Pólo II, 3030-201 Coimbra, PORTUGAL

E-mail: mario.mateus@adai.pt

Abstract. The present study deals with the influence of the sampling parameters on the uncertainty of noise equivalent level in environmental noise measurements. The study has been carried out through the test of different sampling strategies doing resampling trials over continuous monitoring noise files obtained previously in an urban location in the city of Coimbra, in Portugal. On short term measurements, not only the duration of the sampling episodes but also its number have influence on the uncertainty of the result. This influence is higher for the time periods where sound levels suffer a greater variation, such as during the night period. In this period, in case both parameters (duration and number of sampling episodes) are not carefully selected, the uncertainty level can reach too high values contributing to a loss of precision of the measurements. With the obtained data it was investigated the sampling parameters influence on the long term noise indicator uncertainty, calculated according the Draft 1st CD ISO 1996-2:2012 proposed method. It has been verified that this method allows the possibility of defining a general methodology which enables the setting of the parameters once the precision level is fixed. For the three reference periods defined for environmental noise (day, evening and night), it was possible to derive a two variable power law representing the uncertainty of the determined L_{eq} values as a function of the two sampling parameters: duration of sampling episode and number of episodes.

1. Introduction

The current draft document of the ISO 1996-2 standard [1] proposes two distinct sampling methods for estimating environmental noise indicators: one making use of long-term measurements and the other using short-term measurements. In both sampling methods, a measurement campaign may consist of several measurement episodes distributed throughout the assessment period. The details on how to design an appropriate sampling strategy, i.e., how to choose the number and duration of the measurement episodes, are, however, left unspecified in the standard. In addition, there is no guidance as to how the choice of the sampling strategy might influence the overall uncertainty in the determination of environmental noise indicators.

There is, in the literature, some indication as to what extent a required accuracy might be achieved with a particular sampling strategy [2–5]. Given the variability in propagation conditions and source emission characteristics from site to site, however, the sampling strategy that produces the best results in one particular site might not be adequate for a another site. With this problem in mind, the authors have recently proposed a systematic analysis method for assessing the adequacy of a sampling strategy for any given site [6]. The method is based on the bootstrap method, and it produces an estimate of the uncertainty associated with determining environmental noise levels from short-term sampling measurements, as a function of the duration and the number of measurement episodes. In the present

[1] Corresponding author: Tel.:+351 239 708 580; fax: +351 239 708 589
E-mail address: mario.mateus@adai.pt

Content from this work may be used under the terms of the Creative Commons Attribution 3.0 licence. Any further distribution of this work must maintain attribution to the author(s) and the title of the work, journal citation and DOI.
Published under licence by IOP Publishing Ltd

paper, the authors demonstrate the application of this method in quantifying the influence of the sampling strategy on the uncertainty of the long term environmental noise indicator L_{den}.

2. Mathematical formulation

In Ref [7] the expression for calculating the long term environmental noise exposure indicator appears:

$$L_{den} = 10 \times \log_{10}\left[\frac{1}{24}\left(12 \times 10^{\frac{L_d}{10}} + 4 \times 10^{\frac{L_e+5}{10}} + 8 \times 10^{\frac{L_n+10}{10}}\right)\right] \quad (1)$$

The indicators L_d, L_e e L_n are the long term noise levels for the reference periods: day; evening; and night. These, in turn, can be obtained by sampling during the long term period of assessment, taking into consideration that weather conditions can have a significant influence in the way sound energy propagates from source to receiver. ISO1996-2 deals with this by defining four meteorological classes with respect to the propagation of sound energy: M1 – unfavorable; M2 – neutral; M3 – favorable; and M4 – very favorable. If the a number of sample measurements n_{M_i} are performed for a particular weather condition M_i, for which the probability of occurrence p_{M_i} is known or can be estimated, the indicators for each reference period can be estimated from

$$L_{[d],[e],[n]} = 10 \times \log_{10}\left(\sum_i \mathrm{p}_{\mathrm{M}_i,[d],[e],[n]} \times 10^{0.1 \mathrm{L}_{\mathrm{M}_i,[d],[e],[n]}}\right) \quad (2)$$

By the law of propagation of uncertainties, the combined uncertainty of the indicators $L_{[d],[e],[n]}$ is

$$u\left(L_{[d],[e],[n]}\right) = \left\{\sum_{i=1}^{n}\left[\frac{\partial L_{[d],[e],[n]}}{\partial L_{Mi}} \times u(L_{Mi})\right]^2 + \sum_{i=1}^{n}\left[\frac{\partial L_{[d],[e],[n]}}{\partial p_{Mi}} \times u(p_{Mi})\right]^2\right\}^{1/2} \quad (3)$$

Considering the results presented in (3), it is possible to write the expression for the combined uncertainty of the indicator L_{den}, as follows: Assuming independence between the different variables involved, as in Ref [1], the application of the methodology expressed in the GUM [8], considering also due to sampling component, hereafter denoted by u_{samp}, the following expression is obtained for the uncertainty of sound level L_{den}:

$$u(L_{den}) = \left\{\left(\frac{\partial L_{den}}{\partial L_d}\right)^2 \left[u^2(L_d) + u_{samp,d}^2\right] + \left(\frac{\partial L_{den}}{\partial L_e}\right)^2 \left[u^2(L_e) + u_{samp,e}^2\right] + \left(\frac{\partial L_{den}}{\partial L_n}\right)^2 \left[u^2(L_n) + u_{samp,n}^2\right] + u_{loc}^2 + u_{slm}^2\right\}^{1/2} \quad (4)$$

were:

- $\frac{\partial L_{den}}{\partial L_d}$, $\frac{\partial L_{den}}{\partial L_e}$ and $\frac{\partial L_{den}}{\partial L_n}$ are the sensitivity coefficients, i.e. they express the change of L_{den} due to changes in L_d, L_e and L_n, respectively;
- $u(L_d)$, $u(L_e)$ and $u(L_n)$ are the standard uncertainty associated with the measurement performed to characterize the noise in each reference period;
- u_{loc} is the standard uncertainty associated with the location of the measurement point;
- u_{slm} is the standard uncertainty associated with the measurement chain;
- u_{samp} is the standard uncertainty associated with the process of sampling in the period under review.

It is assumed that the components of uncertainty associated with the measuring equipment and the placement of the microphone are constant, i.e., always the same measuring equipment is used and the sampling is performed on the same site. Regarding the uncertainty due to sampling, it is assumed that it should correspond separately to those resulting from the sampling strategy followed in each period.

Generically, the sensitivity coefficients assume the following expression, were w is 0 dB, 5 dB or 10 dB, for the day, evening and the night periods, respectively:

$$\frac{\partial L_{den}}{\partial L_i} = \frac{t_i 10^{0.1(L_i+w)}}{12\times10^{0.1(L_d)}+4\times10^{0.1(L_e+5)}+8\times10^{0.1(L_n+10)}} \tag{5}$$

In Ref. [6], when implementing a fitting process of the observed experimental data with an analytical function of two variables, it was found that a generic power function expression may be used for the three reference periods, varying only the numeric coefficients. The expression is:

$$u_{samp} = c1 \times \Delta T^{\alpha 1} \times N^{-(\alpha 2 + \alpha 3 \times ln(\Delta T))} \tag{6}$$

where α_1, α_2 and α_3 are numerical coefficients found from least squares fitting to locally obtained data.

3. Materials and methods

The site where the study took place is relatively plane with no major obstacles between the source and the adjacent buildings. The dominant noise source is a double lane collecting road following the most recent bridge built over the Mondego river: the Rainha Santa bridge, in Coimbra, a city in the center of Portugal.

Figure 1 shows an aerial photograph of the site, obtained from the web application Google Earth. Superimposed on the photograph is a wind rose centered on the measurement location, showing that the dominant wind direction is from northwest. The most probable velocity is in the range [1.0 m/s; 2.0 m/s].

The annual average daily traffic is about 37,000 vehicles per day, with approximately 10% of heavy vehicles. The traffic is mostly fluid and decelerating, in the west-east direction, from 70 km/h, over the bridge, to 50 km/h, after the bridge. In the east-west direction, the traffic is also fluid and accelerates from 50 km/h to 70 km/h.

The measurement system was built around a National Instruments (NI) data acquisition system and LabVIEW application development platform. The signal from a Brüel&Kjær ½" diameter microphone model 4189, with pre-amplifier model 2671, is converted to digital format by a NI USB-9233 data acquisition board. The microphone was mounted on the exterior of a window, at the third floor level of the Laboratory of Industrial Aerodynamics, of the University of Coimbra. The distance to the road is about 150 m.

Figure 1. Aerial photograph of the noise assessment site, with a wind rose centered on the measurement location.

In order to determine the numerical coefficients in (6), a consecutive 17-day $L_{Aeq,5min}$ time series was selected from a continuous 4-year noise monitoring campaign. Figure 3 compares the 17-day daily average levels with the 4-year daily average levels, showing that the 17-day sample is representative of the expected daily noise level variation at the site. The two dotted lines represent an uncertainty band for the 4-year measurements, considering that, in each period during the day, the noise equivalent level has a Gaussian distribution. A coverage factor equal to 2 (95% probability) has been used. Table 1 shows the numerical coefficients obtained from fitting (6) to the 17-day noise sample.

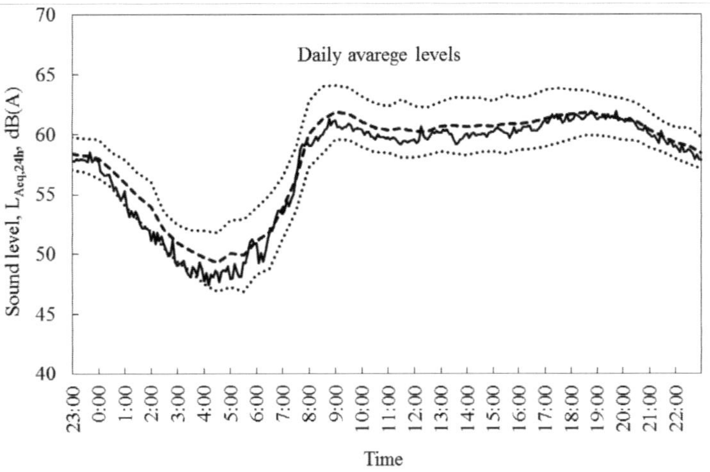

Figure 3. Comparison between the 17-day daily average levels (solid line) and the 4-year daily average levels (dashed line)

Table 1 – Numerical coefficients of equation (4.1) for each reference period.

Coef.	Day	evening	Night
$c1$	2.5224	2.2885	4.6034
$\alpha1$	-0.086	-0.181	-0.068
$\alpha2$	0.408	0.3695	0.4686
$\alpha3$	0.010	0.034	0.002

4. Results and discussion

In Ref [1] the values of uncertainty accounted as fixed components to the sound level meter and microphone placement measurement, should be respectively, 0.5 dB and 0.4 dB. For the latter component its value must be recorded in the case of noise from road noise sources, and whose impact on the mounting location of the microphone is not grazing incidence.

Accounting for only these two components of uncertainty, in which we consider the remaining components in (4) to be zero, leads to a fixed component of the expanded uncertainty $U(L_{den})$ which cannot be less than 1.3 dB.

Following the measurement model assumed in this paper, the accounting of the u_{samp} component will always be affected by the sensitivity coefficients associated with each period. These coefficients have a strong dependence of the noise levels of each period, as shown in Table 2.

Table 2. Sound levels for three different situations, and their respective sensitivity coefficients, used to analyze the influence of sampling in the expanded uncertainty $U(L_{den})$.

Case	L_d (dB(A))	L_e (dB(A))	L_n (dB(A))	$\frac{\partial L_{den}}{L_d}$	$\frac{\partial L_{den}}{L_e}$	$\frac{\partial L_{den}}{L_n}$
A	61	60	50	0.423	0.354	0.224
Ref	61	60	54	0.316	0.264	0.420
B	61	60	58	0.193	0.162	0.645

The accounting of the u_{samp} component, for the three cases considered, the results are presented in Figure 10, providing evidence of the influence of sampling component in the expanded uncertainty $U(L_{den})$, also highlighting its dependence on the number of episodes and duration, as well as differences in the level of indicators L_d, L_e e L_n.

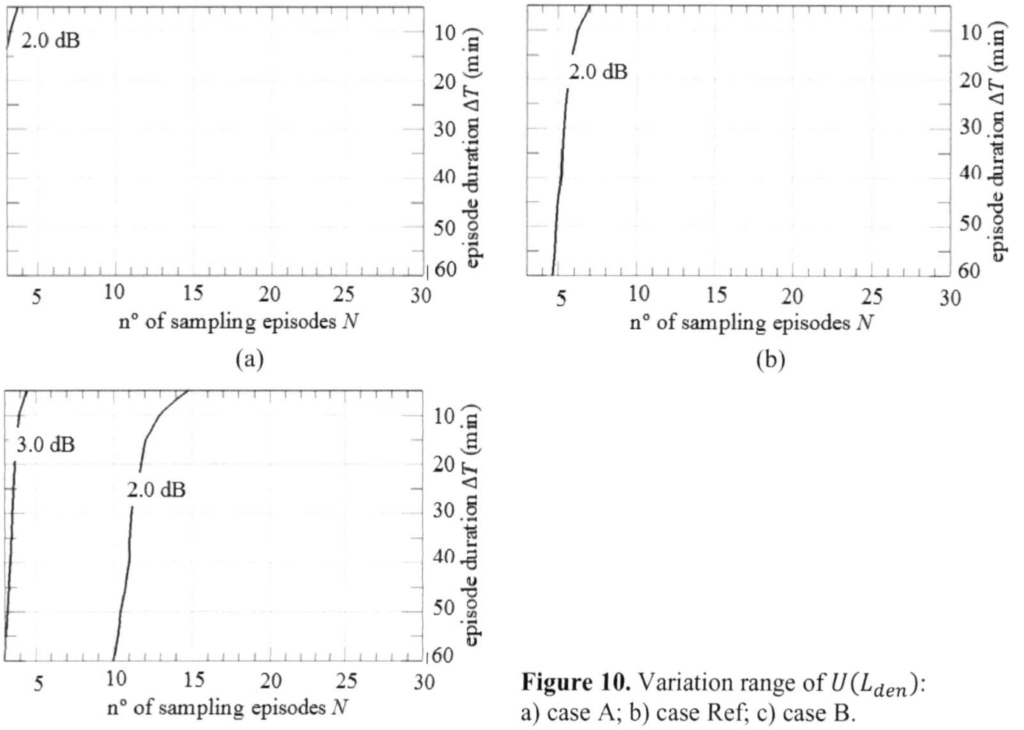

(a)
(b)
(c)

Figure 10. Variation range of $U(L_{den})$: a) case A; b) case Ref; c) case B.

The influence that the sound level of nighttime have on the results presented for $U(L_{den})$, is greater the smaller the difference to the levels of other periods. This feature results from the weight that is given by the respective sensitivity coefficient and also due to the uncertainty of the noise level in this period.

The results presented in these figures refer only to fixed components, not yet part of the uncertainty inherent to the repeatability of the measurements, which will naturally increase further the limits presented.

5. Conclusion

A systematic analysis method, based on the bootstrap method, is proposed by the authors for assessing the quality of sampling strategies for estimating long-term environmental noise indicators at any given site.

The methodology developed allows to determine the best sampling strategy that meets a given requirement regarding measurement uncertainty. Implicitly it follows also that distinct effort should be placed in measurement strategies depending on the contribution that each component has on the final result as well as the precision value associated with it.

Acknowlegment

This work has been partially supported by Fundação para a Ciência e a Tecnologia (FCT) under project Pest-OE/EME/LA-0022/20011. First author acknowledges FCT PhD grant Ref SFRH/BD/37828/2007.

References

[1] Draft 1st CD ISO 1996-2:2012, Acoustics – description, measurement and assessment of environmental noise - Parte 2: determination of environmental noise levels; 2012.

[2] Gaja E., Gimenez A., Sancho S., Reih A. Sampling Techniques for the estimation of the annual equivalent noise level under urban traffic conditions, Appl Acoust 2003; 64(1); 45-53.

[3] Kuehne D. Long-term Leq Errors Expected and how long to Measure (Uncertainty and Noise Monitoring), ForumAcusticum 2005, Budapest (Hungary), Aug 29 - Sep 2; 2005.

[4] Brambilla G., Lo Castro F., Andrea C., Verardi P. Accuracy of temporal samplings of environmental noise to estimate the long-term Lden value, Internoise 2007, Istanbul (Turkey), August 28-31; 2007.

[5] Jagniatinskis A., Fiks B. Assessment of environmental noise from long-term window microphone measurements, Appl Acoust 2014; 76; 2014.

[6] Mateus M, Dias Carrilho J and Gameiro da Silva M. Assessing the influence of the sampling strategy on the uncertainty of environmental noise measurements through Bootstrap method. Manuscript submitted for publication (copy on file with corresponding author).

[7] Directive 2002/49/EC, of the European Parliament and of the Council of 25 June 2002 relating to the assessment and management of environmental noise; 2002.

[8] JCGM 100, Evaluation of measurement data – Guide to the expression of uncertainty in measurement, Paris: Joint Committee for Guides in Metrology (JCGM/WG 1), 2008.

Journal of Physics: Conference Series **588** (2015) 012027 doi:10.1088/1742-6596/588/1/012027

A methodology of healthcare quality measurement: a case study

F Pecoraro[1], D Luzi[1], M Cesarelli[2] and F Clemente[3]

[1] Institute for Research on Population and Social Policies, National Research Council, Rome, Italy

[2] Department of Electrical Engineering and Information Technology and Interuniversity Centre of Bioengineering of the Human Neuromusculoskeletal System, University of Naples, "Federico II", Naples, Italy

[3] Institute of Biomedical Engineering, National Research Council of Italy, Rome, Italy.

Contact person: Fabrizio Pecoraro, f.pecoraro@irpps.cnr.it

Abstract. In this paper we present a comprehensive model for quality assessment taking into account structure, process and outcome dimensions introduced in the Donabedian framework. To test our hypothesis a case study based on the Italian healthcare services is reported focusing on the analysis of the hospital bed management and on the phenomenon of both active and passive patient mobility.

1. Introduction

The demand of high quality healthcare services is constantly increasing due to global demographic changes such as higher life expectancy with greater disease complexity. However, the current budget reductions make it necessary to optimize resources, improve organization and procedures, maintaining and/or improving the quality of healthcare service delivery. Additionally the continuous introduction of administrative and legal constraints (although oriented to improve services) further complicates the consolidation of good practices.

In this framework the identification of quality indicators is essential to assess healthcare services identifying strength and weakness, also to improve their efficiency and effectiveness. To perform this task it is necessary to collect data from different sources, classifying them homogenously and maintaining them over time, thus building a robust model for their acquisition, analysis and presentation. This can help organizations to develop scientific tools, methods and statistics for rapid, accurate and predictive assessment of the efficacy, quality and safety of health. Moreover, this can support the management to improve the results and reduce the time to introduce further innovations (time-to-value). Different kind of indicators have been proposed in the literature [1], generally focused on specific aspects of healthcare delivery (disease-specific, rate-based or sentinel, etc.). In this paper we follow the widely accepted methodology proposed by Donabedian [2, 3] who combines several issues, related to the organizational structure, processes, and outcomes, in a comprehensive model [4]. In particular, the health system is split in four conceptual components: mission, structural capacity, processes and outcomes, all interacting with the macro context [5,6]. This approach has the advantage to measure quality on each dimension of the healthcare system (structure, process and outcome). As a

Content from this work may be used under the terms of the Creative Commons Attribution 3.0 licence. Any further distribution of this work must maintain attribution to the author(s) and the title of the work, journal citation and DOI.
Published under licence by IOP Publishing Ltd

matter of fact research on a suitable methodology to determine proper indicators and the relationship among them is still in progress.

Thus with the final aim of developing a comprehensive model for quality assessment that takes into account these three dimensions and their dynamic relationship, in this paper we test our hypothesis on a case study based on the Italian healthcare services. In particular we focus on the evaluation of efficacy and performance based on the analysis of the hospital bed management.

2. Methods

As said the Donabedian framework has the advantage to measure quality on each dimension of the healthcare system. However, its application needs not only a homogenous classification of indicators related to the healthcare structure, process and outcome, but also a coherent acquisition of data that are usually collected in different information systems and for various purposes [7]. For instance hospital information systems provide information on the staff employed (number of physicians, nurses, technicians) as well as on facilities available (number of beds, diagnostic equipment, etc.), while information on clinical outcome usually derive from hospital discharge forms.

In this paper the proposed methodology is based on the following steps:

- Analysis of available open data sources provided by national authorities and relevant granularity and consistency of data provided.
- Selection of reliable, coherent and suitable set of data according to the purpose of the study.
- Identification of a sound method(s) suitable to match the data available with the three dimensions identified in the Donabedian framework.
- Definition of a case study to verify the reliability of the proposed approach.

3. Materials

For the purpose of our study it is crucial to identify the information sources and flows to select reliable and consistent data to build indicators of each dimensions of the Donabedian framework.

In Italy the Ministry of Health requires a standard set of data to be sent by the hospital information systems that report data on each intervention of a single patient. The Ministry of Health also collects data describing facilities and staff of the healthcare services. This information has recently become freely available for third users following the open data principles, even if with some delay in publishing them.

In this paper we take into account:

- The database of the national healthcare service, which provides statistical data on the territorial healthcare assistance, the personnel employed as well as the healthcare facilities in each local health unit [8].
- The national database of hospital admissions that stores demographic and clinical information about the admissions recorded in Italy. Data are collected based on the discharge report forms [9].
- The National Program of Outcomes Evaluation provides outcome information on the basis of the previous database [10].

Moreover, the data warehouse published by the ISTAT (National Institute for Statistics) [11] has been analyzed to determine overall indicators such as life styles, health conditions and causes of death as well as the population size.

4. Results

4.1. Hospital bed management

To analyse the interaction between structure and process we identify a set of data that meet the requirements of reliability, coherence and adherence related to the efficiency of health structures aggregated at regional level, focusing in particular on the hospital bed management.

To achieve this aim we adopt the Barber and Johnson method [12] to construct a scatter diagram, where data related to structure (number of beds,) and process (in-patient days, , and patient discharges,) are combined to build the following indicators:

- Beds occupancy rate (BOR): percentage of inpatient beds occupied over a specific temporal reference. , where indicates the number of inpatient days available in the given number of days ( ) ;
- Average length of stay (AvLOS): average number of days that inpatients remained in the hospital. ;
- Turnover interval (TOI): average number of days that an available bed remains empty between the discharge of one inpatient and the admission of the next. ;
- Beds turn over (BTO): average number of patients "passing through" each bed during a period. .

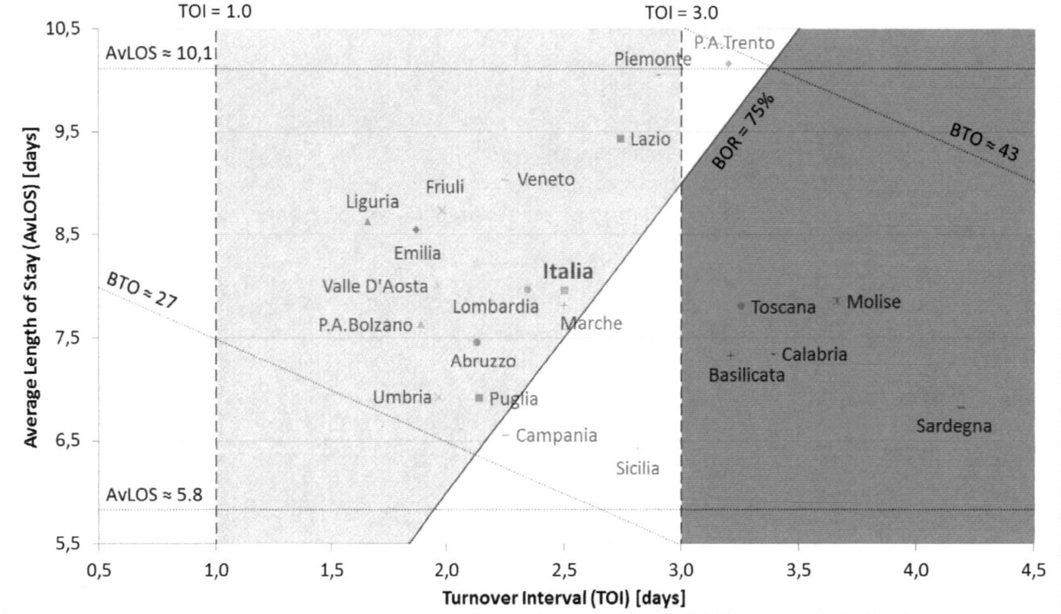

Figure 1. Barber-Johnson diagram of bed level of use distributed by Italian regions. Acceptability area depicted in green is determined considering reference values of turn over interval (TOI = 1.0 and 3.0, green dashed lines) and bed occupancy rate (BOR = 75%, red line) as proposed by national and international guidelines. Moreover, for BTO and AvLOS the average values plus/minus two standard deviations of the overall regions are reported.

Figure 1 shows the hospital bed management distributed by Italian regions, highlighting the reference values of TOI and BOR indicators as proposed by national and international guidelines. Moreover, BTO and AvLOS reference values are also reported considering the average plus/minus two standard deviations of the overall regions, given that these indicators refer to specific hospitalization and diagnosis and do not have general threshold. The reference values of TOI and BTO have been used to identify the following efficiency areas: 1) the red one that identifies regions where both TOI and BOR are outside the reference threshold (e.g. Sardegna); 2) the yellow area that reports regions where either TOI or BOR are outside the threshold (e.g. respectively P.A. Trento and Sicilia) and 3) the green area that identifies regions where both indicators are within the reference thresholds (e.g. Liguria).

4.2. Inter-regional mobility assessment

In order to develop a comprehensive model for quality assessment that takes into account the three dimensions of the proposed framework (i.e. structure, process and outcome dimensions) we further extend the proposed approach integrating the above mentioned indicators with clinical outcome information that take into account the effectiveness of the different healthcare systems in the provision of care services. The analysis of this information has been focused on the 21 regional healthcare systems given that the Italian National Health Service is a regionally decentralized tax-funded system, mainly based on regional budgets thus creating separate and autonomous regional health services.

Thus in the evaluation of singular regional health system a crucial phenomenon is represented by patient mobility that is considered an important marker to assess the supply and demand of services provided. Moreover, this is an appropriate indicator in Italy considering that patients can freely decide the hospital where to be admitted and the choice is essentially determined by distance from home, hospital specialization, waiting lists as well as quality in the supply of care services.

Although it is difficult to classify this indicator in the three pillars proposed in this paper we adopt it to represent the degree of the patient's satisfaction with the specific regional healthcare system focusing the attention on both demand and supply factors in each region and distinguish between pull and push factors. The former describes the ability of a region to attract patients from other regions, whereas the latter describes the ability of a region to restrain the outflows.

To examine the patient flows of each region the mobility index ( ) has been assessed as
, where:

- , the inflow rate is the percentage ratio between non-enrolees admitted and the total number of admissions in a given region and
- , the outflow rate is the percentage ratio between enrolees of a given region admitted in other regions and the total number of admissions of enrolees of the region).

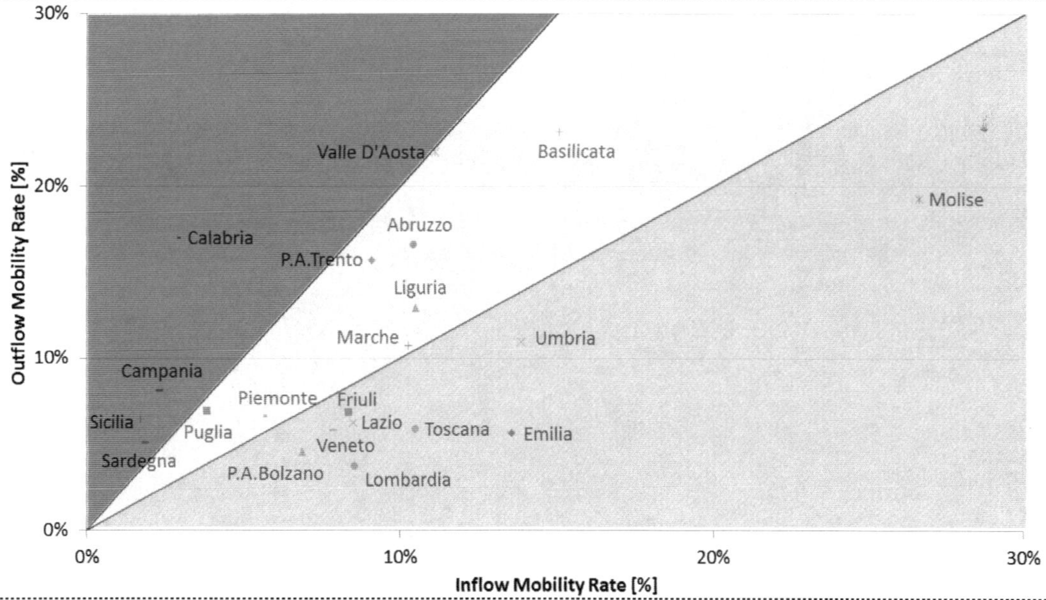

Figure 2. Scattered diagram showing both inflow (abscissa) and outflow (ordinate) mobility rates. Acceptability area depicted in green identifies regions where the inflow mobility rate is higher than the outflow one thus specifying where the mobility index is higher than 1.

Figure 2 shows the distribution of Italian regions considering both inflow and outflow rates. As already reported in the Barber-Johnson diagram, each region has been classified in the following areas: 1) the red one identifies regions where the mobility index is lower than 50% (e.g. Calabria); 2) the yellow area reports regions where the mobility index is higher than 50% but lower than 100% (e.g. Abruzzo) and 3) the green area identifies regions where the mobility index is higher than 100% (e.g. Toscana). In particular regions classified in the green area are "net importers" of patients from other regions, thus being able to offset the outflows with larger inflows.

5. Conclusions

In this paper we propose a methodology based on a comprehensive model for quality assessment integrating structure and process indicators. In particular, for the analysis of healthcare organizational structure we focused on hospital bed management using Barber and Johnson method, while for the analysis of processes we considered the phenomenon of both active and passive patient mobility. The integration of these aspects referred to the 21 Italian regions allows a more detailed analysis of supply and demand of healthcare services, providing on the one hand a comparative analysis of the regional systems and on the other highlighting some important aspects that could help managers to introduce changes and improvements in the organisation of hospital services.

In the future we intend to extend our analysis integrating other outcome indicators to capture the impact of healthcare service provision on health and wellbeing of patients and population. Moreover, we plan to use other information, such as demographic and geo-referenced data to investigate patient mobility also considering the speciality healthcare services, such as cardiology and gynaecology.

References.

[1] Winkel P and Zhang NF 2007 Statistical Development of Quality in Medicine, Wiley.
[2] Handler A, Issel M and Turnock B 2001 A conceptual framework to measure performance of the public health system *Am. J. Public Health* 1235-9.
[3] Donabedian A 1988 The quality of care: how can it be assessed? *JAMA* **260** 1743-8
[4] Clemente F, D'Arco M and D'Avino E 2014 The use of a conceptual model and related indicators to evaluate quality of healthcare in ICUs *Quality Engineering* **26** (2) *196-205*
[5] Hensher M and Keogh B 2009 Quality metrics, Elsevier Ltd. 393-6.
[6] De Vos M, Graafmans W, Keesman E, Westert G, van der Voort PHJ 2007 Quality measurement at intensive care units: which indicators should we use? *J. Critical Care* **22** 267-4.
[7] Pecoraro F, Luzi D, Ricci FL 2013 Secondary uses of EHR systems: A feasibility study *IEEE E-Health and Bioengineering Conference (EHB)*.
[8] www.salute.gov.it/portale/documentazione/p6_2_8_1_1.jsp?lingua=italiano&id=6
[9] www.salute.gov.it/portale/temi/p2_4.jsp?lingua=italiano&area=ricoveriOspedalieri
[10] http://95.110.213.190/PNEed13/
[11] http://dati.istat.it/
[12] Yates J 1982 Hospital Beds: A problem for diagnostic and management, Heinemann Medical Books.

Validating Quantitative Measurement Using Qualitative Data: Combining Rasch Scaling and Latent Semantic Analysis in Psychiatry

Rense Lange[1,2]
[1] Integrated Knowledge Systems, Inc., 107 Carefree Drive, Chatham, IL 62629, USA
[2] ISLA, Rua Cabo Borges, 55, 4430-646 Vila Nova de Gaia, Portugal

E-mail: renselange@earthlink.net

Abstract. An extension of concurrent validity is proposed that uses qualitative data for the purpose of validating quantitative measures. The approach relies on Latent Semantic Analysis (LSA) which places verbal (written) statements in a high dimensional semantic space. Using data from a medical / psychiatric domain as a case study - Near Death Experiences, or NDE - we established concurrent validity by connecting NDErs qualitative (written) experiential accounts with their locations on a Rasch scalable measure of NDE intensity. Concurrent validity received strong empirical support since the variance in the Rasch measures could be predicted reliably from the coordinates of their accounts in the LSA derived semantic space ($R^2 = 0.33$). These coordinates also predicted NDErs age with considerable precision ($R^2 = 0.25$). Both estimates are probably artificially low due to the small available data samples (n = 588). It appears that Rasch scalability of NDE intensity is a prerequisite for these findings, as each intensity level is associated (at least probabilistically) with a well-defined pattern of item endorsements.

1. Overview

Issues of validity have a long-standing history in measurement within the social sciences, as it is may not be clear that questionnaires and tests indeed capture what they are intended to measure, a requirement often called construct validity (Messick, 1989). A test is said to show face validity if test-takers' (respondents') subjective hypotheses concerning a tests's purpose correspond to the designers' intentions. While important in tests' public acceptance (Holden, 2010), face-validity is not considered to be authoritative. Instead, most serious validation research relies on outside statistical criteria: predictive validity reflects the possibility of predicting particular behavioral outcomes, and concurrent validity revolves around a measure's correlation with existing measures of the same construct. Borsboom and Markus (2013) have criticized these approaches for confusing consistency and truth, and others have argued that validity should revolve around the suitability of an instrument for application in actual practice (Fisher, 2003). Rasch scaling aids in this respect as it provides a variety of fit statistics to identify items that fail to show construct invariance and persons for whom the items do not combine to form linear measures (Michell, 2003).

The preceding notions regarding validity stand in stark contrast to the aims of qualitative psychology which focuses on understanding behavior from informants' perspective within a dynamic

Content from this work may be used under the terms of the Creative Commons Attribution 3.0 licence. Any further distribution of this work must maintain attribution to the author(s) and the title of the work, journal citation and DOI.
Published under licence by IOP Publishing Ltd

and negotiated reality (see e.g., Minchiello, 1990). According to Miles and Huberman (1984, p. 225) *"In qualitative research, numbers tend to get ignored. After all, the hallmark of qualitative research is that it goes beyond how much there is of something to tell us about is essential qualities."* Exceptional in this respect are qualitative studies (Lange, et al., 2014) that encode their data in a format suitable for multi-faceted Rasch scaling (Linacre, 2013). However, most qualitative research efforts report their findings in the informants' subjective language thereby making it difficult to judge their validity, as it is difficult to connect the findings to those obtained by other researchers.

Using data from a medical domain (psychiatry) - specifically, the reporting of near-death experiences (NDE) - this paper aims to show that purely qualitative data can be transformed into a form that is suitable for the validation of quantitative measures using unsupervised learning. The approach is based on algorithms that were recently developed in Artificial Intelligence and Natural Language Processing to perform Latent Semantic Analysis (LSA) using large data sets (see below). LSA takes respondents' qualitative (written) accounts of their own subjective opinions, insights, and interpretations of events, and places these inside a high-dimensional semantic space. Also available is a Rasch scaled questionnaire that assesses important aspects of this domain. Standard statistical techniques will then be used to predict respondents' locations on this dimension. Thus, the preceding can be seen as a special case of establishing concurrent validity by connecting qualitative and quantitative data. The approach involves considerable simplification, as LSA reduces the richness and complexity of data valued by qualitative researchers. It is not clear if this simplification actually enhances or detracts from the overall effectiveness of the approach.

2. Latent Semantic Analysis

Assuming that we have d documents and t relevant terms one can construct a matrix X whose rows $1, \ldots, i, \ldots, t$ represent a dictionary of selected terms (or "tokens"), and whose columns $1, \ldots, j, \ldots d$ represent the documents containing these terms. Each entry X_{ij} reflects the weighted frequency with which dictionary term i occurs in document j. The weighting does not affect the following overview, and the topic is further addressed in Section 3.

It has been known since last century that any matrix X can be written as the product of three matrices:

$$X = TSD', \tag{1}$$

where T has size $t \times d$ and D has size $d \times d$, both with orthonormal columns (i.e., $T'T = I$ and $D'D = I$), while S is a $d \times d$ diagonal matrix. The matrices T, D, and S are referred to as the left and right singular vectors and the diagonal matrix of singular values, respectively, and together these define X's Singular Value Decomposition (SVD). By convention, the diagonal elements of S are positive and ordered in decreasing magnitude.

The utility of SVD lies in the fact that keeping just the k largest singular values (i.e., setting $S_{k+1} = S_{k+2} = \ldots S_d = 0$) yields a matrix X^* which increasingly approximates X for greater k. Doing so produces two useful results: First, X^* is a matrix of rank k which is closest to X in a least-square sense. Second, since the corresponding columns in T and D will be multiplied by 0, they can simply be deleted, yielding T^* and D^* with sizes $t \times k$ and $d \times k$, respectively.

LSA (se, e.g., Landauer, Folz, and Laham, 1998) uses SVD to reduce the dimensionality of the semantic space needed to represent a set documents from d to k. The choice of k is important as it should be chosen such that X^* captures mainly the "real" data in X, while omitting the sampling error. There are no obviously proper ways of achieving this goal, and researchers therefore experiment with different choices of k to achieve best results. In practice, the initial dimensionality t is in the order of several thousands, whereas the reduced space has a dimensionality of $k = 300\text{-}500$.

Of course, the approach would be of little use if Eq. 1 has to be recomputed each time a new document y is encountered. However, assuming that the model is correct (i.e., $X = X^*$), a document's reduced coordinates D_y (with k dimensions) is a simple function of it's original coordinates X_y (in d dimensions, $d > k$) as follows (Deerwester et al., 1990, p. 399):

$$D_y = X_y\, T\, S^{-1} \tag{2}$$

While the mathematics behind SVD has long been known, the computation of the T, S, and D of sufficient sizes needed for actual text analysis applications has become feasible only with the advent of modern computers and the development of efficient algorithms. Fortunately, recent years have seen a rapid development of extremely powerful methods based on the use of sparse matrices, and capable of distributed and/or parallel processing. For instance, Řehůřek and Sojka (2010) describe the Python based Gensim software capable of analyzing enormous data sets on a standard personal computer.

Table 1: Rasch Locations (logits) of NDE items for "True NDErs"

Question	Location
Did scenes from the future come to you?	*1.69*
Did scenes from your past come back to you?	*1.30*
Did you seem to be aware of things going on elsewhere?	*1.17*
Did you see deceased or religious spirits?	*0.68*
Were your thoughts speeded up?	*0.64*
Did you suddenly seem to understand everything?	*0.32*
Were your senses more vivid than usual?	*-0.06*
Did you come to a border or point of no return?	*-0.11*
Did you seem to encounter a mystical being or presence?	*-0.20*
Did you see, or feel surrounded by, a brilliant light?	*-0.30*
Did you feel a sense of harmony or unity with the universe?	*-0.55*
Did you seem to enter some other unearthly world?	*-0.61*
Did you have feelings of joy?	*-0.70*
Did you feel separated from your body?	*-0.74*
Did time seem to speed up or slow down?	*-1.07*
Did you have a feeling of pleasantness?	*-1.47*

3. Term Weighting

It is possible to base LSA directly on the frequency by which term i occurs in document j. However, in practice better results are obtained by transforming the between term-document co-occurrence matrix into a locally/globally weighted term frequency - inverse document frequency (*tf-idf*) matrix with real-valued entries. Such transformed values increase proportionally with terms' occurrence within a document, but - in order to control for the fact that some words are more common than others - these values decrease proportional to the logarithm of the frequency of the term across the entire corpus. For instance, assume that there are just three documents d_1, d_2, and d_3 and that some term z occurs 3 times in d_1 and once in d_2. Then the term frequency $tf = 3+1+0=4$ and the logarithm of the inverse document frequency $idf = \ln 3/2 = 0.41$ (using natural logarithms), and hence $tf\text{-}idf_z = tf*idf = 4*0.41 = 1.64$.

4. Case Study: Near-Death Experiences

It is well-known to physicians and mental health professionals that people (adults and children alike) when suddenly faced with their own death experience, a distinctive state of consciousness in which their existence is seemingly unbounded by a physical body or earthly environs. Such *near-death experiences* (NDE) are among the most potent of psychological episodes (for an overview, see e.g., Holden, Greyson & James, 2009). In fact, the *Diagnostic Manual of Statistical and Mental Disorders* contains the V-Code category "Religious or Spiritual Problem" in part to acknowledge and guide clinicians in addressing the impact and aftereffects of NDEs and related experiences (Lukoff, 1998).

Greyson(1983) quantified NDE intensity by means of a series of questions derived from statements of individuals who shared their accounts following such experiences. Specifically, NDErs (i.e., those reporting an NDE) were asked to rate the occurrence of 16 different experiences in terms of three ordered categories which generically represent 'not present,' 'mildly or ambiguously present' or 'definitively present.' Indicative of its measurement validity, the NDE proved to be Rasch scaleable (Lange, Greyson, Houran, 2004).

The questions constituting Greyson's NDE measure are listed in Table 1, together with their item difficulty in logits. Note that items at the lower end of this scale (i.e., with the highest endorsement) refer to experiences of peace, joy and harmony, followed by finding insight and mystical or religious experiences at intermediate levels, while items with the highest values (least endorsed) refer to an awareness of events occurring in a different place or time. By the very nature of Rasch scaling, this sequence defines a true hierarchy such that events mentioned higher in this sequence become salient only after of those lower in the sequence have already been reported - at least probabilistically. In other words, there exists a well-defined relation between NDE intensity and the verbal, i.e., qualitative, contents of the NDE.

Naturally, those who have not had near death experiences cannot be expected to provide meaningful information about NDE. Therefore, Table 1 shows the item locations as computed for those that were classified psychiatrically as having had "True NDE" only (cf, Greyson, 1983).

Table 2: Summary of multiple linear regression runs

Number of dimensions	1-150	1-200	1-250	1-300	1-350
R^2	0.49	0.56	0.64	0.70	0.75
Adjusted R^2	0.29	0.29	0.33	0.31	0.28

An independent sample of 833 NDE accounts was available based on which a semantic space was constructed using Řehůřek and Sojka's (2010) Gensim software. This space used a dictionary with 1500 tokens (terms) that were deemed relevant, and the resulting 1500 x 833 token x document matrix was decomposed according to Eq. 1. A total of 588 of the 833 people also had completed Greyson's NDE scale, and their Rasch NDE intensity estimates were combined with the their accounts' coordinates in the semantic space as computed using Eq. 2. Using the *glm* procedure provided by the R language, the first 50, 100, …, 400 coordinates (ordered by importance) were used to predict respondents' estimated Rasch NDE intensity using standard multiple regression.

Not surprisingly, Table 2 shows that using increasingly complex semantic spaces adds predictive information, as R^2 increases continuously when more predictors are added. Consistent with other research, the corresponding R^2 values when adjusted for sample size and number of predictor variables reach a maximum (i.e., 0.33) for about 250 predictor variables and this value is perhaps a more appropriate estimate of the quality of prediction. Nevertheless, the finding that at least 1/3 of the variation in respondents' quantitative Rasch NDE measures can be predicted from their qualitative NDE accounts strongly supports the notion that qualitative data can be used to validate quantitative variables. The multiple correlation between variance explained by the first 250 factors and their weights in the multiple regression equation is essentially zero ($r = -0.06$), i.e., the semantic factors do not contribute to prediction in proportion to the variance they explain in the SVD context.

It is noted that respondents' age could reliably be predicted from the semantics from their

NDE accounts ($R^2 = 0.25$). However, the multiple correlation with NDErs gender proved negligible.

6. Conclusion

A standard approach to establishing content validity is to correlate the responses to one set of questions to sets of questions purporting to measure the same construct. This approach need not be limited to comparisons between quantitative measures since that Latent Semantic Analysis can be used to capture the meaning of free style written descriptions in order to establish a correlation with other a Rasch scaled variable. It is my conjecture that one variable's Rasch scalability may well prove crucial in this context, as Rasch scalability defines an "item hierarchy" which essentially imposes an orderly progression on the semantic labels corresponding to variables' quantitative properties.

It should be noted that analysis of textual data is not limited by the size of the text corpus, economic factors, or hardware constraints. For instance, Řehůřek and Sojka's (2010) Gensim used here is freely available under the OSI-approved GNU LGPL license. This software is capable of analyzing the complete English Wikipedia by decomposing a sparse matrix with 3.9 million rows x 100,000 dictionary terms with about 760 million non-zero entries in about 13 hours on a standard PC.

References

Borsboom, D. and Markus, K.A. (2013). Truth and evidence in Validity Theory. *Journal of Educational Measurement*, Vol. 50, No. 1, pp. 110–114.

Deerwester, S., Dumais, S.T., Furnas, G.W., Landauer, T.K. and Harshman, R. (1990). Indexing by Latent Semantic Analyis. *Journal of the American Society for Information Science*. 41, 391-407.

Fisher, W. P., Jr. (2003) Mathematics, measurement, metaphor, metaphysics: Part I. Implications for method in postmodern science. *Theory & Psychology*, 13(6), 753-90.

Miles M.B. and Huberman A.M. (1984). *Qualitative Data Analysis: A Sourcebook of New Methods*. Newbury Park, CA: Sage.

Greyson, B. (1983). The Near-death Experience Scale: Construction, reliability, and validity. Journal of Nervous and Mental Disease, 171, 369-375.

Holden, J. M., Greyson, B. and James, D. (eds.). (2009). Handbook of Near-Death Experiences: Thirty Years of Investigation. Santa Barbara, CA: Praeger/ABC-CLIO

Holden, Ronald B. (2010). "Face validity". In Weiner, Irving B.; Craighead, W. Edward. *The Corsini Encyclopedia of Psychology* (4th ed.). Hoboken, NJ: Wiley. pp. 637–638.

Landauer, T. K., Foltz, P. W. and Laham, D. (1998). Introduction to Latent Semantic Analysis. Discourse Processes, 25, 259-284.

Lange, R., Dorsey, K., Wells, N., Roberts, N. and Verhulst, M. (2014). *Rasch analysis of students' qualitative evaluations of faculty professionalism*. Manuscript submitted for publication. Southern Illinois School of Medicine.

Lange, R., Greyson, B. and Houran, J. (2004). A Rasch scaling validation of a 'core' near-death experience. *British Journal of Psychology*, 95, 161-177.

Linacre, J. M. (2013). *Facets computer program for many-facet Rasch measurement, version 3.71.2*. Beaverton, Oregon: Winsteps.com.

Lukoff, D. (1998). From Spiritual Emergency to Spiritual Problem: The Transpersonal Roots of the New DSM-IV Category. Journal of Humanistic Psychology, 38, 21-50.

Messick, S. (1989). Validity. In R. L. Linn (Ed.), *Educational measurement* (3rd ed., pp. 13– 103). Washington, DC: The American Council on Education and the National Council on Measurement in Education.

Michell, J. (2003) Measurement: A beginner's guide. *Journal of Applied Measurement*, 4(4), 298-308.

Minichiello, V. (1990). *In-Depth Interviewing: Researching People*. Longman Cheshire.

Řehůřek, R. and Sojka, P. (2010). Software Framework for Topic Modelling with Large Corpora. *Proceedings of the LREC 2010 Workshop on New Challenges for NLP Frameworks*, La Valetta, Malta, p. 45–50.

Uncertainty evaluation in the self-alignment test of the upper plate of a press

Alexandre S. Lourenço[1] and J. Alves e Sousa[2]

[1]CATIM – Centro de Apoio Tecnológico à Industria Metalomecânica, Porto, Portugal
[2]LREC - Regional Laboratory of Civil Engineering, Funchal, Madeira, Portugal

E-mail: alexandre.lourenco@catim.pt

Abstract. This paper describes a method to evaluate uncertainty of the self-alignment test of the upper plate of a press according to EN 12390-4:2000. The method, the algorithms and the sources of uncertainty are described.

1. Introduction

The testing machinery used to carry out tests on hardened cement are required to comply with a standard specifying their own calibration. In terms of the calibration associated with the measuring force system, the main difference between the two most used standards is the existence of three classes in standard EN 12390-4:2000 instead of four classes in standard ISO 7500-1:2004, with the elimination of class 0,5 .

Despite great emphasis in the application of force in both standards, standard EN 12390-4:2000 sets three compulsory tests in order to guarantee that the force applied on the specimen acts uniaxially.

2. Testing equipment

Annex A of standard EN 12390-4:2000 specifies that the execution of the test should be implemented with the use of a cylinder made of chromium-nickel steel. Its dimensions include a diameter of 100 ± 1 mm and a height of 200 ± 1 mm. It must be instrumented with four complete Wheatstone bridges and with thermal compensation. The placement of the four bridges must be at half the height of the cylinder and at opposite sides. This transducer should be used together with an amplifier display unit, promoting a balanced output signal in a position without any force applied.

The standard specifies a maximum admissible error for the set transducer/display unit of 0.1 % or 5x10-6 mV/V. The calibration of this transducer should be carried out with traceability to national or international standards every two years (at most).

It is usual the certificates for these transducers to be conclusive, stating that the set transducer/display unit is able to perform tests according to a particular standard.

Content from this work may be used under the terms of the Creative Commons Attribution 3.0 licence. Any further distribution of this work must maintain attribution to the author(s) and the title of the work, journal citation and DOI.
Published under licence by IOP Publishing Ltd

3. Test according to EN 12390-4:2000

In order to qualify a press used in the hardened cement test, it is required to perform three tests to quantify the self-alignment of its upper plate, the alignment of the components of the press and the restrictions to the movement of the upper plate.

3.1. Self-alignment of the upper plate

The self-alignment of the upper plate is evaluated by the application of a force between 200 kN and 220 kN. The values of the four bridge positions are registered simultaneously when the plate is tilted to point A and then to point C, in the axis BD, and to point B followed by point D, in the axis AC.

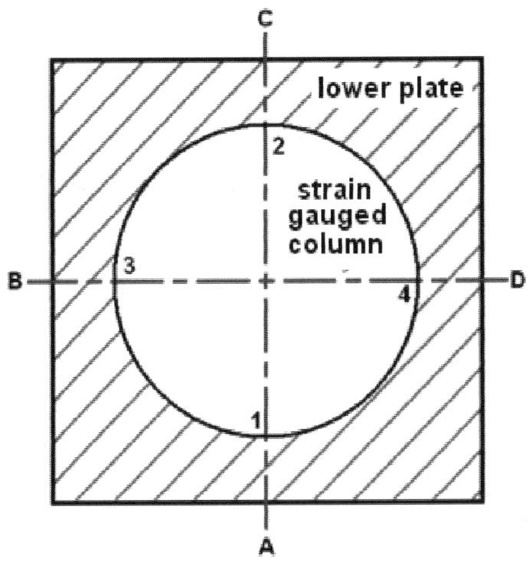

Figure 1
Placement for the verification points of the equipment.
The numbers represent the location of the four bridge positions and the letters relate to the axis of the test equipment.

The extension ratios are determined according to expression (1) for each strain gauge and for each position

$$r_n = \frac{e_n - e_m}{e_m} \quad (1)$$

The self-alignment is determined in relation to the strain gauge bridge by the difference between the highest value of r_n and the lowest value of r_n on the four positions of the upper plate, using expression (2)

$$\Delta r_n = r_{n\,máxino} - r_{n\,mínimo} \quad (2)$$

With this procedure, to evaluate the self-alignment of the upper plate four force applications are required, one for each upper plate tilting, in all four directions (A, B, C e D), and for each force application four values are registered with respect to the four strain gauge bridges of the force transducer (1,2,3 and 4).

3.2. Uncertainty evaluation

The main sources of uncertainty associated with this test, arise from the four bridges force transducer. This transducer is classified by the manufacturer as being class 1, according to standard ISO 376:2011.

During the four repetitions of force application, it is difficult to stabilize the cement test machine in exactly the same value of force, thus the source of uncertainty considered was reproducibility instead of repeatability.

As the reading is in electrical signal and not directly in units of force, the interpolation error was not considered.

Table 1: Sources of uncertainty considered

Source of uncertainty	Type and Distribution	Value	Origin
Maximum error	B-R	0.10 %	EN 12390-4:2000
Reproducibility	B-R	0.20 %	ISO 376:2011 – class 1
Zero	B-R	0.05 %	ISO 376:2011 – class 1
Creep	B-R	0.10 %	ISO 376:2011 – class 1
Uncertainty of calibration	B-N	0.05 %	ISO 376:2011 – class 1
Temperature effect during test	B-R	0.10 %	technical documentation ($\Delta t \leq 10$ K)

It is expected that temperature variations occur during the test, and they might affect each amplifier in different ways. Therefore it was considered the maximum variation in electrical signal for a 10 K variation, as given by the manufacturer.

Table 2: Self-alignment of upper plate test results

Força / kN	e_1	e_2	e_3	e_4
200	0.12644	0.18912	0.17060	0.14917
200	0.11857	0.20090	0.17499	0.14845
200	0.11976	0.19239	0.17356	0.14855
200	0.12084	0.19011	0.16387	0.14103

A full measurement was done and the results are given on table 2. The transducer was centred on the lower plate of the press using a digital calliper and its estimate to be centred at ± 0.05 mm.

The extension ratios and associated uncertainty are shown in Table 3. The uncertainty was calculated using the law of propagation of uncertainty according to GUM 2008 [1].

Table 3: Values of extension ratios and uncertainty

Δr_1	Δr_2	Δr_3	Δr_4
0.058	0.059	0.030	0.023

$U\Delta r_1$	$U\Delta r_2$	$U\Delta r_3$	$U\Delta r_4$
± 0.0048	± 0.0065	± 0.0068	± 0.0054

To perform a validation of results a Monte-Carlo simulation was carried out, for a number of trials of 10^6. The number of trial was select according to JCGM 101:2008 to deliver a 95 % coverage interval for the output quantity such that this length is correct to one or two significant decimal digits[2].

The extension ratios and associated uncertainty, considering the output distribution as Gaussian (k=2), are shown in Table 3.

Table 4: Values of extension ratios and uncertainty using Monte-Carlo simulation

Δr_1	Δr_2	Δr_3	Δr_4
0.058	0.059	0.030	0.023

$U\Delta r_1$	$U\Delta r_2$	$U\Delta r_3$	$U\Delta r_4$
± 0.0049	± 0.0067	± 0.0062	± 0.0053

Using the procedure described by point 8 of JCGM 101:2008 it's possible to validate the uncertainty results obtained early. It was considered only one significant digit on the uncertainty so the δ is equal to 0,0005.

On table 5 are listed the values of y_{low} and y_{high} obtained by the Monte-Carlo simulation and the limits d_{low} and d_{high} used to assess the validity. These absolute values must be not larger than δ calculated earlier.

Table 5: Values y_{low}, y_{high}, d_{low}, d_{high}

	Δr_1	Δr_2	Δr_3	Δr_4
y_{low}	0.0536	0.0527	0.0238	0.0183
y_{high}	0.0631	0.0657	0.0365	0.0287
$\lvert d_{low}\rvert$	0.0001	0.0001	0.0004	0.0005
$\lvert d_{high}\rvert$	0.0001	0.0000	0.0005	0.0002

4. Conclusions

An estimate of the measurement uncertainty using the law of propagation of uncertainty was performed for the self-alignment of the upper plate of a press in accordance with standard EN 12390-4:2000.

It was used also a Monte-Carlo simulation. The uncertainties obtained for both methods were similar. To validate the results it was used the procedure described by point 8 of JCGM 101:2008. The uncertainty must be rounded to one significant digit in order to validate the procedure.

The values of uncertainty obtained are of a magnitude several dozens of times lower than the maximum admissible value of that standard (0.10).

However, only the effects attributable to the force transducer were taken into account.

The correct positioning of the transducer on the bottom plate is of vital importance. Those effects were not analysed in this paper, however it is strongly suggested to include a source of uncertainty to account for them.

References

[1] BIPM, IEC, IFCC, ISO, IUPAC, IUPAP, and OIML. JCGM 100:2008 Evaluation of measurement data — Guide to the expression of uncertainty in measurement, 2008

[2] BIPM, IEC, IFCC, ISO, IUPAC, IUPAP, and OIML. JCGM 101:2008 Evaluation of measurement data — Supplement 1 to the "Guide to the expression of uncertainty in measurement" — Propagation of distributions using a Monte Carlo method, 2008

[3] ISO. ISO 7500-1:2004, Metallic materials -- Verification of static uniaxial testing machines -- Part 1: Tension/compression testing machines -- Verification and calibration of the force-measuring system

[4] ISO. ISO 376:2011, Metallic materials -- Calibration of force-proving instruments used for the verification of uniaxial testing machines

[5] EN 12390-4:2000, Testing hardened concrete -- Compressive strength. Specification for testing machines.

Journal of Physics: Conference Series **588** (2015) 012030 doi:10.1088/1742-6596/588/1/012030

Technical quality assessment of an optoelectronic system for movement analysis

R Di Marco[1], S Rossi[2], F Patanè[1] and P Cappa[1]

[1] Department of Mechanical and Aerospace Engineering, "Sapienza" University of Rome, Italy.
[2] Department of Economics and Management – Industrial Engineerng, University of Tuscia, Viterbo, Italy.

E-mail: roberto.dimarco@uniroma1.it

Abstract. The Optoelectronic Systems (OS) are largely used in gait analysis to evaluate the motor performances of healthy subjects and patients. The accuracy of marker trajectories reconstruction depends on several aspects: the number of cameras, the dimension and position of the calibration volume, and the chosen calibration procedure. In this paper we propose a methodology to evaluate the effects of the mentioned sources of error on the reconstruction of marker trajectories. The novel contribution of the present work consists in the dimension of the tested calibration volumes, which is comparable with the ones normally used in gait analysis; in addition, to simulate trajectories during clinical gait analysis, we provide non-default paths for markers as inputs. Several calibration procedures are implemented and the same trial is processed with each calibration file, also considering different cameras configurations. The RMSEs between the measured trajectories and the optimal ones are calculated for each comparison. To investigate the significant differences between the computed indices, an ANOVA analysis is implemented. The RMSE is sensible to the variations of the considered calibration volume and the camera configurations and it is always inferior to 43 mm.

1. Introduction

The Optoelectronic Systems (OS) are largely used in gait analysis to evaluate the motor performances in clinical analysis [1]. The accuracy of tracking marker trajectories is dependent on several aspects: (i) the number and position of the cameras [2], (ii) the dimension and position of the calibration volume [3, 4], (iii) the calibration procedure [5, 6] and (iv) the algorithm which converts the 2D coordinates of markers acquired by each camera in the 3D space [7]. The reconstruction phase is performed by OS software, therefore it is not possible to specifically evaluate the effect of the implemented algorithm and the other causes on the interval of uncertainty associated to the measurements. In literature several methods to evaluate the accuracy of OS are proposed. Weng *et al.* [6] designed a novel static calibration procedure; they imposed known static input to the OS by means of a custom-made calibration stand. Then, the actual positions of a set of points on the calibration stand (addressed as testing points) are compared with the measured position. Vander Linden *et al.* [3] calculated the error on marker position similarly to Weng *et al.* [6] by testing the accuracy of the OS in nine points of the calibration volume ($54.3 \times 72.0 \times 42.3$ cm^3); Branca and Cappa [4] evaluate the metrological performances via an *ad-hoc* system capable to impose known marker trajectories. Ehara *et al.* [1] provide an estimation of the accuracy on the dynamic reconstruction, comparing the actual

Content from this work may be used under the terms of the Creative Commons Attribution 3.0 licence. Any further distribution of this work must maintain attribution to the author(s) and the title of the work, journal citation and DOI.
Published under licence by IOP Publishing Ltd

value of the distance between two markers on a rigid wand and the values calculated by the OS while moving the wand in the capture volume. In order to provide repeatable trajectories of markers as input for the OS, Windolf *et al.* [2] proposed a *xyz*-robot which moved a L-wand equipped with reflective markers inside a volume of $18.0 \times 18.0 \times 15.0$ cm^3.
In this paper we propose a novel methodology to evaluate the accuracy of the OS. We consider all the above reported causes of uncertainties, enlarging the tested volume to a value comparable with the ones normally used in clinical gait analysis.

2. Methods

The optoelectronic system analyzed was the Vicon MX (Oxford Metrics) composed by eight IR cameras. The OS is installed at the Movement Analysis and Robotics Laboratory of the Children Hospital 'Bambino Gesu̇' (Palidoro, Rome IT). The reconstruction of marker trajectories was conducted by means of the OS manufacturer software.
The experimental protocol consists of a calibration, validation and post-processing phases.
In the calibration session, as also proposed by Vander Linden *et al.* [3], we divided the capture volume ($2.4 \times 3.6 \times 1.6$ m^3) into 12 sub-volumes ($1.2 \times 1.2 \times 0.8$ m^3), six at the floor level and six at a higher level (see Figure 1). The calibration procedure was performed moving a 5-markers

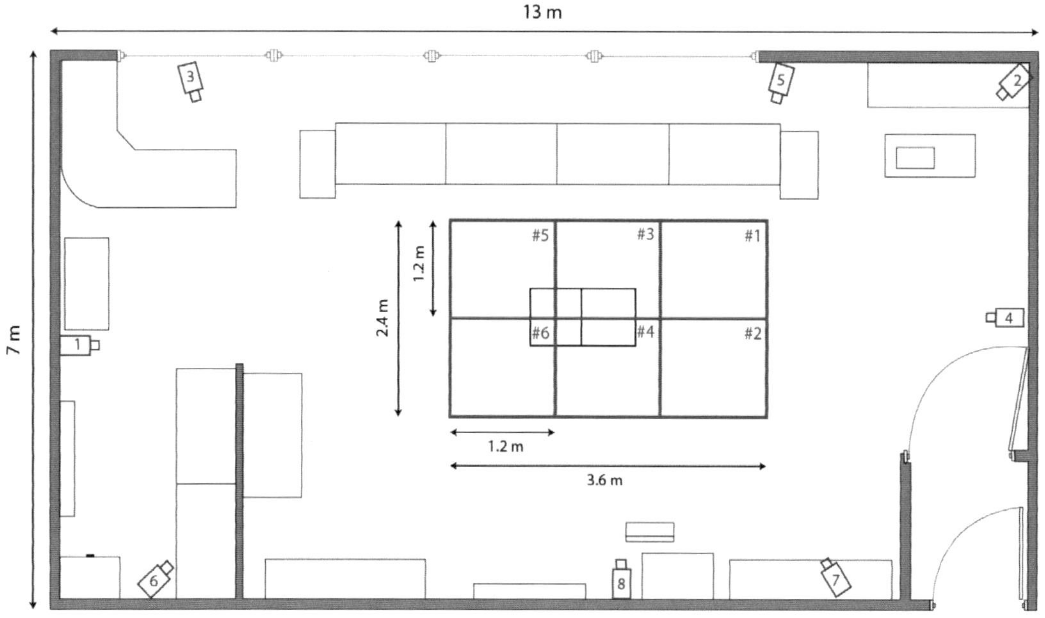

Figure 1. The Movement Analysis and Robotics Laboratory of the Children Hospital 'Bambino Gesù in Palidoro. The green numbers marked the label for the volume at the floor level, while the red numbers are the label for the Vicon cameras.

wand into each selected sub-volume. Each calibration procedure was acquired at 200 Hz for 15 s at least, and it was automatically stopped after each camera has acquired at least 5,000 frames. At the end of the calibration session 13 calibration files have been acquired: 12 for each sub-volume and one for the overall volume.
In the subsequent validation session, the tip of 1 m length wand was equipped with a reflective marker. The operator moved it into the overall volume.
Finally, in the post-processing phase, it was possible to apply to each trial acquired in the

validation session the 13 calibration files acquired during the calibration phase.
In order to evaluate the influence of the redundancy of the camera numbers, we decided to consider five different camera configurations: the first one with all the eight cameras (addressed as all-camera test) and the other four with only three cameras switched on at the same time (addressed as 3-camera test). Even though the minimum number of cameras necessary to track a marker trajectory is, as well known, equal to two we decided to limit it to three since it is the value provided by the manufacturer as optimal value.
The accuracy of the OS was evaluated by means of two indices. The first index used in the present study is the RMSE_C, i.e. the RMS of the difference between the i-th trajectory tracked using one of the '3-camera' configurations and the optimal one ('all-camera') assumed as reference; this index quantifies the effect of the camera redundancy. The second index is the RMSE_V, which gives the effect of the variation of the calibration volume. It is evaluated as the RMS of the distance between the i-th trajectory processed with one of the 12 sub-volume calibration files and the trajectory processed with the file processed taking into account the whole volume calibration.
Significant differences in the indices were evaluated by means of two-way ANOVA (13×5, i.e. volumes $\times$ camera configurations) (p=0.05). Tukey tests were considered as post-hoc test with the same alpha level.

3. Results and Discussion

Considering the whole calibration volume, the mean value of the reconstruction error due to different camera configurations was estimated as equal to 3 mm (max value 9 mm). Considering different volumes with 'all-camera' configuration, the mean value of the reconstruction error was 17 mm (max value 43 mm). The Figure 2 shows the marker trajectory reconstructed by using one of the '3-camera' configuration and four different calibration volumes, those are highlighted in the figures by a cube. Instead, the Figure 3 shows the reconstructed trajectory by using one of the sub-volume calibration files and four different '3-camera' configurations. The marker size in the pictures is equal to the distance of the trajectory from the optimal one (that is the one tracked by using the all-camera configuration in the case of Figure 2 and the whole calibration volume for the Figure 3).

Changing the camera configuration or the calibration volume, the reconstructed trajectory can be unavailable in any parts of it because of the algorithm, which compute the reconstruction. This is not known due to the manufacturer, but probably the unavailability of the trajectories occurs because the algorithm does not find a unique solution for the marker positioning at those frames. In fact, the calibration parameters (i.e. the positions of each camera relative to each other and to the absolute reference frame) are strongly dependent on the calibration volume; therefore the reconstruction of marker trajectory is up to the volume as well.
Unlike what we expected, applying the calibration file related to a specific sub-volume, the reconstruction of the trajectory that is performed in this volume is not more accurate than the one conducted in the other volumes. Both in the cases of RMSE_C and RMSE_V the two main effects give significant differences ($p < 0.01$). Significant differences are always found between sub-volumes and whole calibration volume ($p < 0.01$) and between the 'all cameras' configuration and the other '3-camera' tests ($p < 0.01$). Thus, the calibration volume has to be close to the chosen measurement volume; in fact, changing the dimension of calibration volume, the RMSE_C decrease from a maximum mean value of 4 mm to 1 mm, and RMSE_V decreases from a maximum mean value of 30 mm to 4 mm. The obtained results not only confirm the already stated relationship between the number of cameras and system accuracy, but also allowed a further quantification of the previously mentioned relationship. In fact, considering the 'all-camera' configuration, in the central sub-volumes the RMSE_V decrease from a maximum of 30 mm on bounds of the whole calibration volume to a minimum value of 6 mm.

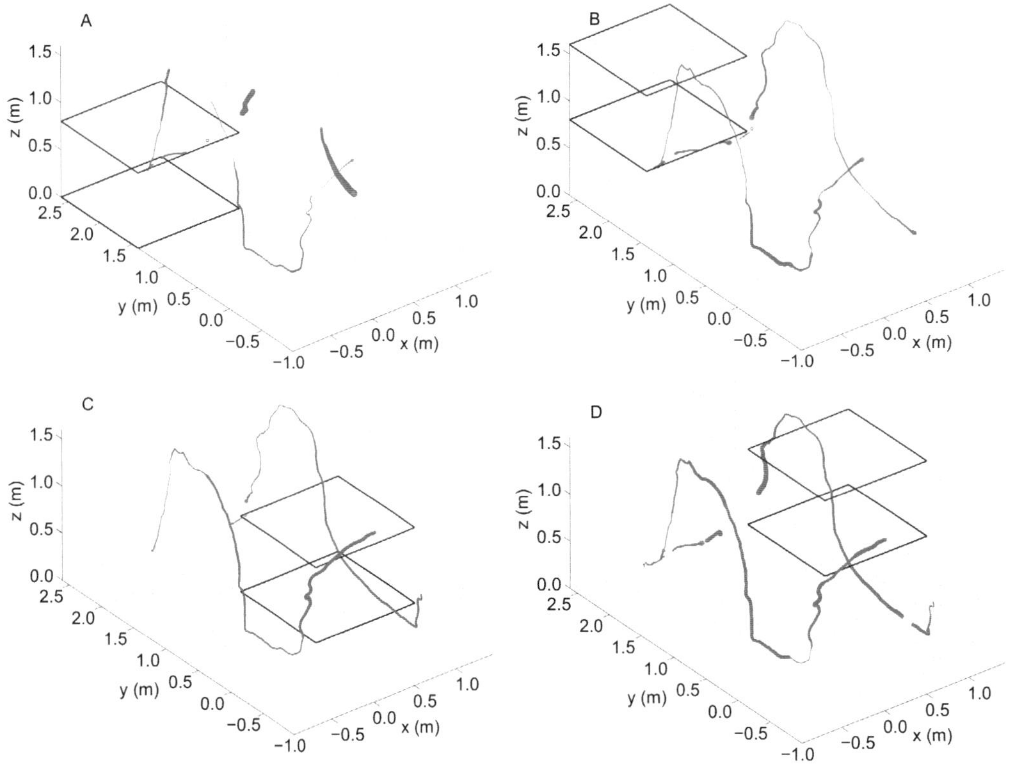

Figure 2. The marker trajectory reconstructed by means of different calibration files: same set of cameras and **(A)** volume 1 floor level, **(B)** volume 1 high level, **(C)** volume 4 floor level and **(D)** volume 4 high level.

The Table 1 shows the RMSE_C values which are related to the number of cameras used for the reconstruction of the trajectory when the whole calibration volume is considered. As expected,

Table 1. The obtained RMSE_C values considering the whole calibration volume and changing the number of camera used to reconstruct the marker trajectory.

Number of cameras	RMSE_C (mm)
3 cameras	7
4 cameras	2
5 cameras	6
6 cameras	1
7 cameras	1

the RMSE_C decrease by increasing the number of cameras. The trend is broken in the case of '4-cameras' configuration, because the chosen cameras for this configuration are positioned in an optimized manner to see the marker. Therefore, not only the number of the camera is an issue for the trajectory reconstruction, but the their position has to be taken into account as

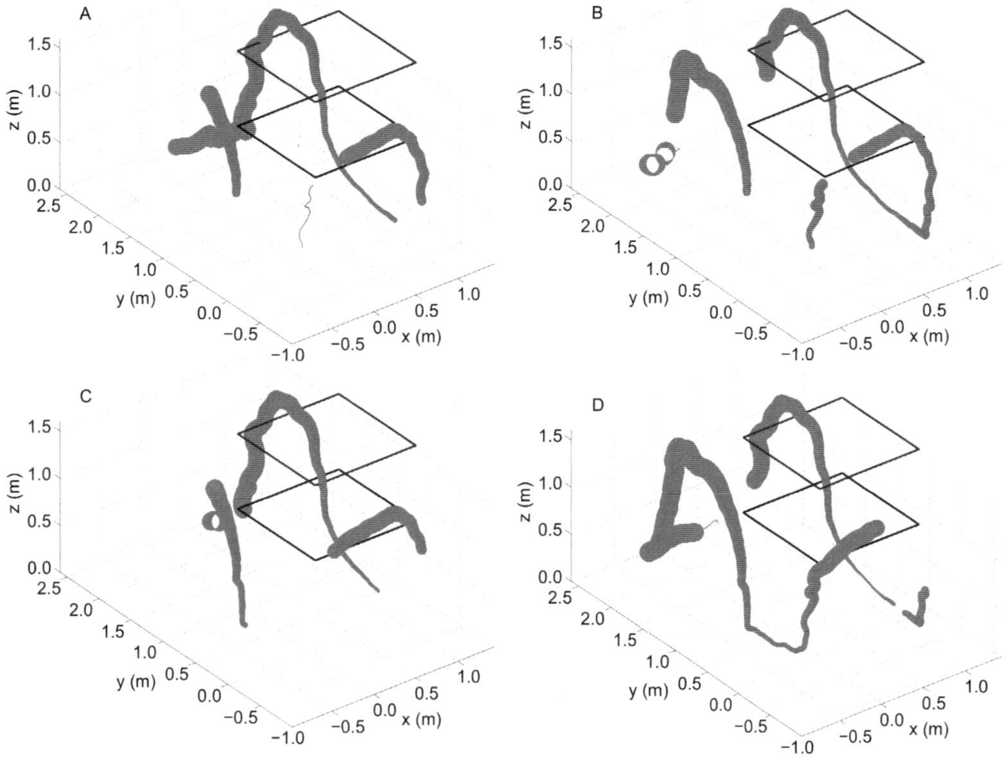

Figure 3. The marker trajectory reconstructed by means of different calibration files: same calibration volume (#4 high level) and **(A)** the camera 1, 3 and 5, **(B)** the camera number 5, 2 and 4, **(C)** the camera 6, 4 and 3, **(D)** the camera 7, 1 and 2.

well. By using six cameras or seven, the RMSE does not change significantly.

4. Conclusion

In the present study we evaluated the accuracy of the optoelectronic system keeping into account the main causes of inaccuracies. From the obtained results, we can assert and quantify the relationship between overall accuracy of the optoelectronic system and the selected calibration volume together with the number of used cameras.

Acknowledgments

The present study was supported by FP7 (FP7-ICT-2011-9; co-PI: Paolo Cappa) and the MIUR Italian Ministry of Instruction, Research and University (PRIN 2012 - 20127XJX57; PI: Paolo Cappa).

References

[1] Ehara Y, Fujimoto H, Miyazaky S, Mochimaru M, Tanaka S, Yamamoto S 1997 *Gait & Posture* **5** 251-255
[2] Windolf M, Gotzen Nils, Morlock M 2008 *Journal of Biomechanics* **41** 2776-2780
[3] Vander Linden DW, Carlson SJ, Hubbard RL 1992 *Physical Therapy* **72** 300-305
[4] Branca FP, Cappa P 1991 *12th Triennial World Congress IMEKO* (Pechino) 1580-1586

[5] Allard P, Blanchi JP, Aissaoui R 1995 *Three-Dimensional Analysis of Human Movement* Human Kinetics Pub. (Champaign) 19-40
[6] Weng J, Cohen P, Herniou M 1992 *IEEE Trans Patt Anal Mach Intell* **14** 965-980
[7] Abdel-Aziz YI, Karara HM 1971 *Proc. of the ASP Symp. on Close-Range Photogram* Urbana 1-18

Ontological Knowledge Base of Physical and Technical Effects for Conceptual Design of Sensors

V M Zaripova and I Yu Petrova

Department of CAD Systems, State Autonomous Educational Institution of Astrakhan Region of Higher Professional Education "ASTRAKHAN CIVIL ENGINEERING INSTITUTE", Astrakhan, Russia

E-mail: vtempus2@gmail.com

Abstract. This article discusses design of the knowledge base of physical phenomena based on domain-specific ontology. Classification of various physical phenomena in the knowledge base is based on energy-information model of circuits (EIMC) suggested by the authors. This model is specially aimed at design of new operating principles of sensing elements (sensors). Such a knowledge base can be used to train intended engineers, specialists in sensors design.

1. Introduction

According to many researchers the global market of sensors annually increases by 7-9%. And the number of physical phenomena and processes applied in sensor design is growing rapidly. But different physical properties that are the basis of sensor operation principle are described through the physical and mathematical tools corresponding to the given phenomena class only. This impedes automation of the design process, increases the period of new sensor development and their cost. This is why it is critical to use an automated data retrieval system, intended for centralized data collection of physical effects, to design the sensor operation principle and train engineers

A number of methods of conceptual design based on the knowledge bases of physical effects and phenomena are well known. For example:

1) Theory of Inventive Problem-Solving - TRIZ or TIPS), the author is Altshuller (Russia) [1],[2];
2) Systematic approach to design engineering, authors are Pahl и Beitz (Germany) [3];
3) Method of conceptual design, the author is Koller (Germany) [4].

All these methods are based on classified knowledge of physical effects. But all of them are aimed at support of conceptual design of any technical devices regardless of the sphere of its application. Physical effects in the knowledge bases of these systems have a worded description as a cause and effect relation and can be followed by an analytic formula of output-input connection. This systems operation results in the synthesis of several alternative physical principles of technical device operation in the form of circuits of serial conversions from input to output. But current difficulties in building mathematical models for description of synthesized variants of the physical principle of operation and evaluation of their performance characteristics do not allow quantitative comparison of these variants. Description of physical effects is not followed by a design study of technical realizations.

Content from this work may be used under the terms of the Creative Commons Attribution 3.0 licence. Any further distribution of this work must maintain attribution to the author(s) and the title of the work, journal citation and DOI.
Published under licence by IOP Publishing Ltd

Conceptual design of sensing elements (sensors) for data measurement and control systems has some peculiar features. On the one hand it is possible to design fairly simple structural diagrams reflecting the principle of the sensor operation; on the other hand input of at least one new phenomenon to the knowledge base generates many new principles of sensor operation.

This paper discusses a conceptual model of knowledge classification based on energy-information models of circuits of different physical nature [5], [6]. The case of the model application for synthesis of multifunctioning sensors is described in [7].

2. Energy-Information Model of Circuits (EIMC)

Energy-information model of circuits (EIMC) enables switching to structural and formalized description of processes of sensor elements by using parametric structural schemes [5], [6]. The operation principle of any element of data measurement and control system is based on interaction of circuits of different physical nature, which is simulated in EIMC by the sequence of physical and technical effects. The EIMC uses the following notions:

Circuit of a certain physical nature is an idealized material medium having defined geometrical dimensions and described in terms of physical constants inherent in phenomena of a given physical nature only.

Generalized circuit values of the same physical nature are varying within wide limits and describe external action on the circuit of the given physical nature and the circuit's response to this action. Main generalized values of EIMC: Q – charge, P – pulse, I – response, U – action.

Generalized circuit parameters describe relative stability of a material medium in which physical process proceeds, these parameters are determined by geometrical dimensions, physical and chemical properties of the materials. Main generalized parameters of EIMC: R – resistivity, $G=1/R$ – conductivity, C – capability, $W=1/C$ – tension, L – inductivity, $D=1/L$ – deductivity (the value opposite to inductivity)

EIMC criteria is a system of equations showing the relations between the generalized values and generalized parameters. The simplest set of parameters for systems with lumped parameters includes six equations [5].

The authors have discovered the systems of generalized-values and generalized-parameters to describe the processes in the circuits of different physical nature (mechanical, thermal, electromagnetic, hydraulic, optic, moisture transfer).

In addition to elementary links which reflect conversions inside the circuit corresponding to the EIMC criteria, links to physical and technical effects (PTE) to represent interaction of circuits of different physical nature with each other are introduced.

Physical and technical effect (PTE) is a real cause and effect relationship, that reflects dependence between the input and output values of different physical nature through the conversion factor.

The whole variety of interactions between values and parameters can be represented in the form of a complex graph (figure 1) using energy-information models for description of circuits of different physical nature. The figure 1 shows the graph of physical and technical effects and in-circuit dependencies for n-circuits: mechanical, magnetic, electric and i-th physical nature circuit.

In the form of circuits of generalized values conversion based on EIMC, an expert describes existing physical phenomena and develops PTE passports [6] for a formal description of processes in technical devices.

Development of a PTE passport is a time-consuming process:

- it requires processing large amounts of information,
- it requires coordinated work of experts in different fields of knowledge,
- complex conversions of known physical equations to formulas containing EIMC values and physical constants,
- involves collection of expert evaluations of performance characteristics for multiple technical implementations of each PTE.

To solve this problem, we can use the ontological knowledge base of physical and technical effects.

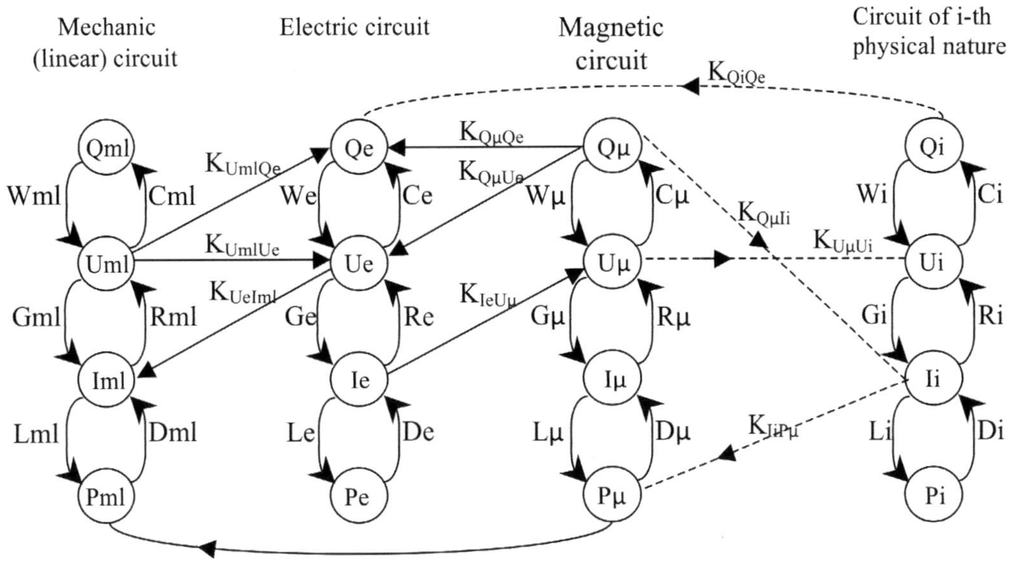

Figure 1. Bond graph for n circuits of the various physical nature.

3. Ontological Knowledge Base to Support Expert Analysis

Ontology is a system of fundamental concepts that provides the developer with an opportunity to model and represent a certain area of the world from the point of view of axiomatic definitions and taxonomic structures [8]. In the context of our task, it is the system of the above EIMC notions.

We suggest using the following types of ontologies to describe the energy-information model:

1. **EIMC model ontology** determines notions of this model (a circuit of a certain physical nature, values, parameters, inter-circuit physical effects, EIMC criteria). See figure 2.
2. **Physical effects ontology** describes classification of physical effects, represents correspondence between EIMC values and parameters and physical values and constants.

3.1. Example of Transformation of Physical Phenomena through EIMC

The task to make an energy-information model of PTE is rather time-consuming, since a proper combination of formulas in a broad list of the existing ones has to be found to bring the initial dependence of physical values to a well-defined form of interrelation of the EIMC analogue-values.

Let us look at the example of classification through the EIMC of a physical phenomena – the Hall effect (production of a voltage difference (the Hall voltage) as a result of a current flow curvature in metals in the magnetic field).

If a conducting plate (made of metal or semi-conductor) of thickness d has a direct current I_e passed in one direction, and magnetic field B (induction density) applied in another direction, then voltage U_e (the Hall voltage) can be measured in the third direction; this voltage is changing, as formula (1) shows, proportional to the magnetic force and current density:

$$U_e = R_H \cdot d^{-1} \cdot B \cdot I_e \,, \qquad (1)$$

where R_H is the Hall coefficient equal to the ratio of carrier mobility to the material specific conductivity.

To implement the energy-information model of this effect, formula (1) has to be converted so as to turn magnetic field B into magnetic flux Q_{mg}, which is the EIMC value. To achieve this, equation (1) can be converted in the following ways:

$$U_e = R_H \cdot d^{-1} \cdot B \cdot I_e \cdot S \cdot S^{-1} = R_H \cdot (d \cdot S)^{-1} \cdot (B \cdot S) \cdot I_e = K_{Q_{mg} I_e U_e} \cdot Q_{mg} \cdot I_e \,, \qquad (2)$$

where S is the sensing area of Hall effect sensor (1-50 mm^2).

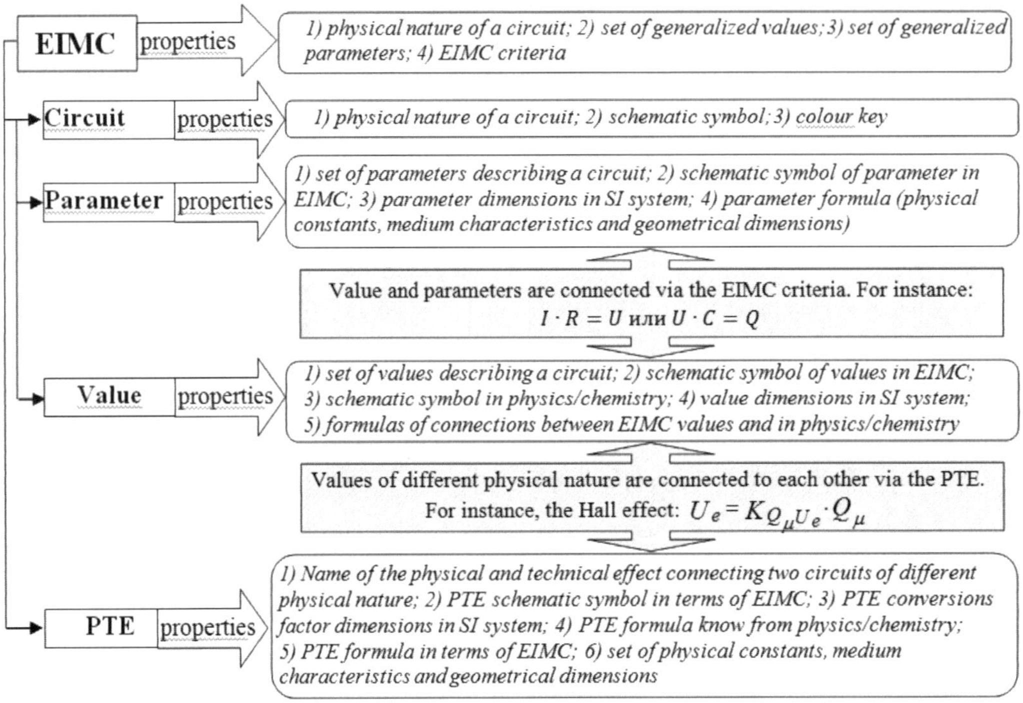

Figure 2. EIMC Model Ontology.

$Q_{mg} = B \cdot S$ - a magnetic flux perpendicular to the sensing area is a generalized EIMC value.

We obtained formula (3) of the conversion ratio for the energy-information model of the Hall effect. The effect model has two inputs – Q_{mg} and I_e, and one output – U_e.

$$K_{Q_{mg} I_e U_e} = R_H \cdot (d \cdot S)^{-1} \,, \qquad (3)$$

3.2. Ontology Knowledge Base and Software to Support the Expert Analysis

We have developed the knowledge and correspondence base of physical values and their EIMC equivalents to reduce expert labor input in classification of physical and technical effects. This knowledge base can be used to search for formulas which enable conversion of formulas known in physics into the EIMC adopted form.

We have proposed the following algorithm to solve this task. The user inputs the request in the form of values and constants that are used in the initial formula. For instance, for the reviewed Hall effect the request will be as follows:

$$U_e, R_H, d, B, I_e \,, \qquad (4)$$

Then the request is processed by the system and then the result of its interpretation is displayed to the expert:

U_e "voltage // physical value",
R_H "Hall constant // constant value",

d "plate thickness // constant value",
B "magnetic field induction // physical value",
I_e "current intensity // physical value".

Each request component is interpreted as a physical value or a constant value. If unambiguous request interpretation is found impossible, the user will be offered the choice of options found in the data base, for instance: "speed // physical value" or "volume // physical value".

The task is to convert the initial list of physical values and constant values into the list containing EIMC analogue-values only and the values having constant values within the physical dependences considered. For the example in review we have:

U_e "EIMC analogue-value",
R_H "constant value",
d "constant value",
B <conversion required>,
I_e "EIMC analogue-value".

The system identifies values $(V_1, V_2, \dots, V_n)$, that have to be converted and selects proper set of formulas $(F_1, F_2, \dots, F_m)$ using the knowledge base of physical value dependencies and EIMC analogue-values. The conversion results in the set of EIMC analogue-values $(E_1, E_2, \dots, E_p)$ and the set of constant values $(C_1, C_2, \dots, C_q)$.

$$(V_1, V_2, \dots, V_n) \xrightarrow[F_1, F_2, \cdots, F_m]{} \langle E_1, E_2, \cdots, E_p \mid C_1, C_2, \cdots, C_q \rangle, \tag{5}$$

In this case the task is to convert B magnetic field induction. The knowledge base has information of the following dependence $Q_{mg} = B \cdot S$ (object F3). The reason to consider object F3 as a candidate formula to convert value B is that the knowledge base contains the fact "F3 uses B". Results:

F_3	defines	Q_{mg}
F_3	uses	S
Q_{mg}	is	EIMC value
S	is	Constant value

"EIMC value" and "Constant value" are the classes of the ontology knowledge base.

Thus, the result of using formula F_3 contains EIMC analogue-values and constant values only. Hereon the search of formulas is completed.

$$\langle U_e, R_H, d, B, I_e \rangle \xrightarrow[Q_{mg} = B \cdot S]{} \langle U_e, I_e, Q_{mg} \mid R_H, d, S \rangle, \tag{6}$$

As seen from equation (6), the conversion factor of the resulting elementary unit of the parametrical structural scheme will contain the following constant values: $\langle R_H, d, S \rangle$.

Figure. 3 shows the fragment of ontology class hierarchy.

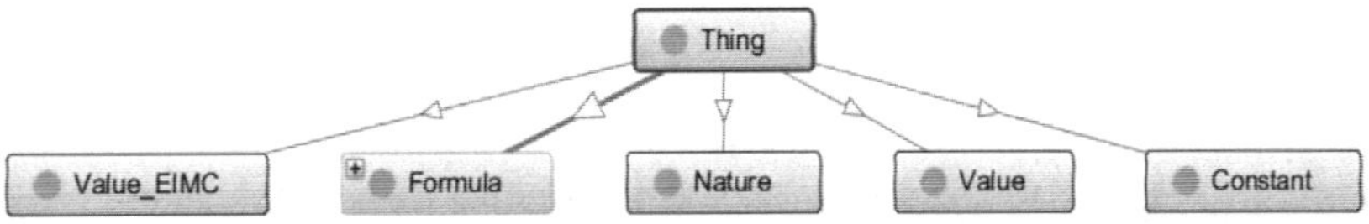

Figure 3. Fragment of the PTE Ontology Knowledge Base.

A fragment of ontology filling with facts is presented in the form of a diagram below in figure 4.

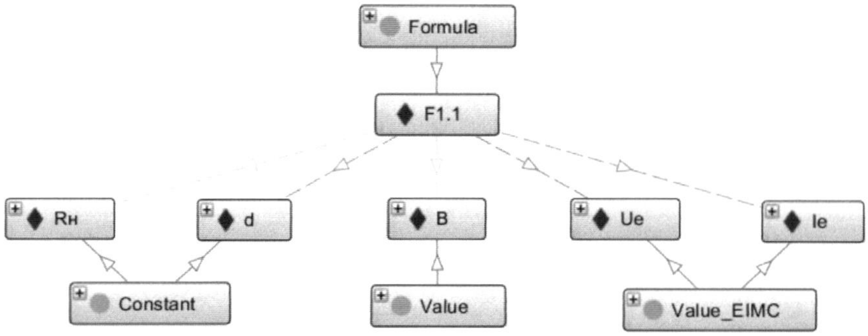

Figure 4. Abstract from Ontology Filling.

Figure 4 shows the following facts (see table 1):

Table 1. The explanatory to figure 4.

F1.1	hasMember RH	d	is Constant
F1.1	hasConstant d	B	is Value
F1.1	hasMember B	U_e	is Value_EIMC
F1.1	is DefinitionOf Ue	Ie	is Value_EIMC
F1.1	hasConstant Ie		
F1.1	is Formula		

The knowledge about nesting object properties in the ontology enables, by the results of ontology classification (using inference engine), adding implicit properties, in this case implicit objects of hasMember property (d, U_e, I_e).

4. Conclusion

The proposed complex of information processing methods in the knowledge base by physical effect provides an opportunity to create unique reference systems and to reduce expert labor input in creation of passports of physical and technical effects and to optimize search for multiple evaluations of performance of technical implementations of these effects.

Ontology and subsystem architecture for ontology management developed by the authors facilitate information search for experts and performance evaluations to be included into each PTE passport. This enables quantitative assessment of each solution and selection of the best one using the complex structured criteria.

References

[1] Altshuller G (Altov H) 2008 *And Suddenly the Inventor Appeared* (Worcester, MA: Technical Innovation Center)
[2] Altshuller G 2000 *The Innovation Algorithm* (Worcester, MA: Technical Innovation Center)
[3] Pahl G, Beitz W 1995 *Engineering Design* (Berlin: Springer-Verlag)
[4] Koller R 1985 *Konstruktionslehre für den Maschinenbau – Grundlagen des methodischen Konstruierens* (Berlin: Springer)
[5] Zaripov M, Petrova I and Zaripova V 2002 *Proc. Joint IMEKO TC-1 & XXXIV MKM Conference Education in Measurements and Instrumentation - Challenges of New Technologies* vol I (Wroclaw: Oficyna Widawnicza Politechniki Wroclawskie) pp 171-176.
[6] Petrova I, Zaripova V 2007 *Int J Information Technologies and Knowledge* **1** 37-43.
[7] Petrova I, Shikulskaya O and Shikulskiy M 2014 *Advanced Materials Research* **875-877** pp. 951 - 956 doi: 10.4028/www.scientific.net/AMR.875-877.951
[8] Guarino N, Giaretta P 1995 *Towards very large knowledge bases—knowledge building and knowledge sharing* (Amsterdam: IOS Press) p 25–32

Metrological Reliability of Medical Devices

E Costa Monteiro[1] **and LF Leon**[2]
[1, 2]Postgraduate Program in Metrology, Pontifical Catholic University of Rio de Janeiro, Brazil

E-mail: beth@puc-rio.br

Abstract. The prominent development of health technologies of the 20th century triggered demands for metrological reliability of physiological measurements comprising physical, chemical and biological quantities, essential to ensure accurate and comparable results of clinical measurements. In the present work, aspects concerning metrological reliability in premarket and postmarket assessments of medical devices are discussed, pointing out challenges to be overcome. In addition, considering the social relevance of the bio-measurements results, Biometrological Principles to be pursued by research and innovation aimed at biomedical applications are proposed, along with the analysis of their contributions to guarantee the innovative health technologies compliance with the main ethical pillars of Bioethics.

1. Introduction

The first discussions concerning traceability in measurements for biosciences occurred during the 20th General Conference on Weights and Measures (CGPM), held in 1995. The theme has been directly addressed in the 21st CGPM (1999), when, recognizing the need of accurate measurements traceable to the International System of Units (SI) in the fields of human health, environmental protection, among others, it was recommended the establishment of an adequate international measurement infrastructure to ensure traceability in biotechnology (11th resolution). Afterwards, in 2000, a Working Group on Bioanalysis (BAWG) was established at the 6th meeting of the Consultative Committee for Amount of Substance: Metrology in Chemistry (CCQM) to address metrological issues in biotechnology and molecular biology. However, to ensure accuracy and comparability of clinical measurement informations, validity not only of biological, but also of physical and chemical measurements are required.

The accelerated evolution of medical devices in the 20th century increased their complexity and capability of measuring physiological parameters essential to diagnosis, prevention, treatment, supporting life and clinical monitoring. On the other hand, the Harvard Medical Practice Study (HMPS), published in 1991, indicated that medical device-related adverse incidents were a common component of hospital care (3.7% of the hospitalizations), with 13.6% of them resulting in death [1].

In the 22nd CGPM (2003), discussions to promote international collaboration led the Consultative Committees to establish strong links, among others, with the World Health Organization (WHO). As a

[1] To whom any correspondence should be addressed.

Content from this work may be used under the terms of the Creative Commons Attribution 3.0 licence. Any further distribution of this work must maintain attribution to the author(s) and the title of the work, journal citation and DOI.
Published under licence by IOP Publishing Ltd

result, in 2004, WHO Consultation on Global Measurement Standards and their use in the in vitro biological diagnostic field was held to discuss metrological traceability and measurement uncertainty.

In 2007, the World Health Assembly adopted its first resolution for health technologies with WHO objectives of improving access and quality, but no approach to metrological reliability was considered.

An important step toward traceability was given in 2009, with the Workshop on "Physiological Quantities and SI Units", at the BIPM, including discussions concerning WHO International Standards and SI, the new ISO/IEC series 80003 standards "Quantities and units used in physiology" under development, besides the role of Consultative Committees in their domain of activities. However, no emphasis has been placed on traceability for medical devices in the subsequent BIPM meeting for bio-measurements in 2011 [2], as well as, during the first and second WHO Global Forum on Medical Devices, held in 2011 and 2013, respectively.

The neologism Biometrology was mentioned in the 24th CGPM meeting in 2011, while announcing a BAWG symposium on "Biometrology in support of Clinical Diagnostics". Considering Biometrology as metrology applied to biosciences, aiming at contributing to the comparability and reproducibility of its wide range of bio-measurements associated, Biometrological Principles should be pursued by research and innovation developments for biomedical application [3].

In the next sessions, metrological reliability in premarket and postmarket assessments of medical devices are discussed and the Principles of Biometrology are presented, along with their contributions to compliance with the main ethical principles of Bioethics.

2. Metrological Reliability of Medical Devices

The earliest initiatives recognizing the need of metrological traceability of health technologies occurred in the area of radiation dosimetry, since the nineteen-sixties [4]. In 1976, the International Atomic Energy Agency (IAEA) together with the WHO established a network of Secondary Standard Dosimetry Laboratories, known as the IAEA/WHO SSDL Network. However, worldwide regulatory systems for medical devices are less developed than those for medicines or vaccines.

In this section, some aspects concerning the metrological reliability, premarket and postmarket evaluation of medical devices are discussed.

2.1. Premarket Evaluation.

Regulatory requirements vary from country to country, but in most of them there is a national authority responsible for implementing medical device regulations. The pre-market approval applied by many of these National Regulatory Agencies requires medical devices to satisfy safety and performance, quality system and labelling requisites, according to a risk-based classification, for manufacturers to launch in the consumer market. Technical standards, published by the International Electrotechnical Commission (IEC) and the International Organization for Standardization (ISO), provide wide-ranging requirements for medical device that are used by regulators for conformity assessment. The 60601 series of standards published by IEC since 1977, contains requirements for medical electrical equipment conformity assessment, with comprehensive general and specific contributions, including requirements for environmentally conscious design (IEC 60601-1-9:2013). Likewise, the International Organization for Standardization (ISO), with a broad range of technical committees targeted at bio-measurement and healthcare issues, contributes for health technology reliability in several aspects that include, for instance, requirements for effective management of the risks associated with the use of medical devices (ISO 14971:2007). However, traceability and uncertainty are not yet emphasized.

The International Organisation of Legal Metrology (OIML) publishes recommendations that may be adopted as national metrological regulations. The OIML metrological characteristics requirements for measuring instruments include the specification of methods and equipment used for conformity assessment and take into account the traceability to SI. In addition, OIML recommendations cover both pre-market and post-market operational life, by means of initial and subsequent verifications, respectively. Among 130 OIML Recommendations already published, only 17 publications consist of a direct contribution to the healthcare sector.

The Quality by Design (QbD) concept, already used in pharmaceutical industry since 2004 [5], consists of a systematic approach that aims to ensure quality by employing statistical, analytical and risk-management methodology in the design, development and manufacturing [6], being a possible tool to be applied to medical device development as well as design the manufacturing processes to meet predetermined quality, safety, and efficacy requisites [7]. The QbD approach implementation aims at improving speed to market and operating efficiency, as well as reducing product variation and costs at all stages of the process [8]. Although being a premarket approach, the QbD proactive quality system is built to reduce post market potential problems [8]. Metrological traceability requirements, however, do not appear in the QbD concept literature and, therefore, should also be incorporated into the predefined objectives considered in the quality approach.

2.2. Post-market evaluation

Aiming at ensuring reliability of devices in use, many regulatory systems adopt post-market surveillance and vigilance, with adverse event reporting. By disseminating the reported informations, it is possible to reduce the likelihood or prevent repetition of adverse events, or alleviate consequences of such repetition [9].

However, with the very few exceptions of medical devices for which there are technical regulations based on OIML international recommendations, the approaches used by each country during post-market phase do not include preventive periodic evaluations to avoid adverse incidents caused by health technologies in use.

An increasing number of reported adverse incidents added to results found in the literature showing non-compliances in safety and performance evaluations of medical devices in use [3, 10, 11, 12, 13, 14, 15, 16], point out to the insufficiency of surveillance and adverse incident schemes to ensure metrological reliability and guarantee safety of the medical equipment in use, underlining the substantial need for post-market periodic evaluation [3, 10, 16].

Voluntary accreditation of health care facility, including periodic calibrations requirements, is an alternative for reliability of biomedical equipment in use, but the benefit is socially restricted. In 2007, IEC Standard 62353 has been published including test procedures, test methods and test intervals which can be used during post-market phase. Another aspect to be considered in order to ensure access to appropriate medical devices, is the proper management and use of medical equipment by means of the inclusion of clinical engineers with expertise in the field of metrology among the members of hospital staff.

Additional challenges to be overcome are the insufficiency of accredited laboratories enable to perform conformity testing of medical devices, the possible inadequacy between maximum permissible measurement error and the precision required in certain diagnostic and therapeutic procedures, the lack of requisites of measurement uncertainty and metrological traceability among the technical standards requirements and the existing gaps in the availability of national or international measurement standards to establish metrological traceability [3, 11, 16, 17].

3. Principles of Biometrology and contributions to Bioethics

Considering the social relevance in healthcare of the application of the science of measurement to biosciences, able to contribute to an ethical orientation of science and technology and to improve global safety and health, the following "Biometrological Principles" should be pursued by research, development and innovations in life-sciences:

- high accuracy, high precision, traceability to SI units and measurement uncertainty;
- low-cost for production/operation;
- low operational complexity;
- social, environmental and economic sustainability;

- non-invasiveness; and
- innocuousness.

The application of these Biometrological Principles contributes to satisfy the requirements of Bioethics, a recent academic inter-disciplinary field consisting of a systematic investigation of human conduct in the area of life sciences and healthcare, encompassing the study of the ethical and moral implications of new biological discoveries and biomedical advances [18].

The context that led to the emergence of bioethics, in the early 1970s, consisted of the advances provided by the scientific and technological development. Although such advances represent hopes of improving the quality of life, there were clear evidence of contradictions that indicated the need for a responsible analysis aiming at ensuring stability and future well-being of the human species and of life itself on the planet.

A commonly used bioethical analysis model is the "principlism", developed in the late 1970s [19]. In this system, ethical frameworks and procedures consider the Four Principles Approach: beneficence, non-maleficence, respect for autonomy and justice. These principles are key points in the life sciences practice. The observation of these principles contributes to the preservation of human dignity, protects the patient from undesirable results and also the professional implications on ethical or legal sphere in view of complications [19].

In this session, the contribution of the here proposed Principles of Biometrology to assure the healthcare technology compliance with the Principles of Bioethics is evaluated by means of the concept association analysis between both principles approaches, structured from the definition and description of the fulfilment of each ethical principle.

The principle of beneficence is characterized by the need to maximise the benefit to the patients. The contribution for an effective and safe medical procedure provided by the biometrological principle of high accuracy, high precision, traceability to SI units and measurement uncertainty, allows for achieving the primary goal of highly accurate diagnosis and therapy. The high accuracy pillar, reducing the uncertainties in the procedures, makes the desired effect preponderate over the foreseeable risks. In addition, the biometrological pillar of low-cost for production and operation guaranties the possibility of same goal expanded to the population at large.

The principle of non-maleficence, universally enshrined by the Hippocratic aphorism: "*Primum non nocere*", which enforces that the benefit is not accompanied by harm, refers to potential risks. Biometrological Principles concerning accuracy, precision, traceability, measurement uncertainty, sustainability, low operational complexity, non-invasiveness and innocuousness of biomedical procedures and technologies contribute for high performance and safety, reducing the occurrence of adverse events such as misdiagnoses, sequelae of treatments, accidents and iatrogenic injuries, evils still common in biomedical practice [20].

The principle of respect for autonomy implies the allocation of power to the patient to make decisions about medical issues, being the basis for the practice of "informed consent" in the physician/patient transaction regarding health care. Before implementing the medical care plan, it is now commonly accepted that the patient must be given an opportunity to make an informed choice. Although still not considered, this information should include explicit mention of the level of uncertainties associated, an essential database for correct decisions and provision of free and informed consent.

The availability of information concerning reliability of healthcare procedures and techniques, with stated accuracy, precision, metrological traceability to the SI, measurement uncertainty and corresponding adequacy for the given purpose, is a contribution provided by the biometrological principle to promote the autonomous decision of the patient. Awareness of risk-benefit associated with the diagnostic and therapeutic orientation, with disclosed information about quality comparability among bio-measurement results achieved by different establishments, contributes to the decision authority of patients.

Finally, the bioethical principle of Justice establishes the pursuit of equity as a fundamental condition. Adopted in 1946, the Constitution of the World Health Organization marks the first formal demarcation of a right to health in international law. In turn, the WHO Constitution defines the right to health as "the enjoyment of the highest attainable standard of health", including, among other principles of this right, the equitable dissemination of medical knowledge and its benefits. Likewise, Article 25 of the Universal Declaration of Human Rights (1948), the first international declaration of fundamental human rights, everyone has the right to medical care. Many people worldwide suffer because they don't have access to high quality, affordable health technology with the problem being more acute in low- and middle-income countries. In 2013, WHO published the Compendium of Innovative Health Technologies for Low-Resource Settings, strengthening the necessity to meet the ethical principle of justice.

The complete group of recommended pillars of Biometrology gives contributions to the universal and equitable access of benefits of life sciences advances, and, consequently, to fulfil the principle of justice.

The biometrological pillars of low-cost and low complexity for production and operation will lead to the development of innovations whose benefits can be widely accessed. On the other hand, metrological performance aspects such as accuracy, precision, metrological traceability to the SI and measurement uncertainty disclosure, also contribute to the principle of Justice, through the reduction of expenses with false positives, represented by costs of misguided treatments or the treatments of the damage to health caused by misguided treatments; with false negatives, including costs of treatments of the complications of not early enough diagnosed diseases; and with the repetition of tests.

By means of the respect to the social, environmental and economic sustainability principle, the benefit provided to those who demand health technology innovation, is not accompanied by damage to others, caused by an unfair distribution of risks, for instance, as a result of non-compliance with the sustainability concepts along the manufacturing process.

In conclusion, concerning health technologies, the Principles of Biometrology provide a technical basis for the practical implementation the ethical foundations of bioethics.

4. Conclusion

In the present work, metrological reliability in premarket and postmarket assessments of medical devices were discussed, and, considering the social relevance of the bio-measurements results, Biometrological Principles to be pursued by research and innovation aimed at biomedical applications were proposed, along with their contributions to meet the main ethical pillars of bioethics.

Although the existence of a regulatory control for medical devices comprising pre-market approval and post-market surveillance based on adverse incident reporting; essential requirements to ensure the reliability, as metrological traceability and periodic evaluation of medical devices in use are not requested. In turn, regulations based on OIML recommendations include metrological requirements and preventive evaluations by means of subsequent verifications of devices in use. However, there are only a few recommendations published for health applications. A possible future implementation of QbD approach to medical device research, development, manufacturing and distribution can give contributions to reliability, but metrological traceability requirements have yet to be incorporated in the quality system. Voluntary health care accreditations have been an alternative to promote metrological reliability of medical devices in use, but with restricted access. An appropriate support of accredited laboratories network sufficient to meet the national demands; the lack of traceability and measurement uncertainty requirements comprised in technical standards and regulations; besides unavailability of national or international measurement standards are challenges to be overcome.

The proposed Principles of Biometrology provide a technical basis for practical implementation of ethical requirements, enabling to achieve complete compliance of healthcare technologies with the ethical foundations of bioethics. Their application aims at providing highly accurate diagnosis and therapy; reducing the occurrence of adverse events and expenses with false positives/negatives and repetition of tests; and contributing to the decision authority of patients and the universal access of benefits of life sciences advances; fully respecting the three aspects of sustainability.

Accuracy, precision, traceability to SI, measurement uncertainty, low-cost, low complexity, full sustainability, non-invasiveness and innocuousness should characterize the innovative health technologies of the new millennium.

5. Acknowledgments

We thank the Brazilian funding agencies CAPES, CNPq, FAPERJ and FINEP for financial support.

References

[1] Baker R G 2004 *Qual. Saf. Health Care* **13** 151-152.
[2] Marriott J, O'Connor G and Parkes H 2011 *Study of Measurement Service and Comparison Needs for an International Measurement Infrastructure for the Biosciences and Biotechnology: Input for the BIPM Work Programme* Rapport BIPM 115 pp.
[3] Costa Monteiro E 2007 Biometrologia: confiabilidade nas biomedições e repercussões éticas *Metrol. Instrum.* **6** 6–12.
[4] IAEA/WHO 1997 *Network of Secondary Standards Dosimetry Laboratories* **37** 1011-2669.
[5] Rathore Anurag S 2009 Roadmap for implementation of quality by design (QbD) for biotechnology products *Trends. in Biotechnology.* **27** 546-553
[6] ICH 2009 *Pharmaceutical development Q8(R2)* (ICH Harmonised Tripartite Guide)
[7] Veljanovski I, 2011 Integrating quality by design (QbD) in medical device manufacturing concept, benefits, and challenges *White paper Stevens Institute of Technology* pp.1–9.
[8] Weinberg S 2011 *Cost-Contained Regulatory Compliance for the Pharmaceutical, Biologics, and Medical Device Industries* (New Jersey: Wiley & Sons, Inc.)
[9] GHTF/SG2/N54R8 2006 *Medical Devices Post Market Surveillance: Global Guidance for Adverse Event Reporting for Medical Devices* (Global Harmonization Task Force)
[10] Costa Monteiro E and Lessa M L 2005 A Metrologia na Área de Saúde: Garantia da Segurança e da Qualidade dos Equipamentos Eletromédicos *Engevista* **7**, p. 51-60.
[11] Goloni C M, Temporão G P and Costa Monteiro E 2013 Metrological reliability of optical coherence tomography in biomedical applications *Journal of Physics*, **459**, p. 012-064.
[12] Palomares R P A, Gouvêa P M P and Costa Monteiro E 2006 Metrological Reliability of High Frequency Surgical Equipment *XVIII IMEKO World Congress* p. 1-4.
[13] Romero J C, Gouvêa P M P and Costa Monteiro E 2007 Avaliação Metrológica de Ventiladores Pulmonares *V Congresso Latino Americano de Metrologia* p. 1-4.
[14] Silva Jr S H and Costa Monteiro E 2008 Incerteza de medição do manômetro usado em esfigmomanômetros *IFMBE Proceedings* **18**, p. 890-894.
[15] Barrientos C P C, Costa Monteiro E, Lessa, L M and Barbosa C R H 2011 Evaluación de la confiabilidad metrológica de esterilizadores a vapor. *VI CBM Proceedings*, p. 1-5.
[16] Ferreira M C, 2011 The role of metrology in medical devices *OIML Bulletin* 4.
[17] Menegotto T, Costa Monteiro E, Lima M S, Almeida G B, Couceiro I B and Grieneisen H P 2011 Measurements of Transfer Standards for Radiometry at Inmetro *VI CBM Proceedings*.
[18] Reich WT 1995 *Encyclopedia of Bioethics* (Macmillan Pub. Co) vol 2 p 2950
[19] Beauchamp T L and Childress J E 2001 *Principles of Biomedical Ethics* (NY: Oxford press)
[20] Lawrence D J, 2007 The Four Principles of Biomedical Ethics: A Foundation for Current Bioethical Debate. *Journal of Chiropractic Humanities.*

Measurement uncertainty consideration in the case of non-linear models for precision length measurement

E. Manske[1], T. Fröhlich, R. Füßl

Technische Universität Ilmenau, Faculty of Mechanical Engineering,
P.O.Box 100 565, 98684 Ilmenau, Germany

eberhard.manske@tu-ilmenau.de

Abstract. The requirements on the precision lengths and precision load measurement rise due to the rapid technical development, in particular of nanotechnologies and precision optics. According to the roadmap "Dimensional metrology for micro- and nanotechnology" of the European National Metrology Institutes (iMERA)[2] in the next few years "traceable 2D (3D) metrology at (sub)-nm accuracy over several 100 mm range" will be required.

1. Introduction

Measurement uncertainties attainable with high precision instruments depend on the measuring ranges, the metrological structure, the sensors and among others on the conditions of use. So, on the one hand, low-error metrological arrangements are necessary for coordinate measurement systems with nanometer precision to achieve small uncertainties in budgeting. On the other hand, a measurement value correction is possible by detailed metrological modelling. These measurement uncertainty budgets can be created on the basis of vector models.

To achieve nanometre accuracy it is very important to use error minimum arrangements. In 3D-nano coordinate measurement machines it is necessary to fulfil the Abbe comparator principle in all three measurement axes [1, 2]. For highest performance we improve this approach by an additional extension: besides the aligned arrangement of measurement object and scale we control all angular deviations to be zero.

On the basis of this approach, we have realized a nanopositioning and nanomeasuring machine with a measuring range of 25 mm x 25 mm x 5 mm and meanwhile an improved resolution of 20 pm. To determine the measurement capability of the machine, we have analysed the measurement uncertainty of the whole machine by complex vector modelling (cf. Fig. 1) [3, 4, 5].

On this basis, error influences can be determined very effectively. We can derive conclusions for the design of the machine or of components as well as for operating the machine or the measuring conditions.

[1] Univ.-Prof. Dr.-Ing. habil. Eberhard Manske, Technische Universität Ilmenau
[2] http://www.technology-roadmaps.eu/doku.php?id=Micro_and_nano

Content from this work may be used under the terms of the Creative Commons Attribution 3.0 licence. Any further distribution of this work must maintain attribution to the author(s) and the title of the work, journal citation and DOI.
Published under licence by IOP Publishing Ltd

$$\vec{r}_{Aa} - \vec{r}_{Pa} + \sum_i \vec{r}_{FRai} - \vec{r}_{IRa} - \vec{r}_{\mathrm{Im}a} - \vec{r}_{\cos a} - \vec{r}_{Abbe-a} - \vec{r}_{MIea} = 0$$

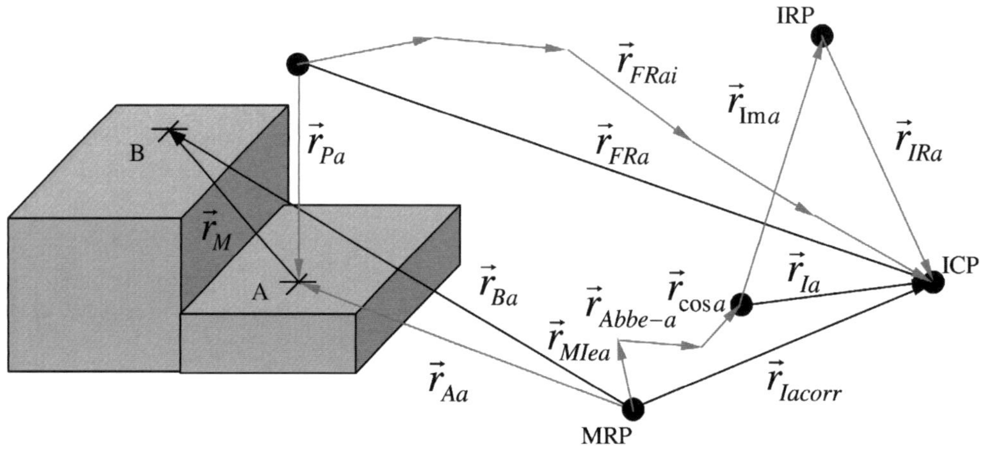

$\vec{r}_P$	probe system vector	$\vec{r}_{\cos}$	cosine error vector
$\vec{r}_{FRi}$	frame system vector	$\vec{r}_{Abbe}$	Abbe error vector
$\vec{r}_{IR}$	interferometer reference path vector	$\vec{r}_{MIe}$	mirror error vector
$\vec{r}_{\mathrm{Im}}$	interferometer measurement path vector		

Figure 1: Vector chain of the measuring point A

Some partial models often lead to non-linear models, for example in the case of geometrical errors. One interesting measurement aspect arises for partial models, where the expectations of all input values are zero, which also leads to an uncertainty of zero in case of the linear GUM approach.

2. Abbe error

In the field of one-dimensional interferometric length measurement it is always possible to minimize the so-called Abbe-error by an aligned arrangement of object to be measured and scale. In precision measurement technology, however, it is well possible for residual sine errors to cause significant errors due to insufficient alignment. For example, if two high-precision laser interferometers are to be compared with each other within a range of, for example, 2m, they cannot be adjusted to each other – in general – better than 500 μm (cf. Fig. 2).

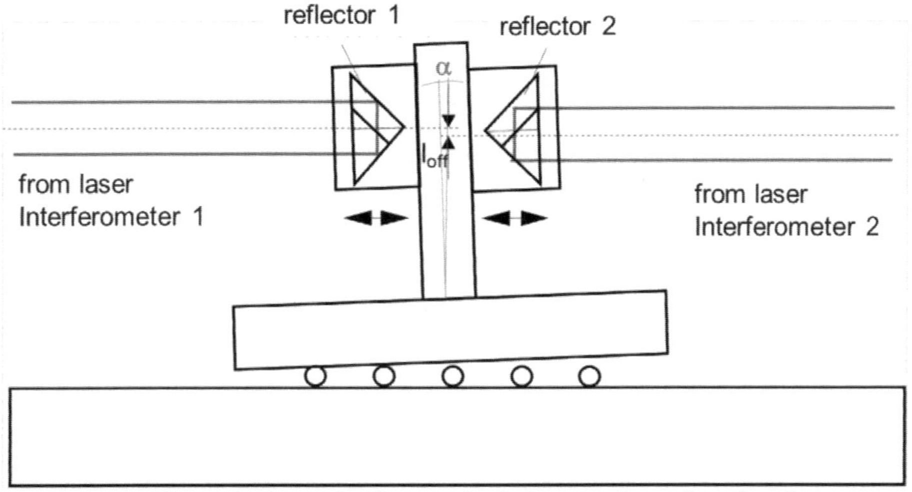

Figure 2: Set-up for the metrological comparison of two interferometers

Thus, for determining the measurement uncertainty contribution, for this set-up the problem arising here is that in the product

$$\Delta l = l_{off} \cdot \sin \alpha$$

both mean values are zero: $\bar{l}_{off} = \bar{\alpha} = 0$

The estimation of the measurement uncertainty according GUM leads to an incorrect solution for the uncertainty:

$$u^2(\Delta l) = \sum_{i=1}^{N} \left(\frac{\partial f}{\partial x_i}\right)^2 u^2(x_i) = \bar{\alpha}^2 u^2(l_{off}) + l_{off}{}^2 u^2(\alpha) = 0$$

In this case, the residual Abbe error can be derived with Steiners theorem:

$$u^2(\Delta l) = \mathrm{VAR}\{\Delta l\} = \mathrm{E}\{(\Delta l - \mathrm{E}(\Delta l)^2\} = \mathrm{E}\{\Delta l^2\} - \mathrm{E}^2\{\Delta l\}$$

With $\mathrm{E}^2\{\Delta l\} = 0$ we get:

$$u^2(\Delta l) = \mathrm{E}\{\Delta l^2\} = \mathrm{E}\{l_{off}^2\} \cdot \mathrm{E}\{\bar{\alpha}^2\}$$

and:

$$u^2(\Delta l) = (\mathrm{VAR}\{l_{off}\} + \mathrm{E}^2\{l_{off}\}) \cdot (\mathrm{VAR}\{\bar{\alpha}\} + \mathrm{E}^2\{\bar{\alpha}\})$$

Finally, because of $\mathrm{E}^2\{l_{off}\} = \mathrm{E}^2\{\bar{\alpha}\} = 0$ we get in contrast to the GUM approach the result:

$$u^2(\Delta l) = \mathrm{VAR}\{l_{off}\} \cdot \mathrm{VAR}\{\bar{\alpha}\} = u^2(l_{off}) \cdot u^2(\bar{\alpha})$$

If we consider under compliance of the Abbe comparator principle realistic values of 500 μm Abbe offset and 60 arcsec angular deviation for a high precision linear stage we can calculate 100 nm as expanded combined uncertainty (at k = 2).

For high precision measurements ($U \leq 10\mathrm{nm}$ at k = 2) this leads to the following deductions:

Journal of Physics: Conference Series **588** (2015) 012033 doi:10.1088/1742-6596/588/1/012033

a) a diminution of the Abbe offset to less than 50 µm is necessary or
b) an additional angular measurement of the linear stage with an uncertainty of less than 6 arcsec is necessary.

In [6], a method for determining the Abbe error as well as for correcting it is described. Here, the measuring table is tilted in a specific way, and the Abbe error induced by this tilting is determined. From this, the Abbe-offset can be specified and finally minimized through iterative adjustment and measurement. However, in the verification experiments some non-linearity that should not be neglected was found. The cause of this phenomenon has not been detected so far (cf. Fig. 4, left-hand side).

The same non-linear behaviour has also been observed by the authors. In Fig. 3, right-hand side, an iteration process can be recognised in the case of which the Abbe error has been reduced through the determined tilting by a constant angle from originally 5 nm to about 1 nm.

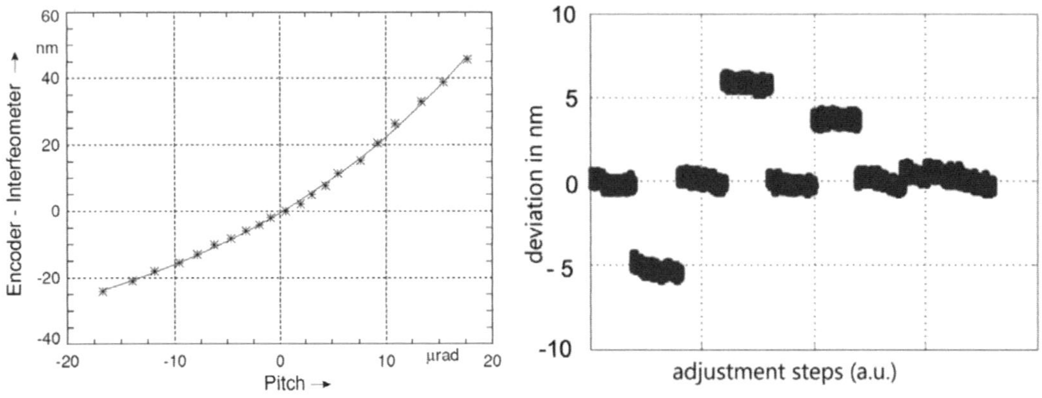

Figure 3: Measurement uncertainty of two measuring systems occurring due to an Abbe error [6] (left-hand side)/ Lowering of the Abbe error through gradual reduction of Abbe offset (right-hand side)

Under certain circumstances, the iterative adjustment process can be very time-consuming. If routine comparisons of interferometers are to be made, this approach will not be suitable. Such a test procedure (Fig. 2) can be extended by an additional angular measurement made by means of an autocollimator and a piezo – angular adjustment unit so as to control in a permanent manner the angular deviations through a closed-loop control circuit.

3. Cosine error

In length measurement, the cosine error is regarded as error of 2nd order. However, it becomes significant for precision measurements. For a detailed analysis of this error, the measuring arrangement must be inspected very carefully. While tactile length measurements lead in general to a positive measuring uncertainty, interference-optical measurements mostly present negative uncertainties:

$$\Delta l_{mech} = l(\frac{1}{\cos\alpha} - 1) \qquad \text{and} \qquad \Delta l_{interf} = l(\cos\alpha - 1)$$

Journal of Physics: Conference Series **588** (2015) 012033 doi:10.1088/1742-6596/588/1/012033

In both cases, the uncertainties represent unbalanced distributions. Figure 4 shows the frequency distribution for an interferometric relative cosine error. As for the angular deviation, it is a rectangle-distributed random error with a half-width of 200 arcsec. In this case, the expected value of the relative length measurement deviation amounts to $-1.2 \cdot 10^{-7}$.

The relative dispersion amounts to $1 \cdot 10^{-7}$, with the expected value being determined according to:

$$\Delta l_{cos} = -1/2 \cdot u^2(\alpha)$$

and the dispersion of the length measurement deviation according to:

$$u(\Delta l) = \sqrt{\frac{4}{5}} \cdot \Delta l_{cos}$$

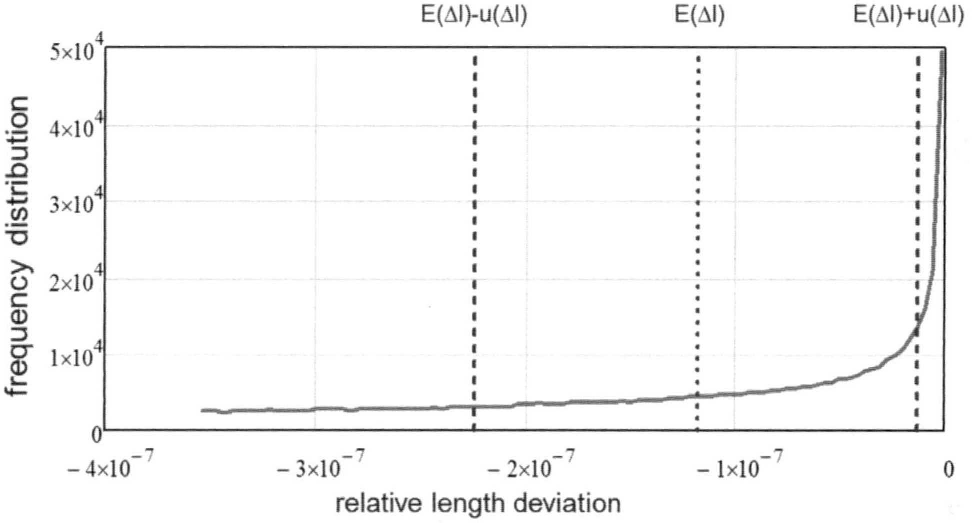

Figure 4: Frequency distribution for an interferometric cosine error

Here, the relation of the fourth central moment to the fourth power of the standard measuring uncertainty amounts to 9/5 (cf. [7]). These theoretically determined values have been checked with view to their correctness by means of the Monte-Carlo-method.

Assuming symmetrical probability (p = 95 %), measurement uncertainty can now be specified only by means of an asymmetrical overlapping interval (e.g. for a displacement path of 1m [-0.34µm, -0.0002µm]).

For the usual indication of 3σ-values, an upper limit of $2 \cdot 10^{-7}$ results erroneously, i.e., of positive cosine errors which, however, cannot occur in practice. Furthermore, it is quite unusual for the expected value of the cosine error to depend on the value of the standard deviation of the angular deviation, which is usually estimated. If this standard deviation is wrongly estimated, not only a change of the measurement uncertainty will result, but also a change of the expected value of the path measured.

4. Summary

In the present paper, some theoretical investigations on the two most important measurement uncertainty contributions of precision length measurement – the Abbe error and the cosine error – have been carried out. They provide non-linear sub-models in the measurement uncertainty analysis. Some practical conclusions concerning measurement arrangements with minimum errors have been drawn, and various proposals for minimizing the Abbe error in practice have been made. A possible method to improve the accuracy of linear length measurement is to additionally measure the pitch and yaw angles, combined with a stable control of the angular deviations in a closed-loop control circuit, or also by a computational correction of the measurement deviations conditioned by the tilt angle. The special feature of the cosine error consists in the asymmetrical distribution and in the correction of the expected value of the length measurement, with this correction being dependent on the value of the occurring angular deviation. Through the intensive work on the measuring uncertainty budget it is possible to derive interesting and important conclusions for further developments of new measurement strategies, arrangements and technology in the field of nanometrology.

5. References

[1] G. Jäger; E. Manske; T. Hausotte; H.-J. Büchner: The metrological basis and operation of nanopositioning and nanomeasuring machine NMM-1. In: Technisches Messen 05, pp. 227 – 234, 2009

[2] E. Manske; G. Jäger; T. Hausotte; R. Füßl: Recent developments and challenges of nanopositioning and nanomeasuring technology. In: Measurement Science and Technology, Vol. 23, 074001 (10pp), 2012

[3] R. Füßl; R. Grünwald; I. Rahneberg; et al.: Uncertainty Consideration of the Coordinate System in Nanopositioning- and Nanomeasuring Machines. In: Technisches Messen, Volume: 76 Issue: 5 pp. 235-238, 2009

[4] R. Füßl; R. Grünwald; I. Schmidt: Analysis of the measurement uncertainties of Nanopositioning and Nanomeasuring Machines (NPM-Machines) by means of a new vectorial model approach. In: Technisches Messen, Volume: 73 Issue: 9, pp. 465-471, 2006

[5] P. Kreutzer; N. Dorozhovets; E. Manske; et al.: Monte Carlo simulation to determine the measurement uncertainty of a metrological scanning probe microscope measurement. In: SPIE Conference: Scanning Microscopy, Monterey, CA, USA, Proceedings of the SPIE - The International Society for Optical Engineering, Vol. 7378, 737816 (12 pp.) DOI: 10.1117/12.821794, 2009

[6] R. Köning; J. Flügge; H. Bosse: A method for the in situ determination of Abbe errors and their correction. In: Measurement, Science and Technology 18, S. 476-481, 2007

[7] W. Kessel: Messunsicherheit - einige Begriffe und Folgerungen für die messtechnische Praxis. In: PTB-Mitteilungen 111, Heft 3, S.220-38, 2001

IMEKO IOP Publishing
Journal of Physics: Conference Series **588** (2015) 012034 doi:10.1088/1742-6596/588/1/012034

A comparison of measurement concepts across physical science and social science domains: instrument design, calibration, and measurement

M Wilson[1#], L Mari[2*], A Maul[3], D Torres Irribarra[1]
[1] Graduate School of Education, University of California, Berkeley, CA, USA
[2] School of Engineering, Università Cattaneo – LIUC, Castellanza (VA), Italy
[3] School of Education, University of Colorado, Boulder, CO, USA

E-mail: MarkW@berkeley.edu

Abstract. Is there a framework common to measurement in both physical and social sciences? The answer to this question would determine to an important extent the possibility of building a shared measurement-related body of knowledge across these traditionally separate domains. In this paper, we outline a framework of the processes involved in the construction and use of measures that includes instrument design, instrument calibration, and ultimately measurement using the instrument. A comparison of these steps across the two domains reveals both (a) formal parallelism, and (b) important differences in the way calibration is intended and implemented. We examine the similarities and differences to determine whether this is a case of irreducible difference, or whether the similarities are such that measurement in the two domains can be viewed within a single conceptualization.

1. Introduction

Measurement results are expected to convey information on the measurand in terms of the adopted scale (e.g., the numerical quantity value reported in the measurement result, the ratio measurand/unit), and to be independent of the adopted measuring instrument and of quantities other than the measurand.

The exclusive dependence of measurement results on the measurand would guarantee [a] the *objectivity* of measurement information, i.e., its object-relatedness, and the reference to a universally accessible scale would guarantee [b] the *intersubjectivity* of measurement information, i.e., its subject-independence, where object-relatedness and subject-independence can be considered characterizing features of measurement itself [1], [2]. How can they be obtained?

The answer to this question has nothing to do with the (possibly quantitative) form of the measurement results [3], but is grounded instead on the experimental structure of measurement systems: it is the suitable (i) design, (ii) calibration, and (iii) operation of such systems that gives measurement results the features of object-relatedness and subject-independence to an acceptable degree.

(i) *Design*. In the design stage the two basic requirements are:

[#] To whom any correspondence should be addressed.

[*]One of the authors is a member of the Joint Committee on Guides in Metrology (JCGM) Working Group 2 (VIM). The opinion expressed in this paper does not necessarily represent the view of this Working Group.

Content from this work may be used under the terms of the Creative Commons Attribution 3.0 licence. Any further distribution of this work must maintain attribution to the author(s) and the title of the work, journal citation and DOI.
Published under licence by IOP Publishing Ltd

[a] measuring instruments are designed so to have a stable behavior, sensitive to the measurand and insensitive to external influences ('influence quantities'), thus aiming at obtaining object-related information;
[b] measurement standards are designed to allow the realization of scales and units that are stable and socially widespread, thus aiming at obtaining subject-independent information.
(ii) *Calibration*. The outcome of the design stage is generally a class of measuring instruments or measurement standards, which must be then realized, i.e., instanced in specific devices that can be operated here-and-now. This requires establishing the values in the here-and-now conditions of the parameters left unspecified in the design, an outcome obtained by the calibration stage. Calibration is a process aimed at:
[a] removing instrument influence from the measurement results, and thus obtaining object-related information;
[b] making measurement results traceable to a given scale/unit, and thus obtaining subject-independent information.
(iii) *Measuring*. Finally, a calibrated measuring system can be operated in the measurement stage, where the operation can proceed:
[a] without interference from outside influences, and
[b] with consistency across subjects.

This framework justifies the structure of the meta-process: design first, then calibrate, and finally measure. The question we are addressing in this paper is the following: *Is this structure common to measurement in both physical and social sciences?* A positive answer to this question would seem to be a precondition for building a shared measurement-related body of knowledge across these different domains.

Some historical background in the social sciences on this can be gathered by considering the views of O.D. Duncan: "Let us note the clear separation between the two tasks: first, to scale the statements, or to infer their location on a linear attitude continuum defined in terms of the polar contrast of "strongly affirmative" with "strongly negative"; and, second, to place respondents on that same attitude continuum insofar as their locations on it can be inferred from the selection of items they choose to endorse. Not only are the two problems logically distinct and operationally separated, but also the former – the scaling of statements – precedes and is presupposed by the latter." [4: p.174].

Despite the differences in language and content (the description above is about Thurstone's scaling, a basic measurement method in the social sciences [5]), the reference is manifestly to a situation that is usual to physical measurement too: first, the measuring instrument is designed and calibrated; second, it is used in measurement, where calibration and measurement are indeed, in Duncan's words, "logically distinct and operationally separated" processes and "the former precedes and is presupposed by the latter". On the other hand, Duncan continued: "By contrast, psychometric or "test theory" methods of attitude measurement either attempt to accomplish the two tasks simultaneously and jointly or even to bypass the scaling of items altogether." [4: p.174].

Here the pivotal role of calibration is under discussion: is it really possible to measure and to calibrate at the same time, or to measure by means of an uncalibrated instrument? Duncan has demonstrated that at least some social science measurement approaches do not seem to conform to the 3-stage system described above. Now, our task is to investigate whether this is a case of irreducible difference (i.e., the same term, "measurement", is used with two irreconcilably different meanings), or whether the differences may instead be merely due to differences in commonly-used terminology and conventional practices between the two domains.

2. A black box approach to measuring instrument design, calibration, and measurement

With the aim of introducing what may constitute the simplest sharable basis for a measurement-related conceptual framework, we propose a purely structural view of the meta-process under consideration as follows (an even simpler description would be about measurement by direct comparison, as performed by a ruler or an equal arm balance, but this case is much less interesting for our purposes). The reader can interpret what follows as the design, calibration, and operation process of, say, a simple thermometer.

1. One observes that a property **Y** of an object *b* changes (say, the length **Y**(*b*) of *b*). From further observations one hypothesizes that the changes of **Y**(*b*) are systematically and regularly caused by the changes of a property **X** of another object *a* that somehow interacts with *b* (say, the temperature **X**(*a*) of *a*, being it possibly the environment surrounding *b*). "Systematically and regularly" only means here that, *ceteris paribus*, if in different times, t_1 and t_2, one observes $\mathbf{Y}(b(t_1))$ is [not] equal to $\mathbf{Y}(b(t_2))$ (the length of *b* is [not] the same at the times t_1 and t_2), then she hypothesizes that also $\mathbf{X}(a(t_1))$ is [not] equal to $\mathbf{X}(a(t_2))$ (the temperature of *a* is [not] the same at the times t_1 and t_2). In this situation one can write F(**X**(*a*)) to denote the property **Y**(*b*) caused by **X**(*a*), where then F is a causation mapping that describes the *transduction* $\mathbf{X} \rightarrow \mathbf{Y}$ performed by *b*. One also observes that the same behavior of *b* is obtained by its interaction with a whole set of objects $\{a_i\}$, i.e., all a_i have the property **X** and all $\mathbf{X}(a_i)$ systematically and regularly produce corresponding changes of **Y**(*b*).

2. A socially widespread enough and stable enough object a^* in $\{a_i\}$ is available, so that $\mathrm{F}(\mathbf{X}(a^*))$ can be easily obtained and generally $\mathbf{X}(a^*(t_1)) = \mathbf{X}(a^*(t_2))$ (how can the relationship 'having the same temperature' be determined without recurring to *b*? let us omit the discussion of this subject here).

3. Given the object a^* the relation 'having the same **X** as a^*' (having the same temperature as a^*) can be experimentally determined by means of the object *b*, according to the procedure:

[P] for each candidate object *a*, in $\{a_i\}$, *a* has the same **X** as a^* if $\mathrm{F}(\mathbf{X}(a)) = \mathrm{F}(\mathbf{X}(a^*))$

(note that this procedure does not require the comparison, nor the comparability, of **X**(*a*) and $\mathbf{X}(a^*)$). The information that *a* has the same **X** as a^* can be written $\mathbf{X}(a) = \mathbf{X}\text{-of-}a^*$ for short.

Hence:

– *a* is the object under measurement;
– a^* is an object used as a measurement standard;
– *b* is used as a measuring instrument;
– **X** is the property subject to measurement;
– **Y** is the instrument indication;
– **X**-of-a^* is the value realized by the measurement standard,

and:

– Stage 1 corresponds to the *design* of the measuring instrument *b*;
– Stage 2 corresponds to the *selection of a measurement standard* and the (first) *calibration* of *b*;
– Stage 3 corresponds to a *measurement* performed by means of *b* according to the procedure [P].

Note also that:

– only the causal relationship between **X** and **Y**, as realized by *b*, is assumed: no known mathematical functions or models are required;
– despite this simplicity, model-ladenness is unavoidable, at least in the assumptions of ceteris paribus comparison (i.e., all influence properties did not change), and stability of the standard.

Unfortunately, something else must be added to this description.

4. The transduction behavior $\mathbf{X} \rightarrow \mathbf{Y}$ of *b* could be recognized to be not perfectly stable, i.e., $\mathrm{F}(\mathbf{X}(a^*(t_1))) \neq \mathrm{F}(\mathbf{X}(a^*(t_2)))$ even if $\mathbf{X}(a^*(t_1)) = \mathbf{X}(a^*(t_2))$ (the caused lengths are different even if the causing temperature remains the same). This shows that in general the mapping F could be (for example) time dependent in its turn (where the causes of this time dependence might be in the failure of the ceteris paribus hypothesis, but again we will not develop this subject here). Hence $\mathrm{F}(t, \times)$ must be established so that from $\mathbf{Y}(b(t)) = \mathrm{F}(t, \mathbf{X}(a(t)))$ (the length of *b* obtained by its interaction with the temperature of *a* at the time *t*) one can obtain whether $\mathbf{X}(a) = \mathbf{X}\text{-of-}a^*$ or $\mathbf{X}(a) \neq \mathbf{X}\text{-of-}a^*$. The idea is, trivially, of identifying $\mathrm{F}(t', \mathbf{X}(a^*(t')))$ at a time *t*' "close enough" to the measurement time *t*, so that one can suppose that the following time-dependent version of the procedure [P]:

[P'] for each candidate object *a*, in $\{a_i\}$, *a* has the same **X** as a^* if $\mathrm{F}(t, \mathbf{X}(a(t))) = \mathrm{F}(t', \mathbf{X}(a^*(t')))$

can be applied. This stage corresponds then to a (re)*calibration* of *b*.

This shows the most fundamental reason why instrument calibration and measurement are (complementary but) distinct processes: the former operates on an object – a measurement standard – whose quantity under consideration is assumed to be known, whereas the latter operates on an object – the object under measurement – whose quantity under consideration is assumed to be unknown and in fact looked for by means of measurement itself.

3. Social science measurements: the case of norm-referenced measurement

Now, the question can be asked: does the black box approach described in the previous Section apply to (at least some) social science measurements? As the quotation from Duncan above shows, it is not straightforwardly clear that the response is always "yes". In fact, it must be said that social science measurement does not usually allow for the concept of a measurement standard (i.e., the object a^*) in the straightforward way that it is used above. There are, for example, no actual persons who are generally agreed to be the measurement standards of 100 IQ (or 115, 130, etc.), and this is also true in other social science applications such as the measurement of attitudes, and academic knowledge and skills. Instead, the most common form of standardization is based on a so-called "norm-referenced" perspective where, say, the mean response in a population is labeled with a certain numeral (e.g., 100 in the case of IQ tests), and other points are given labels depending on their location in the frequency distribution of the responses (such as 115 for 1.0 standard deviations above the mean, 101 for $1/15^{th}$ of a standard deviation above the mean, etc.). Clearly, this technique depends on not only the instrument as it is usually conceptualized (usually as a set of items to which the persons respond in some way, which are then categorized according to a particular theory), but also on the sample of persons that were used to calculate the statistics on which the labels were based (i.e., the mean and standard deviation of the population in the IQ example). Thus, if one extends the concept of an instrument to include the particular sample to which it was administered (or perhaps more broadly, to a definition of a "sample representative of a population"), then the status of this as a measurement standard can be (at least abstractly) claimed. In fact, there are historical cases where this dependence on a specific sample has been maintained over extended periods of time (one example is the commonly used scale of scores for the SAT, which were originally established on a particular sample of college entrance candidates in the 1940s, and then used as a consistent framework through the 1990's, although the link has been subsequently abandoned). The manner in which this historical link is maintained (or is not maintained) can have a range of interesting consequences. For example, Flynn's [6] showed that, according to the results of typical IQ tests, the IQ of the populations of many countries has risen considerably throughout the 20th century, yet this phenomenon was not generally apprehended by IQ researchers during the 20th century precisely because they continually updated their IQ scales to match the current population, and the norm-referenced standardization described above concealed the change.

The details of how the norm-referenced approach can be conceptualized as a calibration are given in [6]. The essential insight is to see that the specific set of response vectors that correspond to an IQ score of, say, 100, are indeed the measurement standards here. The correspondence of this example to the black box approach given above is given by the following:

– a is the object under measurement (a person);
– $\mathbf{X}(a)$ is the intelligence of a;
– b is the IQ test (including both content and technical specifications);
– $\mathbf{Y}(b)$ is the vector of responses to b;
– y is the indication value, the so-called "score" of $\mathbf{Y}(b)$,

and for each measurement standard:

– a^* is the set of persons whose response vectors have the same score as that which has been assigned for that measurement standard (e.g., the set of persons who have a score that is sufficiently close to the mean).

Now to make this a bit more concrete, take a (non-typical, but small enough to be clear) case where there are 10 dichotomous items in an IQ test, and a score of, say, 3 is found to be the score closest to the mean, and hence given a numeral of 100. In this case, there are $2^{10} = 1024$ distinct response vectors, of which 252 correspond to the measurement standard (the mean), labeled as 100.

We can then see how the three steps of the instrument design, calibration, and operation are carried out in this situation. First, the instrument development would be constituted by the selection of a set of items whose responses were believed to be related (positively) to a person's intelligence. Second, the calibration would consist of gathering the item response vectors of a sample of persons from a suitable population, and calculating the required statistics for the calibration (such as the item difficulties, the mean person score, etc.), and the assignment of the response vectors to the labeled measurement standards (such as the mean, with a score of 100). Third, the operation of the instrument occurs when a person now takes the IQ test, and thus generates a response vector, the indication value of which is compared to the various measurement standards (e.g., if her response vector has a score greater than the mean, she will have a label greater than 100, etc.). Thus, in a formal sense, this procedure of norm-referenced calibration does indeed have the same structure as the black box approach presented above.

However, what one observes from this situation is that these measurement standards are not very satisfactory: as noted above regarding historical examples, they are sensitive to the particular sample that was used to calculate the mean and other statistics used, which *prima facie* seems to suggest that the results obtained will not be independent of the particular instrument used, and will thus not be objective. In addition, a second disadvantage is that the connection from the basic and interpretable parts of the instrument (i.e., the items and the coded responses) to the standard (i.e., the specific set of response vector that corresponds to a particular norm-referenced measurement standard) is very complex, and hence difficult to use for any interpretational purpose. To see this, consider the example mentioned where an IQ score of 100 corresponds to 252 distinct response vectors! It is not at all clear how one would make an interpretation of such a score based on the individual 10 items and their respective scoring codes, which were the initial constructive pieces of the test. Of course, with more items (as there typically are), the number of distinct response vectors becomes much higher, and even more so when polytomous scores allowed on the individual items. Thus, these measurement standards are not of much practicable use, and hence, it is our interpretation that it is for this reason that their conceptualization as "measurement standards" has been ignored by the very psychologists and others who have implicitly used those standards so frequently.

With this in mind, one can interpret the comment of Duncan that "psychometric or 'test theory' methods of attitude measurement either attempt to accomplish the two tasks simultaneously and jointly or even to bypass the scaling of items altogether" as being both incorrect and insightful at the same time. He has not recognized the distinct parts of calibration and measurement in the standard norm-referenced approach, and has not acknowledged the role of the items in the creation of the measurement standards. But, he has also correctly, and insightfully, observed that the calibration itself is not very satisfactory, as it renders the results dependent on the particular sample of persons that is used, and that the (crucial) role of the items is rendered all but invisible in the process.

The question that remains is whether this unfortunate situation can be remedied. That is, can one replace the problematical technique of norm-referenced calibration with a better one--one that reduces (or, ultimately, eliminates) the dependence of the measurements on the sample of persons, and also makes the role of the items more transparent in the calibration? In pursuit of this question, Wilson [7] has described several of the major approaches to measurement that have been developed in the last century, including classical test theory (equivalent to the norm-referenced approach described above), Guttman's scalogram technique [8], Rasch scaling [9], [10], and Construct Modeling [11], and shown how these last three can each be seen as making improvements on the preceding one. We will not repeat that account here, but it will suffice to say that the improvements have not led to the elimination of all problems, but that they are important for the future of measurement in the social sciences, and also form a basis for establishing the structural equivalence of measurement in the physical and social sciences.

4. Conclusion

In this paper we have sketched a comparison between formalized versions of measurement in physical and social science measurement: in sum, they can be seen to have important similarities. We have illustrated this for one of the major approaches to social science measurement, (norm-referenced measurement), noting some points where the comparison seems somewhat tenuous, or, at least, not straightforward. In next steps, we will add a comparison to a second approach to social science measurement (construct modeling), which we see as being closer in its logic to physical science measurement, though differences still remain.

5. References

[1] Mari L 2000 Beyond the representational viewpoint: a new formalization of measurement *Measurement* **27** 71-84
[2] Mari L, Carbone P and Petri D 2012 Measurement fundamentals: a pragmatic view *IEEE Trans. Instr. Meas.* **61(8)** 2107-2115
[3] Mari L, Maul A, Torres Irribarra D and Wilson M 2013 Quantification is neither necessary nor sufficient for measurement *IOP J. Phys.: Conf. Ser.* **459** 012007
[4] Duncan OD 1984 *Notes on social measurement. Historical and critical* (New York: Russell Sage Foundation)
[5] Thurstone LL 1927 The method of paired comparisons for social values *Journal of Abnormal and Social Psychology* **21** 384-400
[6] Flynn JR 1987 Massive IQ gains in 14 nations: What IQ tests really measure *Psychological Bulletin* **101** 171-191
[7] Wilson M 2013 Using the concept of a measurement system to characterize measurement models used in psychometrics *Measurement* **46(9)** 3766-3774
[8] Guttman LA 1950 The basis for scalogram analysis *Studies in Social Psychology in World War Two* ed SA Stouffer, LA Guttman, FA Suchman, PF Lazarsfeld, SA Star and JA Clausen (Princeton: Princeton University Press) vol 4 Measurement and prediction pp 60-90
[9] Rasch G 1960 / 1980 *Probabilistic models for some intelligence and attainment tests* (Chicago: University of Chicago Press)
[10] Wright B and Stone M 1979 *Best Test Design* (Chicago: MESA Press)
[11] Wilson M 2005 *Constructing measures: An item response modeling approach* (Mahwah, NJ: Erlbaum)

Increasing the Safety in Recycling of Construction and Demolition Waste by Using Supervised Machine Learning

P Kuritcyn[1], K Anding[1], E Linß[2] and S M Latyev[3]

[1]Faculty of Mechanical Engineering, Department of Quality Assurance and Industrial Image Processing, Ilmenau University of Technology, Ilmenau, Germany

[2]Faculty Civil Engineering, Bauhaus-University of Weimar, Weimar, Germany

[3]Faculty of Photonics and Optical Information Technology, Department of Computer Photonics and Videomatics, National Research University of Information Technologies, Mechanics and Optics, St. Petersburg, Russia

E-mail: petr.kuritcyn@tu-ilmenau.de, katharina.anding@tu-ilmenau.de

Abstract. This paper discusses the possibility of the optical identification of recycled aggregates of construction and demolition waste (CDW) using methods of image processing, spectral analysis and machine learning. The classification performances in colour images shown, that we have to use other added spectral information to solve the recognition task in a satisfactory manner. In addition to investigations on a large colour image dataset first investigations in visible (VIS) and infrared (IR) spectrum were done for analysing significant characteristics in spectrum, which are useful for classification the C&D aggregates.

1. Introduction

Construction and demolition waste (CDW) are with around 80 million tons per year the biggest waste flow in Germany. Certainly the rate of the recyclable amount depends on the composition and heterogeneity of material. For recycled masonry aggregates and recycled mixed aggregates the lowest recycling rates are found because of the high heterogeneity and the mineral admixtures. Therefore the reuse of these materials is very difficult.

The recycling industry of building materials is dominated by simple technologies. For instance the single-stage crushing is used with advance sieving and separation of reinforcement steel by over belt magnetic separator. For the processing of building materials sorting processes are only used for the separation of light components until now. These technologies are not able to separate the incidental mixed aggregates. They are suitable in no way for "new building materials" including connected building materials, which will use more and more in building industry.

The heterogeneity of C&D aggregates prevents the profitable reuse. Therefore it is necessary to reduce the heterogeneity. But this is possible only by a multi-stage process with several classify and sorting steps.

The investigations are focused on the optical differentiation of the CDW classes according to the standard specification DIN 4226-100, which describes recycled aggregates for concrete and mortar [1].

Content from this work may be used under the terms of the Creative Commons Attribution 3.0 licence. Any further distribution of this work must maintain attribution to the author(s) and the title of the work, journal citation and DOI.
Published under licence by IOP Publishing Ltd

First investigations, which showed the feasibility of automatic recognition of a selection of CDW classes on the basis of colour image information, were published in [2] and [3]. In the further investigations in this publication the amount of subclasses and instances are significantly extended for evaluation the necessity of different wavelength information for solving such complex recognition task. So the observed wavelength range for solving this problem was extended from information out of colour images over VIS spectrum information to the point of IR spectrum information.

The used 50 subclasses of the 5 superordinated classes of DIN 4226-100 show in many cases a high phenotypic similarity for example concrete / aerated concrete / lightweight concrete and also porous and dense brick. The investigations were done on new, not used building materials, which were crushed for obtaining homogeneous samples.

The aim is to develop an optical solution for determination of building material classes by using methods of image processing, spectral analysis and machine learning for realize a real closed cycles and a high standard of quality in recycling. Several optical and spectral attributes were found, which have discriminatory power to classify the chosen materials. Several classification algorithms of supervised machine learning were tested on different feature vectors as a numerical representation of objects of the given dataset.

2. State of the art

As in other sectors of recycling, for example the glass or plastic recycling, the sensor-based sorting has been becoming more interesting in the recycling of building materials and sorting of minerals in the last years. There are mainly used methods with optical, magnetic, NIR or X-Ray sensors.

The application of automatic sensor sorting in the areas of mining and recycling is successful in Europe and will increase in the following years. The benefits are the increase of the end product value and the cost reduction of downstream handling steps in the processing [4], [5].

3. Image analysis – object recognition in colour images

The first trial to solve the given recognition task was to use only phenotypic characteristic in VIS for classifying C&D aggregates. So the goal was to acquire colour images of all given CDW classes and use the visual contour, colour and texture features for training supervised classifier algorithms.

3.1. Image acquisition

For this research in the field of image based recognition of C&D aggregates in the visible spectrum (VIS) a specialized system for feeding, separation and image acquisition was developed and used. It is possible to handle the sample components sized from 8 till 16 mm in diameter with the developed device. The sample is filled in the charging bin, gets separated via a chute and tracks over a belt conveyor. The aim is to get non-touching and non-overlapping objects in the field of view of the colour line scan camera resulting in single object images after segmentation.

A large dataset was generated with different sample objects by using the described image acquisition device. Totally nearly 32.000 objects were given for building up training and testing dataset by using cross-validation. This dataset was used to compare several supervised classifiers and tune classifier parameters for example the SVM parameter via cross-validation.

3.2. Feature extraction

After an accurate segmentation of the foreground region the images was transformed from RGB to HSI colour space. Then a feature vector with 234 features (several contour, colour and texture features) has been calculated. The used feature extraction algorithms are part of the machine vision software Halcon and were described in the release notes for MVTec HALCON 8.0.3 [6]. These specific feature vectors provide the basis of the classifier training independent from the chosen classifier algorithm.

3.3. Comparison of classification algorithms

Several classification algorithms of supervised machine learning were tested on the given colour image dataset to solve the recognition task with an automatic recognition routine. The tested classification algorithms are part of the machine learning library Weka [7], such as support vector machines like the LibSVM as C-SVM and nu-SVM, decision tree classifiers like J48 and Random Forest classifier, distance-based classifiers like k-nearest neighbour and statistical classifiers like the Naive Bayes classifier by using a 10-fold cross validation.

3.3.1. Recognition rates of different classifiers. The classifiers with the best performance on the given problem are Random Forest, the *Nu-* and *C*-LibSVM with obtained TRR of 94.7% up to 97.2%. Statistical classifiers like Naive Bayes (TRR = 44.8%) and other simple distance-based classifiers like Nearest Neighbour (TRR = 84.8%) fail or show worse performances on the given dataset (TRR=total recognition rate).

Table 1. Comparison of different classifiers.

	Classifiers											
	NNeighbour		C-SVM		Nu-SVM		Naive Bayes		RandomForest		J48 Tree	
Classes	IRR (%)	TRR (%)	IRR (%)	TRR (%)	IRR (%)	TRR (%)	IRR (%)	TRR (%)	IRR (%)	TRR (%)	IRR (%)	TRR (%)
concrete and aggregates	78.5		95.4		97.2		79.7		91.1		81.4	
clinker and dense brick	76.5		99.0		94.3		97.4		95.5		91.1	
sand-lime brick	82.0	84.8	97.8	97.2	96.4	95.8	71.3	44.8	93.7	94.7	84.3	88.0
other mineral components	89.5		97.6		95.9		12.2		96.6		92.3	
foreign components	82.4		95.7		98.1		77.2		92.0		81.5	

IRR= individual recognition rate; TRR=total recognition rate

If we only use colour image analysis good recognition rates could be reached. But there are also a bigger amount of false classifications between critical and uncritical classes for recycling. So the results of colour image analysis are not enough for using so classified C&D aggregates as high quality building materials.

4. Spectral analysis – enlargement of spectral wavelength range

The results of object recognition in colour images shown, that we have to use other added spectral information to solve the recognition task in a satisfactory manner. So investigations in VIS and infrared (IR) Spectrum were done for analysing significant characteristics in spectrum on the same datasets, which are useful for classification the C&D aggregates.

4.1. Used spectrometer devices

We used two different spectrometers for analysing the CDW classes, the VIS spectrometer Ocean Optics USB2000+ with a linear silicon CCD array as detector and a detector range of 200 to 1100 nm and the Polytec PSS 2120 with an InGaAs detector and a range of 1100 to 2100 nm.

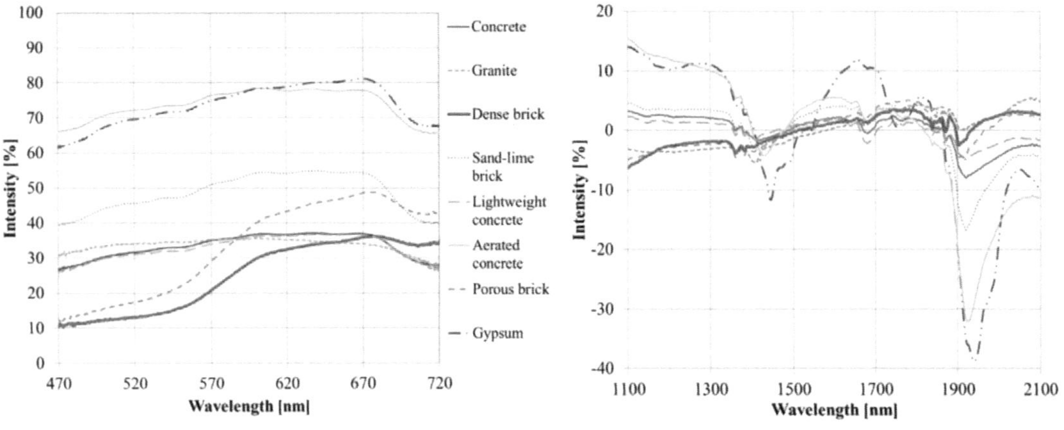

Figure 1. Averaged VIS (left) and averaged and centered IR spectrums (right) of several CDW classes.

4.1.1. Analysis of VIS Spectrum of several CDW classes. First investigations of the specific VIS spectrums are shown in Figure 1 (left side). We used a light source, which shows a stable emitted lightning till 720 nm. In the spectrum range of 470-720 nm some materials show significant features, for example the classes of brick show significant characteristics in VIS for identifying in context of the classes of concrete. Other materials show only differences in intensity like concrete, gypsum, aerated concrete etc. During the capturing process of the spectrums it is difficult to reach the same intensity for all samples (subclasses) out of one class. This depends on the different object contours and on the different behaviour of reflecting the light. Classes with similar chemical substances, for example lightweight concrete and concrete, show very similar spectral characteristics without significant differences in intensity. And also phenotypic very similar classes like sand-lime brick and aerated concrete, which shown very similar VIS spectrum characteristics with the exception of differences in intensity, needs the use of other wavelength ranges (like IR spectrum information) for auspicious results. This causes the necessity for the additional using of colour image information and other spectral information.

4.1.2. Analysis of IR Spectrum of several CDW classes. Within a research project at the Bauhaus University Weimar were carried out studies on the spectral characteristics of building materials forming pure minerals in cooperation with the firm LLA Instruments GmbH [8]. Many pure minerals have spectral features (absorption bands) in the near infrared wavelength range between 1100-2500 nm regard to their active near-infrared absorptions of the minerals can be divided into the following groups:

- Water containing minerals,
- Minerals with characteristic hydroxide,
- Carbonate minerals.

In particular, the hydroxide groups can be in the wavelength range between 1350 - 1430 nm evaluate well and are mineral-specific. Because of different conditions in the crystal lattice varying the wavelength at which absorbs hydroxide. This variation is between different groups of minerals only a few nanometres, but is sufficient for their identification. The background spectra are characterized by broad absorption bands especially in the wavelength range between 1350 - 1450 nm and 1800 - 2100 nm, which are composed of the superposition of the absorption of various minerals and water.

The characters appearing in construction materials are minerals, especially the mineral classes carbonates, silicates (including layered silicates), and sulphates (especially gypsum products) and oxides (e.g. hematite) assigned.

In further investigations a bigger amount of samples (nearly 1100 samples out of the 9 superordinated classes: lightweight concrete, concrete, aerated concrete, sand-lime brick, dense and porous brick, gypsum, asphalt and granite) was used by capturing and analysing their IR spectrum.

Figure 1 (right side) shows that the class specific characteristics in IR is much better for classification between the different classes as in VIS (see Figure 1 (left side)). From Figure 1 (right side) follows a good distinction of some of the defined categories of building materials in the IR-spectrum. Concrete, lightweight concrete and aggregates are similar in the IR-spectrums. The mineral composition of normal concrete and also lightweight concrete varies depending on the used aggregates. However the mineral calcite is detectible by near infrared at 1911 nm. A differentiation of normal and lightweight concrete is not easy (only differences in intensities of several wavelengths).

Aerated Concrete can be seen very well on the basis of the formed tobermorite phase in the infrared spectrum. During the steam curing in the autoclave of aerated concrete tobermorite formed between 30 and 40 percent by mass. The spectrum of the aerated concrete show adsorption bands at 1430 and 1920 nm and additional a less intensive band at 1680 nm.

All sulphates like gypsum, plasters and plasterboard show high characteristic absorption bands at 1440, 1750 and 1930 nm, which is very good detectable. There is a small shifting about 10 to 15 nm in the spectrum of gypsum and plasterboard, which can be used for the differentiation of these materials. Sulphates are foreign material which is to separate completely by sorting if the secondary aggregates will be used for the production of new concrete for instance.

The examined sand lime bricks consist mainly of quartz, which is not detectable by IR. As a vapour cured product the sand lime brick also contains the mineral tobermorite. The characteristic absorption bands are at 1410, 1680 and 1920 nm.

The bricks are very variable in its mineralogical composition depending on the used clay for the production. The peak in the spectrum is at 1900 nm and is different from the concrete peak at 1911 nm, which is the feature for differentiation. To keep apart the dense and porous bricks we should be using the differences in intensities of several wavelengths by near infrared. From the recycling point of view a brick with a bulk density of 2.0 g/cm³ and more can be used for the production of recycling concrete so that a separation makes sense.

In previous studies it could be shown that the application of optical classifiers is predestined for the recognition of dense and porous brick [2], [3]. Also the recognition of the different concrete types like normal and lightweight concrete can be solved better by adding the colour image analysis in the visible spectrum and the use of classifiers.

After analysing the IR spectrums a principal component analysis was done. The first 5 and 10 principal components with highest information content were used for the application of a supervised classifier for differentiation between the given classes. Figure 2 shows the first 3 principal components of the IR spectrums. The classes lightweight concrete, concrete, aerated concrete and sand-lime brick build a main cluster with an extensive size. There are small, well-defined clusters like dense and porous brick, gypsum, asphalt and granite. The classes aerated concrete and granite build ellipsoidal clusters with good homogeneity. The class gypsum shows a plane ellipsoidal cluster, which has a long distance to the main cluster and is relatively homogeneous. The two classes of brick build two closely spaced ellipsoidal clusters, which are separable between each other and the main cluster. The class aerated concrete lie compactly on the border to the main cluster and shows a marginal overlapping with the cluster of sand-lime brick. In contrast the classes concrete, lightweight concrete and sand-lime brick (especially the first two classes) show significant overlapping and the class concrete builds an inhomogeneous cluster. This fact causes a difficult classification between the three classes. The total variance by using the first 3 principal components is 99%, this shows that we can visualize the spectral information of the dataset without information loss by using only the first three components. The visualization of the first three components allows the conclusion that the classes aerated concrete, dense and porous brick, gypsum, asphalt and granite are good distinguishable.

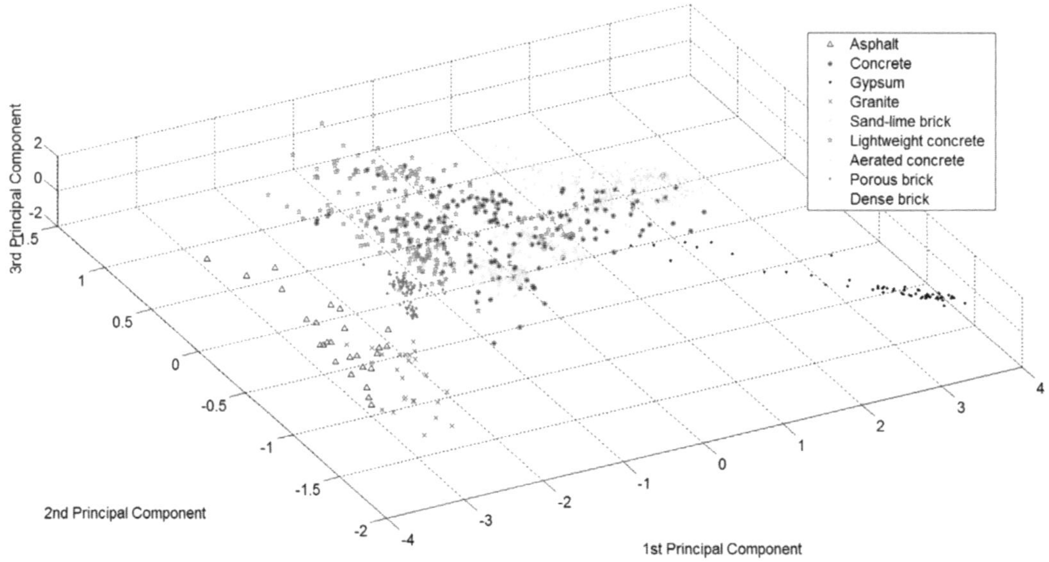

Figure 2. Principle component analysis (PCA) of IR spectrums of several CDW classes.

The application of only the first 10 principal components instead of the 501 dimensional wavelength-specific information accelerated the classification time significantly without producing a loss of information. A total recognition rate of 98.4% by using only the first 10 components (cumulative variance = 99.995%) and of 90.8% by using only the first 5 components (cumulative variance = 99.95%) and by using a 10-fold cross validation. The individual recognition rates are 100.0% for asphalt, 97.4% for concrete, 98.3% for gypsum, 100.0% for granite, 98.5% for sand-lime brick, 97.5% for lightweight concrete, 99.1% for aerated concrete, 100.0% for dense brick and 98.1% for porous brick. The most false classifications appear between the classes lightweight concrete and concrete. Also critical are the false classifications between the classes dense and porous brick. If the goal is the recycling of C&D aggregates to win high quality building materials, these false classifications have to be avoided. The conclusion is that the best way for solving the classification and sorting task is to combine the IR information and the information of the colour image analysis (in VIS) and to use supervised classifiers.

The reached high recognition rates by using only IR spectrums are based on the relatively low intra class variability of the nine C&D aggregate classes. The cause of the relatively low intra class variability is the use of new building materials, which were crushed for obtaining homogeneous recyclates. Real C&D aggregates show a higher intra class variability because of contaminations and dust on their surfaces and a higher variability in shape and size. Further investigations are planned for combining feature vectors of colour image analysis and IR spectrum analysis and using used real C&D aggregates as sample dataset.

5. Results

The further investigations on the extended colour image dataset by using supervised machine learning algorithms showed that the obtained total recognition rates (TRR) of the best classifiers (Random Forest, Nu- and C-LibSVM) were 94.7% up to 97.2% for using the given 5 superordinated classes in context of the DIN standard.

The first investigations in the VIS spectrum range showed that some materials show significant spectrum characteristics for identifying, for example the classes of brick. Other materials show only differences in intensity like concrete, gypsum, aerated concrete. Classes with similar chemical substances, for example lightweight concrete and concrete, showed very similar spectral characteristics without significant differences in intensity. This causes the necessity for the additional using of colour image information and other spectral information.

The first investigations in the IR spectrum range showed that concrete and brick in the infrared spectrum are well distinguishable in principle. Gypsum as impurity in the C&D waste is very well detectable by IR. Aerated concrete and sand lime brick can also be very well recognized by IR sensors. Lightweight and normal concretes and dense and porous brick show a little amount of false classifications in the IR spectrum. A total recognition rate of 98.4% by using only the first 10 principal components (cumulative variance = 99.995%) and of 90.8% by using only the first 5 principal components (cumulative variance = 99.95%) and by using a 10-fold cross validation. So the best way for solving the analysing task is to complete the IR information by information of the colour image analysis (in VIS) and to use supervised classifiers.

Further investigations are planned for combining feature vectors of colour image analysis and IR spectrum analysis and using used real C&D aggregates as sample dataset.

References

[1] DIN 4226-100: 2002-02 Aggregates for concrete and mortar - Part 100: Recycled aggregates, *DIN standard*

[2] Anding K, Linß E, Träger H, Rückwardt M and Göpfert A 2011 Optical Identification of Construction and Demolition Waste by Using Image Processing and Machine Learning Methods *14th Joint Int. IMEKO Symp. (Jena, Germany)*

[3] Linß E, Träger H, Ludwig H M, Anding K and Linß G 2012 Study of the Identification of Aggregates of Construction and Demolition Waste by Using Object Recognition Methods, *IALCCE 2012: 3rd Int. Symp. on Life-Cycle Civil Engineering (Vienna, Austria)*

[4] Dehler M. 2003 Optical Sorting of Mineral Commodities (Optische Sortierung von mineralischen Rohstoffen), *online publication*

[5] COMMODAS, We sort it out, 2007, *Catalog*. 7 p. http://www.commodas.de

[6] Halcon, *Reference Manual* - Halcon 8.0.3, MVTec Software GmbH, www.mvtec.com 2009

[7] Hall M, Eibe F, Holmes G, Pfahringer B, Reutemann P and Witten I H 2009 The WEKA Data Mining Software: An Update, *SIGKDD Explorations* **11**(1)

[8] NIR-Sorting of CDW. Final report of AiF scientific project. AIF-KF 2155712 Bauhaus-University Weimar 2013

[9] Linß E, Anding K, Schnellert T and Ludwig H M 2012 Identification of Construction and Demolition Waste by Using Image Processing in the Visual and Near-infrared Spectrum, and Machine Learning Methods *Proc. 18th iBausil (Weimar, Germany, 12-15 September2012)*

Food Safety by Using Machine Learning for Automatic Classification of Seeds of the South-American Incanut Plant

Thomas Lemanzyk[1], Katharina Anding[1], Gerhard Linss[1], Jorge Rodriguez Hernández[2] and René Theska[3]

[1]Faculty of Mechanical Engineering, Department of Quality Assurance and Industrial Image Processing, Ilmenau University of Technology, Ilmenau, Germany

[2]Departamento Académico de Ingeniería, Pontificia Universidad Católica del Perú (PUCP), San Miguel, Lima, Peru

[3]Faculty of Mechanical Engineering, Department of Precision Engineering, Ilmenau University of Technology, Ilmenau, Germany

E-mail: thomas.lemanzyk@tu-ilmenau.de

Abstract. The following paper deals with the classification of seeds and seed components of the South-American Incanut plant and the modification of a machine to handle this task. Initially the state of the art is being illustrated. The research was executed in Germany and with a relevant part in Peru and Ecuador. Theoretical considerations for the solution of an automatically analysis of the Incanut seeds were specified. The optimization of the analyzing software and the separation unit of the mechanical hardware are carried out with recognition results. In a final step the practical application of the analysis of the Incanut seeds is held on a trial basis and rated on the bases of statistic values.

1. Introduction

Through increasing globalization, the compliance with international quality standards is necessary. The quality demand of food is related for example to avoidance of uncleanness inside or on food products, contamination through environmental influences, foreign material or degeneration of the product itself. To avoid this inadvertent decrease of quality, control- and counteractive measures in the handling process of the end-product are necessary.

This paper particular handles the quality control in the production process of seeds of the so called Sacha Inchi, also known as Incanut because of its South-American origin.

As a natural product, the seeds of the Incanut are open to different climatic conditions. These nature given influences have a direct effect on the end-product, which can appear for example in discolouration, rests of shells, rottenness or missing seed-parts. All the unwanted parts in an amount of seeds are called impurities, as it is also done with other grocery, especially in the case of grain. Seeds which fit to in design and colour to the demand of the producer are named as proper seeds or seeds in a proper condition. To avoid common contamination as good as possible, they have to be recognized and identified initially. In general there are two different ways of identification. On the one hand there

Content from this work may be used under the terms of the Creative Commons Attribution 3.0 licence. Any further distribution of this work must maintain attribution to the author(s) and the title of the work, journal citation and DOI.
Published under licence by IOP Publishing Ltd

is the manual identification of unwanted impurities by human sighting (the state of the art) and on the other hand there is the automatic identification with the help of industrial image data processing through the mechanical systems and production engineering. The automatic identification of the impurities means the automatic classification of objects in the given sample into their related classes by using supervised machine learning algorithms. The main attention in this paper is spotted on the identification and the classification of the appearing seeds through machine help under the regard of the environmental conditions in the manufacture processing of the Incanut [1].

With an automatic control and handling of the Incanut, health risks and quality degradation of the end-product should be avoided, especially in comparison to manual methods of quality control.

2. Classes of the Incanut

The evaluation to relate the seeds to their fitting classes is usually made manually by sighting and sorting. These manual classified sorted samples build the a priori knowledge for training the classifiers.

Figure 1. Constituent parts of the Incanut seeds

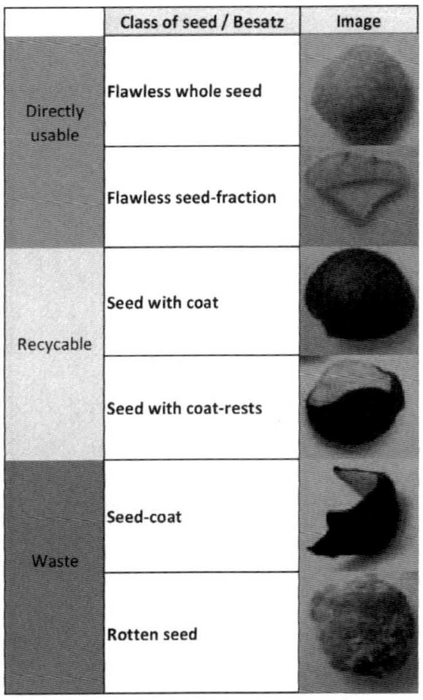

Figure 2. The 6 subclasses and the 3 superordinated classes of the Incanut

Figure 1 illustrates the constituent parts of the Incanut seeds and figure 2 illustrates the different six subclasses and the three superordinated classes of the Incanut, which are important for production process and food safety. Four of the six different classes of seeds can be utilized or rather be leaded back to the process. Just two classes of the main class "waste" are classified as useless and are being disposed. Therefore in the daily routine of production, the main classes "directly usable", "recyclable" and "waste" are mainly distinguished and accordingly sorted.

3. Image analyses - object recognition in colour images

The automatic classification of Incanut seeds based on the image processing of colour images and the using of an adapted trained classifier. The manual classification and sorting of the seeds, this means their fitting in the defined classes, builds the basis of the developing of an automatic recognition routine. The investigations are based on the grainspector, an analysing software and a mechanical hardware device for the automatic analysis of grain, which were developed in a prior project of the department of Quality Assurance and Industrial Image Processing, Ilmenau University of Technology (see [2], [3], [4], [5]). So the task was the optimizing of the classifier and the separation unit of the mechanical hardware device. The colour images are captured on a conveyor belt with a 3-CCD colour line scan camera with divisor prism. A combination of incident light and transmitted light was chosen. A wide sample dataset is generated, which contains all object classes in a large number of object samples. Our research is based on a dataset of the 6 object classes (seen in figure 2) with totally nearly 6500 objects.

Randomly chosen two-thirds of the images of the image-dataset are used for the classifier training and randomly chosen one-third of the images are used for the testing. The classifier is being learned through the training process with the attribute-vectors of the learning-dataset.

Thereby altogether 227 contour, colour and texture attributes of the software Halcon are used [6]. Three of the used features should described in detail. As for example the contour feature *area_center* computes the area and the center of an input region. The contour feature *contlength* calculates the total contour-length of the input region and the contour feature *roundness*, which is a specific shape factor from the contour and is determined as the relation between the mean value of the distance (*Distance*, see equation (1)) and the standard deviation of the mean distance (*Sigma*, see equation (2)) [6]. The Roundness is calculated in equation (3) with p is the center of the area, p_i the pixels and F the area of the contour.

$$Distance = \frac{1}{F}\sum\left\|p - p_i\right\| \tag{1}$$

$$Sigma^2 = \frac{1}{F}\sum\left(\left\|p - p_i\right\| - Distance\right)^2 \tag{2}$$

$$Roundness = 1 - \frac{Sigma}{Distance} \tag{3}$$

4. Classification with Support Vector Machines (SVM)

The used analysing software [2], [3], [4], [5] allows the training of several SVM classifiers after feature extraction with different kernel types (for example Radial Basis Function (rbf) or homogeneous and inhomogeneous polynomial kernel functions) and different internal parameters. In these investigations the implemented SVM of [6] was optimized on the given recognition task by optimising the parameters of the SVM classifier. The classification with SVM needs an optimal parameter selection for a good classification performance.

The SVM was first introduced by [7] and is mentioned as being one of the most powerful classifiers today. It is derived from the statistical learning theory [8]. The algorithm is motivated by the structural risk minimization which says that not only the training error but also the complexity of the model influences its generalization ability. The SVM was designed to solve binary classification problems but there are different strategies to solve multi-class problems, too. The SVM executes a non-linear projection of data in a higher dimensional feature space. The classes are separable in a linear way in this higher dimensional feature space. During the training process an optimal hyperplane is constructed. Optimal means that it leaves a maximal margin between the hyperplane and the closest training point on both sides.The kernel function $k(x,x_i)$ extends the linear discriminant SVM to a nonlinear machine. The given decision function [9] is given in equation (4) and the used radial basis function kernel (rbf) as one of the most popular kernel functions is defined in (5).

$$f(x) = \operatorname{sgn}\left(\sum_{i=1}^{m}\alpha_i y_i k(x, x_i) + b\right). \tag{4}$$

$$k(x, x') = e^{-\gamma\left\|x - x'\right\|^2} \tag{5}$$

After optimizing the given hard- and software and after elimination of images with foreign-particles as wrong apriority-knowledge in manual expert-sorted samples or images with artefacts caused by belt-defects the individual recognition rates (RR) for the 6 subclasses obtain are for flawless seed-fractions 95.8 %, flawless whole seeds 94.4 %, seeds with coat 98.2 %, seeds with coat-rests 89.2 %, seed-coats 93.4 % and rotten seeds 94.3 %. The total recognition rates over all 6 classes obtain 94.4 % (see figure 3).

Under application of the three main classes (superordinated classes), which make sense for the production process, significantly higher recognition rates are achieved (see figure 4). Thereby the individual recognition rates for the main classes are for directly usable 97.3 %, Recyclable 97.8 % and Waste 96.1 % with a total recognition rate of 97.0 %.

		Classified as							
	Classes	[1]	[2]	[3]	[4]	[5]	[6]	Individual RR [%]	Total RR [%]
Real	Flawless seed fractions [1]	680	9	0	0	17	4	95.8	94.4
	Flawless whole seeds [2]	10	252	0	0	2	3	94.4	
	Seeds with coat [3]	0	0	321	1	3	2	98.2	
	Seeds with coat rests [4]	1	0	24	273	8	0	89.2	
	Seed coats [5]	10	0	4	1	470	18	93.4	
	Rotten seeds [6]	18	1	0	0	2	348	94.3	

Figure 3. Optimised SVM – recognition performance for 6 subclasses

		Classified as				
	Main class	Directly usable	Recycable	Waste	Individual RR [%]	Total RR [%]
Real	Directly usable	951	0	26	97,3	97,0
	Recycable	1	619	13	97,8	
	Waste	29	5	838	96,1	

Figure 4. Optimised SVM – recognition performance for 3 superordinated classes

5. Results

An informative basic research on the plant and plant-components of the South-American Incanut and their use was carried out through initial extensive research. Although the plant is generally not well-known in Europe, a definition of the classes of the Incanut-seeds was done on basics of experiences with the plant in South-America.

With the help of these information it was possible to create the basics for the realization of the classification of the Incanut-seeds for the production of Incanut-products. The software and hardware of the grainspector [2], [3], [4], [5] were modified and optimized to the Incanut-seeds. With this configuration, initial practical tests with samples of material of Incanut-seeds from Ecuador could already be analyzed. The test results and the function of the hard- and software are very satisfying. After optimizing the hard-and software and after the elimination of the pictures with belt-defects and foreign-particles (wrong apriority-knowledge in the actual correctly sorted samples) the individual recognition rates for the classes are: flawless seed-fractions 95.8 %, flawless whole seeds 94.4 %, seeds with coat 98.2 %, seeds with coat-rests 89.2 %, seed-coats 93.4 % and rotten seeds 94.3 %. The total recognition rate of all classes was 94.4 %.

Under application of the three main classes, which make sense for the production process, significantly higher recognition rates are achieved. Thereby the total recognition rates for the main classes are: directly usable 97.3 %, recyclable 97.8 % and waste 96.1 % with a total recognition rate of 97.0 %.

The next steps for these investigations will be the increasing of datasets and the further optimizing of the hardware device for decreasing of image artefacts and in-contact objects.

References

[1] Lemanzyk T 2013 Basic Research on automatic optical Classification of Sacha Inchi Seeds *master thesis* Ilmenau University of Technology Ilmenau Germany

[2] Brückner P, Anding K, Dambon M and Garten D 2010 Measuring Wheat Quality, *Vision Systems Design Magazine – Vision and Automation Solutions for Engineers and Integrators Worldwide* **15** Issue 6 Nashua USA

[3] Anding K and Garten D 2008 Comparison of Different Classification Algorithms at the Application of Automatical Quality Assurance of Grain *Proceedings of the 53rd International Scientific Colloquium (IWK)* of Ilmenau University of Technology, Faculty of Mechanical Engineering p 349-350

[4] Anding K, Brückner P and Linß G 2010 Influence of Dataset Character on Classification Performance of Support Vector Machines for Grain Analysis, *Artificial Intelligence and Applications* Innsbruck Austria p. 75 – 81

[5] Garten D, Anding K, Lerm S, Linß G and Brückner P 2013 Image Analysis of Natural Products, *Conference on Optical Characterization of Materials* OCM 2013 Karlsruhe

[6] Halcon, *Reference Manual* - Halcon 8.0.3, MVTec Software GmbH, www.mvtec.com 2009

[7] Boser B, Guyon I and Vapnik V 1992 A Training Algorithm for Optimal Margin Classifiers *Proceedings of the Fifth Annual Workshop on Computational Learning*

[8] Vapnik V 1995 The Nature of Statistical Learning Theory (New York: Springer)

[9] Guyon I, Elisseeff A 2006 Feature extraction: foundations and applications (Berlin: Springer)

Multispectral Image Correction for Geometric Measurements

M Rosenberger[1], G Linß[1]

[1]Ilmenau University of Technology, Department Quality Assurance and Industrial Image Processing, Gustav Kirchhoff Platz 2, 98693 Ilmenau, GER

maik.rosenberger@tu-ilmenau.de

Abstract. Multispectral- and hyperspectral imaging technologies enable new possibilities in industrial measurement applications. Based on the knowledge of remote sensing a lot of investigations were made in the last decades of years. Nevertheless the demands on remote sensing versus technical multi spectral image processing are quite different. In the field of precise geometric measurement technics it is necessary to correct the image data between different spectral channels with a high accuracy, normally in the micron range. Otherwise the geometric absolute value of fail detection on edges can be become very large. State of the art in industrial imaging and detection of geometric features is the calibration of only one imaging channel. In this paper, the studies on a twelve channel multi spectral imager were presented. For the applied filter wheel system, investigations on the improvement of lens aberration as well as for the defocus problem were made. Therefore a calibrated high precision geometric test chart was used to calibrate the system geometrically. To correct the geometric errors on the image plane a special moving filter approach, based on linear convolution, was developed. For every channel a calibration matrix were calculated and applied on the image system output.

1. Introduction

The analysis of space resolved spectral information is only possible to a certain limit with an ordinary colour camera [1],[2].

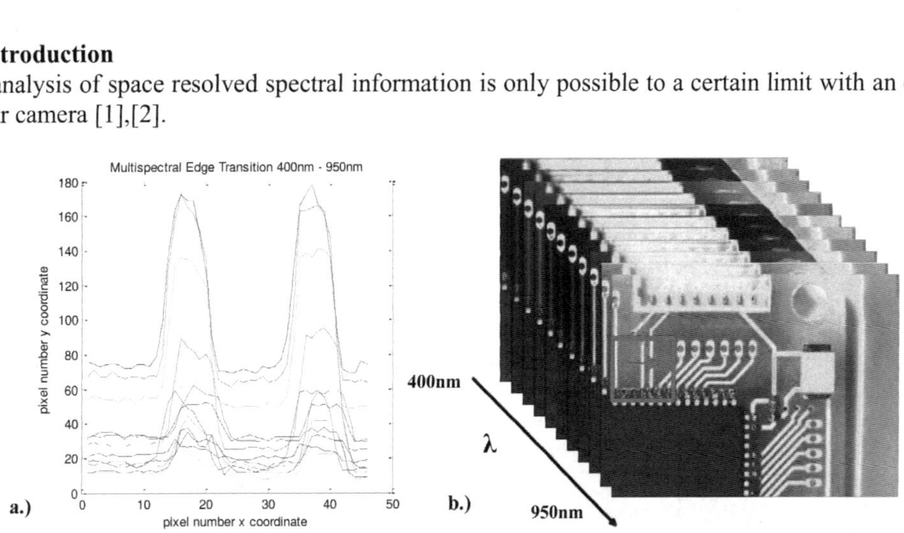

Figure 1: a.) multi spectral edge transition, along the search line (red arrow) of a twelve channel spectral cube (b.); (blue graphs 400-500nm, green graphs 550-650nm, red graphs 700-800 nm, grey scaled graphs 850-950nm);

The technical function principle of ordinary Red Green Blue (RGB) cameras with a maximum of three colour channels can dissolve only a limited range in the colour space and loose spectral image

Content from this work may be used under the terms of the Creative Commons Attribution 3.0 licence. Any further distribution of this work must maintain attribution to the author(s) and the title of the work, journal citation and DOI.
Published under licence by IOP Publishing Ltd

information. The comparatively large sensitivity of the three individual colour channels results to lose some spectral information. A multispectral camera operates with several colour channels and reconstructs a complete spectrum with specific numerical methods. Filter wheel cameras present an effective and affordable alternative in multispectral image acquisition Figure 2 [3],[4].

Figure 2: multispectral imager 3D-model and inside view on the left, assembled twelve channel multispectral imager on the right;

Their modular and versatile structure, as well as the consistent enhancements in filter and sensor technology offer many pros for use of these systems. Actually a lot of investigations were made to increase the spectral resolution for space resolved spectral imaging systems. The simultaneously data acquisition of the complete two dimensional field is not solved already. Only push broom scanners can capture spectral images like a line camera. Filter wheel cameras do not have that problem but during the filter exchange it loss information of fast moving object scenes. Every technology has its pros and cons depending on the application. As well as the resolution in wavelength, for geometric measurement an accurate correction of aberrations are necessary to achieve lowest measurement uncertainties. That leads to a correction of every spectral channel. The need for the transfer of multispectral imaging technologies in industrial application is to make features visible that actually cannot be detected with greyscale or RGB –imaging. Figure 1 shows a multispectral cube captured with a filter wheel camera [3] and the information along a search line in all channels. There is lot of complicated measurement tasks for industrial measurement especially in the field of injection moulded parts and composite technologies. In the following the imaging setup will be presented partly and a short introduction of wavelength depended aberrations for spectral imaging is given.

2. Challenges in Multispectral Image Acquisition

The two major challenges in multispectral imaging are the increase of wavelength stability and wavelength resolution as well as the increase in spatial resolution. Therewith it is essential to calibrate the system. In many publications a model based correction is described mostly for applications in bio science and remote sensing. In the case of precision measurements the goal should be a very small amount of residue. With that the problem is that the most lenses are corrected in three wavelengths in the RGB range. Based on the longitudinal aberration in diffractive optics the focus point of every wavelength channel is located on a different position and must be corrected. The spot size in the image plane differs from the focused wavelength (minimum spot size) up to10 microns (Figure 4 a.) This effect is based on the different propagation wavelength in materials. A possibility to correct this aberration is to enlarge the optical path of light specifically. Therefore the sellmeier equation with the sellmeier coefficients were used to calculate the thickness of the correction plates [5]. First the wavelength dependent refractive index can be calculated by equation 1.

$$n(\lambda) = \sqrt{1 + \frac{B_1\lambda^2}{\lambda^2 - C_1} + \frac{B_2\lambda^2}{\lambda^2 - C_2} + \frac{B_3\lambda^2}{\lambda^2 - C_3}} \qquad (1)$$

With the knowledge about the wavelength dependent refractive index and the assumption of an additional way for the propagating light the following equation 2 leads to the thickness of the correction plates dl_λ.

$$dl_\lambda = \frac{df_\lambda n_{\lambda,BK7}}{n_{\lambda,BK7} - n_{\lambda,air}} \tag{2}$$

To calculate the thickness of the correction plates the focus deviations have to be measured in the optical path or have to be simulated. To consider the tolerances in the complete optical path like the tolerances in the lens, the tolerance in the lens mountings, the tolerances in the filter mountings and as well as the tolerances in the sensor mounting , the focus shift were measured inside the system on all twelve filter wavelengths. For this reason a calibration chart were used to get sharp edge transitions. The contrast values on the edge transition were evaluated. On the contrast maximum the optimal focus position will be expected. To realize the focus procedure a high precision linear bearing with a measurement system which resolves up to 10 microns were adapted into the camera system. The adjustable motion system carries the image sensor and can be moved for the measurement very precisely. The results of the measurements of the wavelength depended focus shift is given in Figure 3. The start position of the focus position was placed in the infrared region.

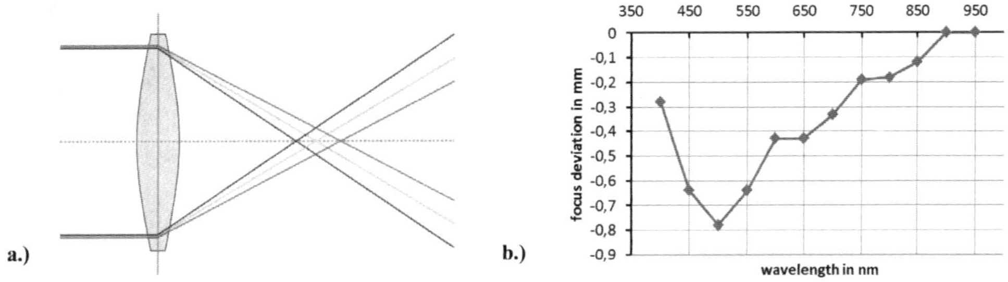

Figure 3: a.) schematic of wavelength dependent focus deviation, b.) measured focus deviation on the image plane (image sensor ICX 415Al lens: Distagon 2,8/25 ZF IR);

From that position a linear trend were expected. This trend can be observed until 550nm (Figure 3). Below 550 nm the trend distance to the start position increases. This effect can be traced back to a deviation of thickness of the color filters as well as of the chromatic lens optimization in the RGB-Range. At the end a range between 150 microns up to 2000 microns were calculated for the correction of the entire wavelength range for the used lens.

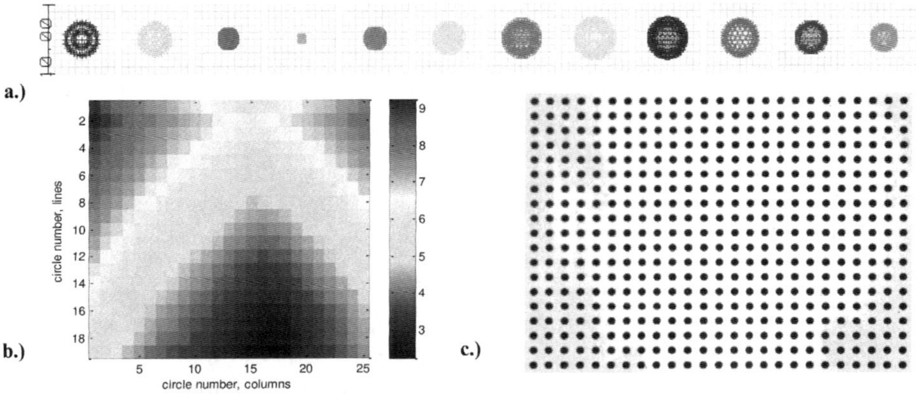

Figure 4: a.) optical simulation of longitudinal aberration of an apochromat, beginning from the 400nm band in steps of 50nm, base wavelength 550nm; b.) maximum field aberration between all channels of the test target (the colour value is the aberration in pixel); c.) calibrated test chart with detected centre points;

With the knowledge of the thickness for the different correction plates the focus shift were compensated. The mounted correction plates generating an additional optical lateral shift which have to be corrected on the image plane. This optical shift and the distortion of the lens give the absolute error on the image plane. With the calibration test chart (pitch 1000µm, Figure 4 c.) the maximum measured deviation between the different wavelength ranges is nine pixels. The spatial error distribution is displayed in Figure 4 b.). To calculate the complete distortion the following procedure was developed. After the correction with the calculated correction plates, the focus shift is strong minimized, so every filter were evaluated separately. To extract the circles in the image a Hough-Transformation was used. At the end of the applied transformation a table with x and y coordinates were generated. After that the approach of an minimal distortion in the middle of the lens were assumed. With this assumption the five circles in the idle of the image were used as the reference. Furthermore the distance of the different circles of the calibration target in the image are known. With the knowledge of the reference positions of the calibration targets an reference matrix of circle points were calculated. The difference between the x- and y-coordinates leads to the deviations in the separate directions. Furthermore the pixel factor can be calculated with this data. At the end an deviation cube of 19 x 25 x 12 data points were processed for the correction algorithm. In case of another lens only the size have of the matrix have to be changed.

3. Approach for Correction

Based on the measured centre points (Figure 3. b-c.) the deviations from the expected positions of the centre points can be calculated for every channel. For this operation a polynomial with an grade of three were used to fit the discrete values into an continuous function. With x the measured position in the column direction (x) and a_x-d_x the polynomial coefficients leads to the approach (equation 3):

$$f(x_c) - x = a_x x^3 + b_x x^2 + c_x x + d_x \rightarrow min \quad (3)$$

After the optimization the four coefficients for every column in 19 lines were done. That leads to an matrix with nineteen a_x, b_x, c_x, d_x coefficients. Furthermore the calculated coefficients a_x-d_x can be described also as continuous functions in the line direction (y) (equation 4-7):

$$f(a_x) - a_x = a_{y3} y^3 + b_{y3} y^2 + c_{y3} y + d_{y3} \rightarrow min \quad (4)$$
$$f(b_x) - b_x = a_{y2} y^3 + b_{y2} y^2 + c_{y2} y + d_{y2} \rightarrow min \quad (5)$$
$$f(c_x) - c_x = a_{y1} y^3 + b_{y1} y^2 + c_{y1} y + d_{y1} \rightarrow min \quad (6)$$
$$f(d_x) - d_x = a_{y0} y^3 + b_{y0} y^2 + c_{y0} y + d_{y0} \rightarrow min \quad (7)$$

At the end of the calculation of all coefficients the deviation can be described with sixteen coefficients for every channel. For the calculation of the corrected position the following equation 8 was derived:

$$\begin{aligned} dx = {} & (a_{y3} y^3 + b_{y3} y^2 + c_{y3} y + d_{y3}) x^3 + \\ & (a_{y2} y^3 + b_{y2} y^2 + c_{y2} y + d_{y2}) x^2 + \\ & (a_{y1} y^3 + b_{y1} y^2 + c_{y1} y + d_{y1}) x + \\ & (a_{y0} y^3 + b_{y0} y^2 + c_{y0} y + d_{y0}) \end{aligned} \quad (8)$$

The same procedure has to be applied for the y coordinate correction of the image. The approximations of the discrete coefficients versus the discrete coefficients are graphical displayed in the following Figure 5. For every pixel the two correction values gives information about the direction and the amount of deviation. With the approximation for the y and x direction of the deviations of the circles on the calibration target a very precise correction can be calculated for every pixel. Therewith up to four pixels have to be weighted to generate the result pixel.

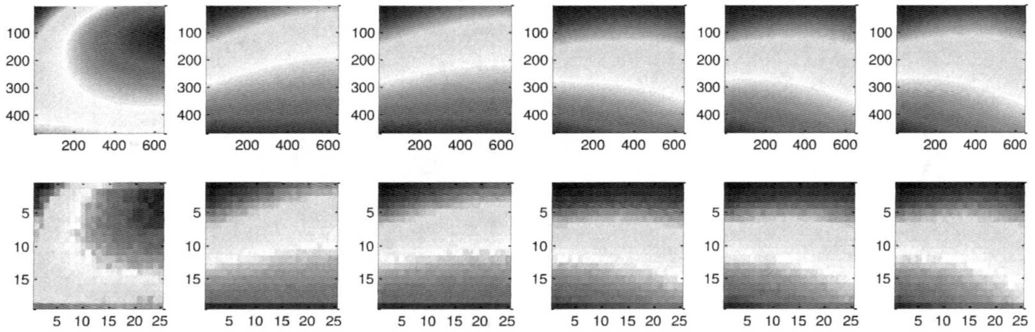

Figure 5: approximation of the deviation versus discrete sampling points beginning from the 400nm channel up to 650nm channel, upper pictures displaying the approximation on the complete image lower images displaying the discrete sampling points of the detected circles;

In order to implement this algorithm in a smart spectral imager, the correction values are placed in a digital filter core which moves over the distorted picture and is actualized with for every new incoming pixel. Formal a linear folding operation with changing coefficients is applied to correct the distorted images. The big advantage is that only the filter core has to be updated during an incoming image. The size of the core depends on the absolute value of error and has an odd core size. For the used lens the absolute value of the deviation is nine pixels. The deviation is direction dependent that leads to the effective size of the filter core. In this case the dimension is nine. The deviation based on the optical distortions, leads to a non-raster synchron movement. For this reason the integral assumption of four pixels were applied. Therefore it is assumed that the pixel intensity is divided in four pixels hugely (Figure 6).

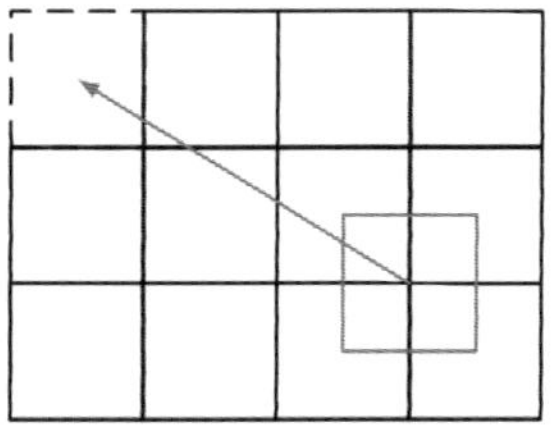

Figure 6: example for an uniformly distributed pixel, in this case a quarter intensity of the four pixel were used;

To move the gray value on the correct position two pixels in x and two pixels in the y direction which were weighted with the position values. That leads to the usage of a maximum of four pixels.

4. Results

After the analysis and the processing of the data, Figure 7 shows the results of the correction on the calibration target. With the applied method a deviation less than a half pixel can be achieved. In Figure 7 part d.) the uncorrected data along the search line is displayed. The difference between the edge transitions over all channels has a deviation of nine pixels between the different wavelength channels.

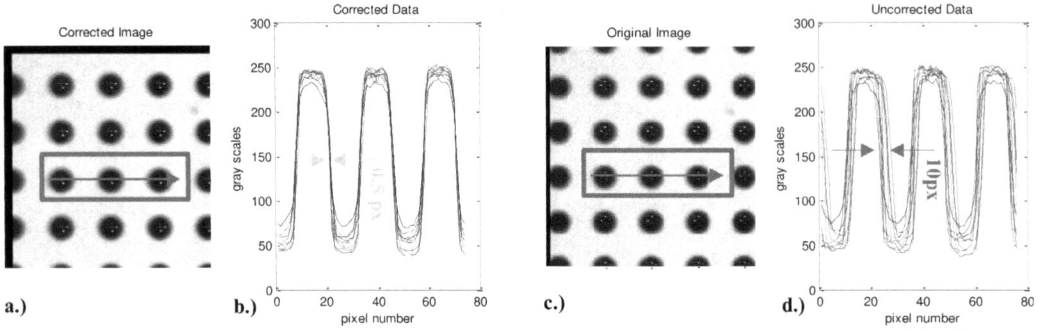

Figure 7: a.) corrected image the green crosses (expected position without errors) are in the centre of the circle after correction (channel 4, 550nm), b.) edge transition of the corrected images (left), c.) detected centre of the circles (red crosses) (channel 4, 550nm), d.) edge transition uncorrected; (blue graphs 400-500nm, green graphs 550-650nm, red graphs 700-800 nm, grey scaled graphs 850-950nm);

With the used lens and the applied correction algorithm a deviation of less than 20 microns at a field of 20000x26000 microns in all channels is achievable. For geometric detection a decrease of a factor of ten in uncertainty is possible with the correction. Furthermore multi-channel edge detection algorithms will deliver better results.

References

[1] F. Konig and P. Herzog, "On the limitations of metameric Imaging", Proceedings of IS&T PICS, 1999, pp. 163-168.

[2] M. Yamaguchi, H. Haneishi, and N. Ohyama, "Beyond Red-Green-Blue (RGB): Spectrum-based color imaging technology," J. Imag. Sci. Technol., vol. 52, no. 1, pp. 010 201-1–010 201-15, Jan. 2008

[3] M. Preißler, M. Rosenberger, M. Correns, M. Schellhorn and G. Linss, „Investigation on a modular high speed multispectral camera", Proceedings of the 20th IMEKO TC2 Symposium on Photonics in Measurement : May 16 - 18, 2011, [Linz Austria], pp.59-62, Aachen : Shaker, 2011.

[4] M. Preißler, M. Rosenberger, G. Linß: "Smart Spectral Imager", 2012 IEEE Photonics Conference (IPC 2012) : Burlingame, California, USA, 23 - 27 September 2012. - Piscataway, NJ : IEEE. - 2012, S. 591-592

[5] H. Haferkorn: "Bewertung optischer Systeme", Berlin : Dt. Verl. der Wiss., 1986, ISBN 3-326-00000-6

Colour based fire detection method with temporal intensity variation filtration

K Trambitckii[1], K Anding[2], V Musalimov[1], G Linß[2]

[1]Faculty of Fine Mechanics and Technology, Department of Mechatronics, National Research University of Information Technologies, Mechanics and Optics, Saint-Petersburg, Russia

[2]Faculty of Mechanical Engineering, Department of Quality Assurance and Industrial Image Processing, Ilmenau University of Technology, Ilmenau, Germany

E-mail: tr-kostya@ya.ru

Abstract. Development of video, computing technologies and computer vision gives a possibility of automatic fire detection on video information. Under that project different algorithms was implemented to find more efficient way of fire detection. In that article colour based fire detection algorithm is described. But it is not enough to use only colour information to detect fire properly. The main reason of this is that in the shooting conditions may be a lot of things having colour similar to fire. A temporary intensity variation of pixels is used to separate them from the fire. These variations are averaged over the series of several frames. This algorithm shows robust work and was realised as a computer program by using of the OpenCV library.

1. Introduction

For the early fire detection in areas of high fire risk, it is necessary to monitor such areas regularly. Improvement of monitoring systems can go in several directions: increasing possibility of fire detection at the earlier steps, reducing the cost of monitoring, reducing the influence of the human factor, etc.

Development of video, computing technologies and computer vision gives a possibility of automatic fire detection on video information. Such a system was implemented as a computer program, which uses and processes video information from a video camera with the goal of fire detection. Only task of computer operator in that process—to react to the signals of alarm system in proper time.

Different algorithms was implemented to find a more efficient way of fire detection. Fire detection speed was selected as one of the most important criteria. It gives the possibility to work in a real time.

That work was held under the project called “Program-technical system of detection and prediction of large-scale wildfires”.

2. State of the art

Detection of moving objects are often used to find possible fire regions. For this purpose, mainly two ways are used: consecutive frames subtraction and background subtracting [1]. In the first method changes are calculated in the images during the transition from one frame to another. The problem of this method is, that the overlapping area of the images can be mistakenly taken as a background. In the process of background subtraction dynamic regions are extracted from a static background image.

Content from this work may be used under the terms of the Creative Commons Attribution 3.0 licence. Any further distribution of this work must maintain attribution to the author(s) and the title of the work, journal citation and DOI.
Published under licence by IOP Publishing Ltd

The problem of this method is, that the extracted region can be quite inaccurate if the background image cannot be updated on time or at all. For example, for the forest fire detection this method is not applicable because of moving of trees, which is caused by wind. In that case the most parts of the image becomes dynamic.

Colour selection is another way of finding fire areas. Celik et al. [2] described the colour model of fire in RGB colour space. Decision whether colour pixel belongs to fire is made on the basis of the rules of two groups of inequalities. Chen et al. [3] allocated fire area in RGB colour space using three rules within which each possible fire pixel intensity have to be greater than threshold value. Celik et al. [4] proposed a model to determine the colour of fire pixels in the colour space YCbCr. To decide whether the pixel is a fire pixel a fuzzy inference system is used. System parameters obtained as a result of processing a large number of images with fire. Phillips et al. [5] uses the test data in which the fire manually selected for machine learning to create the necessary colour table. So it is possible to increase the detection accuracy for a specific fire type.

In addition to colour information, the dynamic characteristics of fire are important features, which are usable to separate fire from similar coloured objects.

Phillips et al. [5] calculated temporary intensity variations over several frames. If the temporary variation of definite pixel is more than a certain threshold value (which is typical for flame) the pixel is defined as a fire pixel. Zhang et al. [6] used the fact that the height of the fire varies with time due to movements of the flames, so that feature is taken as the main dynamic characteristics of fire. Torein et al. [7] kept track of the history of changes in the red channel of each RGB pixel of the image belonging to the contour of the fire within a short period of time and uses this data as the input wavelet method.

3. Tools and computing system

As the main tool for video data processing OpenCV was used. It is a popular open source library of computer vision, image processing and general-purpose numerical algorithms.

The fire detection software was implemented in C++ language. Open source integrated development environment (IDE) Eclipse was used as a development environment.

4. Algorithm

The first step of the implemented algorithm described below is a pixel based colour filtration. The system uses stabilized camera. Therefore described algorithm does not use background subtraction as a separate step. Methods with a similar realization have the following advantages:

- high processing speed, which makes it possible to use the algorithm in real-time video system;
- the lack of a background image subtraction remove the need of updating it, and thus, gives the possibility of using the algorithm for the detection of forest fires;
- relative ease for implementation.

Next steps include temporary intensity variation filtration and making decision step. The complete block scheme of the algorithm is shown in figure 1. Figure 2 and figure 3 show some examples of correct detected fire situations by using the described algorithm.

4.1. Colour filtration

At the first step the colour analysis is applied. An analysis of the space of abstract mathematical colour models is used, which are sets of 3 or 4 channels. As a colour model the RGB colour space is used. At the first step algorithm searches pixels which have typical fire colour in RGB colour space.

It is determined whether definite pixel is a possible fire pixel by the following equations [2]:

$$\left.\begin{array}{c} R(x,y) > R_{mean}, \\ R_{mean} = \frac{1}{K}\sum_{i=1}^{K} R(x_i, y_i) \\ R(x,y) > G(x,y) > B(x,y) \end{array}\right\} \quad (1)$$

where K is the total pixels number, R_{mean} is the mean value of the red colour at the whole image, $R(x, y)$, $B(x, y)$ and $G(x, y)$ are the values of the red, blue and green channel of the definite pixel (x, y).

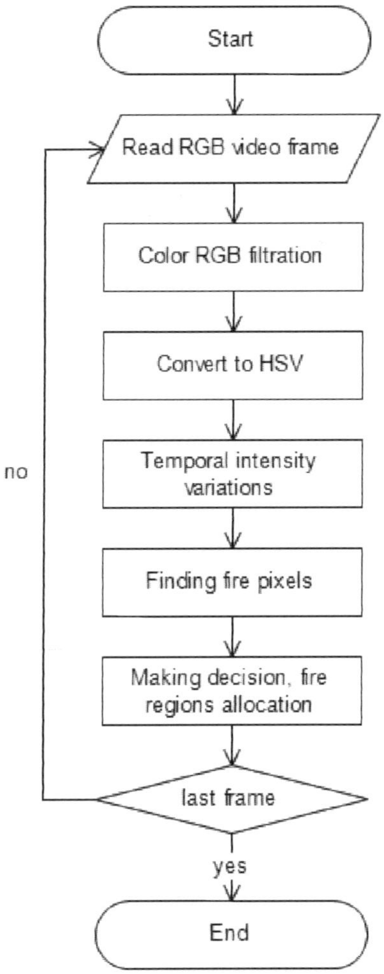

Figure 1. Algorithm block scheme.

Figure 2. Example 1 of fire detection.

Figure 3. Example 2 of fire detection.

These conditions work for most of the fire pixels. Thus if the RGB value of the pixel is within these three conditions, it is supposed that the current pixel is a fire pixel. That information is stored in the fire pixels separate image. Its dimension is equal to the size of one frame of the input video. When the pixel belongs to the RGB colour specified range, the filtered image is calculated as follow:

Journal of Physics: Conference Series **588** (2015) 012038 doi:10.1088/1742-6596/588/1/012038

$$I_{fire}(x, y, i) = 1 \tag{2}$$

where $I_{fire}(x, y, i)$ is the pixel of the *i*-th frame with the coordinates of *x* and *y*. Otherwise the pixel is equal to zero.

Figure 4 shows the result image after the first step of algorithm applied on the colour pixels with conditions described above. White pixels are image elements which are equal to one and black ones are equal to zero.

Figure 4. Original snapshot from video (left) and colour filtered image (right).

4.2. Averaging, fire pixels calculation

By nature fire is translucent. Therefore, the estimation of fire colour is averaged for a short period of time. First, each frame is estimated by colour accessories. The last *n* frames are processed to make a decision whether the pixel is a pixel of fire.

To calculate the probability that a colour pixel belongs to the fire, the average value for the last n frames is calculated:

$$I_{av}(x, y) = \frac{\sum_{i=1}^{n} I_{fire}(x,y,i)}{n} \tag{3}$$

with n as the number of averaged frames.

If a pixel value of I_{av} is more than a threshold value k_1, then it is supposed that the pixel is a fire pixel. Thus the new image is calculated:

$$I_{colour}(x, y, i) = \begin{cases} 1 \; if \; I_{av}(x, y, i) > k_1 \\ 0 \; if \; I_{av}(x, y, i) \leq k_1 \end{cases} \tag{4}$$

where k_1 is a definite threshold value.

Figure 5 shows a comparison of images I_{fire} and I_{colour}. For our example $n = 5$ and $k_1 = 0.2$. Thus if one pixel of I_{av} image appeared in the last 5 frames at least twice times, then that pixel is a fire pixel and the corresponding element of image is equal to one. As seen from the comparison, the areas of the new image are more saturate than in I_{fire} image because of the data averaging in the series of frames. That step is also useful for the reduction of noise pixels.

4.3. Temporary intensity variations

Colour information is not always enough to properly detect fire on video information. The main reason is that there can be many objects in observation conditions, which have a colour similar to fire. To distinguish them from fire, temporary changes of intensity of pixels calculated in a series of shots are used.

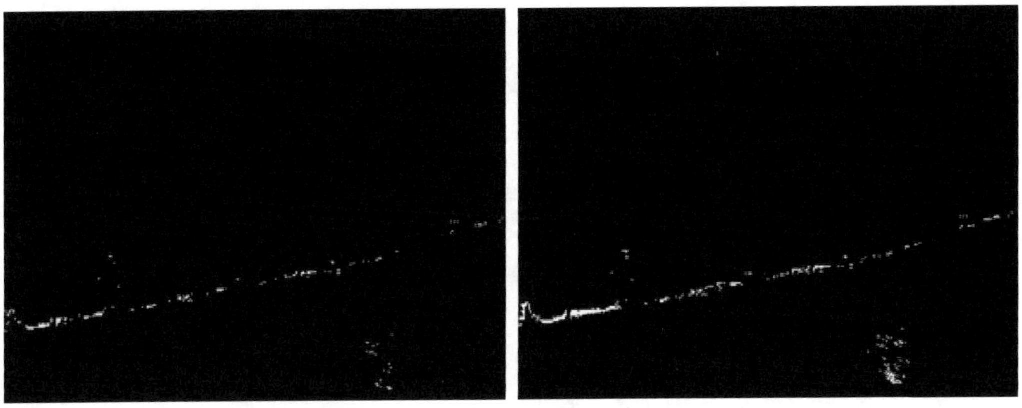

Figure 5. Colour filtered matrix (left) and averaged matrix (right).

In RGB colour space all three channels are colour values and there are no channels with information about the intensity of pixels. One way to solve the problem is to change the colour space to HSV. In this colour space the V channel has information about intensity.

The first using information from V channel of HSV colour space in the form of the temporary intensity variations over n shots, which are calculated for each pixel:

$$I_{diff}(x,y) = \frac{\sum_{i=2}^{n} |V(x,y,i) - V(x,y,i-1)|}{n-1} \quad (5)$$

where $V(x, y, i)$ and $V(x, y, i-1)$ is the intensity (V channel of HSV colour space) of pixel with x and y coordinates of current and previous frame respectively.

However, this change of intensity can be wrong, for example in the case where in addition to the intensity flickering of fire pixels overall image intensity changes because of changing lighting conditions. Therefore average intensity change is also calculated for each non-fire pixel of image:

$$I_{diff_a} = \frac{\sum_{x=1}^{m} \sum_{y=1}^{p} I_{diff}(x,y)}{\sum_{x=1}^{m} \sum_{y=1}^{p} 1} \quad (6)$$

where m and p are numbers of horizontal and vertical pixels respectively for which $I_{colour}(x, y) = 0$. At the next step the overall average intensity change I_{diff_a} is subtracted from each pixel of the matrix $I_{diff}(x, y)$:

$$\Delta I(x,y) = I_{diff}(x,y) - I_{diff_a} \quad (7)$$

Matrix ΔI is the actual temporal variation of the intensity of pixels in video information. The influence of environment light condition changes is crucially reduced.

4.4. Fire pixel detection

At that stage it is possible to definitively identify the pixels of fire. If the pixel is a possible fire pixel (using colour condition) and the actual temporal intensity variation $\Delta I(x, y)$ is larger than the threshold value k_2, it means that the pixel is a pixel of fire:

$$I_f(x,y) = \begin{cases} 1, if\ I_{colour}(x,y) = 1\ and\ \Delta I(x,y) > k_2 \\ 0, otherwise \end{cases} \quad (8)$$

Figure 6 shows the final image I_f of fire detection on a video information. In our example $k_2 = 10$. It can be seen from the example that the described algorithm correctly removes pixels with colour which is similar to fire but do not belong to fire. And there are only fire pixels in the final image.

Figure 6. Original source image (left) and final result of detection (right).

4.5. Making decision
The final step in fire detection is to decide whether there is a fire in this video image or not. The easiest way is to count the number of fire pixels detected at previous steps and a further comparison of the amount counted from the threshold value. If quantity of fire pixels is above the threshold, then this means that there is a fire on the video with a high probability. In this case, using the decision, it is possible to display an alarm message on the screen of the computer.

Figure 2 and figure 3 show some examples of correct detected fire situations by using the described algorithm.

5. Conclusion
Development of computer systems and computer vision technologies has made it possible to use an alternative method for monitoring areas where the probability of fire occurrences is high. That algorithm does not require expensive high performance equipment and it can works in real time on a regular personal computer by using a cheap video camera. This algorithm shows robust work and was realised as a computer program by using the OpenCV library.

References
[1] Toreyin B U , Cetin A E 2007 Online detection of fire in video *IEEE Conference on Computer Vision and Pattern Recognition* (Minneapolis) pp 1–5
[2] Celik T, Demirel H, Ozkaramanli H, Uyguroglu M 2006 Fire detection using statistical colour model in video sequences *IEEE Intern. Conf. on Acoustics, Speech and Signal Proc* pp 213–216
[3] Chen T H , Wu P H , Chiou Y C 2004 An early fire-detection method based on image processing *IEEE International Conf. on Image Processing (ICIP)* pp 1707–1710
[4] Celik T, Ozkaramanli H, Demirel H 2007 Fire and smoke detection without sensors: image processing based approach *European Signal Processing Conference*
[5] Philips W, Shah M and Lobo N V 2007 Flame recognition in video (Istanbul)
[6] Zhang J H, Zhuang J, Du H F 2006 A new flame detection method using probability model *Int. Conf. on Computational Intelligence and Security* pp 1614–1617
[7] Toreyin B U, Dedeoglu Y, Gudukbay U, Cetin A E 2006 *Pattern Recognition Letters* Computer vision based method for real-time fire and flame detection **27** pp 49–58

IMEKO IOP Publishing
Journal of Physics: Conference Series **588** (2015) 012039 doi:10.1088/1742-6596/588/1/012039

Probabilistic inversion: a preliminary discussion

Giovanni Battista Rossi and Francesco Crenna
Laboratorio di Misure
Università degli Studi di Genova - DIME/MEC, Genova, Italy

Abstract.
We continue the discussion on the possibility of interpreting probability as a logic, that we have started in the previous IMEKO TC1-TC7-TC13 Symposium. We show here how a probabilistic logic can be extended up to including direct and inverse functions. We also discuss the relationship between this framework and the Bayes-Laplace rule, showing how the latter can be formally interpreted as a probabilistic inversion device. We suggest that these findings open a new perspective in the evaluation of measurement uncertainty.

1. Introduction

In the last Joint IMEKO TC1-TC7-TC13 Symposium (2013) we proposed a logical approach to probability [3-5;7-10] that proved to be quite effective in treating measurement representations accounting for uncertainty as an inherent feature [1-2;12-13][1].

We now proceed in this research line by providing tools for effectively describing the measurement process. It has been noted that the measurement process can be modelled in fairly general terms as the concatenation of an observation and a restitution phase. The former considers the "physical" behaviour of the measuring system, the latter includes the data processing required to produce a measurement result including a statement on measurement uncertainty [6]. If a probabilistic model is assumed for describing the observation phase, restitution may be interpreted as a "probabilistic inversion" of observation [11].

Yet, so far this interpretation has been provided on an intuitive basis mainly, assuming that the Bayes-Laplace rule can be understood as providing some kind of probabilistic inversion. Here we reconsider this question in a more formal way, following the logical approach to probability that we have recently developed, at least in an initial way.

2. Probabilistic functions

Let us briefly recall the general reference framework [13]. Consider a finite universe A, a finite collection $\mathcal{E} = \{\mathcal{A}_1, \mathcal{A}_2, ...\mathcal{A}_M\}$ of structures over A and a probabilistic distribution over $\mathcal{E}$, $P(\mathcal{A}_i)$, such that

$$P(\mathcal{E}) = P\{\mathcal{A} \in \mathcal{E}\} = \sum_{i=1}^{M} P(\mathcal{A}_i) = 1. \tag{1}$$

Let ϕ be any statement whose truth can be assessed on any element (structure) of $\mathcal{E}$ [5]. Then

[1] References are listed in chronological order to give a feeling of the historical development of the subject

Content from this work may be used under the terms of the Creative Commons Attribution 3.0 licence. Any further distribution of this work must maintain attribution to the author(s) and the title of the work, journal citation and DOI.
Published under licence by IOP Publishing Ltd

Table 1. An illustrative example of a probabilistic function from A to B.

Structure	Function	$f_i(a)$	$f_i(b)$	$P(\mathcal{A}_i)$
$\mathcal{A}_1$	f_1	c	c	0.2
$\mathcal{A}_2$	f_2	d	d	0.2
$\mathcal{A}_3$	f_3	c	d	0.5
$\mathcal{A}_4$	f_4	d	c	0.1

$$P(\phi) = P\{\mathcal{A} \in \mathcal{E}|\phi\} = \sum_{\mathcal{A}_i|\phi} P(\mathcal{A}_i), \tag{2}$$

where $\mathcal{A} \in \mathcal{E}|\phi$ denotes a structure where the statement ϕ is true.

In Refs. [12, 13] we have amply discussed the application of this framework to order and to generic m-ary relations. Let us now focus to *functions*. To this purpose, note that the statement

$$v = f(u), \tag{3}$$

referring to a function $f : A \to B$, may be seen as denoting a binary relation on $A \times B$, such that

$$\forall u \in A \exists v \in B(v = f(u)), \tag{4}$$

$$\forall u, v, z \in A(v = f(u) \wedge z = f(u) \Rightarrow v = z). \tag{5}$$

Therefore we can consider a finite collection of structures $\mathcal{E} = \{\mathcal{A}_1, \mathcal{A}_2, ...\mathcal{A}_M\}$, where $\mathcal{A}_i = (A, B, f_i)$, and an associated probability distribution $P(\mathcal{A}_i)$ over $\mathcal{E}$, that satisfies condition (1). Then the probability that the elements $a \in A$ and $b \in B$ are linked by the function f is given by

$$P(f(a) = b) = P\{\mathcal{A} \in \mathcal{E}|f(a) = b\} = \sum_{\mathcal{A}_i|f(a)=b} P(\mathcal{A}_i). \tag{6}$$

Consider, as an example, $A = \{a, b\}$ and $B = \{c, d\}$. There are thus four possible functions from A to B, as shown in Table 1.

Suppose that a probability is assigned to each of them, as shown in the same table. Then it is possible to calculate $P(f(u) = v)$, for each $u \in A$ and each $v \in B$. We obtain, in this case:

- $P(f(a) = c) = 0.7$,
- $P(f(a) = d) = 0.3$,
- $P(f(b) = c) = 0.3$,
- $P(f(b) = d) = 0.7$.

3. Probabilistic inversion

Consider now the probabilistic inverse function to the above, $g : B \to A$. Note that in the deterministic case, only two of the four functions above are invertible, namely f_3 and f_4. Instead, in the probabilistic case the notion of inversion becomes more general as we will see in the following development.

To approach the notion of probabilistic inversion, consider first that we can calculated directly the probabilities associated to each value of g, on the basis of the (complete) knowledge

Table 2. Probabilistic inverse function, from B to A.

Structure	Function	$g_i(c)$	$g_i(d)$	$P(\mathcal{B}_i)$
$\mathcal{B}_1$	g_1	a	a	0.21
$\mathcal{B}_2$	g_2	b	b	0.21
$\mathcal{B}_3$	g_3	a	b	0.49
$\mathcal{B}_4$	g_4	b	a	0.09

associated to the (direct) probabilistic function f, through the very definition of inverse function. In fact we can establish the following rule:

$$P(g(b) = a) \propto P\{\mathcal{A} \in \mathcal{E} | f(a) = b\} = \sum_{\mathcal{A}_i | f(a) = b} P(\mathcal{A}_i), \tag{7}$$

where anyway a proportionality, rather than an equality, holds between the first and the second member of the equation. In fact this rule does not provide directly a probability distribution; yet the proportionality factor can be calculated by imposing the following closure condition:

$$\sum_{v \in A} P(g(v) = a) = 1. \tag{8}$$

If we apply this rule to our numerical example, we obtain[2]:

- $P(g(c) = a) = 0.7$,
- $P(g(c) = b) = 0.3$,
- $P(g(d) = a) = 0.3$,
- $P(g(d) = b) = 0.7$.

Under a different perspective, since g is a function from B to A, we recognise that we have four such functions, as shown in Table 2, where the corresponding structures, $\mathcal{B}_k = (B, A, g_k)$, are also defined. There probabilities can thus be calculated by the rule

$$P(g_k) = P(\bigwedge_{b \in B} g(b) = g_k(b)). \tag{9}$$

As an example, we can apply this rule to our example and we obtain:

- $P(g_1) = P(g(c) = a \wedge g(d) = a) = P(g(c) = a)P(g(d) = a) = 0.7 \cdot 0.3 = 0.21$,
- $P(g_2) = P(g(c) = b \wedge g(d) = b) = P(g(c) = b)P(g(d) = b) = 0.3 \cdot 0.7 = 0.21$,
- $P(g_3) = P(g(c) = a \wedge g(d) = b) = P(g(c) = a)P(g(d) = b) = 0.7 \cdot 0.7 = 0.49$,
- $P(g_4) = P(g(c) = b \wedge g(d) = a) = P(g(c) = b)P(g(d) = a) = 0.3 \cdot 0.3 = 0.09$,

These results are summarized in the table. So rules (7-9) allow us to define the probabilistic inverse function g.

[2] In this example the proportionality becomes an equality, but this is due to the simplicity of the example and does not happen in the general case.

Table 3. Probabilistic function f, from X to Y.

Structure	Function	$f_i(1)$	$f_i(2)$	$f_i(3)$	$P(\mathcal{A}_i)$
$\mathcal{A}_1$	f_1	4	4	4	0.1
$\mathcal{A}_2$	f_2	4	4	5	0.4
$\mathcal{A}_3$	f_3	4	5	4	0.1
$\mathcal{A}_4$	f_4	4	5	5	0.2
$\mathcal{A}_5$	f_5	5	4	4	0.0
$\mathcal{A}_6$	f_6	5	4	5	0.1
$\mathcal{A}_7$	f_7	5	5	4	0.0
$\mathcal{A}_8$	f_8	5	5	5	0.1

4. Relationship with the Bayes-Laplace rule

Consider now the special but important case, where A and B are two finite sets of integers, that is $A = X \in \mathbb{N}$ and $B = Y \in \mathbb{N}$. Then the function $f : X \to Y$ can now be regarded as linking two probabilistic variables, $x \in X$ and $y \in Y$. The corresponding probabilistic inverse function $g : Y \to X$ can still be obtained by formulae (7-9).

From another standpoint, note that the *conditional distribution* $P(y|x)$ can be defined by

$$P(y = j|x = i) = P\{\mathcal{A} \in \mathcal{E}|f(i) = j\} = \sum_{\mathcal{A}_k|f(i)=j} P(\mathcal{A}_k). \tag{10}$$

So there is a close link between the probabilistic function f and the conditional distribution $P(y|x)$.

Now, the (inverse) conditional distribution $P(x|y)$ can be obtained by the Bayes-Laplace rule, assuming a uniform distribution for x. We obtain

$$P(x = i|y = j) \propto P(y = j|x = i), \tag{11}$$

where the proportionality factor can be fixed through the closure condition

$$\sum_y P(x|y) = 1. \tag{12}$$

On the other hand, such distribution can be obtained from the inverse function g through rule (8), now rewritten as

$$P(x = i|y = j) = P\{\mathcal{B} \in \mathcal{E}|g(j) = i\} = \sum_{\mathcal{B}_k|g(j)=i} P(\mathcal{B}_k), \tag{13}$$

where $\mathcal{B}_k = (B, A, g_k)$ and the two paths yield consistent results.

An example is provided in Table 3, where $X = \{1, 2, 3\}$ and $Y = \{4, 5\}$,.

The corresponding inverse function, calculated through rules (7-9), is reported in Table 4. Alternatively, we can proceed by following the Bayes-Laplace rule. We thus calculate the (direct) conditional probability distribution $P(y|x)$ and we obtain:

$$P(y = 4|x = 1) = P(f(1) = 4) = 0.8$$
$$P(y = 5|x = 1) = P(f(1) = 5) = 0.2$$
$$P(y = 4|x = 2) = P(f(2) = 4) = 0.6$$
$$P(y = 5|x = 2) = P(f(2) = 5) = 0.4$$
$$P(y = 4|x = 3) = P(f(3) = 4) = 0.2$$

Table 4. Probabilistic inverse function g, from Y to X.

Structure	Function	$g_i(4)$	$g_i(5)$	$P(\mathcal{B}_i)$
$\mathcal{B}_1$	g_1	1	1	0.07
$\mathcal{B}_2$	g_2	1	2	0.14
$\mathcal{B}_3$	g_3	1	3	0.29
$\mathcal{B}_4$	g_4	2	1	0.05
$\mathcal{B}_5$	g_5	2	2	0.11
$\mathcal{B}_6$	g_6	2	3	0.21
$\mathcal{B}_7$	g_7	3	1	0.02
$\mathcal{B}_8$	g_8	3	2	0.04
$\mathcal{B}_9$	g_9	3	3	0.07

$$P(y = 5|x = 3) = P(f(3) = 5) = 0.8$$

Then we calculate the inverse probability distribution $P(x|y)$, through rules (9-10), and we obtain

$$P(x = 1|y = 4) = P(g(4) = 1) = 0.500$$
$$P(x = 2|y = 4) = P(g(4) = 2) = 0.375$$
$$P(x = 3|y = 4) = P(g(4) = 3) = 0.125$$
$$P(x = 1|y = 5) = P(g(5) = 1) \cong 0.143$$
$$P(x = 2|y = 5) = P(g(5) = 2) \cong 0.286$$
$$P(x = 3|y = 5) = P(g(5) = 3) \cong 0.571$$

Where some rounding of numerical values has been done to obtain a greater readability of the results. On the other hand, we can also calculate this distribution through (13), and the two procedures yield consistent results, apart from rounding effects.

To sum up,

- a probabilistic function, f, can be defined by formulae (3-6);
- the corresponding probabilistic inverse function, g, by formulae (7-9);
- the conditional distribution, $P(y|x)$, corresponding to function f, can be obtained according to formula (10)[3];
- the inverse conditional distribution, $P(x|y)$, corresponding to the inverse function g, can be obtained thanks to the Bayes-Laplace rule, reported in formulae (11-12);
- the same inverse conditional distribution, $P(x|y)$, can also be obtained directly from the corresponding inverse function g through the rule expressed by (13), which thus established a linked between the Bayes-Laplace rule and the notion of probabilistic inverse function, here obtained after considering probability as a logic.

5. Conclusion

We have shown how a probabilistic logic can be developed up to including direct and inverse functions. In particular, we have shown that in a framework where probability is interpreted as a (first order) logic it is possible to formally define the notion of *probabilistic function* and of

[3] Here the (special) case where $y = f(x)$ and x and y are probabilistic variables, i.e., numerical entities, is considered, yet a more general formulation is also possible in a very similar way.

probabilistic inverse function, and that the Bayes Laplace rule provides an alternative consistent way for performing *probabilistic inversion*.

Therefore, the Bayes-Laplace rule can be *formally*, and not only intuitively, interpreted as a probabilistic inversion device. This opens a new perspective in the development of a probabilistic approach to measurement, that frees it from unnecessary links with subjective interpretations of probability [13]. The full exploitation of these ideas will be the object of future studies.

References

[1] Leaning M S and Finkelstein L 1980 *ACTA IMEKO 1979* ed G Streker (Amsterdam:Elsevier) p 73
[2] Falmagne J C 1980 *Philosophy of Science* **47** 277
[3] Fenstad J E 1980 *Studies in inductive logic and probability Vol. 2* ed R C Jeffrey (Berkeley: University of California Press) p 251
[4] Regenwetter M 1996 *Journal of mathematical psychology* **40** 219
[5] Rigamonti G 2005 *Corso di logica* (Torino: Bollati Boringhieri)
[6] Rossi G B 2006 *Measurement* **39** 34
[7] Accardi L 2008 La matematica: problemi e teoremi ed C Bartocci and P Odifreddi (Bologna: Einaudi) p 793
[8] Costantini D 2011 *Verso una visione probabilistica del mondo* (Padova: GEM)
[9] Haenni R, Romeijn J W, Wheeler G, Williamson J 2011 *Probabilistic Logics and Probabilistic Networks* (Dordrecht: Springer)
[10] Benoit E 2011 *Uncertainty in fuzzy scales based measurements* Int. IMEKO TC7 Symp. August 31th - September 2nd Jena, Germany
[11] Rossi G B 2012 *IEEE Transactions on Instrumentation and Measurement* **61** 2097
[12] Rossi G B, Crenna F 2013 *Probability as a logic for measurement representations* Joint IMEKO TC1-TC7-TC13 Symp. September 3-5 Genova, Italy
[13] Rossi G B, Crenna F 2014 *A first order probabilistic logic with application to measurement representations* submitted for publication.

A proposal for the measurement of graphical statistics effectiveness: Does it enhance or interfere with statistical reasoning?

M Agus[1], M P Penna[1], M Peró-Cebollero[2], J Guàrdia-Olmos[2]

[1]Department of Pedagogy, Psychology, Philosophy, Faculty of Humanistic Studies, University of Cagliari, Via Is Mirrionis, 1, 09123 Cagliari, Italy

[2]Department of Methodology of the Behavioural Sciences, Faculty of Psychology, University of Barcelona, Pg. Vall d'Hebron, 171, 08035 Barcelona, Spain

E-mail: mirian.agus@unica.it

Abstract. Numerous studies have examined students' difficulties in understanding some notions related to statistical problems. Some authors observed that the presentation of distinct visual representations could increase statistical reasoning, supporting the principle of graphical facilitation. But other researchers disagree with this viewpoint, emphasising the impediments related to the use of illustrations that could overcharge the cognitive system with insignificant data. In this work we aim at comparing the probabilistic statistical reasoning regarding two different formats of problem presentations: graphical and verbal–numerical. We have conceived and presented five pairs of homologous simple problems in the verbal numerical and graphical format to 311 undergraduate Psychology students (n=156 in Italy and n=155 in Spain) without statistical expertise. The purpose of our work was to evaluate the effect of graphical facilitation in probabilistic statistical reasoning. Every undergraduate has solved each pair of problems in two formats in different problem presentation orders and sequences. Data analyses have highlighted that the effect of graphical facilitation is infrequent in psychology undergraduates. This effect is related to many factors (as knowledge, abilities, attitudes, and anxiety); moreover it might be considered the resultant of interaction between individual and task characteristics.

1. Introduction

Over the last decades many studies have been carried out to evaluate the influences on problem solution produced by the overall disposition of the matter. Probabilistic statistical problems are a particular type of enigmas, that persons are requested to solve both in academic contexts and in everyday life. Some authors affirm that the application of some visual aids could appear as a capital way to enlighten statistical reasoning [e.g. 1]. In contrast, Knauff and Johnson-Laird [2] sustain that illustrations could inhibit reasoning in these dilemmas because they could burden the cognitive system with supplementary evidences, beside the point.

Content from this work may be used under the terms of the Creative Commons Attribution 3.0 licence. Any further distribution of this work must maintain attribution to the author(s) and the title of the work, journal citation and DOI.
Published under licence by IOP Publishing Ltd

In the light of bibliography, our study attempts at evaluating and comparing the performance in some pairs of analogous problems, presented in verbal-numerical and graphical format. By this, we investigate the existence of the effect of 'graphical facilitation' versus 'graphical impediment', considering also the influence of others important dimensions identified in literature (visuo-spatial ability, numerical skill, attitude towards statistics and statistical anxiety) [e.g. 3,4].

2. Aim and method

The aim of this investigation is to propose an experimental protocol in order to explore whether graphical representations can facilitate or impede probabilistic statistical reasoning in psychology undergraduates. Therefore we assessed and compared the student's performance in problem solving with two homologous formats for each problem: verbal-numerical and graphical.

2.1. Participants, instrument and procedure

We studied 311 undergraduates in Psychology, enrolled in the first year of the degree course in three universities: University of Cagliari (Italy) (n=64), University of Trieste (Italy) (n=92) and University of Barcelona (Spain) (n=155). The mean age of the participants was 20.8 ± 5.9 years of age (mean ± standard deviation) and consisted of 85 males (27.3 %). Data collection was acquired from September to October 2013. The samples involved students who voluntarily participated in the research (accidental sampling). We submitted our questionnaires in one session, in papery form, to a large group in a lecture room; every group of undergraduates compiled the questionnaires respectively in their native language. By means of several pilot studies [4,5,6,7] we constructed a specific questionnaire in order to evaluate probabilistic statistical reasoning, comprising five pairs of homologous problems in two formats (verbal numerical – N - and graphical - G). The problems were presented in different orders (verbal-numerical and graphical -NG -vs. graphical and verbal-numerical -GN) and different sequences. We chose to inspect the reasoning in "inexperienced" students (without statistical skills) in order to assess the authentic impact of the format for the problem presentation. For each problem (in both formats), we offered a brief explanation and then four closed response options (when only one was correct).

With the purpose of estimating the numerical and visuo-spatial prerequisites for the problem solution, we applied two related scales of the Intermediate Form of Primary Mental Abilities (P.M.A.) [8,9]. Furthermore, we applied the Survey of Attitudes Towards Statistics (SATS-28) [10] and the Statistical Anxiety Scale (SAS) [11] to inquire the relationships among attitudes and emotional aspects and the problem solving (both these questionnaires demonstrated the cross-country validity between Italian and Spanish samples) [12,13].

In this work we evaluated the differences in problem solving between countries, the potential effect of the order of presentation between in N and G format. But, especially, we attempted to assess the potential 'graphical facilitation' effect. For these purposes we computed the Bivariate Chi Square, the Multiple Correspondence Analysis and the Hierarchical Cluster Analysis.

3. Results

We evaluated the descriptive statistics in relation to the answers given to each pair of problems. In order to specifically evaluate the supposed 'graphical facilitation' effect, we recoded the answers in each pair of problems in a new variable, devising a typological index [14]. In fact, we attempted to identify more precisely the students that solved the problems only in the graphical format and not in the other ways of presentation. For these reasons we constructed a new categorical variable for each pair of problem and we distinguished between: subjects who didn't give correct answers in correspondence to both presentation formats, subjects who gave correct answers only in the verbal-numerical format, subjects who gave correct answers only in the graphical format, subjects who gave correct answers in both presentation formats. These new indices classified the answers in two formats in a single categorical variable for each pair of problems, individuating a profile of all potential arrays of answers in the problem presentation.

In relation to these new indices, we applied the Bivariate Chi Square to detect the potential effect of the country (Italy and Spain); this analysis didn't show a significant effect ($p > .05$) then we decided to analyse all the students as a unique sample.

In Figure 1 we can observe the distribution frequencies for each index, in relation to all five pairs of problems; these indices are evaluated regarding the different orders of problems presentation (NG versus GN). Afterwards we computed a Bivariate Chi Square for each profile to distinguish the effect of the presentation order (NG versus GN) in the answers given in all five problems. The problem presentation order did not have an effect on the responses in relation to profile 1, profile 3 and profile 5. Indeed we found significant effects in relation to profile 2 ($\chi^2 = 11.166$; df=3; p=.011) and 4 ($\chi^2=18.348$; df=3 ; p= .0001). Especially in these two profiles we observed a higher number of subjects that did not solve in both forms in the GN order and in the NG order.

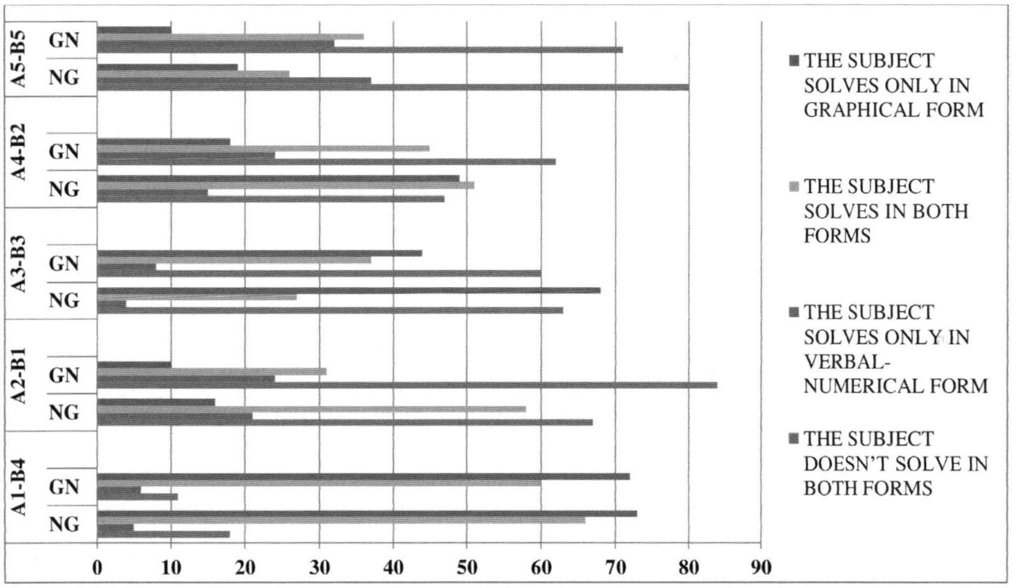

Figure 1. Frequencies for response profile in relation to each five pairs of problems

We decided to evaluate all the categorical indices related to five response profiles as extensive dimensions applying a Multiple Correspondence Analysis (MCA) [e.g. 15], using the software SPAD 5.5. We referred to the *Ecole française d'analyse des données* [e.g. 16,17]. This method allows exploring and resuming the categorical responses in relation to few dimensions, investigating the configuration of some categorical variables. It can be a powerful statistical tool because it reveals groups of variables and modalities in the dimensional spaces, without requiring fulfilling the stringent assumptions demanded by other statistical techniques largely used. This statistical method allows distinguishing among "active" and "supplementary" variables. The first ones define the space solution and the factor construction; the second ones do not affect the axis orientation, but complete and sustain the interpretation of dimensions identified by active variables.

Then in this technique we used as 'active' variables the categorical indices obtained in relation to each pair of probabilistic problems (five categorical variables associated to a four modalities each one, as previously indicated). As 'supplementary' variables we used the categorical variables related to personal data (gender, age, country), to the order of problem presentation (Numerical Graphical NG / Graphical Numerical GN), to the levels of abilities (numerical and visuospatial), to the levels of attitudes towards statistics and to the levels of statistical anxiety.

In the MCA we chose to consider two factors. The first could be identified as the dimension of 'probabilistic statistical reasoning' (eigenvalue = 0.4514; % intertia= 15.05). It was characterised by the active modalities of good solution in both formats for all problems. Moreover, as supplementary variables, it was associated in the positive semi-axis with a high level of visuo-spatial abilities, high time solution in both formats, low levels of statistical anxiety and high positive attitudes towards statistics. In the negative semi-axis the active and supplementary modalities show the opposite trend (see Table 1).

Table 1. MCA – Factor 1

Positive semi axis				
active modalities	**V. test**	**Factorial coordinates**	**Contributions**	**Cos^2**
P2 SOLVE N & G	13.01	1.17	17.2	.54
P4 SOLVE N & G	11.32	.96	12.6	.41
P5 SOLVE N & G	10.76	1.23	13.2	.37
P1 SOLVE N & G	9.23	.64	7.2	.27
P3 SOLVE N & G	7.96	.89	7.2	.20
Negative semi axis				
active modalities	**V. test**	**Factorial coordinates**	**Contributions**	**Cos^2**
P4 NOT SOLVE N & G	-10.39	-.79	9.9	.35
P2 NOT SOLVE N & G	-9.97	-.58	7.2	.32
P5 NOT SOLVE N & G	-9.88	-.57	7.1	.31
P3 NOT SOLVE N & G	-8.45	-.59	6.1	.23
P1 NOT SOLVE N & G	-7.50	-1.28	7.2	.18

These results should be read as an evidence that lead us to suppose that the 'graphical facilitation' effect could subsist particularly when the subject know the structure of problem (previously presented and solved correctly in the verbal numerical format).

The second factor could be recognised as the dimension of 'graphical facilitation' vs 'graphical impediment' (eigenvalue = 0.2764; % intertia = 9.21). Indeed it is characterised in the positive semi-axis by the active modalities for good solution only in the graphical format of all problems. As supplementary variables it is related with a low time for solution in the graphical format. The active and supplementary modalities display the reverse tendency in the negative semi-axis (see Table 2).

On the two factors resulting from the MCA, we applied the Hierarchical Cluster analysis to find undergraduate groups having analogous features, to group subjects, referring to the factors identified [e.g. 18]. This method allows us to classify the students' groups based on their performance in all pairs of problems, in order to identify the meaningful taxonomies. The 5-membership clustering was deliberated to offer the best cluster solution. This solution individuated students' groups that are good solvers in both formats of all problems (cluster 5, n=99) and bad solvers in both formats of all problems (cluster 2, n=133). But moreover three clusters were characterised by a minor number of students that experiment the 'graphical facilitation' effect in all problems (cluster 4, n=29) and the 'graphical impediment' (cluster 1, n=41 and cluster 3, n=11). Especially cluster 3 is interesting because it consists in a small number of students that experiment a 'very strong graphical impediment'. Generally, we could observe that the number of students that experience 'graphical impediment' is higher than the number of subjects that check out the 'graphical facilitation' effect.

Table 2. MCA – Factor 2

active modalities	V. test	Factorial coordinates	Contributions	Cos^2
Positive semi axis				
P5 NOT N, SOLVE G	9.54	1.69	19.1	.29
P2 NOT N, SOLVE G	8.52	1.60	15.4	.23
P4 NOT N, SOLVE G	8.13	.88	12.0	.21
P3 NOT N, SOLVE G	3.23	.24	1.6	.03
P1 NOT N, SOLVE G	2.65	.16	.9	.02
Negative semi axis				
active modalities	**V. test**	**Factorial coordinates**	**Contributions**	**Cos^2**
P3 SOLVE N, NOT G	-8.56	-2.43	16.4	.24
P1 SOLVE N, NOT G	-8.39	-2.49	15.8	.23
P4 SOLVE N, NOT G	-6.77	-1.02	9.3	.15
P5 SOLVE N, NOT G	-4.30	-.46	3.3	.06

4. Discussion

In our work we aim at evaluating the effect of 'graphical facilitation' vs 'graphical impediment' in probabilistic statistical reasoning, in psychology undergraduates without statistical expertise. For this reason we devised an experimental protocol characterised by five pairs of analogous probabilistic problems in verbal numerical and graphical format. The presentation of pairs of problems and their data analyses highlighted that only few students could advantage of the 'graphical facilitation' effect in the solution of these homologous of problems on probabilistic statistical reasoning [19]. It appears infrequent for our psychology undergraduates to experiment the effect of 'graphical facilitation'. Our results suggest the relevance of knowledge of the problem structure in facilitating the problem solving in both formats. Also the role of prerequisites, attitudes, and anxiety is shown, because bad solvers in both formats are characterised by high levels of statistical anxiety and low level of visuo-spatial ability. Moreover we observed that the good solvers in both formats had low levels of anxiety and high positive attitudes towards statistics. These facts are important because these students did not have a statistical education.

Our results might be also related to the subject's learning style; particularly we refer to the 'visual-spatial learners' that are characterised by strong visual-spatial abilities and weaknesses in verbal information sequential processing [e.g. 20]. In fact, some authors recognise that the application of visual-spatial representation could be useful for the 'visual-spatial learners', but not for subjects with different learning styles [e.g. 20]. This aspect might explain both the effect of graphical facilitation [19] and the effect of graphical impediment [2]. Moreover our results support the theoretical concept of the individual-task combination [3] which emphasised that the probabilistic statistical reasoning could be considered consistent with a multifactorial relation among many features (the tasks organisation in relation to the individual idiosyncrasy). The outcomes of this work could be useful regarding the intention to develop statistical teaching and to elaborate appropriate learning experiences to improve statistical reasoning. At this point, the assessment of subjects' characteristics seems to be advantageous for the enhancement of statistical reasoning, in relation to the application of graphical methods. In fact, the effect of 'graphical facilitation' cannot be considered as neglected by interaction with this multiplicity of factors.

References

[1] Brase G L 2009 Pictorial representations in statistical reasoning *Appl. Cogn. Psychol.* **23** 369-81. Doi:10.1002/acp.1460
[2] Knauff M and Johnson-Laird P N 2002 Visual imagery can impede reasoning *Mem. Cognition* **30** 363–71. Doi:10.3758/BF03194937
[3] Zhu L and Gigerenzer G 2006 Children can solve Bayesian problems: The role of representation in mental computation. *Cognition* ***98*** 3 287–308. Doi: http://dx.doi.org/10.1016/j.cognition.2004.12.003
[4] Penna M P, Agus M, Peró-Cebollero M, Guàrdia-Olmos J and Pessa E 2014 The use of imagery in statistical reasoning by university undergraduate students: a preliminary study *Qual. Quant.* **48** 1 173-187. Doi:10.1007/s11135-012-9757-5
[5] Agus M, Penna M P, Peró-Cebollero M, Guàrdia-Olmos J and Pessa E 2014 The application of graphical representations in estimation of probabilistic events *J. Th. Res. Ed.* **9** 1 235-252. Doi: 10.6092/issn.1970-2221/4299
[6] Agus M, Peró-Cebollero M., Guàrdia-Olmos J and Penna MP 2013 The measurement of statistical reasoning in verbal-numerical and graphical forms: a pilot study. *J.of Phys.Conf. Ser.* **459** 12023. Doi:10.1088/1742-6596/459/1/012023
[7] Agus M, Peró-Cebollero M, Penna M P and Guàrdia-Olmos J. 2014 Towards the development of problems comparing verbal-numerical and graphical formats in statistical reasoning. *Qual Quant.*1–19. Doi: 10.1007/s11135-014-0018-7
[8] Thurstone LL and Thurstone TG 1981a *PMA: abilità mentali primarie: manuale di istruzioni K-1 - Batteria fattoriale delle abilità mentali primarie* (Firenze: Organizzazioni Speciali)
[9] Thurstone LL and Thurstone TG 1981b *TEA: tests de aptitudes escolares [niveles 1, 2 y 3]: manual* (Madrid: Tea).
[10] Schau C, Stevens J, Dauphinee TL and Vecchio AD 1995 The Development and Validation of the Survey of Attitudes toward Statistics. *Educ. Psychol. Meas.* **55** 5 868–875.
[11] Colet AV, Seva UL and Condon L 2008 Development and validation of the Statistical Anxiety Scale. *Psicothema* **20** 1 174–180.
[12] Carmona J, Primi C, Chiesi F. 2008 Testing for measurement invariance of the Survey of Attitudes Toward Statistics: A comparison of Italian and Spanish students. III European Congress of Methodology, Oviedo, Spain.
[13] Chiesi F, Primi C, Carmona J. 2011 Measuring Statistics Anxiety. *J Psychoeduc Assess* **29(6)** 559–69.
[14] Marradi A 2007 *Metodologia delle scienze sociali* (Bologna: Il Mulino)
[15] Greenacre M 2010 *Correspondence analysis in practice*. (NW: Chapman and Hall/CRC).
[16] Benzécri J P. 1992 *Correspondence analysis handbook.* (CRC Press).
[17] Lebart L, Morineau A and Piron M. 1995 *Statistique exploratoire multidimensionnelle*. (Paris: Dunod).
[18] Di Franco G. 2006 *Corrispondenze multiple e altre tecniche multivariate per variabili categoriali.* (Milano: FrancoAngeli).
[19] Moro R and Bodanza G A 2010 El debate acerca del efecto facilitador en problemas de probabilidad condicional: ¿Un caso de experimentación crucial? *Interdiscipl.* **27** 163-74
[20] Silverman L K. 1989 The Visual-Spatial Learner. *Prev Sch Fail Altern Educ Child Youth* **34(1)** 15–20.

IMEKO IOP Publishing
Journal of Physics: Conference Series **588** (2015) 012041 doi:10.1088/1742-6596/588/1/012041

The measurement of enhancement in mathematical abilities as a result of joint cognitive trainings in numerical and visual-spatial skills: A preliminary study

M Agus, M L Mascia, M C Fastame, V Melis, M C Pilloni, M P Penna

Department of Pedagogy, Psychology, Philosophy - Faculty of Humanistic Studies, University of Cagliari, Via Is Mirrionis, 1, 09123 Cagliari, Italy

E-mail: mirian.agus@unica.it

Abstract. A body of literature shows the significant role of visual-spatial skills played in the improvement of mathematical skills in the primary school. The main goal of the current study was to investigate the impact of a combined visuo-spatial and mathematical training on the improvement of mathematical skills in 146 second graders of several schools located in Italy. Participants were presented single pencil-and-paper visuo-spatial or mathematical trainings, computerised version of the above mentioned treatments, as well as a combined version of computer-assisted and pencil-and-paper visuo-spatial and mathematical trainings, respectively. Experimental groups were presented with training for 3 months, once a week. All children were treated collectively both in computer-assisted or pencil-and-paper modalities. At pre and post-test all our participants were presented with a battery of objective tests assessing numerical and visuo-spatial abilities. Our results suggest the positive effect of different types of training for the empowerment of visuo-spatial and numerical abilities. Specifically, the combination of computerised and pencil-and-paper versions of visuo-spatial and mathematical trainings are more effective than the single execution of the software or of the pencil-and-paper treatment.

1. Introduction

Mathematics is one of the basic school subjects that involves different psychological factors [1, 2, 3]. The most relevant factors seem to be motivation [for a review see 4], maths self-confidence [5] and cognitive components, such as working memory [e.g. 2, 4, 5]. A body of studies identifies the cognitive processes underlying the development of numerical knowledge, showing the key role played by semantic, syntactic, lexical, and counting processes [e.g. 6, 7]. Working memory is involved in controlling, regulating, processing and actively maintaining related information to carry out different cognitive tasks (e.g. mathematical processing) [7]. In this regard, visual-spatial working memory efficiency is an important factor in understanding individual differences in mathematic achievement in children [3]. Furthermore, there is evidence that visual-spatial skills are crucial in the development of mathematical skills since pre-school age [e.g. 8, 9, 10]. Indeed, visual mental images and non verbal reasoning are strictly related to higher performance in mathematical problem solving tasks [e.g. 9, 10]. In this regards, the use of visual aids favours the development of the quantity concept in early primary

Content from this work may be used under the terms of the Creative Commons Attribution 3.0 licence. Any further distribution of this work must maintain attribution to the author(s) and the title of the work, journal citation and DOI.
Published under licence by IOP Publishing Ltd

school grades. Holmes and colleagues [11] state that when children are presented with mathematical tasks, they convert visual-spatial data into verbal mental representations (i.e., symbols). From a developmental perspective originally pre-schoolers use the visuo-spatial format to manipulate quantities, later on they learn to use verbal representations (e.g. digit numbers, algebraic signs) [11,12].

During the last decade, a new trend of research has been developed to propose educational solutions aimed at favouring the scholastic achievements of children with typical and atypical developmental paths. In this regard, several studies suggest the use of trainings to enhance specific cognitive processes in childhood. Nowadays, many technological tools have been introduced in educational environments and there has been a strong increase in research on the potentialities of computer-assisted interventions to support educational activities [13,14]. ICT is however an excellent tool to develop computer-assisted psycho-educational interventions that can be proposed at school. There is also a body of studies suggesting that visuo-spatial pencil-and-paper and/or computerised trainings can contribute to the development of mathematical skills [13]. Räsänen et al. [15] show the positive effect of computer-assisted interventions in the development of numeracy skills in pre-schoolers; and also Van Garderen [10, 11] found a similar pattern of results in students with learning disabilities, average achievers and gifted sixth graders.

2. Aim and method

The aim of the current investigation was to evaluate the impact of single and combined visuo-spatial and mathematical trainings, respectively presented in computer-assisted or pencil-and-paper modalities on the development of numeracy skills in second graders. Therefore, at pre-test and post-test children were collectively presented with a battery of tests assessing mathematical and visuo-spatial problem solving abilities, respectively.

2.1. Participants, instrument and procedure

One hundred and forty six 7 year old children were recruited in several Italian primary schools located in Sardinia (Italy). The participants (47.9% female) were divided into seven experimental groups that followed one or combined type of trainings, in pencil-and-paper and/or computerised formats.

In order to assess respectively the numerical and visuo-spatial abilities, at pre and post-test, our participants compiled a battery of standardised tests (Calculation Ability – MT Group - AC-MT 6-11 [16]; Coloured Progressive Matrices – CPM [17]). After that, children were presented the numerical training "Sviluppare l'intelligenza numerica vol. II" developed by Lucangeli et al. [18] and/or the visuo-spatial psycho-educational programme "Recupero in ... abilità visuo-spaziali" by Fastame and Antonini [19] for 10 weekly sessions. Specifically, the experimental groups carried out the following treatments: G1 was presented with the mathematical training in computerised format (n=24); G2 carried out the mathematical training in computerised format and visuo-spatial training in computerised format (n=32); G3 was given the mathematical training in the pencil-and-paper format (n=17); G4 carried out the visuo-spatial training both in the pencil-and-paper and computerized formats (n=17); G5 carried out only the visuo spatial training in the pencil-and-paper modality (n=13); G6 was presented with the mathematical training in pencil-and-paper format and computer-assisted modalities (n=13); G7 was presented the visuo spatial training in the pencil-and-paper format and the mathematical training in computerized modality (n=30). The teachers and their pupils voluntarily participated in the research (non-probabilistic sampling). Overall, one hundred and thirty seven children completed the battery of cognitive tests that was presented at pre-test and post-test.

We applied the MANCOVA and a Two Step Cluster Analysis in order to evaluate and describe the gain in the numerical and visuo-spatial abilities in each experimental condition.

Journal of Physics: Conference Series **588** (2015) 012041 doi:10.1088/1742-6596/588/1/012041

3. Results

We analysed the data on participants from the seven groups. The mean and standard deviations for the scales of AC-MT (*Written calculation, Accuracy, Speed, Semantic and Syntactic numerical knowledge*) and CPM were examined; then we evaluated the Pearson *r* correlation to investigate the linear relationship between dimensions inquired in two moments (pre and post training); as expected, we also observed positive correlations among mathematical and visuo-spatial abilities.

We calculated the gain between post and pre evaluations, in relation to CPM and to each scale of AC-MT (*Written calculation, Speed, Accuracy, Semantic and Syntactic numerical knowledge*); this index is the difference between post and pre measurement of each dimension. Moreover we applied a MANCOVA on the gain scores of the AC-MT and CPM with the 'training group' as the factor, using the pre-test measures in numerical and visuo-spatial abilities as covariates. The multivariate tests were significant for the covariates and for the factor 'training group' (Wilks' lambda $_{[30, 490]}$ = .678, p=.016). Then univariate tests indicated a significant effect of 'training group' in terms of *Written calculation* ($F_{(6, 126)}$ = 2.251, MSE = 180.719, p=.042, partial ηp^2 = .096) and *Speed* ($F_{(6, 126)}$ = 3.892, MSE = 267.066, p = .001, partial ηp^2 = .156). With the application of Post Hoc Comparison of Fisher's Least Significant Difference, we observed specifically the differences between the training groups. Table 1 summarises the main results.

Table 1. Main results of Post Hoc Comparison LSD

group		group	Dependent variable
G1 mathematical training in computerised format	>	G3 mathematical training in paper format	*Written calculation, Speed*
G2 mathematical training in computerised format and visuo spatial training in computerised format	>	G1	*Written calculation*
G6 mathematical training in paper format and mathematical training in computerised format	>	G1	*Written calculation*
	>	G4 visuo spatial training in paper format and visuo spatial training computerised format	*Written calculation, Speed*

Furthermore to describe the potential effects of different trainings, we performed the Two Step Cluster Analysis (combination of the hierarchical and partitioning methods) [20]. This statistical method applies a two-stage approach: Firstly, the algorithm is similar to the application of k-means cluster analysis; then, on the base of these results, the successive analysis is based on the application of a procedure similar to the hierarchical agglomerative methods. Thus, we have identified 2 clusters. The first cluster was constituted by the groups realizing a single or combined training to the same matter (G1, G3, G4, G5, G6); the second cluster was established by the pupils that assisted the combined trainings in mathematical and visuo spatial abilities (G2, G7) (Table 2).

Table 2 . Two Step Cluster Analysis – Cross tabulation between Cluster and training groups

TRAINING GROUP	**Cluster 1**	**Cluster 2**	**Total**
G1 mathematical training in computerised format	24	0	24
G2 mathematical training in computerised format and visuo spatial training in computerised format	0	32	32
G3 mathematical training in paper format	17	0	17
G4 visuo spatial training in paper format and visuo spatial training computerised format	17	0	17
G5 visuo spatial training in paper format	9	0	9
G6 mathematical training in paper format and mathematical training in computerised format	13	0	13
G7 visuo spatial training in paper format and mathematical training in computerised format	0	25	25
Total	80	57	137

In relation to each dimension, examining the cluster centroids, we observe a stronger improvement in numerical abilities related to *Written calculation* and *Speed* in the second cluster (Table 3). These results are consistent with our MANCOVA results.

Table 3. Two Step Cluster Analysis - Centroids

		Written calculation	**Numerical knowledge**	**Speed**	**Accuracy**	**Visuo-spatial ability**
		Mean (sd)	Mean (sd)	Mean (sd)	Mean (sd)	Mean (sd)
Cluster	**1**	-1.63 (12.78)	-1.08 (6.87)	-1.07 (8.43)	.58 (11.43)	1.12 (9.29)
Cluster	**2**	2.85 (12.02)	1.81 (8.49)	2.87 (10.02)	-1.45 (9.10)	-1.33 (6.68)
	Combined	.23 (12.62)	.11 (7.69)	.56 (9.30)	-.26 (10.54)	.10 (8.37)

4. Discussion

The main goal of the present research was to investigate the part played by single or combined visuo-spatial and mathematical trainings in favouring the enrichment of numerical knowledge. Therefore, computerised-assisted and/or paper-and-pencil single or combined treatments were proposed to 146 second graders, attending several Italian primary schools.

A large group of studies suggests that the psycho-educational programmes for the empowerment of cognitive functions can contribute massively to school achievements, both for typically developing children and those showing learning disabilities [e.g. 21, 22].

Overall, the current findings are consistent and extend previous results. Indeed, our outcomes suggest at least five different conclusions. First, the use of computer-assisted mathematical trainings is more effective than the paper-and-pencil mathematical treatments, whereas the paper-and-pencil mathematical training combined with the software for the mathematical skills is more effective than the use of the single computer-assisted mathematical training. Furthermore, although the use of computer-assisted software is very effective in empowering cognitive abilities in primary school [23], our study suggests that the use of the computer *per se* is not sufficient to enrich numeracy knowledge in second graders. Indeed, we found that children trained by a computer-assisted and pencil-and-paper mathematical treatment outperformed children trained by a combined computer-based and pencil-and-paper visuo-spatial trainings in terms of processing speed for number manipulation. Therefore, the 'pure' computerised and pencil-and-paper mathematical training is more effective than the 'pure' computerised and pencil-and-paper visuo-spatial treatment in empowering number processing in second graders. However, regardless the modality (i.e., computer-assisted versus pencil-and-paper) in which the trainings were presented, the combination of visuo-spatial and mathematical materials was as much effective as the 'pure' combined (i.e., computer-assisted and pencil-and-paper) mathematical treatment.

Therefore, we can conclude that our findings are consistent with those by previous researches [e.g. 13, 15, 23] sustaining that the use of a computer-assisted intervention supports the development of numerical knowledge in children, especially when the training is combined with mathematical and visuo spatial abilities.

In summary, the current results suggest that the combined computerised and pencil-and-paper mathematical treatment developed by Lucangeli and collegues [18] is very effective for the empowerment of numeracy skills in primary school. Moreover, a similar positive effect is also found when the computerised version of the training developed by Lucangeli et al. [18] is combined both with the computerised version or the pencil-and-paper visuo-spatial trainings proposed by Fastame and Antonini [19].

References

[1] Kyttälä M, Aunio P, Lehto JE, Van Luit J, Hautamäki J. 2003 Visuospatial working memory and early numeracy. *Educ Child Psychol.* **20(3)** 65–76.

[2] Cornoldi C and Lucangeli D 2004 Arithmetic Education and Learning Disabilities in Italy. *J Learn Disabil* **37(1)** 42–9

[3] Raghubar KP, Barnes MA and Hecht S A 2010 Working memory and mathematics: A review of developmental, individual difference, and cognitive approaches. *Learning and Individual Differences* **20(2)** 110-122

[4] Middleton JA, Spanias PA. 1999 Motivation for achievement in mathematics: Findings, generalizations, and criticisms of the research. *J Res Math Educ* **30** 65–88.

[5] Casey MB, Nuttall RL, Pezaris E. 1997. Mediators of gender differences in mathematics college entrance test scores: a comparison of spatial skills with internalized beliefs and anxieties. *Dev Psychol* **33 (4)** 669.

[6] De Hevia MD, Vallar G and Girelli L. 2008 Visualizing numbers in the mind's eye: the role of visuo-spatial processes in numerical abilities. *Neurosci Biobehav Rev* **32(8)** 1361–72
[7] De Smedt B, Janssen R, Bouwens K, Verschaffel L, Boets B and Ghesquière P. 2009. Working memory and individual differences in mathematics achievement: A longitudinal study from first grade to second grade. *J of Exp Child Ps* **103(2)** 186-201.
[8] Van Garderen D. 2006 Spatial Visualization, Visual Imagery, and Mathematical Problem Solving of Students With Varying Abilities. *J Learn Disabil* **39(6)** 496–506
[9] Heathcote D. 1994 The role of visuo-spatial working memory in the mental addition of multi-digit addends. *Cah Psychol Cogn - Psychol Cogn. CPC* **13(2)** 207–45
[10] Van Garderen D and Montague M. 2003 Visual-Spatial Representation, Mathematical Problem Solving, and Students of Varying Abilities. *Learn Disabil Res Pract* **18(4)** 246–54
[11] Holmes J, Adams JW and Hamilton CJ. 2008 The relationship between visuospatial sketchpad capacity and children's mathematical skills. *Eur J Cogn Psychol* **20(2)** 272–89
[12] Holmes J and Adams JW. 2006 Working Memory and Children's Mathematical Skills: Implications for mathematical development and mathematics curricula. *Educ Psychol* **26(3)** 339–66
[13] Praet M and Desoete A. 2014 Enhancing young children's arithmetic skills through non-intensive, computerised kindergarten interventions: A randomized controlled study. *Teach Teach Educ* **39** 56–65
[14] Penna MP and Stara V 2010. Opinions on computers, and efficacy of a computer-based learning: A pilot study. *Educ Inform Tech* **15(3)** 181-204.
[15] Räsänen P, Salminen J, Wilson AJ, Aunio P and Dehaene S. 2009 Computer-assisted intervention for children with low numeracy skills *Cogn Dev* **24(4)** 450–72
[16] Cornoldi C, Lucangeli D and Bellina M *AC-MT 6-11 Test: Test per la valutazione delle difficoltà di calcolo [The AC-MT arithmetic achievement test].* Trento, Italy: Erickson. 2012.
[17] Belacchi C, Scalisi TG, Cannoni E and Cornoldi C *Coloured Progressive Matrices Standardizzazione Italiana [CPM-Coloured Progressive Matrices Italian Standardization].* Firenze, Italy: Giunti OS 2008
[18] Lucangeli D, Poli S and Molin A *Sviluppare l'intelligenza numerica II.* Trento, Italy: Erickson; 2011
[19] Fastame MC and Antonini R. *Recupero in... abilità visuo-spaziali. Percorsi e attività per la scuola primaria e secondaria di primo grado.* Trento, Italy: Erickson; 2011
[20] Chiu T, Fang D, Chen J, Wang Y, Jeris C. 2001. A robust and scalable clustering algorithm for mixed type attributes in large database environment. *Proceedings of the seventh ACM SIGKDD international conference on knowledge discovery and data mining.* 263–8.
[21] Dunning DL, Holmes J, Gathercole SE. 2013. Does working memory training lead to generalized improvements in children with low working memory? A randomized controlled trial. *Dev Sci.* **16(6)** 915–25.
[22] Beck SJ, Hanson CA, Puffenberger SS, Benninger KL and Benninger WB. 2010 A controlled trial of working memory training for children and adolescents with ADHD. J *Clin Child Adolesc Psychol* **39(6)** 825–36.
[23] Chen Y-N, Lin C-K, Wei T-S, Liu C-H and Wuang Y-P. 2013 The effectiveness of multimedia visual perceptual training groups for the preschool children with developmental delay *Res Dev Disabil* **34(12)** 4447–54

Metrological traceability in education: A practical online system for measuring and managing middle school mathematics instruction

D Torres Irribarra[1,2], R Freund[1], W Fisher[1], M Wilson[1]
[1] BEAR Center, Graduate School of Education, University of California, Berkeley, CA (USA)

Abstract. Computer-based, online assessments modelled, designed, and evaluated for adaptively administered invariant measurement are uniquely suited to defining and maintaining traceability to standardized units in education. An assessment of this kind is embedded in the Assessing Data Modeling and Statistical Reasoning (ADM) middle school mathematics curriculum. Diagnostic information about middle school students' learning of statistics and modeling is provided via computer-based formative assessments for seven constructs that comprise a learning progression for statistics and modeling from late elementary through the middle school grades. The seven constructs are: Data Display, Meta-Representational Competence, Conceptions of Statistics, Chance, Modeling Variability, Theory of Measurement, and Informal Inference. The end product is a web-delivered system built with Ruby on Rails for use by curriculum development teams working with classroom teachers in designing, developing, and delivering formative assessments. The online accessible system allows teachers to accurately diagnose students' unique comprehension and learning needs in a common language of real-time assessment, logging, analysis, feedback, and reporting.

1. Introduction

Traceability to consensus standard units is a primary means of coordinating quantitative communications in science [1]. In any given field, theory, data, and instruments focused on phenomena expressed in common terms improve the likelihood of effectively applying new results, of successfully advancing the state of the art, and of obtaining significant economic efficiencies [2]. The third edition of the International Vocabulary of Metrology [3] therefore asserts the general value of standardized terms and concepts, including psychological and social measurement for the first time.

Of course, many more questions than answers follow from the suggestion that some form of recognizably metrological concepts and methods might be achievable in psychology and the social sciences. Many of the problems center on connecting measures of additive effects in a common framework that supports data- and theory-based checks on the meaning and quality of the unit. Building on advances in theory and method [4], research conducted over the last several decades has identified, modelled, scaled, and described the substantive properties of a number of invariant psychological, social, and physical constructs [5-7]. But a variety of reasons, many involving assumptions about the nature of quantification and the culture of measurement practices in education

[2] To whom any correspondence should be addressed.

Content from this work may be used under the terms of the Creative Commons Attribution 3.0 licence. Any further distribution of this work must maintain attribution to the author(s) and the title of the work, journal citation and DOI.
Published under licence by IOP Publishing Ltd

and psychology, have prevented development and tests of viable methods for effectively and efficiently maintaining a shared metrological frame of reference for the constructs of interest in these fields. Work that can be broadly construed as relevant to this purpose has largely involved computerized administration of item banks, which lends itself to implementation in networked contexts.

However, online test administration alone is an insufficient response to the challenges that must be faced. Recent experience has decisively demonstrated the dysfunctional effects caused when summative accountability purposes are imposed on educators who do not have access to the tools they need if they are to take responsibility for their students' outcomes [8]. Combining summative and formative applications is complex [9]. Effective assessment of learning readiness, and putting the results of those assessments to use in the classroom, can be incorporated in systematic routines to limited extents. The most challenging difficulties are those involving close and ongoing (formative) attention to individual differences in students, curricula, teachers, schools, which also require the monitoring of overall (summative) progress. The online environment provides a natural way to provide the tools needed to guide teachers and others in the performance of complex operations they would not otherwise attempt. Framing the cognitive context for asking the right questions is important for measurements to inform immediate instructional goals, as well as to assess longer term outcomes. Innovative technologies and advances in predictive theoretical models are now rapidly changing the paradigm for what is possible in instructionally-oriented assessment [10-11].

2. Sound measurement for quality assessment

The capacity to bring the right questions, technologies, and information together in the education context is the focus of the BEAR Assessment System [12], a measurement approach built on the idea that quality assessment demands sound measurement. This system emphasizes a cyclical process of construct (measurand) characterization, item development, outcome definition and evidence collection based on analysis of student responses (Fig. 1). This process is aimed at providing meaningful interpretations of student work relative to the cognitive and developmental goals defined by a curriculum.

The BEAR Assessment System is built on four principles:

1. A developmental perspective that specifies what is being assessed and how it will be identified, from emerging perspectives to the developing views of higher order proficiency.
2. A match between instruction and assessment that ensures what is taught is what is measured, and that teaching, learning and assessment support each other in improving student learning.
3. Appropriate feedback, feed forward, and follow-up that engage teachers and the students in the assessment process and that brings metacognition fully into play to improve outcomes.
4. Generation of quality evidence, so that teachers, parents, schools and other stakeholders can have confidence in reported outcomes, and can readily interpret and use assessment findings to improve instruction and student learning outcomes.

The Berkeley Assessment System Software (BASS; Fig. 2) is a web application implementing the BEAR Assessment System. It allows researchers and teachers to design, develop, and deliver formative assessments and to monitor and report student progress within an interpretive context. BASS is being developed to support the use of the BEAR Assessment System. Accordingly, BASS is designed to be used in close alignment with educational goals and instructional content. This allows teachers to understand the assessment results in terms of the learning progressions defined by each curriculum, and to diagnose students' comprehension levels and learning needs according to their placement along these progress variables.

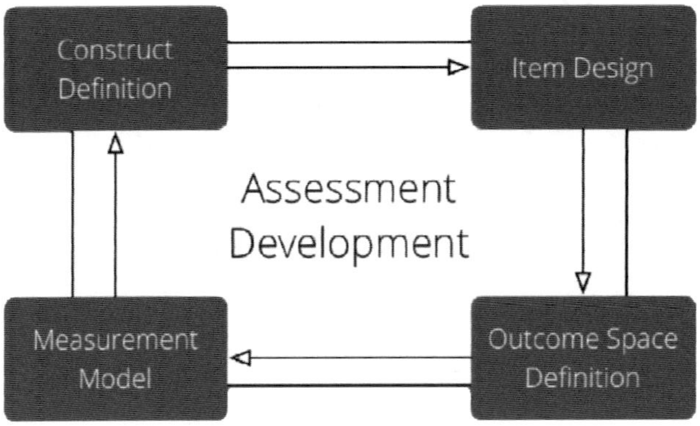

Fig. 1. The BEAR Assessment System

This system allows easy management of all kinds of assessment, including in-class activities, homework assignments, end-of-unit tests, and more. BASS builds on more than a decade of research in the design of assessment tasks and measurement techniques to support evidence-based assessment in the classroom [13-15]. In order to make the system usable to instructors in classrooms, teachers are placed at the center of the design process. The development of the teacher tools in BASS emphasizes (1) giving teachers flexibility in designing and assigning activities for students, (2) simplifying the scoring process, and (3) providing them with useful reports to share with students and parents and to use in planning instruction. The assessments include computerized modifications of existing paper-and-pencil tasks; new computer-based assessment tasks (e.g., card-sort problems, interactive graph problems) are planned for future development. An estimation engine under development will automate current manually conducted comparisons of student performances on paper-and-pencil and computer-based tasks at both the task and test level.

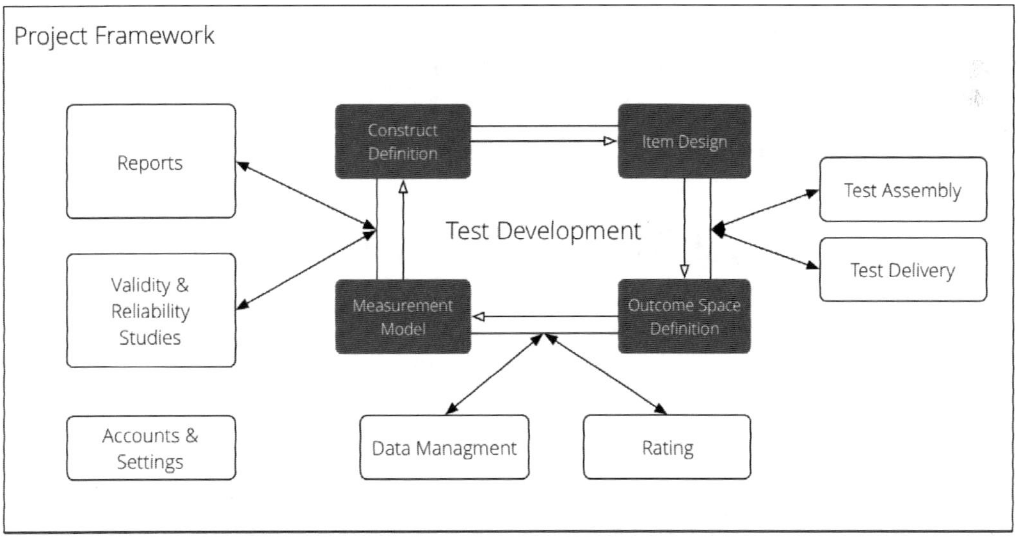

Fig. 2. Structure of the BEAR Assessment System Software

3. A sample application: The Analysis and Data Modeling Project

The Analysis and Data Modeling (ADM) project is a middle school mathematics curriculum designed by researchers at Vanderbilt University (led by Rich Lehrer) in collaboration with the educational assessment team at the BEAR Center at Berkeley [16]. The assessments being designed and tested are intended to support the development of middle school students' understandings of concepts and practices of data modeling. Data modeling refers to the invention and revision of models of chance to describe the variability inherent in particular science processes. The construction of the assessments involved analysis of core concepts and practices of data modeling that were useful to students for learning, to teachers for developing their instructional practices, and to assessment developers for developing assessments. In all, seven constructs span the range of the curriculum. The seven constructs are: Data Display, Meta-Representational Competence, Conceptions of Statistics, Chance, Modeling Variability, Theory of Measurement, and Informal Inference.

Defining the construct to be measured involves specifying a logical sequence of increasing evidence of ability, where later levels build on previous accomplishments. Fig. 3 shows an example from the ADM project for the Theory of Measurement construct. Each level of the construct corresponds to a specific level of student understanding. Each construct map characterizes the expected levels of student understanding and the main learning performances associated with each level.

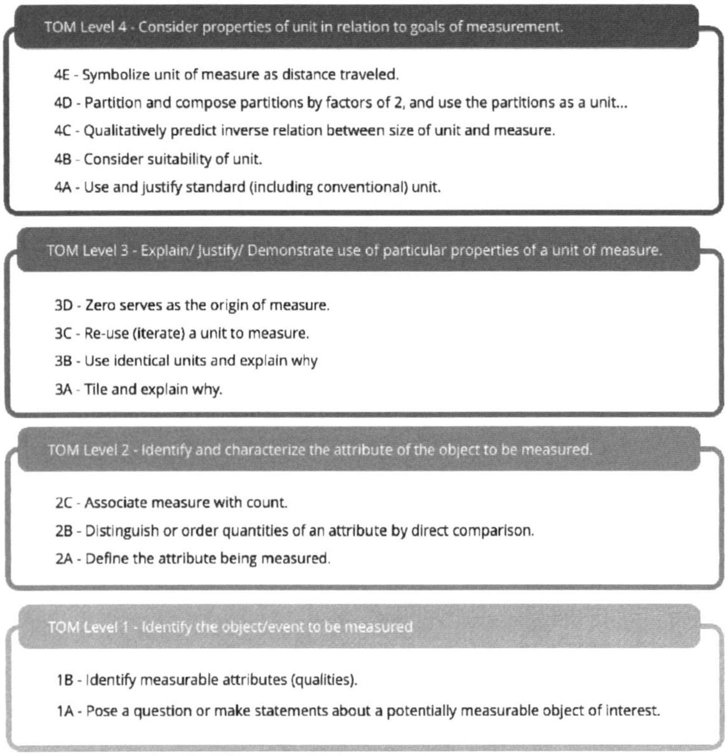

Fig. 3. Levels in the Theory of Measurement construct

Assessment items are developed with the intention of targeting specific construct levels (Fig. 4). For each item, an exemplar, or scoring guide, describes the learning performances and sample responses. Each item is explicitly related to one or more constructs through the scoring guide. Each level of a scoring guide has a description, taken directly from the description of the level of the construct, a rationale, which outlines in detail how responses to this item can correspond to that level of the construct, and one or more examples of student responses at this level. Whenever possible, these responses are taken from actual student productions.

4. Future directions

The formative assessment application BASS is built on the idea of measurement scales that stand for generalized learning structures. As the stability of these structures becomes better understood in the context of education practice, they will naturally be linked together in common systems, not unlike the metrological systems of weights and measures we take for granted in commerce and the natural sciences [17-18]. Common languages like these will eventually impact human, institutional, and information resources in ways that will make educators and employers better able to work together to identify and meet human resource and a wide variety of other needs.

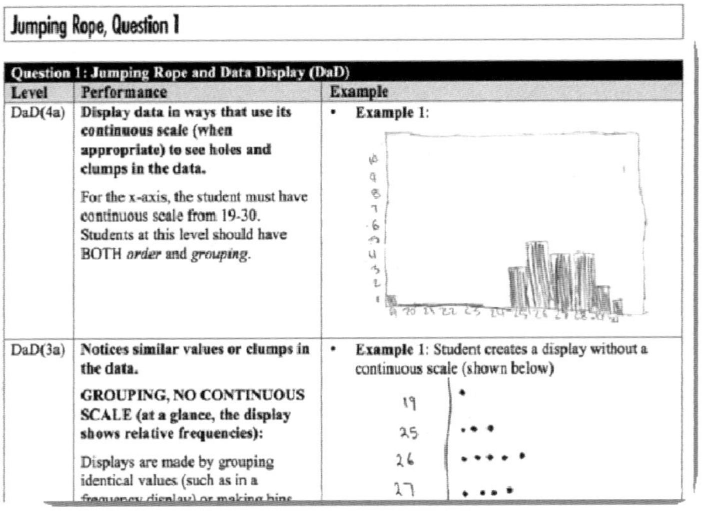
Jumping Rope, Question 1

Question 1: Jumping Rope and Data Display (DaD)		
Level	**Performance**	**Example**
DaD(4a)	**Display data in ways that use its continuous scale (when appropriate) to see holes and clumps in the data.** For the x-axis, the student must have continuous scale from 19-30. Students at this level should have BOTH *order* and *grouping*.	• **Example 1:**
DaD(3a)	**Notices similar values or clumps in the data.** **GROUPING, NO CONTINUOUS SCALE (at a glance, the display shows relative frequencies):** Displays are made by grouping identical values (such as in a	• **Example 1:** Student creates a display without a continuous scale (shown below)

Fig. 4. Sample ADM items from the Data Display construct

5. References

1. Mari L 2007 *Measurement* **40** 233-242
2. National Institute for Standards and Technology 2009, 20 July *Outputs and outcomes of NIST laboratory research* http://www.nist.gov/director/planning/studies.cfm
3. Joint Committee for Guides in Metrology 2008 *International vocabulary of metrology: Basic and general concepts and associated terms, 3rd ed* (Sevres, France: International Bureau of Weights and Measures)
4. Rasch G 1960 *Probabilistic models for some intelligence and attainment tests* (Reprint, with Foreword and Afterword by B. D. Wright, Chicago: University of Chicago Press, 1980) (Copenhagen, Denmark: Danmarks Paedogogiske Institut)
5. Dawson T L 2004 *Journal of Adult Development* **11** 71-85
6. Andrich D and Styles I M 1998 *Psychological Methods* **3** 454-469
7. Pelton T and Bunderson V 2003 *Journal of Applied Measurement* **4** 269-281
8. Mintrop H and Sunderman G L 2009 *Educational Researcher* **38** 353-364
9. Wilson M 2004 *Towards coherence between classroom assessment and accountability (National Society for the Study of Education Yearbooks vol 103, part II)* (Chicago: University of Chicago Press)
10. De Boeck P and Wilson M (Eds) 2004 *Explanatory item response models* (New York: Springer-Verlag)

11. Stenner A J, Fisher W P Jr, Stone M H and Burdick D S 2013 *Frontiers in Psychology: Quantitative Psychology and Measurement* **4** doi: 10.3389/fpsyg.2013.00536
12. Wilson M 2005 *Constructing measures: An item response modeling approach* (Mahwah, New Jersey: Lawrence Erlbaum)
13. Wilson M 2004 Towards coherence between classroom assessment and accountability (National Society for the Study of Education Yearbooks vol 103, part II) (Chicago: University of Chicago Press)
14. Wilson M and Sloane K 2000 *Applied Measurement in Education* **13** 181-208
15. Wilson M 2009 *Journal of Research in Science Teaching* **46** 716-730
16. Lehrer R, Kim M J, Ayers E and Wilson M 2014 Toward establishing a learning progression to support the development of statistical reasoning *Learning over time: Learning trajectories in mathematics education* ed A Maloney, J Confrey and K Nguyen (Charlotte: Information Age Publishers) in press
17. Wilson M and Draney K 2002 A technique for setting standards and maintaining them over time *Measurement and multivariate analysis* ed S Nishisato, Y Baba, H Bozdogan and K Kanefugi (Tokyo: Springer-Verlag) pp 325-332
18. Fisher W P Jr 2009 *Measurement* **42** 1278-1287

Acknowledgments

This work was supported by a grant from the US Institute for Education Sciences (award R305A120217). The development of the constructs and classroom implementation of the ADM curriculum is being led by Richard Lehrer and Ryan Seth Jones (Vanderbilt University). Technical staff contributing to the software portion of the project include Ana Maria Albornoz Reitze, Shazi Khan, Daniel Stanfield, Sevan Tutunciyan, Richard Vorp and Mike Kendall (all with the University of California, Berkeley). Additional recent work on psychometric analyses was contributed by Perman Gochyyev and Shonte Stephenson (University of California, Berkeley).

The role of metrology in mediating and mobilizing the language and culture of scientific facts

W P Fisher Jr[1,2] **A J Stenner**[3]
[1] BEAR Center, Graduate School of Education, University of California, Berkeley, CA (USA)
[3]MetaMetrics, Inc., Durham, NC (USA)

Abstract. The self-conscious awareness of language and its use is arguably nowhere more intense than in metrology. The careful and deliberate coordination and alignment of shared metrological frames of reference for theory, experiment, and practical application have been characteristics of scientific culture at least since the origins of the SI units in revolutionary France. Though close attention has been focused on the logical and analytical aspects of language use in science, little concern has been shown for understanding how the social and historical aspects of everyday language may have foreshadowed and influenced the development and character of metrological language, especially relative to the inevitably partial knowledge possessed by any given stakeholder participating in the scientific enterprise. Insight in this regard may be helpful in discerning how and if an analogous role for metrology might be created in psychology and the social sciences. It may be that the success of psychology as a science will depend less on taking physics as the relevant model than on attending to the interplay of concepts, models, and social organization that make any culture effective.

1. Introduction

Language plays a widely recognized and visible role in culture and cultural identities. Geographic associations—sometimes quite specific ones—can often be easily inferred from linguistic clues, such as dialects or accents. Subtle matters of decorum, economic status, and social mores, as well as more obvious connections with clothing, hair styles, and adornments, may also be closely allied with variations in language.

But where language is often used in a relatively unreflective way in everyday life, language in science and engineering is explicitly oriented toward carefully crafted precision and clarity [1]. New words for systematically implemented and detailed component processes, methods, phenomena, effects, etc. need to convey very specific meanings to be useful. Consistent and routine usage of standardized metrological terms in educational, laboratory, industrial, and other contexts inevitably contributes to the shaping of social organization and cultural identity [2-11].

Conceptual complexity has increasingly well-documented interactions with social organization, with demand for metrological uniformity in science and commerce having a long history of consequences in government, academic, research, and business institutions [6]. Recent research on the

[2] To whom correspondence should be addressed.

Content from this work may be used under the terms of the Creative Commons Attribution 3.0 licence. Any further distribution of this work must maintain attribution to the author(s) and the title of the work, journal citation and DOI.
Published under licence by IOP Publishing Ltd

intertwined processes of concept formation and social organization [7-11] suggests a basis for resituating scientific language, concepts, and thinking relative to their origins in everyday processes. New possibilities for informing the theory and methods of psychology and the social sciences emerge as the salience of existing work into invariance, instrument calibration and equating, and the creation of shared frames of reference becomes apparent [12-15].

2. Text and technology

Over many millennia, a wide variety of technologies, from prehistoric discoveries of fire and the wheel to contemporary telecommunications and computing devices, have made it possible for persons lacking advanced skills and resources to take advantage of difficult and complex operations that would otherwise be unavailable to them. Technologies embody concepts and things in ways that make their associations accessible to end users unable to perform or invent for themselves the operations involved in producing a desired effect. As Whitehead [16] put it in 1911,

> It is a profoundly erroneous truism, repeated by all copy-books and by eminent people when they are making speeches, that we should cultivate the habit of thinking of what we are doing. The precise opposite is the case. Civilization advances by extending the number of important operations which we can perform without thinking about them.

In performing this function, technique and technology show their conceptual origins in the ancient Greek term, *techne*, meaning "to make" [17-18]. Other words sharing this root include text and textile, which conceptually overlap in the way they entail related spheres of weaving, yarns, threads, spinning, etc. Though the concrete media required for written text are obvious forms of technology, spoken language's phonemes, phonetics, syntax, semantics, etc. also exhibit the features of *techne*.

Technique and technology embody the advance work done by language in the sense of codifying elaborate routines in a compact and portable method or tool. People who learn a word in a language typically have no or very limited knowledge of the experiences that led to the formation of the relevant concept but nonetheless accept the association without wondering very much about it. Just as is the case with technologies like electrical appliances, automobiles, or computers, we are born into and enculturated within social groups with long traditions of capitalizing on the practical knowledge of language and tool use while knowing little or nothing of the technical knowledge that makes the concepts and methods involved effective. Written text is in a sense the paradigmatic definition of technology in that it is the earliest obvious example of how learning was codified in a form later generations could access, whether or not they knew anything about the origins of the concepts, tools, or methods learned.

Whenever a newly salient event in the world is noted and remarked upon, an analogous word, sound, image, or sign of some kind (a text, broadly applied) is metaphorically situated in relation to other signs within a pre-existing referential system. The text of language is a model for everything technical in the sense that semiotic associations of words-things-concepts give rise to analogous associations of tools-things-models. The making of meaning via these kinds of associations is the fundamental creative act, and sets up the process by which anything else is made. The root pattern of methodical thinking lies in the way words come into language as the medium embodying the unity of an ideal concept and something in the world.

Hence we have the Greek root of method as *meta-odos*, a following along after (*meta-*) on the path (*odos*) [19] traced by the "activity of the thing itself" that thought experiences [20]. The word "method" and the concept of methodical thinking converge here, making

> an explicit claim to return for evidence to 'the things themselves' or 'phenomena,' i.e., to things as they show themselves before the work of abstraction and theorizing has carved out a language of fixed essences for them removed from human praxis, history and culture [21].

When a name is given by an individual to something notable or remarkable repeatedly experienced and re-cognizable in the world, that thinking process methodically and logically coordinates and aligns a word, a concept, and a thing. This semiotic coordination might be shared with another person, and the word, or a related sign, and its position in a larger system of shared signification may then become distributed throughout a community as a part of that society's cultural fabric [2-11, 22-23].

If that happens, the word and its associations effectively pre-think the world for those who learn what the word means from others. Those who learn the word from others do not have to acquire enough experience with the phenomenon to conceive an idea of it, do not have to give birth to new meaning embodied in a word of their own, and do not then have to translate words across personal idioms relating the same concepts and things.

This process is implicated by Gadamer [20] when he writes of the way formal logic inevitably begins from "the logical advance work language has done for it." The advance work performed by language pre-thinks the world for us to the point of bringing about new efficiencies in the making of meaning. The to-and-fro play of conversation relieves speakers of the burden of taking the initiative by building on pre-existing shared meanings to the point of facilitating fluid, satisfying and pleasurable shared experiences. Not only does the back-and-forth of question and answer absorb interlocutors into elucidating the object of the conversation, but this may occur to the extent that their horizons or frames of reference fuse in relation to that object. This fusion is itself the moment at which both a common language and an awareness of different perspectives are created, and mutual understanding is achieved, however provisional and circumstantial it may be.

3. Advancing civilization

The prototypical metrological processes performed in the standardizations of spoken and written language pave the way for science's deliberately arranged shared cultural frames of reference. That is, the reasoning processes used in scientific problem solving are not qualitatively different from the reasoning processes people use in other areas of life [9]. The question that arises, then, is how civilization might be advanced via psychology and the social sciences: how might we increase the number of important operations in these areas that we can perform without thinking of them? That is, instead of taking physics as the standard of success in the conduct of science, might it not make more sense to understand the social and linguistic processes through which physics has succeeded, with the aim of extending those processes into psychology and the social sciences?

Almost all state-of-the-art measurement in education, health care, performance assessment, etc. plainly depends almost entirely on the active participation of people able to think about the important operations that must be performed. Absent skilled experts, state of the art measurement simply does not usually happen in psychology or social sciences research and practice. Even when experts are involved, the complications and expense of high quality measurement are often enough to prevent it from taking place.

Why? Might it be because, with a very limited number of exceptions [12-15, 24-28], measurement in psychology and the social sciences lacks virtually any methods and traditions concerned at all with metrological traceability? With uniform unit standards? With consensus processes for determining standard product definitions? With the power of metrology for simplifying processes, for reducing costs, for streamlining communication, and for amplifying collective intelligence? Little or no attention is being focused on metrology even in the wake of recent developments that would seem to make its relevance unavoidably evident: networked communications, item banking, instrument equating, adaptive instrument administration, and the predictive control needed for on-the-fly automated item generation [14-15, 24-28]. Inevitably, however, increasing pressure to put two and two together will be applied as the human, economic, legal, moral, social, etc. implications of advancing psychological and social measurement technologies become apparent.

In wondering how to advance civilization by simplifying end use functions, Whitehead is suggesting a model of a different kind of person than the rational Cartesian subject usually assumed as the agent of scientific and economic activity. No one, no matter how brilliant or economically advantaged, has the time and resources to be completely informed about every important factor affecting the decisions of daily life, much less in the complex operations of science. Even the simplest communication would be prohibitively cumbersome if each person was burdened with the tasks of completely re-inventing for themselves every word in a language and always being completely logical in their decision making. If everyone had to formulate their own vocabularies, grammars, etc., and then translate between them to effect any communication at all, the flow of conversation would be so obstructed as to be changed utterly in its basic character.

Even though few speakers of a language seek out any significant degree of understanding of the origins of the alphabet, script, words, grammar, orthography, etc. comprising that language, this does not prevent proper, comprehensible usage and successful communication. The capacity of language to point at things not present and to make intended meanings comprehensible even in the absence of any overt expository skill in etymology and grammar brings efficiencies to communication that make shared knowledge possible. Explaining these efficiencies is, according to Hayek, the

> central question of all social sciences: How can the combination of fragments of knowledge existing in different minds bring about results which, if they were to be brought about deliberately, would require a knowledge on the part of the directing mind which no single person can possess? [29]

The continuing relevance of Hayek's question is noted by Hancock [30], who emphasizes the ways decentralized decision-making processes enable societies to employ multiple methods for exploring alternative approaches to solving problems. Others have similarly lately remarked, quoting Hayek [29], that "the marvel of the market thus resides in 'how little the individual participants need to know to be able to take the right action'" [31]. The qualification of "little" here is, of course, relative. Massive amounts of low quality information may never support right action, and the social investments made in creating high quality scientific information may be a major portion of the total economy [2].

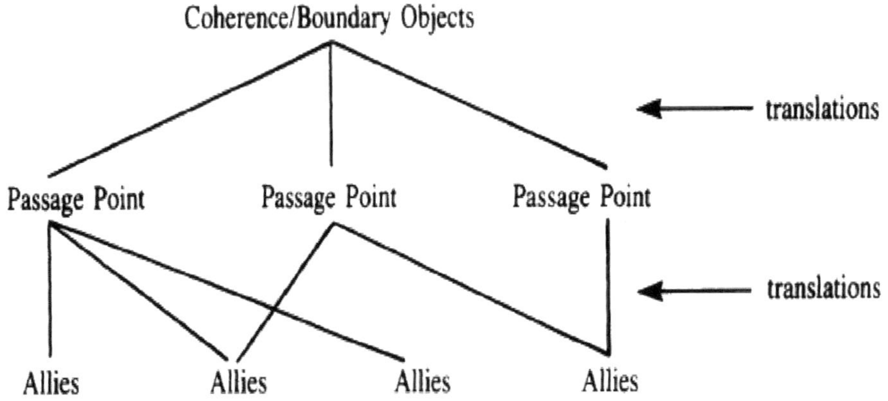

Fig. 1. Translation network schematic [32]

Bringing technical processes and objects into everyday use thus requires the coordination and alignment of a wide variety of domains of expertise, with the added problem of making each domain's technical aspects transparent to all of the others. Without financing, accounting, management, sales, marketing, human resources, metals mining, cable manufacturing, rubber and plastics insulators, and consumers, the electrical industry would be as nonexistent as it would be if inventors, scientists and engineers had never created electrical concepts and tools in the first place.

In the terms of contemporary social studies of science [2-11], the problem is one of translating each area of stakeholders' perspectives on technical boundary objects into the languages of each other area of stakeholders (Fig. 1) [32-33]. Ideally, alliances advance each stakeholder group's interests further than would otherwise be possible, but this is not, of course, always the case. In education, for instance, psychometricians, statisticians, learning theorists, curriculum designers, teachers, parents, students, testing agencies, publishers, principals, researchers, accountants, and others may be allied or alienated depending on whether they successfully define a common boundary object and translate their interests in it into terms incorporated and advanced by each other area of stakeholders. The pace and spread of innovations depends in large part on being able to express technical effects in ways that capture the imaginations and interests of stakeholders across domains enough for them to coordinate and align their processes and outcomes [32-36]. Attending to the semantic role performed by technological objects within and across stakeholder communities may provide productive new paths for research and development in psychology and the social sciences [37-38].

4. Future directions

Rephrasing the question to make the interplay of concepts and social organization plain [2-11], how can we create a world in which the facts of psychological and social measurement can survive? What kind of environment would be required to build networks in which the outside world has the same form as the instruments in the laboratory? What kinds of continuous trails can be created to tie all of the literacy measures together, all of the numeracy measures together, and all of the respective relationship quality, physical functioning, and health status measures together? Can individual differences be appropriately understood and qualified? What opportunities for such networks can be envisioned, and are there any approximations of such networks already in place in education, psychology, or the social sciences? And quite importantly, how can the interests of each group of stakeholders in a given area be satisfied and represented? Can divergent and even conflicting interests be productively mediated within and between various organizations and institutions [35-36]? Can the rules, roles, and responsibilities constituting efficient market economics [34] be brought to bear on exchanges of human, social, and natural capital value [37]? Even negative answers to these questions will give clearer assurances about viable paths for productive science and economics than could be obtained if the questions had never been raised at all.

5. References

[1] Joint Committee for Guides in Metrology 2008 *International vocabulary of metrology: Basic and general concepts and associated terms, 3rd ed* (Sevres, France: International Bureau of Weights and Measures)
[2] Latour B 1987 *Science in action* (New York: Cambridge University Press)
[3] Latour B 2005 *Reassembling the social* (Oxford: Oxford University Press)
[4] Mendelsohn E 1992 The social locus of scientific instruments *Invisible connections: Instruments, institutions, and science* ed R Bud and S E Cozzens (Bellingham, WA: SPIE Optical Engineering Press) pp 5-22
[5] O'Connell J 1993 *Social Studies of Science* **23** 129-173
[6] Ashworth W J 2004 *Science* **306** 1314-1317
[7] Schaffer S 1997 Metrology, metrication, and Victorian values *Victorian science in context* ed B Lightman (Chicago: University of Chicago Press) pp 438-474

[8] Schubert C, Sydow J, and Windeler A 2013 *Research Policy* **42** 1389-1405
[9] Nersessian N J 2012 *Mind, Culture, and Activity* **19** 222-239
[10] Hutchins E 2010 *Topics in Cognitive Science* **2** 705-715
[11] Hutchins E 2012 *Mind, Culture, and Activity* **19** 314-323
[12] Fisher W P Jr 2009 *Measurement* **42** 1278-1287
[13] Rasch G 1960 *Probabilistic models for some intelligence and attainment tests* (Reprint, with Foreword and Afterword by B. D. Wright, Chicago: University of Chicago Press, 1980) (Copenhagen, Denmark: Danmarks Paedogogiske Institut)
[14] Stenner A J, Fisher W P Jr, Stone M H and Burdick D S 2013 *Frontiers in Psychology: Quantitative Psychology and Measurement* **4** doi: 10.3389/fpsyg.2013.00536
[15] Wilson M 2005 *Constructing measures: An item response modelling approach* (Mahwah, New Jersey: Lawrence Erlbaum)
[16] Whitehead A N 1911 *An introduction to mathematics* (New York: Henry Holt and Co)
[17] Heidegger M 1977 The question concerning technology *Basic writings* ed D F Krell (New York: Harper & Row) pp. 283-317
[18] Zimmerman M E 1990 *Heidegger's confrontation with modernity: Technology, politics, art* (Bloomington, Indiana: Indiana University Press)
[19] Heidegger M 1991 *The principle of reason* (Bloomington, Indiana: Indiana University Press)
[20] Gadamer H-G 1989 *Truth and method* (New York: Crossroad)
[21] Heelan P A 1994 Galileo, Luther, and the hermeneutics of natural science *The question of hermeneutics* ed T J Stapleton (Dordrecht, The Netherlands: Kluwer) pp. 363-374
[22] Black M 1962 *Models and metaphors* (Ithaca, New York: Cornell University Press)
[23] Gibbs R W Jr 2011 *Discourse Processes* **48** 529-562
[24] Stenner A J, Burdick H, Sanford E E, Burdick D S 2006 *J Appl Meas* **7** 307-322
[25] Bond T, Fox C 2007 *Applying the Rasch model* (Mahwah, New Jersey: Lawrence Erlbaum)
[26] Embretson S E 2010 *Measuring psychological constructs* (Washington, DC: American Psychological Association)
[27] Engelhard G Jr 2012 *Invariant measurement* (New York: Routledge Academic)
[28] Fisher W P Jr, Wright B D 1994 *International Journal of Educational Research* **21** 557-664
[29] Hayek F A 1948 *Individualism and economic order* (Chicago: University of Chicago Press)
[30] Hancock D J 2009 The triumphs of Mercury: Connection and control in the emerging Atlantic economy *Soundings in Atlantic history: Latent structures and intellectual currents, 1500-1830* eds B Bailyn, P L Denault (Cambridge, MA: Harvard University Press) pp. 112-140
[31] Tadelis S, Williamson O 2010 *Transaction cost economics* (Berkeley, CA: Haas School of Business, University of California, Berkeley http://faculty.haas.berkeley.edu/stadelis/-tce_org_handbook_111410.pdf)
[32] Star S L, Griesemer J R 1989 *Social Studies of Science* **19** 387-420
[33] Fox N J 2011 *Sociology* **45** 70-85
[34] Miller P, O'Leary T 2007 *Accounting, Organizations, and Society* **32** 701-734
[35] Woolley A W, Chabris C F, Pentland A, Hashmi N, Malone T W 2010 *Science* **330** 686-688
[36] Woolley A W, Fuchs E 2011 *Organization Science* **22** 1359-1367
[37] Fisher W P Jr, Stenner A J 2011 A technology roadmap for intangible assets metrology. International Measurement Confederation (IMEKO) TC1-TC7-TC13 Joint Symposium. Jena, Germany. http://www.db-thueringen.de/servlets/DerivateServlet/Derivate-24493/ilm1-2011imeko-018.pdf
[38] Gorur R 2013 *Critical Studies in Education* **54** 132-142

Kinematic measurements for the characterization of the dynamics of hopping

Francesco Crenna and Giovanni B Rossi

Measurement Laboratory, University of Genova DIME/MEC, Via all'Opera Pia 15a, 16145 GENOVA, Italy

E-mail: francesco.crenna@unige.it

Abstract. The study of human movement requires the accurate measurement of the forces and of the kinematic quantities involved. Here we consider hopping, that deserves great attention in the literature, since it combines the activation of a complex sensorimotor control with the relative simplicity of the gesture. The movement of the center of mass and the contact conditions are of special interest. Thus a measuring system that combines digital image acquisition, force measurement based on a force plate and extensive use of inertial sensors, for linear and rotational motion, has been developed and characterized. The results concerning the accurate measurement of contact conditions are specially outlined.

1. Introduction

Hopping is widely considered as a reference gesture when studying lower limb mechanical behavior [1-6]. The gesture is mainly bidimensional, with dominant movements in the sagittal plane. Repeatability and reproducibility can be improved by controlling the hopping rate. Preliminary studies have defined a preferred frequency for humans so the gesture can be repeated at this frequency with the help of a metronome [1]. A satisfactory representation of the gesture can be obtained by a model of the whole body, specifically of the lower limbs, consisting of rigid elements [1,2,3]. Related measurements involve both kinematic and dynamic quantities. Several dynamic issues are involved in hopping, such as the ground reaction force and its relationship with centre-of-mass acceleration, the elastic contribution of the lower limbs and the behaviour of the moment they express, as a function of position or angles. In this paper we are mainly interested in kinematic quantities since they are directly involved in the asset control required to maintain a macroscopic constant position of the body in the sagittal plane, during the performance [8]. In this scenario, among all kinematic quantities, we focus mainly on two: the centre of mass (CM) of the whole body and the point where the forces and moments are exchanged through the feet with the floor (Centre of Pressure, COP). Generally the quality and reliability of the measurement result is taken as grated, while some attention is given to the processing algorithms [7]. This paper presents a measurement system designed to obtain reliable measurement results for a set of kinematic and dynamic quantities, based on a video system operating with markers, inertial sensors and a force platform. The kinematic of the lower limbs may be obtained from two independent methods, so it is possible to provide a result verification by comparison of the two measurement results. The positions of the points of interest on the sagittal plane can be obtained directly from video or force data, or they can be evaluated

Content from this work may be used under the terms of the Creative Commons Attribution 3.0 licence. Any further distribution of this work must maintain attribution to the author(s) and the title of the work, journal citation and DOI.
Published under licence by IOP Publishing Ltd

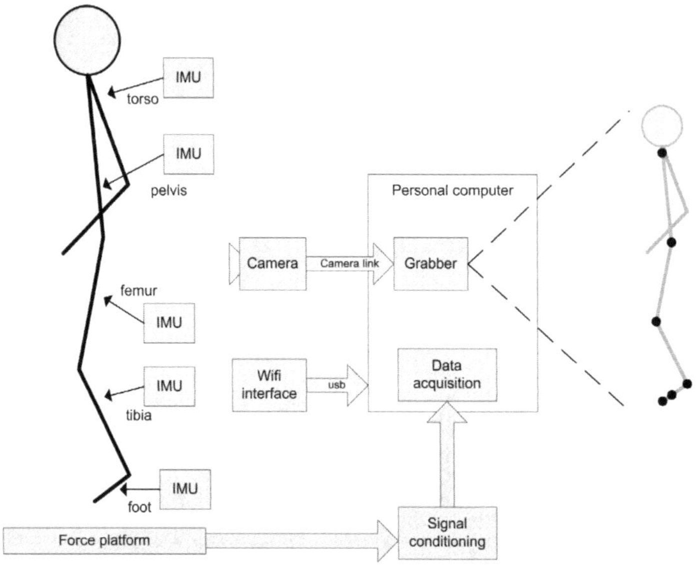

Figure 1. Schematic of the measurement system

through the use of a rigid model of the limb, by using the inertial data [9]. In the paper we will present the measurement system in detail, including the processing of the measurement signals; we will introduce an experimental procedure that enables the evaluation of repeatability and reproducibility of the gesture and finally discuss the measurement results with particular focus on kinematic data for the CM and COP and their verification.

2. The gesture and its model

Hopping is an interesting gesture from the biomechanical point of view since it provides a large amount of information about the lower limbs behaviour, under natural, relaxed, conditions or when the task requires a higher level of performance. Beside that, since the hopping is performed on site, with rather small body displacements in the horizontal plane, it presents interesting biomechanical aspects as regards the position control required to maintain the equilibrium in a fairly constant horizontal position, while hopping and moving the whole body in the vertical direction. In literature some studies about the dynamics of hopping are available. They are mainly focused on dynamics aspects combining kinematic and force data to evaluate the moment required and the limbs joints, studying the adjustment of the lower limbs behaviour, according to the required performance, or due to the geometric nature of the ground, such as flat, inclined or uneven, or to ground's dynamic nature such as rigid or compliant. Rarely the kinematics of the CM and COP are deeply investigated. In this case our interest is to develop a model of an hopping man, considering a schematization of lower limb behaviour, but introducing also the problem of stability, so considering the whole body behaviour. Of course such a model has to be based on reliable experimental kinematic data. As regards the measurement problem, hopping is rather favourable since, on one hand it is a simple gesture occurring, in first approximation, in the sagittal plane, on the other it offers the possibility to carry on repeated and reproducible measurements. The former aspect influences the hardware set up that is simpler as compared with three dimensional gestures, while the latter enables measurement procedures including estimation of repeatability and reproducibility uncertainties.

Hopping frequency may vary according to the task required, such as maximum height or test duration. In our case the subject is required to hop at a fixed frequency of 2,2Hz, as presented by a metronome. Such frequency corresponds to the expected hopping frequency when a natural task is required, without particular effort or performance [1]. Each subject has to perform at least two sets of tests: hopping naturally and at maximum height [6] maintaining in both cases the same pace as suggested by the metronome.

3. The measurement system

The measurement setup is presented in figure 1. The video system is based on a BASLER camera (details in table 1) connected via IEEE1394. Optics has been selected to have a complete view of the subject during hopping, considering also the maximum height case. Active markers consisting in high intensity white led, are sewn on an elastic running suit. Subject performing the test wears the instrumented suit and markers are adjusted in proper positions.

Table 1. Camera technical details.

Resolution	640 (vertical) x 480 (horizontal)
Pixel size	9,9 μm $\times$ 9,9 μm
Sensor dimension	1/2 in
Maximum frame rate	60 Hz
Optics	16mm C mount

The subject is instrumented also with 5 inertial sensors (Inertial Motion Unit, IMU) from Xsens Technology placed on limb's segments and torso. Each unit is equipped with triaxial accelerometer, gyroscope and magnetometer and wireless connected to a base station, some technical details are available in table 2. IMU signals are internally pre-processed by Xsens software, obtaining as a result the three orientation angles in the space for each unit, together with acceleration data. Synchronization between frame acquisition, force data acquisition and IMU sensors is made possible by acquiring a synch signal from the IMU base station. The synch signal enables off line time synchronization of all the data.

Table 2. IMU - Xsens MTW - technical specifications.

Dynamic range	$\pm$ 200 $°/s$,$\pm$ 160 m/s^2
Static accuracy (roll-pitch)	<0,5 °
Static accuracy (yaw)	1 °
Angular resolution	0,5 °
Internal sampling rate	1800 Hz
Bandwidth	>120 Hz
Wireless sampling rate (6 sensors)	75 Hz

After camera calibration, during video capture, optics diaphragm is closed, to obtain dark images with light markers, then a threshold is applied to obtain black or white (binary) pixels in each frame. The position on the image of the white spots, corresponding to each marker, is computed and converted in positions by using the calibration constant, determined, for the specific set up, before and after each test.

4. Data processing

Markers and IMU sensors are placed on the leg facing the line of view of the camera, as presented in table 3. By processing position data it is possible to obtain either segment angles in the sagittal plane, or, with the support of a simple rigid model of the lower limbs and torso, segment extremities positions. In both cases these data can be compared with IMU measurement values, validating the experimental results.

Table 3. marker positions definition.

marker	1-3	4	5-6	7	8-9	10	11
position	foot	ankle	tibia	knee	femur	hip	shoulder

Center of mass position in the sagittal plane is measured considering the position of the marker corresponding the iliac crest, that is located at about the same height from the ground of the center of mass of the whole body, or it is calculated on the basis of a rigid segment model of the lower limb and torso, considering the measurement of the orientation of each segment. Center of pressure is defined as the point, in the transversal plane, where the overall force is exchanged with the ground. It's position can be determined considering the force data from the platform. It can be interesting the measurement of the contact point also, the point around which the foot rotates during contact, due to the movement of the ankle. As for the COM, it can be estimated using video system (markers located on the heel, malleolus and metacarpus) or considering the IMU data in the hypothesis of a foot consisting of two rigid parts articulated in the metacarpus.

5. Measurement procedure

To guarantee results reproducibility we have developed a detailed measurement procedure. The video system is calibrated using a vertical reference with marks every 100 mm located at the same distance from the camera of the instrumented right hand side of the subject, before and after the hopping test. Generally we obtained a resolution of about 3,2 mm/pixel, differences between before and after the test were below the resolution of the system and even the reproducibility due to different measurement setup was very low and less than 0,2 %. In order to compare video and COP data, it is necessary to calibrate the systems together. We have proceeded recording force and video data with a 20 kg weight placed in the center of the force platform and at at a known distance from it. By processing force data it is possible to measure the COP due to the weight and compare its position in frontal direction with the one obtainable from the video. In such a way video and COP measurements in the frontal direction refers both to the same origin.

IMU sensors are placed near the test site and, after activation, they rest quiet for at least one minute of warm up, while stability of the readings is verified. Subject wears the instrumented running suit and IMU sensors are placed on site by using elastic straps. Markers positions are finely adjusted in the reference points as in table 3. Before and after the hopping the subject stands still in front of the camera and a complete still measurement of about 45 s is performed. The hopping test is performed with a reference sound giving the 2,2 Hz pace. The test starts when the subject feels a good accordance with the pace, and it lasts for at leat 10-15 s, then the subject rest for some time, while data are checked. The experimental session consists of several test at both normal and high height, and it ends with a static recording. After the test all the measurement data are verified and if they present inconsistencies, such as large differences between the two calibrations or the two still measurements, rhythm anomalies and so on, the hopping test is discarded.

6. Preliminary experimental results

The measurement system is still under development, in particular as regards the video acquisition system, since we are going to use a higher resolution and higher speed camera, nevertheless we can present here an example of the obtainable results, demonstrating the measurement system potentialities and some of the difficulties we are going to face in such kind of measurements. Figures refer to hopping tests, performed by a 65 kg, 1,72 m, male subject.

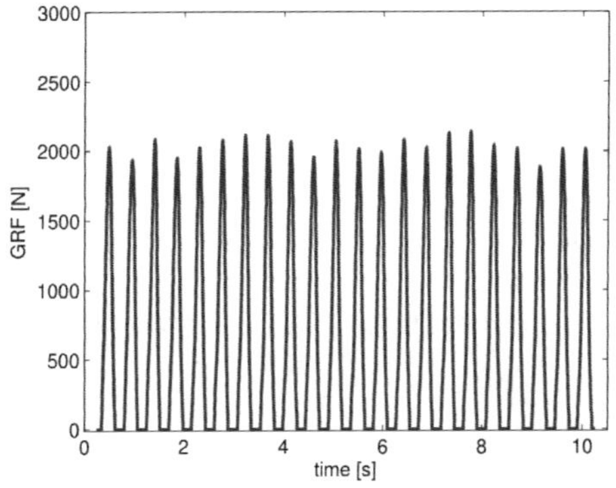

Figure 2. GRF for normal height, 2,2 Hz hopping.

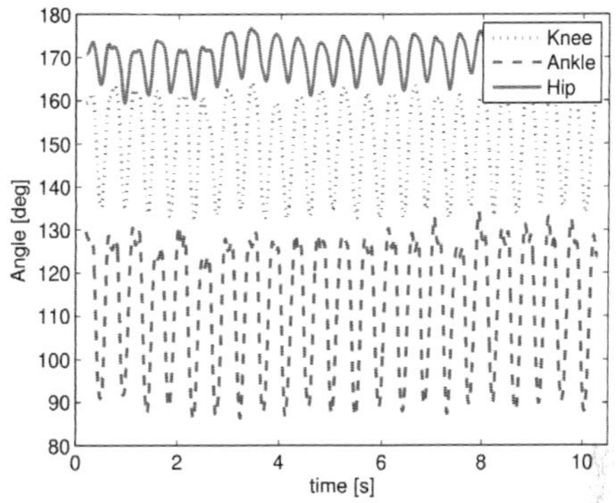

Figure 3. Angles movements for the leg and torso.

Figures 2 and 3 presents the time histories of the ground reaction force (GRF) and of the angles involved in the gesture, for a subset of hops in the same test: the good repeatability is apparent. A reference contact signal may be obtained by thresholding the Ground Reaction Force signal. It is now possible to analyse values of interest at touchdown and their variations during the contact phase as presented in table 4, together with some reference values from the literature [2].

Table 4. Experimental results. Mean values and standard deviation of the measurements in multiple tests. Literature references from [2].

	Normal height			Maximum height		
Quantity	Experiment	σ	Liter.	Experiment	σ	Liter.
Hopping rate	2,2 Hz	1,7%		2,2 Hz	2 %	
time spent at ground	260 ms	9 ms	308 ms	198 ms	9 ms	260 ms
Ground reaction Force	2180 N	3,6 %	1740 N	2800 N	3,6 %	2179 N
Touchdown angles [deg]						
Hip	172	1	171	175	2	173
Knee	160	2	150	160	2	154
Ankle	126	2	125	126	2	130
Excursions at ground [deg]						
hip	10	1	10	8	1	3
Knee	24	2	23	23	2	13
Ankle	39	2	32	40	3	26

Mean values show a rather good agreement with literature values considering that results from this study refers to several tests by a single subject while in the literature case a set of subjects was considered, and it had balanced gender and age composition. Beside that it is worth noting that absolute angles depends on marker alignment on the subject's limb.

Moving now to the main aim of this study we consider frontal movements of the center of mass (CM) of the whole body and of the center of pressure, COP. The CM position may be measured through the markers data, considering that it is about 15 cm above the hip marker. Otherwise its position can be evaluated considering the CM of each body segment and its position in the space. This second procedure may use marker video data, or orientation data from the inertial sensors, and it is still under development. Figures 3 presents a time history data for the frontal movement in the sagittal plane of the hip marker, the foot markers, the COP and the center of mass of the whole body, together with the contact signal.

We can briefly discuss this graph considering that since the CM is located above the hip marker, and the torso is a bit inclined and not fully upright, the CM position is a bit displaced from the hip. From a dynamical point of view the COP at touchdown tends to the back and returns in the front of the foot, while the CM present a smoother behaviour in accordance with the COP.

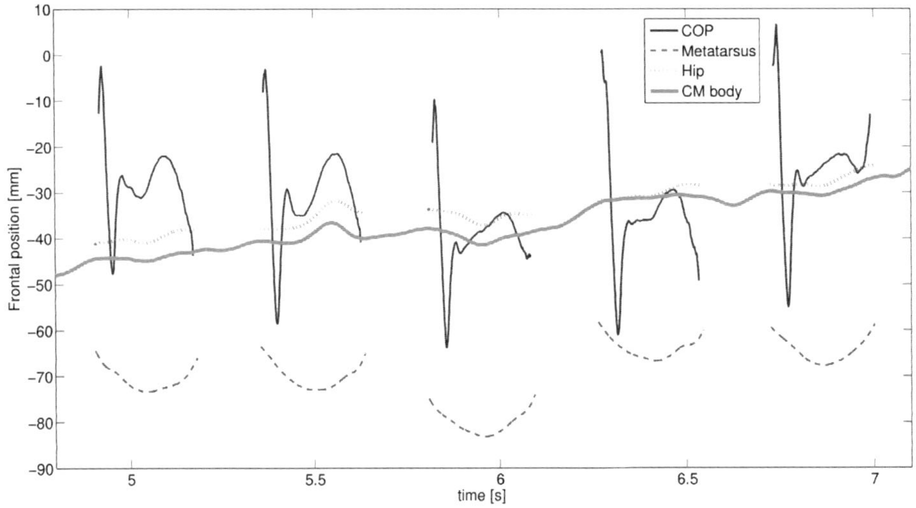

Figure 4. Frontal movements of COP, CM and other keypoints during the contact phase.

In the study of the contact phase, in addition to the points previously described, the instantaneous centre of rotation (CR) plays an important role. Generally it is assumed to be coincident with foot metacarpus, but of course it is interesting to measure its position. This rather complex goal can be archived assuming foot top as a rigid element and considering velocity vectors measured at malleolus and metacarpus markers. Since we are considering a rotation of a rigid segment, at every instant the two straight lines by the extremes of the velocity vectors intersect in the instantaneous centre of rotation. Considering marker positions direct measurements we have developed an indirect CR measurement procedure that is currently under validation.

The reliability of the measurement results depends on several influence quantities related to both the measurement system and the measurand itself. In the particular case of the tests we have presented here, we had good repeatability and reproducibility but they also depend on subject performance, that, if necessary can be improved by training. At the interface between the measurand, or the subject performing the gesture, and the measurement system, there are markers and IMU sensors. It is possible to validate the results by comparing the angles as measured by IMU or from markers' positions, as presented in the following figure 5. The limited differences between them is on one side a validation of the results, on the other it might be due to a slight misalignment between the sagittal plane, assumed coincident with the plane of the image, and the IMU coordinate system. Another validation possibility is to verify the length of body segments that can be assumed as rigid. Figure 6 presents the femur and tibiae lengths as a function of time during the hopping. Segments' lengths are evaluated considering the markers distance between malleolus and knee (tibiae) and between knee and hip (femur). They are not rigourously constant, probably because markers move from their reference positions during performance. Markers are sewed to the elastic fabric adherent to the subject's leg with the help of some elastic tape also. Nevertheless, during the performance, due to movements mainly at ankle and knee and to muscular activation, they move causing a segment length variation. This is a typical problem in biomechanics, but both the rather small variation in segment length which is less than 1,5 % for tibiae, the worst case, and the good agreement on the angles computed from video and inertial data gives confidence to the reliability of the results.

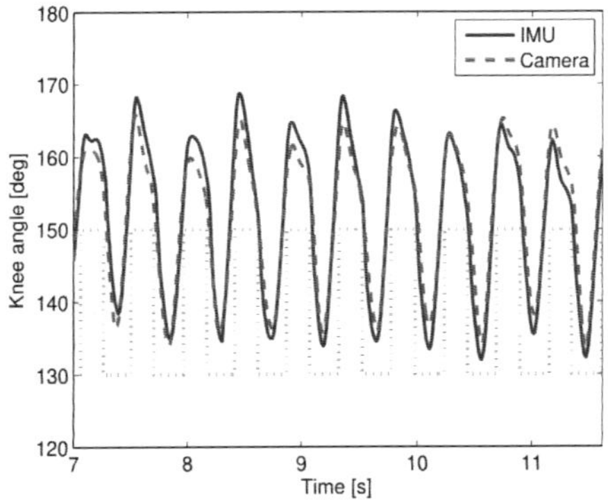

Figure 5. Knee angle as measured by video and IMU systems.

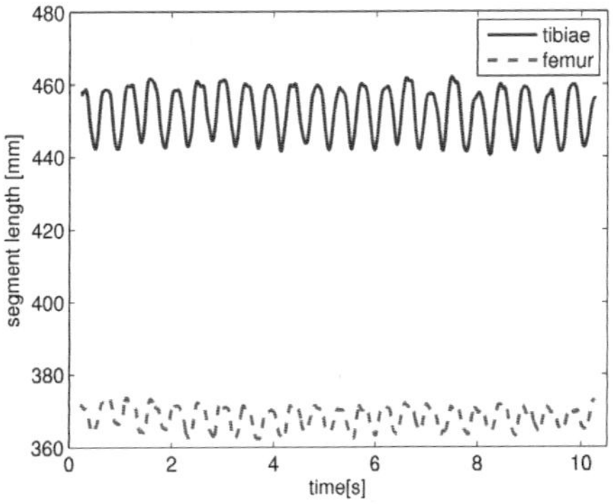

Figure 6. Tibiae and femur segments' lengths.

7. Conclusions

We have presented a redundant measurement system for the kinematic characterization of hopping. Several studies of this gesture are available in literature, mainly focused on the general dynamics of lower limbs. To understand the dynamics of hopping it is necessary to better characterise the contact conditions through proper kinematics quantities, than currently available in the literature.

A combination of marker, video and inertial sensors was used to get the possibility to validate the results by comparison and internal validation. In this case our interest was mainly focused on the contact phase characterised by the center of pressure, the center of mass of body segments and whole body, the foot position and its instantaneous center of rotation.

The paper presents and discusses some measurement results which are in accordance with the available results in literature, show good repeatability and reproducibility and can be verified by comparison between different measurement methods. Nevertheless results present some measurement problems that emphasize the common difficulties to be faced when dealing with measurements in biomechanics. In this particular situation the possibility to cross validate the results obtained from different measuring methods is a useful opportunity.

References

[1] Farley C T, Blickhan R, Saito J, Taylor C R 1991 *Journal Applied Physiology* **71**, 2127-2132
[2] Farley C T, Morgenroth D C 1999 *J.Biomechanics* **32**, 267-273
[3] Rapoport S, Mizrahi J, Kimmel E, Verbitsky O, Isakov E 2003 *J. Biomechanical Engineering* **125** 507-514
[4] Hobara H, Kimura K, Omuro K, Gomi K, Muraoka T, Iso S, Kanosue K 2008 *J. Biomechamics* **41** 506-514
[5] Moritza C T, Farley C T 2006 *J. Biomechanics* **39** 1030-1038
[6] Hobara H, Muraoka T, Omuro K, Gomi K, Sakamoto M, Inoue K, Kanosue K 2009 *J. Biomechanics* **42** 1768-1771
[7] Losier K H, Eriksson A 2014 *J. Biomechanics* **47** 115-121
[8] Riesen S, Seyfarth A, Grimmer S 2013 *J. Biomechanics* **46** 2207-2212
[9] Zhoua H, Stone T, Huc H, Harris N 2007 *Medical Engineering Physics*

Evaluation of the uncertainty of electrical impedance measurements: the GUM and its Supplement 2

Pedro M Ramos[1], Fernando M Janeiro[2] and Pedro S Girão[1]
[1] Departamento de Engenharia Electrotécnica e de Computadores, Instituto Superior Técnico, Universidade de Lisboa and Instituto de Telecomunicações, Av. Rovisco Pais, n. 1, 1049-001 Lisbon, Portugal
[2] Instituto de Telecomunicações and Universidade de Évora, Departamento de Física, Rua Romão Ramalho, n. 59, 7000-671 Évora, Portugal

E-mails: pedro.m.ramos@tecnico.ulisboa.pt, fmtj@uevora.pt, psgirao@tecnico.ulisboa.pt

Abstract. Electrical impedance is not a scalar but a complex quantity. Thus, evaluation of the uncertainty of its value involves a model whose output is a complex. In this paper the comparison of the evaluation of the uncertainty of the measurement of the electrical impedance of a simple electric circuit using the GUM and using a Monte Carlo method according to the Supplement 2 of the GUM is presented.

1. Introduction

The publication, in 1993, of the Guide to the Expression of Uncertainty in Measurement (GUM) was the first significant result of the work of more than 15 years developed by individuals and organizations to produce an internationally accepted procedure for expressing measurement uncertainty and for combining individual uncertainty components into a single total uncertainty. The Guide, revised in 1995 and 2008 [1], materializes Recommendation INC-1 (1980) by the CPIM Working Group on the Statement of Uncertainties by providing a conceptual framework allowing a consistent treatment of uncertainty contributions. Complemented by the contribution of many [2-7], the impact of GUM in Metrology was increased when, in 2005, standard ISO/IEC17025 [8] established that accredited test laboratories were required to estimate the uncertainty and to report it. Notwithstanding, the original GUM soon evidenced some shortcomings, namely: (a) dealing with non-symmetrical measurement uncertainty distributions, with non-linearity of the measurement system, with input dependency and systematic bias; (b) it mainly concerns univariate measurement models, namely models having a single scalar output quantity.

To cope with these limitations, work has been developed by many authors (e.g. [9-18]) and two supplements to the GUM – Supplement 1 and Supplement 2 – were issued. Based on the same basic ideas of the GUM – measurement model and probability distribution of the input quantities – Supplement 1 [19], which deals also only with models having a single scalar output quantity, introduces numerical simulation using the Monte Carlo method (MCM) as alternative to analytical or numerical integration for calculating the combination of the probability distributions of the input quantities involved in the propagation of distributions required for the evaluation of measurement

Content from this work may be used under the terms of the Creative Commons Attribution 3.0 licence. Any further distribution of this work must maintain attribution to the author(s) and the title of the work, journal citation and DOI.
Published under licence by IOP Publishing Ltd

uncertainty. Supplement 2 [20] extends the use of GUM and Supplement 1 to multivariate measurement models, namely models with more than one output quantity. It is true that *the GUM includes examples, from electrical metrology, with three output quantities [JCGM 100:2008 H.2], and thermal metrology, with two output quantities [JCGM 100:2008 H.3]* [20], but Supplement 2 *describes a generalization of that Monte Carlo method to obtain a discrete representation of the joint probability distribution for the output quantities of a multivariate model. The discrete representation is then used to provide estimates of the output quantities, and standard uncertainties and covariances associated with those estimates* [20].

Supplements 1 and 2 to the GUM originate some inconsistency between the GUM and the Supplements. This fact, added to some non-solved limitations of the GUM itself, lead to the need of a revision of the GUM that, to the best of our knowledge, is underway [21].

The electrical impedance is a simple example of a complex quantity and the output of its measurement model is a complex number. The uncertainty evaluation of electrical impedance measurements using the GUM and using the Supplement 2 to the GUM are compared.

2. Electrical impedance

Electrical impedance extends the concept of resistance to alternating current (AC) circuits, describing not only the relative amplitudes of the voltage and current, but also their phase difference. However, electrical impedance measurement is important not only in the analysis of electric circuits but also for other purposes. In fact, because the electrical impedance allows the quantification of the behavior of a conducting medium to an electric current, the impedance measurement is used in the determination of the electromagnetic properties of materials and is the basis of various methods of electrical transduction with applications in fields such as chemistry [22,23] and biomedicine [24-27].

Sometimes abusively used in other time varying regimes, electrical impedance, usually represented by Z, is an electric quantity defined in the context of sine wave alternating current. Thus if $v(t)$ is the impedance sinusoidal voltage with constant amplitude V_M, constant frequency f, and constant phase φ_v

$$v(t) = V_M \cos\left(2\pi f t + \varphi_v\right) = \mathrm{Re}\left(V_M \mathrm{e}^{\mathrm{j}2\pi f t} \mathrm{e}^{\mathrm{j}\varphi_v}\right), \tag{1}$$

where j is the imaginary unit and Re(x) is the real part of x, the impedance current is

$$i(t) = I_M \cos\left(2\pi f t + \varphi_i\right) = \mathrm{Re}\left(I_M \mathrm{e}^{\mathrm{j}2\pi f t} \mathrm{e}^{\mathrm{j}\varphi_i}\right). \tag{2}$$

In complex amplitudes (phasors) both can be written as

$$\overline{V} = V_M \mathrm{e}^{\mathrm{j}\varphi_v} \text{ and } \overline{I} = I_M \mathrm{e}^{\mathrm{j}\varphi_i}. \tag{3}$$

The impedance is not a phasor but a complex number defined as the ratio $\overline{V}/\overline{I}$ whose amplitude (module) is Z and whose phase is φ.

$$\overline{Z} = \frac{\overline{V}}{\overline{I}} = \frac{V_M}{I_M} \mathrm{e}^{\mathrm{j}(\varphi_v - \varphi_i)} = \frac{V_M}{I_M} \mathrm{e}^{\mathrm{j}\varphi} = Z\,\mathrm{e}^{\mathrm{j}\varphi} = Z\cos\left(\varphi\right) + \mathrm{j}Z\sin\left(\varphi\right). \tag{4}$$

The principles, methods, equipment and procedures for electrical impedance measurement are diverse [28] depending, namely, on the frequency and on the application. In the following sections, and because the purpose here is to compare the uncertainty evaluation using two different methods, the considered case is that of electrical impedance measurements, based on digital acquisition of sinusoidal voltages and estimation of their amplitudes and phase differences as described in [29].

3. Results

3.1. Measurement Model

Figure 1 depicts the setup for the measurement of impedance Z through the measurement of the voltages at its terminals and the terminals of a reference impedance Z_R. These voltages are buffered with two instrumentation amplifiers (IAs) with unitary gain and simultaneously acquired using two analogue-to-digital converters (ADCs). The series of Z and Z_R is excited by the output of a sine wave generator. Figure 2 shows an example of the sampled signals for a 1 kHz measurement frequency. The signal processing algorithm implemented in the DSP estimates the voltage amplitudes and their phase difference using a seven-parameter sine-fitting algorithm [30].

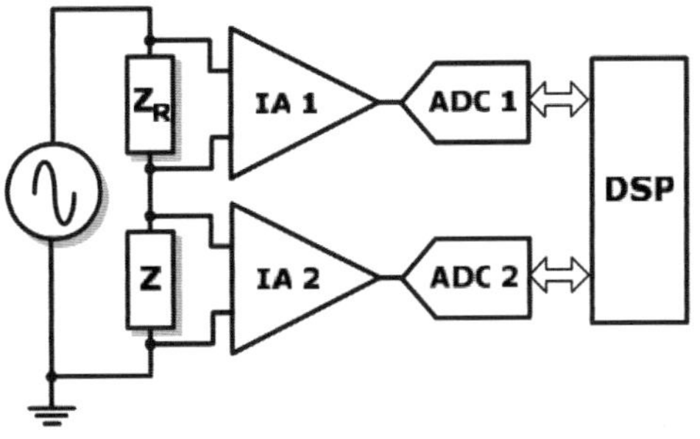

Figure 1. Setup for the measurement of Z.

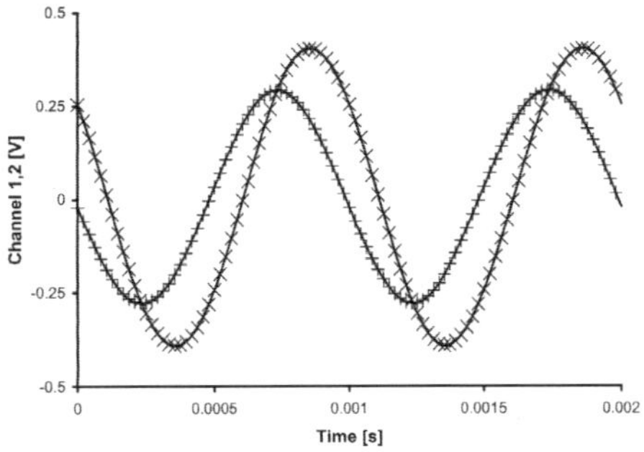

Figure 2. Example of ADC 1 and ADC 2 acquired samples.

From Figure 1, and since the input impedance of the IAs is much higher than the amplitudes of both impedances,

$$\frac{\overline{U_1}}{\overline{Z_R}} = \frac{\overline{U_2}}{\overline{Z}} \tag{5}$$

where $\overline{U_1}$ and $\overline{U_2}$ are the complex amplitudes of the voltage across Z_R and Z acquired by ADC 1 and ADC 2, respectively. The impedance measurement model is

$$\overline{Z} = \frac{\overline{U_2}}{\overline{U_1}} \overline{Z_R} \,. \tag{6}$$

The acquisition process yields an estimation of the complex ratio $\overline{U_2} / \overline{U_1}$. The uncertainty of Z_R (which originates from the measurement of Z_R with another impedance measurement method) must be included to obtain the uncertainty of Z.

3.2. Measurements

The results presented correspond to the 1 kHz measurement of an impedance with about $1\,\text{k}\Omega$ amplitude and a phase of 5°. The reference impedance also has amplitude near $1\,\text{k}\Omega$ and a phase near -85°, which corresponds to a non-ideal capacitor.

In Figure 3, the 2D histogram of the measured $\overline{U_2} / \overline{U_1}$ ratio is shown for 100 000 measurements with 101 bins both for real and imaginary components. The experimental histogram closely resembles a bivariate normal distribution. Notice that the $\overline{U_2} / \overline{U_1}$ ratio is located in the complex plane near the positive imaginary axis. This will result in an impedance near the positive real axis since the reference impedance is a capacitor. The average amplitude ratio value is 0.998619264 while the experimental standard deviation of the mean is 7.9×10^{-8}. The average phase is 89.5464203° with a experimental standard deviation of the mean of 0.0000034°. The coverage regions for 68% and 95% are shown in Figure 4.

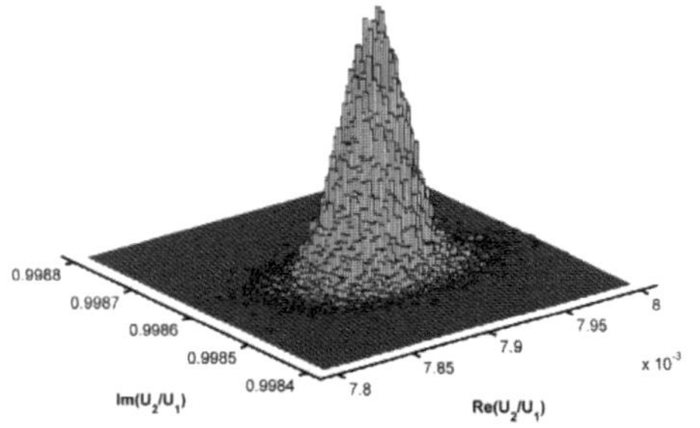

Figure 3. 2D histogram of the measured $\overline{U_2}/\overline{U_1}$ ratio.

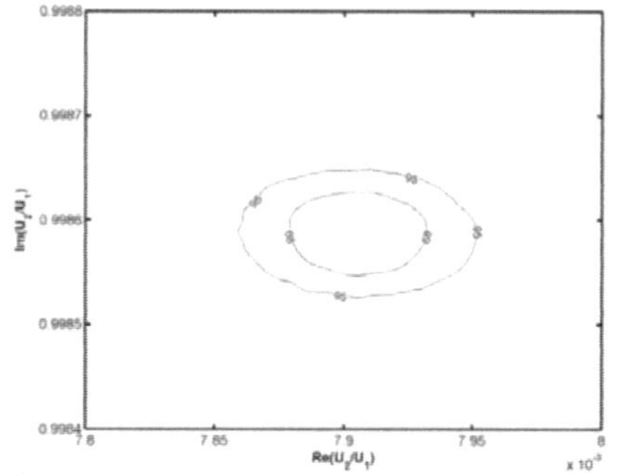

Figure 4. Coverage regions for 68% and 95% of the measured $\overline{U_2}/\overline{U_1}$ ratio.

3.3. Uncertainty evaluation following the law of propagation of uncertainty

The use of the law of propagation of uncertainty (LPU) is explained in the GUM [1], but is restricted to measurement models with one scalar output. This method can be applied to the measurement model (6) if it is divided into two: the impedance amplitude and the impedance phase, according to

$$|Z| = \frac{|U_2|}{|U_1|}|Z_R| \text{ and } \varphi = \Delta\varphi - \varphi_{Z_R}. \tag{7}$$

In this situation, the combined standard uncertainty of the impedance amplitude is

$$u_{|Z|} = \sqrt{\left(\frac{|U_2|}{|U_1|}u_{|Z_R|}\right)^2 + \left(|Z_R|u_{\frac{|U_2|}{|U_1|}}\right)^2}, \tag{8}$$

since the amplitude ratio measurement and the measurement of the reference impedance are independent. The combined standard uncertainty of the impedance phase is

$$u_\varphi = \sqrt{u_{\Delta\varphi}^2 + u_{\varphi_{Z_R}}^2}. \tag{9}$$

The reference impedance is measured using an AGILENT 4294A. The obtained reference impedance amplitude is 1004.721 Ω with a maximum error of 0.76 Ω. As for the phase, the measured value is -84.2528° with a maximum error of 0.043°. To obtain the uncertainty of the Z_R measurements (amplitude and phase), the GUM indicates that a uniform probability density function (PDF) should be considered and therefore the uncertainties are obtained by dividing the maximum error by the square root of 3. Notice, that the assumption that the PDF is uniform was validated with consultations with the manufacturer of the impedance meter. Therefore, the standard uncertainty of the reference impedance amplitude and phase are $u_{|Z|} = 0.44\,\Omega$ and $u_\varphi = 0.025\,^\circ$.

In the complex plane, the uncertainty of the impedance amplitude corresponds to an uncertainty in the distance to the origin of the complex number, while the uncertainty of the impedance phase corresponds to an uncertainty of the complex number phase. The combination of these two uncertainties defines a region with the generic shape depicted in Fig. 8 of [31] and shown in Figure 5. Notice that the considerable distance to origin and the reduced phase uncertainty makes all the lines resemble straight lines when in fact, the real, limits of the coverage phase intervals are arcs.

Another way to interpret the results of the law of propagation of uncertainty is to combine the two outcomes (the amplitude and phase PDFs) considering that they are independent. Since the GUM transforms every uncertainty into a normal PDF distribution, it is possible to combine both output PDFs (amplitude and phase) to obtain a bivariate PDF. The resulting equivalent 2D histogram is shown in Figure 6. Note that, the coverage regions of this histogram (which are ellipses) do not correspond to the intersection areas depicted in Figure 5.

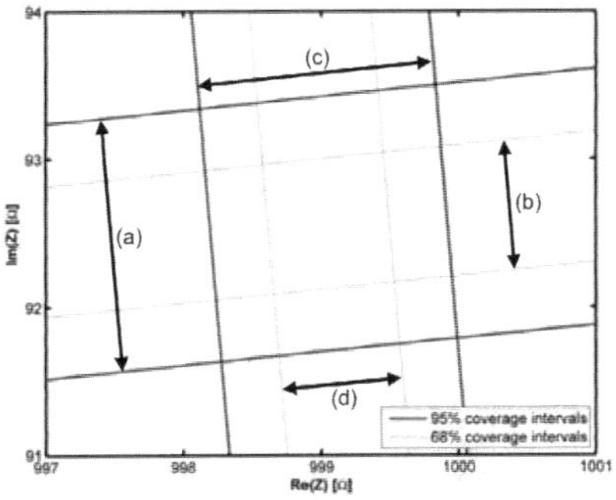

Figure 5. Coverage intervals obtained using the LPU. (a) and (b) are the coverage intervals of the impedance phase for 95% and 68%. (c) and (d) are the coverage intervals of the impedance amplitude for 95% and 68%.

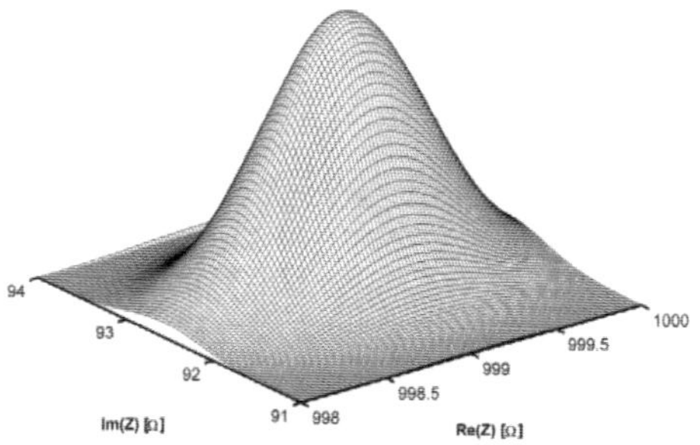

Figure 6. Equivalent 2D histogram obtained using the law of propagation of uncertainty as defined in the GUM [1].

3.4. Uncertainty evaluation using Monte Carlo method

The law of propagation of uncertainty approach, as shown in the previous section, has significant shortcomings in this (and similar) applications since it neglects any phase/amplitude dependence of the measured $\overline{U_2}/\overline{U_1}$ ratio and takes a simplistic interpretation of the uncertainty of the reference impedance (transforming the uniform PDFs into normal ones). These shortcomings can be addressed by evaluating the uncertainty of the measurement model (6) using Monte Carlo simulations. In Supplement 1 [19], the use of Monte Carlo method (MCM) is described for single output measurement models where the law of propagation of uncertainty is not suited due to: (a) the type of PDFs of each measurement model input are substantially different from normal PDFs; (b) the first order Taylor approximation used in LPU is not suited for some measurement models; (c) measurement models that cannot be described by a closed algebraic equation (to determine the partial derivates). Supplement 2 [20], goes further by expanding the use of MCM for measurement models with any number of output quantities. One significant evolution in this supplement is the definition of coverage regions and in particular that of the smallest coverage region (and interval which is also suited for measurement models with a single output).

In the considered case of impedance measurements, the PDF of the reference impedance amplitude is uniform as is the PDF of the reference impedance phase. In order to obtain the histogram of the measured impedance, for each of 100 000 $\overline{U_2}/\overline{U_1}$ ratio measurements, 100 000 values of the reference impedance amplitude and phase are randomly generated in accordance to their uniform distributions. The values of the resulting estimated impedance are then used to generate the 2D histogram (Figure 7) and from the histogram, the coverage region with 95% is obtained (Figure 8).

It can be seen that the 2D histogram is basically a plateau due to the uniform distribution of the reference impedance amplitude and phase. On the edges of the plateau, it decreases quickly (practically following a one-sided normal distribution) due to the relatively narrow 2D histogram of the measured $\overline{U_2} / \overline{U_1}$ ratio. Due to the abruptness of the plateau decline, the coverage region for 68% is basically the same as for the 95% case.

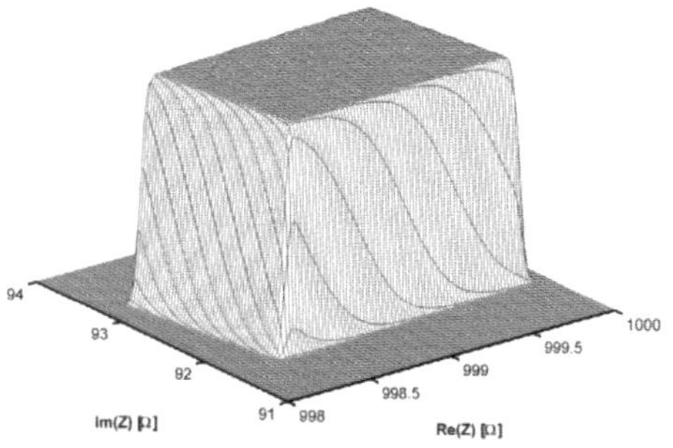

Figure 7. 2D histogram of the measured impedance.

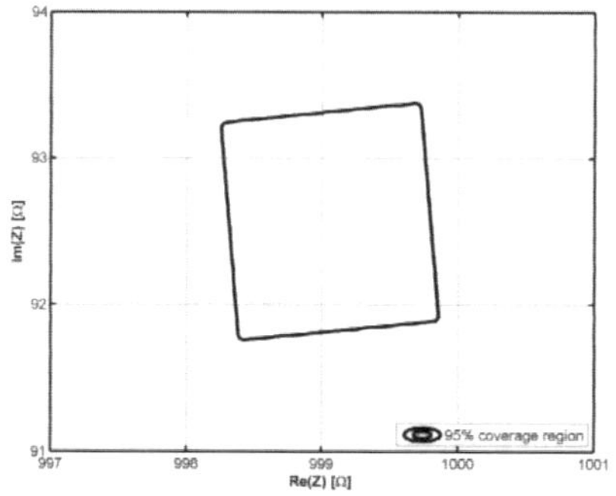

Figure 8. Coverage region for 95% of the measured impedance.

One important aspect that should be pointed out, concerns the numerical implementation procedure used to estimate the histogram of Figure 7. The total amount of estimated impedance values is 10^{10}, which would require 160 GBytes of data using a straightforward brute-force approach (herein called the traditional method). This amount of data is unmanageable to process and estimate the final histogram.

The new proposed method consists in processing the data in segments, estimate the 2D histogram contribution of each segment, discard the data and proceed to the next segment storing only the cumulative histogram contribution. In Figure 9, relative simulation times are shown for the new method and that of the traditional method as a function of the number of measured points M – which is considered to be the same as of the estimated reference impedance values for each measurement – i.e., the total number of estimated impedance values is M^2. Processing is done in MATLAB on a PC with an i7 processor and 16 GB of RAM.

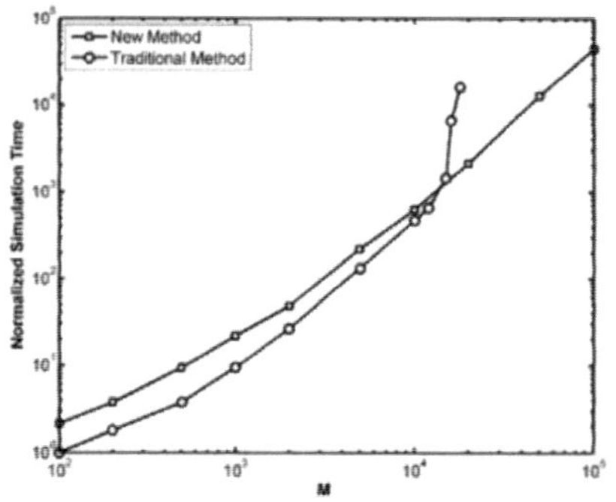

Figure 9. Normalized simulation time (ratio between simulation time and the lowest value) to obtain the output histogram as a function of the number of measurements for the traditional method and the new method.

Although the new method is slower for $M < 10^4$, due to the multiple histogram assessments, for $M > 10^4$, the simulation time of the traditional method increases substantially and for $M > 18\,000$, Matlab can no longer conclude the process due to the use of virtual memory and the corresponding increase in access time to the data. On the other hand, the new method has no such constrain and could be used even for $M > 10^5$. Note that the amount of required memory for the traditional method is $16M^2$ Bytes while for the new method it is only $16M$ Bytes. For $M = 10^4$, the traditional method requires 1.6 GBytes, while the new method requires only 160 kBytes.

The drawback of the new method is that the histogram range and its number of points must be defined in the beginning of the process and cannot be changed during the process. This issue requires that the process is started with a low value of M to define the histogram range, which is refined before the final simulation is executed.

4. Conclusions

The example presented in this paper highlights some of the known limitations of the law of propagation of uncertainty. For this particular case where the estimated output is a complex number, its separation into two measurement models removes the dependency of the estimated complex ratio amplitude and phase. If the real and imaginary components were considered instead, the resulting ellipsoid coverage regions would be different. The Monte Carlo method shows that the actual coverage region is not an ellipsoid but instead a rotated rectangle mainly due to the uncertainty of the reference impedance and its considerable area when compared with the coverage region of the measured complex voltage ratio.

5. References

[1] BIPM, IEC, IFCC, ILAC, ISO, IUPAC, IUPAP and OIML 2008 Guide to the Expression of Uncertainty in Measurement, JCGM 100:2008, GUM 1995 with minor corrections, http://www.bipm.org/utils/common/documents/jcgm/JCGM1002008E.pdf

[2] Phillips SD, Eberhardt KR, Parry B 1997 Guidelines for Expressing the Uncertainty of Measurement Results Containing Uncorrected Bias *NIST Journal of Research* Institute of Standards and Technology

[3] Bell S 1999 *A Beginner's Guide to Uncertainty of Measurement* (London: NPL)

[4] Lira I 2002 *Evaluating the Uncertainty of Measurement. Fundamentals and Practical Guidance* (Bristol, UK: Institute of Physics)

[5] Kirkup L and Frenkel RB 2006 *An Introduction to Uncertainty in Measurement. Using the GUM (Guide to the Expression of Uncertainty in Measurement)* (Cambridge University Press)

[6] NASA HDBK-8739.19-3 – Measurement Uncertainty Analysis Principles and Methods, July, 2010

[7] Willink R, Hall BD 2013 An extension to GUM methodology: degrees-of-freedom calculations for correlated multidimensional estimates", arXiv:1311.0343 [physics.data-an]

[8] ISO/IEC 17025:2005 General requirements for the competence of testing and calibration laboratories

[9] Yeung H and Papadopoulos CE 2000 Natural gas energy flow (quality) uncertainty estimation using Monte Carlo simulation method *International Conference on Flow Measurement*, FLOMEKO

[10] Khu ST, and Werner MG 2003 Reduction of Monte-Carlo simulation runs for uncertainty estimation in hydrological modelling *Hydrology and Earth System Sciences* **Vol. 7** No. 5 680-690

[11] Andrae ASG, Mller P, Anderson J and Liu J 2004 Uncertainty estimation by Monte Carlo simulation applied to life cycle inventory of cordless phones and microscale metallization processes *IEEE Transactions on Electronics Packaging Manufacturing* **27** No. 4 233-245

[12] Rossi GB, Crenna F, Harris PM, Cox MG 2006 Combining direct calculation and the Monte Carlo method for the probabilistic expression of measurement results *Advanced Mathematical and Computational Tools in Metrology VII*, Series on Advances in Mathematics for Applied Sciences vol. 72 221-228, (World Scientific Publishing Co.)
[13] Cox MG and Bernd Siebert RL 2006 The use of a Monte Carlo method for evaluating uncertainty and expanded uncertainty *Metrologia* **43** 4 178-188
[14] Silva Ribeiro A 2006, *Avaliação de Incertezas de Medição em Sistemas Complexos Lineares e Não-Lineares*. PhD. thesis (in Portuguese)
[15] Alves e Sousa J, Silva Ribeiro A, Forbes AB, Harris PM, Carvalho F, Bacelar L 2007 The Relevance of Using a Monte Carlo Method to Evaluate Uncertainty in Mass Calibration *Proc. IMEKO TC3, TC16 & TC22 International Conference* Merida México
[16] Jing H, Huang M-F, Zhong Y-R, Kuang B and Jiang X-Q 2007 Estimation of the measurement uncertainty based on quasi Monte-Carlo method in optical measurement *Proceedings of the International Society for Optical Engineering*
[17] Schueller GI 2007 On the treatment of uncertainties in structural mechanics and analysis, *Journal Computers and Structures* **85** 5-6 235-243
[18] Koch KR 2008 Evaluation of uncertainties in measurements by Monte-Carlo simulations with an application for laser scanning *Journal of Applied Geodesy* **2** 67-77
[19] BIPM, IEC, IFCC, ILAC, ISO, IUPAC, IUPAP and OIML 2008, Supplement 1 to the 'Guide to the Expression of Uncertainty in Measurement'—Propagation of distributions using a Monte Carlo method JCGM 101:2008, http://www.bipm.org/utils/common/documents/jcgm/JCGM.101.2008E.pdf
[20] BIPM, IEC, IFCC, ILAC, ISO, IUPAC, IUPAP and OIML 2011, Supplement 2 to the 'Guide to the Expression of Uncertainty in Measurement'—Extension to any number of output quantities JCGM 102:2011, http://www.bipm.org/utils/common/documents/jcgm/JCGM.102.2011.E.pdf
[21] Bich W *et al.* 2012 Revision of the 'Guide to the Expression of Uncertainty in Measurement *Metrologia* **49** 6 702-705
[22] McNeil CJ, Athey D, Ball M, Ho WO, Krause S, Armstrong RD, Wright JD and Rawson K 1995 Electrochemical Sensors Based on Impedance Measurement of Enzymel Catalyzed Polymer Dissolution: Theory and Applications *Anal. Chem.*, **67**, 3928-3935
[23] Roy P, Tsai T-C, Liang C-T, Wua R-J and Chavali M 2011 Application of Impedance Measurement Technology in Distinguishing Different Tea Samples with Ppy/SWCNT Composite Sensing Material *Journal of the Chinese Chemical Society* **58** 714-722
[24] Guimerà A, Calderón E, Los P and Christie AM 2008 Method and device for bio-impedance measurement with hard-tissue applications *Physiological Measurement* **29** 6 15-26
[25] Tang WH and Tong W 2009 Measuring impedance in congestive heart failure: Current options and clinical applications *Am Heart J.* **157** 3 402–411
[26] Gupta AK 2011 Respiration Rate Measurement Based on Impedance Pneumography *Texas Instruments Application Report SBAA181*
[27] Marquez JC, Rempfler M, Seoane F and Lindecrantz K 2013 Textrode-enabled transthoracic electrical bioimpedance measurements – towards wearable applications of impedance cardiography *J Electr Bioimp.* **4** 45–50
[28] Callegaro L 2013 *Electrical Impedance: Principles, Measurement and Applications* (Boca Raton Fl. USA: CRC Press)
[29] Ramos PM, Silva MF, Serra AC 2004 Low Frequency Impedance Measurement Using Sine-Fitting *Measurement* **35** 89-96
[30] Ramos PM, Serra AC 2008 A new sine-fitting algorithm for accurate amplitude and phase measurements in two channel acquisition systems *Measurement* **51** 135-143
[31] Ramos PM, Janeiro FM, Tlemçani M, Serra AC 2009 Recent Developments on Impedance Measurements with DSP-Based Ellipse-Fitting Algorithms *IEEE Transactions on Instrumentation and Measurement* **58** 1680-1689

Microwave Doppler radar in unobtrusive health monitoring

P Silva Girão[1], O Postolache[2], G Postolache[3], PM Ramos[1], JM Dias Pereira[4]
[1]DEEC/IST/UL, IT, Avenida Rovisco Pais, 1, 1049-001 Lisbon, Portugal
[2]ISCTE-IUL, IT, Avenida das Forças Armadas, 1649-026 Lisbon, Portugal
[3]Escola Superior de Saúde, Universidade Atlântica, Oeiras, Portugal
[4]EST/IPS, IT, Rua do Vale de Chaves, Estefanilha, 2910-761 Setúbal, Portugal

E-mails: psgirao@tecnico.ulisboa.pt, opostolache@lx.it.pt, gabrielap@uatla.pt, pedro.m.ramos@tecnico.ulisboa.pt, joseper@est.ips.pt

Abstract. This article frames the use of microwave Doppler radar in the context of ubiquitous, non-obstructive health monitoring. The use of a 24GHz CW (continuous wave) Doppler radar based on a commercial off-the-shelf transceiver for remote sensing of heart rate and respiration rate based on the acquisition and processing of the signals delivered by the radar is briefly presented.

1. Introduction

Since described in 1842 by Christian Johann Doppler [1] in the context of star motion, the Doppler Effect has been used in many applications namely for the measurement of physical quantities such as speed, flow, temperature with impact in areas such as cosmology and meteorology. Because Doppler based sensing solutions require no contact of the sensor with the sensed quantity, they are particularly attractive for medical applications and thus are the support of various ultrasound based equipment used primarily for non-invasive, non-intrusive and unobtrusive medical diagnosis of conditions including detection of foetal heartbeat, detection of air emboli, blood pressure monitoring, detection and characterization of blood flow, and localization of blood vessel occlusions (e.g. echocardiogram, Doppler ultrasound). New small size radar integrated devices, developed namely for automotive applications, paved the way to medical applications that require or advise the unobtrusiveness and pervasiveness permitted by Doppler sensing.

As it is well known, the Doppler Effect consists on the shift in frequency emitted by a sound or electromagnetic source and the one sensed by a receiver when the distance between the source and the receiver changes. If the distance between the source and receiver decreases, the acoustic or electromagnetic waves become closer (smaller wavelength) and thus the frequency of the received signal increases; if the distance between the source and receiver increases, the inverse happens, that is, the frequency at the receiver is lower than the emitted by the source. The distance may change either (a) because the velocities of the source and receiver are different (Figure 1 (a)) or (b) because the source and receiver are stationary at the same place but the waves emitted by the source are reflected by a moving body (Figure 1 (b). The shift in frequency and can be expressed by

$$\Delta f = \frac{\Delta v}{c} f_s \qquad (1)$$

Content from this work may be used under the terms of the Creative Commons Attribution 3.0 licence. Any further distribution of this work must maintain attribution to the author(s) and the title of the work, journal citation and DOI.
Published under licence by IOP Publishing Ltd

where $\Delta f = f_r - f_s$ is the frequency shift (difference between the frequency at the receiver and of the source), $\Delta v = v_r - v_s$ is either the velocity of the receiver relative to the source (case (a) above mentioned) or the velocity of the moving body (case (b) above mentioned), c is the velocity of the waves in the medium and f_s is the frequency of the source.

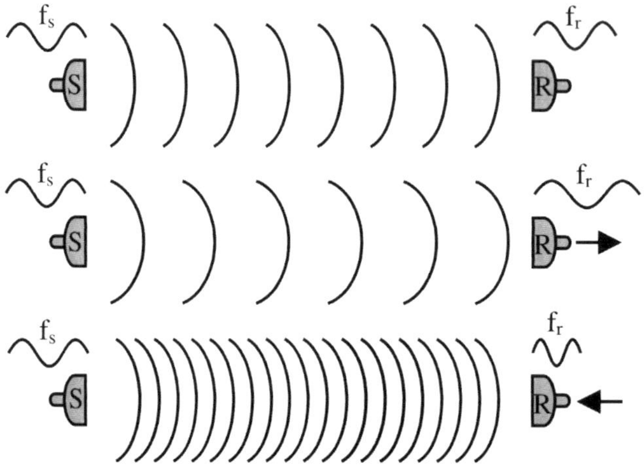

(a)

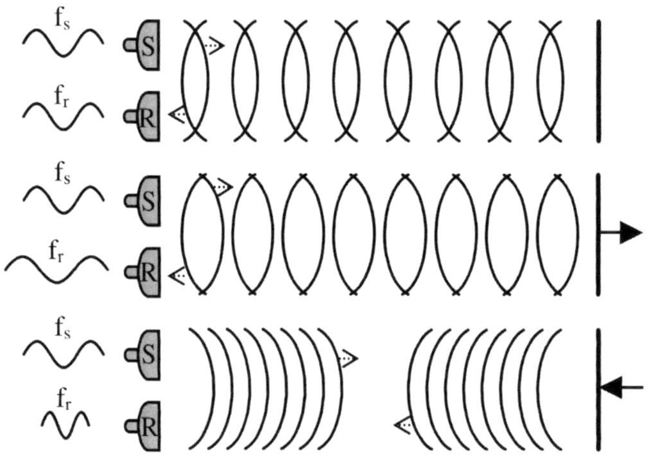

(b)

Figure 1. Doppler Effect: S – source; R – receiver; f_s and f_r frequency of signal at the source and at the receiver, respectively. (a) from top to bottom: distance S-R constant → $f_r = f_s$, distance S-R increasing → $f_r < f_s$, distance S-R decreasing → $f_r > f_s$; (b) from top to bottom: the same three situations but when S-R are stationary and the emitted wave face an obstacle (case that interests most biomedical applications

2. Health monitoring

Medicine – *ars medicina* in Latin, the art of healing - is an applied science and, as such, a never ending subject for research in all the three major aspects involved: diagnosis, treatment, and prevention of disease. Along the time, Medicine and been changing from a treatment centred science to a preventive one. Demography, social changes, and the rising costs of health and social care have driven that tendency: the health quality of people increases and the costs of health systems are reduced if illness is kept as low as possible and the first move in this direction is through prevention and early detection and diagnosis of diseases.

Health monitoring encompasses different activities namely clinical analysis, screening tests, diagnostic tests, medical exams, and self-exams. Image based tests have already reached the possibility of identifying molecules in a laboratorial environment [2] but no one will be surprised if in the future diseases can be detected, identified and perfectly characterized based only on image based common equipment. In the meantime, image equipment based on the Doppler Effect using ultrasound have been assisting health technicians in health monitoring, either in diagnostic but also for supporting continuous health care services.

In a recent work, The College of Fellows, American Institute for Medical and Biological Engineering (CF-AIMBE) considers that two of the six major challenges for medical and biological engineering in the 21st are *engineering personalized health care and engineering solutions to injury and chronic diseases* [3]. Personalized health care is only possible if appropriate data on each individual is available. This is the objective of electronic health records (EHRs) [4]. The authors have been working on this subject in Portugal and under a Portuguese funded project, are developing a platform for EHR filling, storage, and search and query. For several years the authors also share the view of the CF-AIMBE in what concerns the importance of assisting injured and the chronically ill.

Bearing in mind that demographics and costs lead to periods of internment in hospitals increasingly shorter, it is crucial to develop solutions that allow treatment and monitoring of health both in ambulatory and at home, either of the patient's family or in nursing homes. This is particularly pertinent when elderly are at stake.

2.1 Unobtrusive, pervasive sensing
Equipment aimed at acquiring pertinent information of the physiological status of a person in non-hospital environments, namely during daily activity, must be as unobtrusive and pervasive as possible. Nowadays there are several solutions achieving these goals at different levels. Particularly, it is worth referencing here solutions and systems based on instrumented clothing (e-textiles) [5-7] because of the potential they bring to non-intrusive acquisition of physiological signals. The authors have been working and proposing unobtrusive and pervasive (ubiquitous) solutions [8-13] namely based on the use of non-conductivity electrodes [14,15] and ballistocardiography sensing [16-19], where the sensing part is embedded in wheelchairs and crutches [20].

3. Sensing with Microwave Doppler Radar
Sensing using the Doppler Effect is particularly suited for unobtrusive sensing. Ultrasonic medical imaging uses signals of 2 MHz and higher with power densities generally less than 1 W/cm^2. Ultrasound waves tend to reflect almost totally (up to 99% [21]) wherever air meets biological tissue. For this reason ultrasonic sensing requires not only contact of the source and receiver with the skin, but also that no air bubbles are in their interfaces. Sensing requires a probe, a gel that conducts well sound waves and an operator to handle the probe. Ultrasound sensing is thus not an alternative for pervasive sensing of physiological signals, particularly if the purpose is no contact between sensor and the person to monitor. Electromagnetic waves do not have the problem of ultrasound mentioned in the previous paragraph; they penetrate well non-metallic objects and thus the human body, which means that the source and receiver may be placed at some distance from the examinee.

Since de 1970s that it is known that microwave Doppler radar can be used for cardiovascular and respiratory measurements [22-23]. From then on, and until the 1990s, radar technology was slowly introduced in medical applications [24]. With the beginning of the current millennium, and due particularly to technological advances in ultra wide band (UWB) radio, radar technology has been increasingly used both for medical imaging and for heart rate and pulmonary motion monitoring [24-26].

Continuous wave (CW) Doppler radars are very sensitive devices in detecting movement, such as time varying physiological phenomena. But the electromagnetic wave reflected for instance by the chest and heart of a person, contains information namely about person's movement, respiration and blood circulation. To separate them, range information is necessary, which can be achieved using modulated signals such as pulsed signals or frequency modulated (FM) signals [27-31]. UWB radar is exactly based on pulsed signals and several medical applications have already been reported [32] namely for heart rate, respiration rate (RR), and heart movements recording [33,34]. Nevertheless, there is much to be done to take fully advantage of UWB radar for other medical applications, which may include apnoea, allergy, and asthma monitoring, without forgetting the potential negative effects of electromagnetic fields on humans.

In the framework of the work related to the implementation of pervasive, non-intrusive solutions for continuous monitoring vital signs of elderly, the authors have been studying the use of FM-CW radar through the implementation of devices and systems based on off-the-shelf components [35-38] namely as components of health monitoring systems installed in chairs, wheelchairs, and crutches.

3.1 FM-CW Doppler radar

In FM-CW Doppler radar the emitted signal is a constant amplitude sinewave (carrier) whose frequency is modulated by a triangular or sawtooth signal (modulating signal). The received signal is the sum of the signals resulting from reflections when the emitted signal faces a zone of characteristic impedance transition (that we will call an "obstacle"). The frequency of the received signal contains thus information of the position of the reflecting zones and also of their velocity if their positions relative to the emitted signal source are changing. Processing the received signal should then allow obtaining that information. Figure 2 shows an example of a sawtooth modulating signal when the emitted signal suffers a single reflection by an obstacle moving away from the emitted signal source.

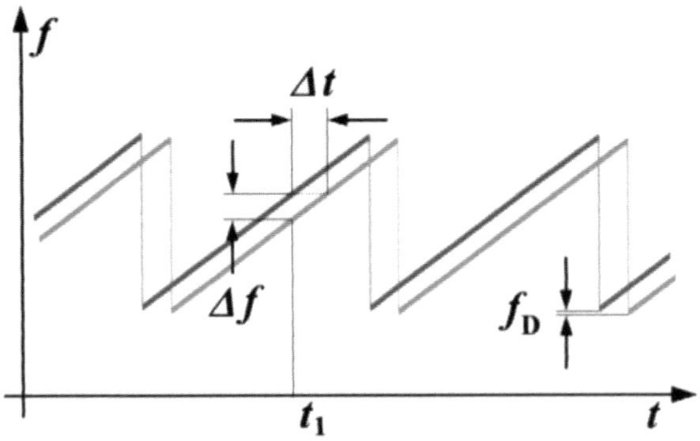

Figure 2. Sawtooth modulating signal of an FM-CW radar (red) and frequency demodulation of the received signal (green). Δt is a measure of the distance between emitter and obstacle and f_D a measure of the velocity at which the obstacle is moving away from the emitter.

When using FM-CW Doppler radar for health monitoring the situation is very different from the one depicted in Figure 2 because there are several sources of reflections, the received signal is a complex one, and thus identification of all reflecting sources is extremely different and only possible using signal processing techniques. The example present in the next section evidences this statement.

Figure 3 shows the simplified block diagram of a transceiver for FM-CW Doppler radar. In the simple case of Figure 2, the output of the mixer – intermediate frequency signal – allows determination of the distance between source and obstacle (range) and obstacle velocity.

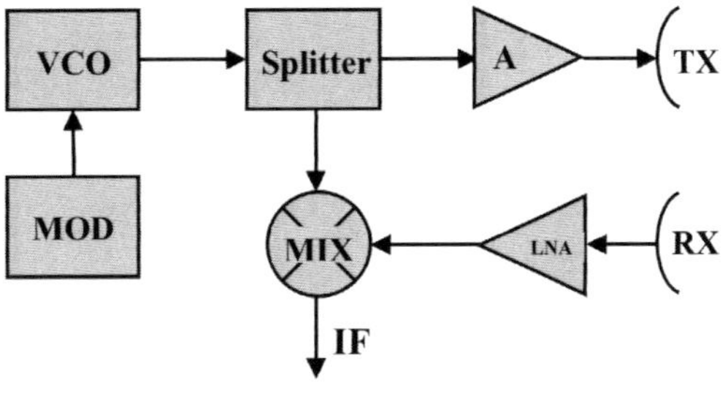

Figure 3. Block diagram of FM-CW radar transceiver: VCO – voltage controlled oscillator; MOD - modulator; A – amplifier; LNA – low noise amplifier; MIX – mixer; LPF – low pass filter; TX – transmitting antenna; RX – receiving antenna. IF – intermediate frequency signal.

The space resolution Δd of FM-CW radar – minimum distance that can be measured - depends only on the bandwidth (BW) of the modulating signal and not on the base frequency of the carrier,

$$\Delta d = \frac{c}{2BW} \tag{2}$$

To detect and measure for instance chest movement due to respiration, which can be as low as 1mm, the modulating signal BW must be of 150 GHz! This value would require carriers of much higher frequency! This means that physiological parameters to obtain using radar must be based on velocity measurement and not on range detection.

4. Radar vital signals monitor (RVSM)

Heart rate and respiration rate are two basic vital signals and its monitoring represents the first step in health care. Figure 4 shows a commercial off-the-shelf (COTS) FMCW/FSK transceiver developed for automotive applications that the authors have been using namely for heart and respiration rate measurement. The transceiver VCO centre frequency is 24 GHz, the low limit of one of the industrial, scientific and medical (ISM) radio bands [39-40].

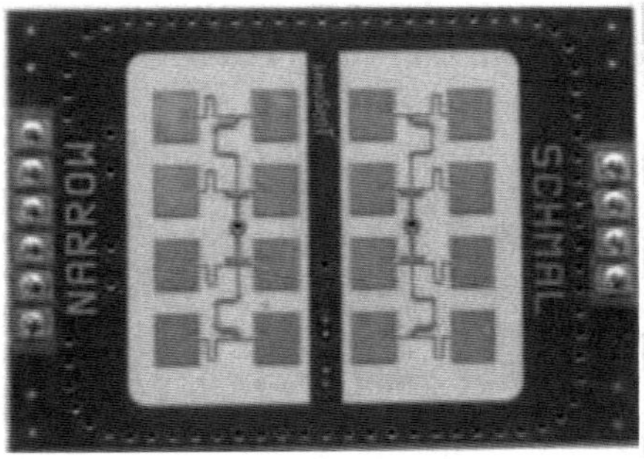

Parameter	Symbol	Min.	Typ.	Max.	Units
transmit frequencies	f		24.000 - 24.250		GHz
varactor tuning voltage	V_{tune}	0.5		10	V
tuning slope			40		MHz/V
temperature drift (frequency)	Δf		-1		MHz/°C
output power (EIRP)	P_{out}		15		dBm
IF-output	offset	1.0	2.2	4.0	V
IF-amplifier	bandwidth		DC - 50k		Hz
	gain		20		dB
full beam width @ -3dB	horizontal		45		°
	vertical		38		°
side-lobe suppression	horizontal		13		dB
	vertical		13		dB
supply voltage	V_{cc}	4.75	5.0	5.25	V
supply current	I_{cc}		35	50	mA
operating temperature	T_{op}	-20		+60	°C

Figure 4. InnoSenT IVS-162 FMCW/FSK transceiver and its main specifications. The golden squares constitute the transmitting and receiving antenna.

The transceiver permits FM or FSK modulation; therefore measurement of distance as well as recognition of stationary objects is possible, depending on modulation. The receiving path includes IQ mixing - stereo (dual channel) operation -, which allows for direction of motion identification. The transceiver includes an IF pre-amplifier whose bandwidth is limited to 50 kHz for lowest noise performance. The transceiver dimensions (40x30x19 mm) and power requirements (5V/50mA) allow ubiquitous installation (e.g. crutches and walkers, Figure 5, for motion detection [38]).

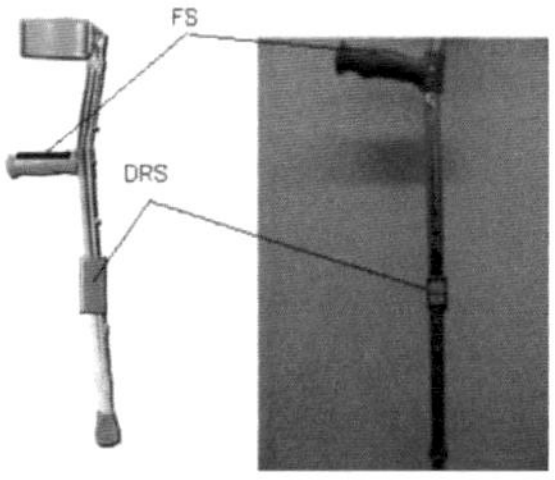

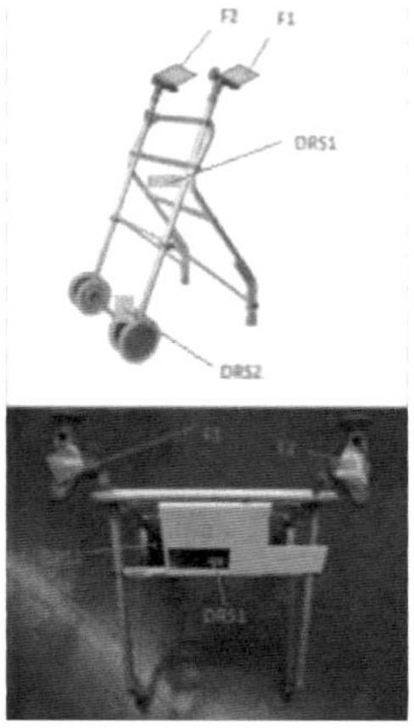

Figure 5. Smart crutch (top) and smart walker (right) (DRS, DRS1, DRS2 - Doppler radar sensors; FS, F1, F2 - piezoresistive force sensors).

Figure 6 shows the overall block diagram of the circuitry that implements the hardware component of a heart and respiration rate monitor based on InnoSenT IVS-162 transceiver developed by the authors [12].

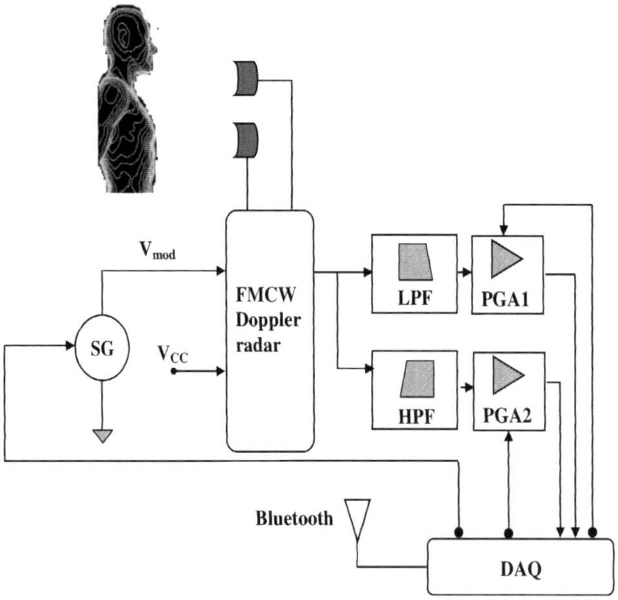

Figure 6. Heart rate and respiration rate non-intrusive measurement based on InnoSenT IVS-162 FM-CW Doppler radar transceiver: SG – signal generator; V_{mod} – modulating signal; V_{CC} – supply voltage; LPF – low pass filter; HPF – high pass filter; PGA1, PGA2 – programmable gain amplifiers; DAQ – data acquisition module.

A signal generator (SG) module, which is controlled by the Bluetooth data acquisition module (DAQ), delivers a voltage V_{mod} that modulates the VCO of the transceiver. The heart rate and respiration rate can be obtained by measuring the velocities of heart movement and of the chest. Thus, the radar is operated at a constant frequency selected by a constant value of V_{mod} and only IF signal in-phase with the transmitted signal (I component) is used to obtain both heart and respiration rates. For that purpose, the I component of the IF signal is filtered using a set of two active filters: an LPF (0.3 Hz), to extract the very low frequency breathing signal, and an HPF (0.9 Hz) to isolate the higher frequency signal associated with cardiac activity. Two programmable gain amplifiers, PA1 and PA2 increase the dynamic range of the filters output to take advantage of the 16bit ADC of a data acquisition module, DAQ (BlueSetry from GridConnect).

Figure 7 shows the signals IF and HPF output signals obtained from a stationary person. As can be seen, further processing is required to extract heart and respiration rate. The output of the LPF and HPF are digitized, transmitted to a remote processing unit (e.g. PC) for the removal of an eventual DC component in the IF signal resulting from reflections on stationary objects and for digital filtering and detrending algorithms, based on digital wavelet transform, applied to the resulting data.

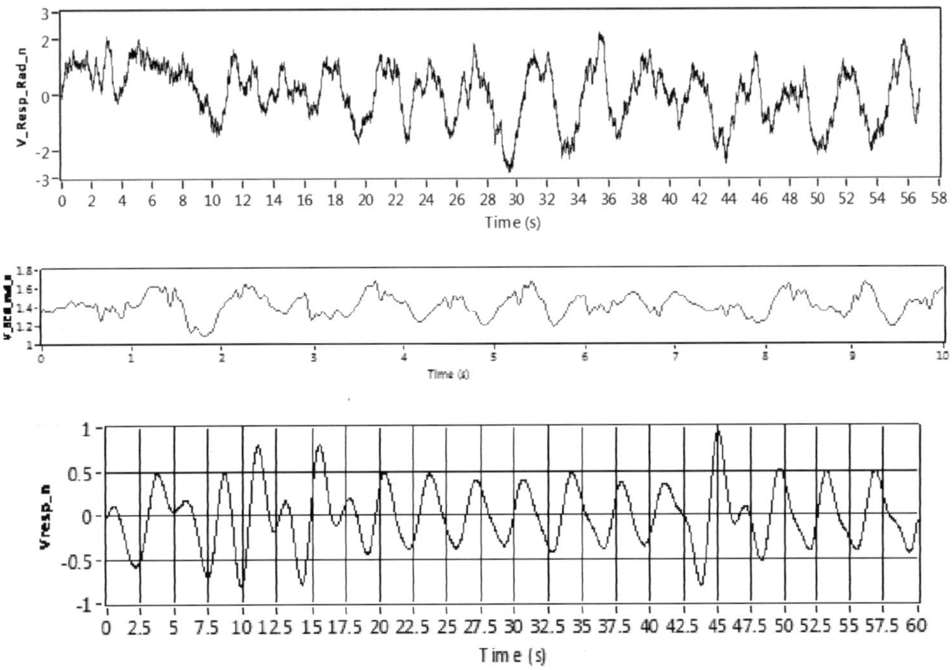

Figure 7. From top to bottom (arbitrary units): I-component of the Doppler radar IF signal taken from a stationary person; output of the HPF; output of the LPF.

The signals shown in Figure 8 were obtained with the same type of Doppler radar installed in a walker used by three different moving persons: one normal (normal gait), one suffering from myopathy (myopathic gait), and another from hemiparesis (hemiparetic gait). The signals are sufficiently different, but further tests and data processing are required to definitely claim the validity of the Doppler radar for people gait identification.

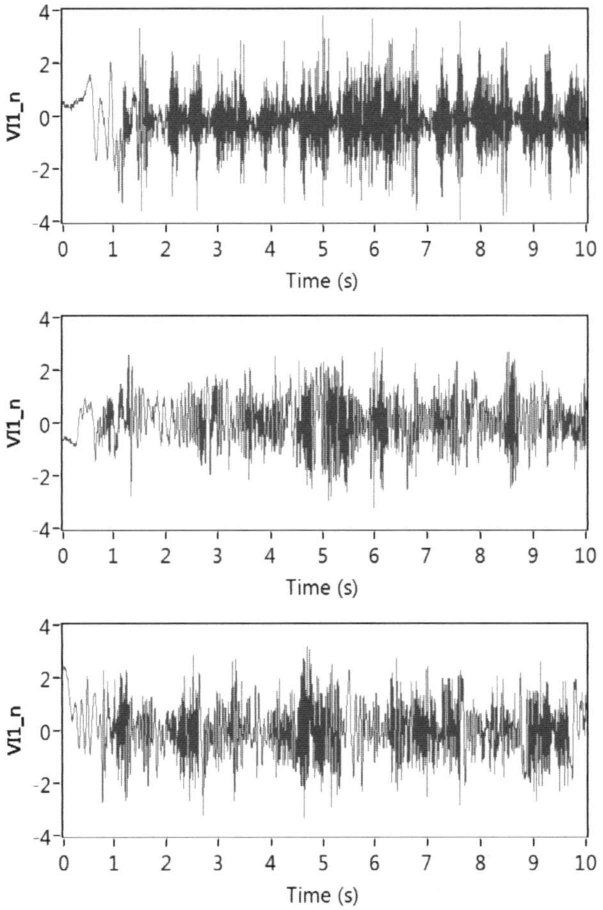

Figure 8. Gait sensed by a Doppler radar embedded in a walker (Vtune – 6 Vpp, f = 10kHz, triangular). From top to bottom (arbitrary units): regular gait; myopathic gait; hemiparetic gait.

5. Conclusion

Pervasive (ubiquitous) healthcare is a research field of increasing importance. Technological progress in areas such as materials, electronics, sensors, telecommunications, computer systems and signal processing have allowed the development of interesting and promising devices and systems appropriate to health care in non-hospital environment. Microwave Doppler radar is a fairly well establish technology, but its use for medical applications is still at a research stage, namely in what UWB radar is concerned. To the best of our knowledge, no device or system is currently in a market stage.

The heart and respiration rates are two vital signals fairly accessible to microwave radar usage. The 24GHz microwave CW (continuous wave) Doppler radar based monitor described in this paper is an example of what is possible nowadays. The results obtained validate the proposed solution, but only when measurements are made on stationary people. We look forward to demonstrating the efficiency of the monitor in environments where several people are in motion recurring to the possibility of frequency modulating the transmitted signal, which is possible with the transceiver used, and additional data processing. The authors will also go on researching the use of microwave Doppler radars in the monitoring of other physiological signals and persons activity and gait identification.

6. References

[1] Doppler C 1842 Über das farbige Licht der Doppelsterne und einiger anderer Gestirne des Himmels *Proc. Royal Bohemian Society of Sciences* Part V **2** 465-482 Prague

[2] Zhang R, Zhang Y, Dong ZC, Jiang S, Zhang C, Chen LG, Zhang L, Liao Y, Aizpurua J, Luo Y, Yang JL, Hou JG 2013 Chemical mapping of a single molecule by plasmon-enhanced Raman scattering *Nature* **498**, 82–86

[3] College of Fellows, American Institute for Medical and Biological Engineering 2013 Medical and Biological Engineering in the Next 20 Years: The Promise and the Challenges *IEEE Trans. on Biomedical Eng.* **60** 7

[4] Hoerbst A, Ammenwerth E 2010 Electronic Health Records. A Systematic Review on Quality Requirements *Methods Inf. Med* **4** 1-17

[5] Zysset C, Kinkeldei T, Cherenack K, Tröster G 2010 Woven Electronic Textiles: An Enabling Technology for Health-care Monitoring in Clothing *Proc. UbiHealth 2010* Copenhagen Denmark

[6] Pereira TP, Carvalho H, Catarino AC, Postolache O, Girão PM 2013 Wearable Biopotential Measurement Using the TI ADS1198 Analog Front-End and Textile Electrodes *Proc. IEEE Int. Symp. Medical Measurements and Applications – MeMeA* **1** 1-6 Gatineau Canada

[7] Catarino A, Carvalho H, Dias MJ, Pereira T, Postolache O, Girão PS 2012 Continuous Health Monitoring Using E-Textile Integrated Biosensors 2012 *Proc. Int. Conf. Exp. Electrical and Power Engineering (EPE 2012)* Iasi Romania

[8] Ribeiro MR, Postolache O, Girão PS 2014 A Novel Smart Sensing Platform for Vital Signs and Motor Activity Monitoring *Sensing Technology: Current Status and Future Trends I* Mason A, Mukhopadhyay SC, Jayasundera KP, Bhattacharyya N (Heidelberg: Springer)

[9] Postolache O, Girão PS and Postolache G 2013 Pervasive Sensing and M-Health: Vital Signs and Daily Activity Monitoring *Pervasive and Mobile Sensing and Computing for Healthcare - Technological and Social Issues* - Smart Sensors, Measurement and Instrumentation **2** Mukhopadhyay SC, Postolache OA 1–49 (Berlin Heidelberg: Springer-Verlag)

[10] Postolache O, Girão PS, Postolache G 2013 Comparative analysis of two systems for unobtrusive heart signal *Proc. Int. Conf. IEEE Engineering in Medicine and Biology Society – EMBC* **I** 1-4 Osaka Japan

[11] Postolache O, Girão PS, Ribeiro M, Guerra M, Pincho J, Santiago F, Pena A 2011 Enabling telecare assessment with pervasive sensing and Android OS smartphone *Proc. IEEE Int. Workshop Medical Measurements and Applications* 288-393 Bari Italy

[12] Postolache O, Girão PM, Pinheiro EC, Postolache G 2010 Unobtrusive and Non-invasive Sensing Solutions for on-line Physiological Parameters Monitoring *Wearable and Autonomous Biomedical Devices and Systems for Smart Environment* Lay-Ekuakille A, Mukhopadhyay SC (Berlin: Springer)

[13] Pinheiro EC, Postolache O, Girão PS 2010 A Survey on Unobtrusive Measurements of the Cardiovascular Function and their Practical Implementation in Wheelchairs *Sensors & Transducers Magazine* **9** 12 182-199

[14] Pinheiro EC, Postolache O, Girão PS 2013 Contactless Impedance Cardiography Using Embedded Sensors *Measurement Science Review* **9** 3 157-164

[15] Pinheiro EC, Postolache O, Girão PM 2012 A Practical Approach Concerning the Capacitive Acquisition of the Electrocardiogram in a Moving Wheelchair *Proc. IMEKO World Congress* **I** Busan, South Korea

[16] Pinheiro EC, Postolache O, Girão PS 2012 Study on ballistocardiogram acquisition in a moving wheelchair with embedded sensors *Metrology and Measurement Systems* **XIX** 4 739-750

[17] Postolache O, Girão PS, Postolache G 2011 Seismocardiogram and Ballistocardiogram Sensing *Int. J. Meas. Tech. Inst. Eng. (IJMTIE)* **1** 3 67-88

[18] Pinheiro EC, Postolache O, Girão PS 2010 Theory and Developments in an Unobtrusive Cardiovascular System Representation: Ballistocardiography *The Open Biomedical Engineering Journal* **4** 1 201-216
[19] Pinheiro EC, Postolache O, Girão PS 2009 Pulse Arrival Time and Ballistocardiogram Application to Blood Pressure Variability Estimation *Proc. IEEE International Workshop on Medical Measurements and Applications* **I** 132-136, Cetraro Italy
[20] Postolache O, Girão PS, Dias Pereira JM, Postolache G 2011 Smart Walker for Pervasive HealthCare *Proc. Int. Conf. Sensing Technology – ICST* **I** 1-5 Palmerston North, New Zealand
[21] Aldrich JE 2007 Basic physics of ultrasound imaging *Crit. Care Med.* **35** 5 S131-S137
[22] Lin JC 1975 Non-invasive microwave measurement of respiration *Proc. IEEE* **63** 1530
[23] Lin JC 1979 Microwave apexcardiography *IEEE Trans. MTT* **27** 618-620
[24] Øyvind Aardal and Jan Hammerstad 2010 Medical radar literature overview *Norwegian Defence Research Establishment (FFI) rapport 2010/00958*
[25] Xu Y, Lu YH, Zhang HX, Wang YQ 2007 An overview of ultra-wideband technique application for medial engineering *Proc. IEEE/ICME Int. Conf. Complex Med. Eng.* 408-411 Beijing PR China
[26] Staderini EM 2002 UWB radars in medicine *IEEE Aerosp. Elec. Syst. Mag.* **17** 1 13-18
[27] Pfanner F, Maier J, Allmendinger T, Flohr T and Kachelrieß M 2013 Monitoring internal organ motion with continuous wave radar in *CT Med. Phys.* **40** 9 091915
[28] Mostov K, Liptsen E and Boutchko R 2010 Medical applications of shortwave FM radar: Remote monitoring of cardiac and respiratory motion *Med. Phys.* **37** 1332-1338
[29] Mazlouman SJ, Tavakolin K, Mahanfar A and Kaminska B 2009 Contact-less Assessment of In-vivo Body Signals Using Microwave Doppler Radar *Biomedical Engineering* **13** 239-260 Mello CAB (InTech)
[30] Droitcour A, Lubecke VM, Lin J and Boric-Lubecke O 2007 A Microwave Radio for Doppler radar Sensing of Vital Signs *IEEE MTTS Int. Microwave Symp. Dig.* **1** 175-178 Phoenix AZ USA
[31] Guohua L, Jianqi W, Yu Y and Xijing J 2007 Study of the ballistocardiogram signal in life detection system based on radar *Proc. 29th IEEE EMBS Ann. Int. Conf.* 2191-2194 Lyon France
[32] Immoreev I and Ivashov S 2008 Remote Monitoring of Human Cardio-respiratory System Parameters by Radar and its Applications *Proc. Int. Conf. Ultrawideband and Ultrashort Imp. Sign.* 34-38 Sevastopol Ukraine
[33] Lazaro A, Girbau D and Villarino R 2010 Analysis of vital signs monitoring using an IR-UWB radar *Progress in Electromagnetics Research* **100** 265–284
[34] Bilich CG 2006 Bio-medical sensing using ultra wideband communications and radar technology: a feasibility study *Proc. Int. Perv. Health Conf. Work.* 1–9
[35] Postolache O, Girão PS, Dias Pereira JM, Postolache G 2011 FM-CW radar sensors for vital signs and motor activity monitoring *ICST Trans. on Ambient Systems* **11** 10-12 1-10
[36] Postolache O, Girão PS, Postolache G, Mendes J 2011 Cardio-Respiratory and Daily Activity Monitor Based on FMCW Doppler Radar Embedded in a Wheelchair *Proc. Int. Conf. IEEE Engineering in Medicine and Biology Society – EMBC* Boston USA
[37] Postolache O, Girão PS, Pinheiro EC, Madeira R, Dias Pereira JM, Mendes J, Postolache G, Moura C 2011 Multi-usage of Microwave Doppler Radar in Pervasive Healthcare Systems for Elderly *Proc. IEEE Int. Inst. Tech.Conf. - I2MTC* 30-34 Hangzhou PR China
[38] Postolache O, Girão PS, Madeira R, Postolache G 2010 Microwave FMCW Doppler radar implementation for in-house pervasive health care system *Proc. IEEE International Workshop on Medical Measurements and Applications* 47-52 Ottawa Canada
[39] Frequency ranges for global or regional harmonization of short-range devices 2011 *Recommendation ITU-R SM.1896*
[40] Limitation of radiation from industrial, scientific and medical (ISM) equipment Recommendation ITU-R SM.1056-1

Metrological Measurement of ^{99m}Tc Half-life Radionuclide by Germanium Detector

R L da Silva[1], J U Delgado[1], R Poledna[1], R N Alves[2], M C M de Almeida[2,3]

[1]Laboratório Nacional de Metrologia das Radiações Ionizantes-LNMRI/IRD/CNEN, Av. Salvador Allende, s/n – Recreio CEP 22780-160, Rio de Janeiro, Brazil;[2] Seção Engenharia Nuclear, IME, Pça Gen Tibúrcio 80, SE7 – Praia Vermelha, CEP 22290-270, Rio de Janeiro, Brazil; [3] Dinor, Comissão Nacional de Energia Nuclear(CNEN), Rua General Severiano 82, Botafogo, Rio de Janeiro, Brazil

E-mail: marcandida@yahoo.com.br, ronaldo@ird.gov.br

Abstract: An approach for half-life determination using gamma spectrometry with Ge detector is presented. This measurement reduces the contribution of the type B component on the total uncertainty. The independence of the method with respect to the instrumental interferences or radiochemical impurities was evidenced. The method shows no limitations for the measurements of samples containing a genetically related impurity of concentrations below than 3% with the same energy or very close to the one of the radionuclide to be measured, e.g.,^{99}Mo in a^{99m}Tc sample. The results of half-lives are in good agreement with the literature and the associated uncertainties lower than 0.1 %.

1. Introduction

In the nuclear medicine, the accurate determination of the half-life of radionuclides is important, therefore it contributes for damages reduction caused to the patient during the diagnosis or treatment, due to an unnecessary exposition, causing loss of time, some discomfort and damages to the health. The accurate knowledge of the half-life will advantage the application of futures radiopharmaceuticals, which will be able to replace or become an alternative for those used nowadays, as much in the area of radiodiagnostic as in the radiotherapy, aiming at better results in the exams and a smaller cost of production of the radiopharmaceuticals.

To determine the activity of a radionuclide, independently of the application or purpose, it is necessary to know its half-life and associated uncertainty.

Content from this work may be used under the terms of the Creative Commons Attribution 3.0 licence. Any further distribution of this work must maintain attribution to the author(s) and the title of the work, journal citation and DOI.
Published under licence by IOP Publishing Ltd

Journal of Physics: Conference Series **588** (2015) 012047 doi:10.1088/1742-6596/588/1/012047

In the recent years, there was an increasing interest in the accurate determination of half-life of gamma emission radionuclides through the spectrometric technique with germanium detector. A Ge detector has a high capability to discriminate energy lines, becoming less dependent of factors as background and instrumental fluctuations. This method has also the advantage to identify qualitatively and quantitatively radiochemical impurities of gamma radiation usually present in the samples to be analyzed.

2. Methodology

Radionuclide to be measured is ^{99m}Tc, it is characterized by gamma emission and has about 6 h of half-life. A procedure for preparation of the liquid sources was adopted. It was used one ampoule source with radioactive solution. One of the spectrum is shown in figure 1.

In half-life determination for ^{99m}Tc the most intense gamma energies was used: 140.5 keV (89.1 %).

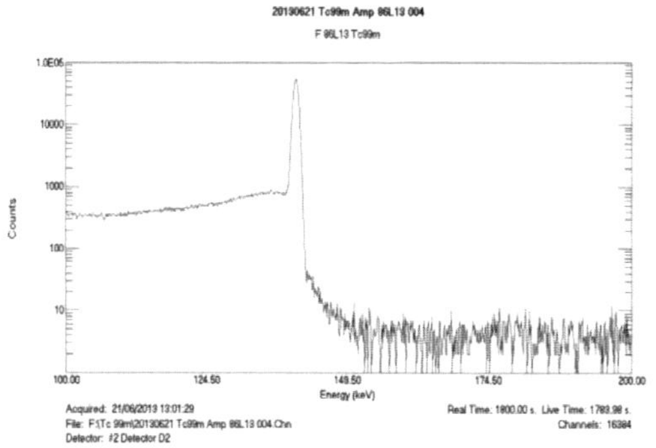

Figure 1: The ^{99m}Tc source spectrum

3. Results

For the set of spectra a graphic ln C versus real time of counting was obtained (C=counting rate). This real time was corrected for the start of counts (figures 2). After computing this data, the half-life value of the radionuclide was determined.

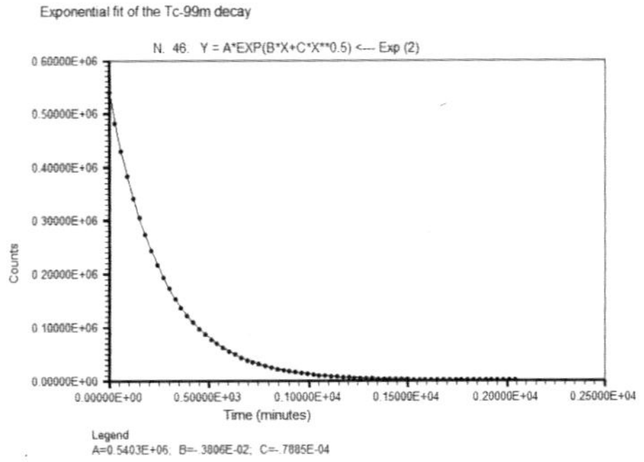

Figure 2: Exponencial fit of ^{99m}Tc decay

After 138 counting set fot the activity decay, it was obtained 6.0088 (5) h of half-life to ^{99m}Tc. The A and B component uncertainties are considered. However, the half-life expected from LNHB is 6.0067 (10) h. Here, it can be evidenced that the determined value of half-life is in good agreement with that published values in the literature.

4. Conclusions

The results obtained show that this method using the gamma spectrometry allows one to determine with accuracy radionuclides half-lives.

With the counting conditions established here for the related method made it possible a reduction in the total value of the uncertainty in the determination of the half-life, while the influence of the B-type component was practically neglected.

It was demonstrated also that when the impurity possess energy line equal or very close to that one radionuclide to be analyzed, the method depends on the previous knowledge of the impurity.

5. References

[1] Morel J., Etcheverry M. and Blanchis P., 1992 *Bulletin* BNM **88** 19.

[2] Woods M. J. and Lucas S. E. M., 1992 *Nucl. Instr. Meth.* **A 312** 346.

[3] Bowles N. E., Woods S. A., Woods D. H., Jerome S. M., Woods M. J., De Lavison P., Lineham S., Keightley J. and Poupaki I., 1998 *Appl. Radiat. Isot.*, **49** 9-11 1345.

[4] Lagoutine F., Coursol N. and Legrand J., 1984 *Table de Radionucléides*. CEA-BNMGif-Sur-Yvette, France.

[5] Nucleíde Gamma and Alpha Library,LNHB, http://laraweb.free.fr/ 2013

[6] Da Silva, C. J., Iwahara, A., Moreira, D. S., Delgado, J. U. and Gomes, R. S., 2012 Appl. Radiat. Isot. **70(9)** 1924.

Validation of VARK learning modalities questionnaire using Rasch analysis

E D Fitkov-Norris and A Yeghiazarian
Kingston University, Kingston-upon-Thames KT2 7LB, UK

E-mail: e.fitkov-norris@kingston.ac.uk

Abstract. This article discusses the application of Rasch analysis to assess the internal validity of a four sub-scale VARK (Visual, Auditory, Read/Write and Kinaesthetic) learning styles instrument. The results from the analysis show that the Rasch model fits the majority of the VARK questionnaire data and the sample data support the internal validity of the four sub-constructs at 1% level of significance for all but one item. While this suggests that the instrument could potentially be used as a predictor for a person's learning preference orientation, further analysis is necessary to confirm the invariability of the instrument across different user groups across factors such as gender, age, educational and cultural background.

1. Introduction

A number of factors such as learner motivation, study skills and ability to assess their own learning needs have been identified as good predictors of student performance [1]. External factors such as classroom climate and aligning teaching style with learners' learning preferences may bring additional benefits for learners [2,3], with students being happier when taught using their preferred learning mode, namely Visual, Auditory, Read/Write or Kinaesthetic [4].

VARK type questionnaires have been used extensively to test the learning mode preference of postgraduate and undergraduate students on a number of degrees such as nursing, psychology and business [5,6]. Self-reported questionnaires are often used by educationalists to evaluate the propensity of learners to use particular learning approaches and tools in order to assess good and bad learning practices and identify adequate learning support mechanisms [7]. These instruments often rely on parametric statistical analysis to build measurement scales and to make overall recommendations, although as the data is often collected using Likhert or binary (Yes, No) scales [8-10], this manipulation may not be suitable. Rash analysis can be used to transform the ordinal or binary response scales into linear interval scales that can be manipulated freely [11]. The analysis can also be used to identify items that do not fit the scale as well as for construct validation and development particularly when construct invariance across different groups of users (for example male and female or younger and older) is required [12]. Furthermore the analysis can be used to analyse the psychometric properties of a scale and identify overlapping response categories that can be merged [13].

Content from this work may be used under the terms of the Creative Commons Attribution 3.0 licence. Any further distribution of this work must maintain attribution to the author(s) and the title of the work, journal citation and DOI.
Published under licence by IOP Publishing Ltd

This paper will use Rasch analysis to examine the internal construct validity of a questionnaire designed by Neil Fleming [14] which assesses the preferences for students to receive and process information using Visual, Auditory, Read/Write and Kinaesthetic modes. It consists of 16 scenarios each asking the respondent to identify all of the information processing modes they would adopt in a particular scenario. The number of ticks given for each category are counted to give an overall score for each learner's preference for this mode of information processing. Based on these scores the overall preference of the learner is identified by using normative [15] comparison to the average scores obtained by subjects who have completed the questionnaire in the past and recommendations for the best mode for learning and information processing are given. Although the VARK questionnaire has been analysed in the past using factor analysis, taking into account the testlet nature of the questions led to an improvement in the factor model fit. However the overall fit to the four learning modalities was still not statistically satisfactory which cast some doubt on its validity [16]. Furthermore, the suitability of the questionnaire for measuring individual learning preferences by combining its scales and computing averages from the binary responses has never been examined. Rasch analysis will be used to carry out a preliminary examination to assess the internal validity of the four sub-scales of the questionnaire and check if response patterns fit the interval scale for each of the four learning preference trait. Response patterns fitting the Rasch model well would be suitable for parametric statistical analysis and increase the confidence in the recommendations made by the instrument.

Rasch analysis has been used extensively for analysing both questionnaire and construct validity and reliability in a number of aspects of healthcare [11,17] and social sciences [18-20]. In the context of education and learning the most prevalent use of Rasch analysis is in assessing the match of question difficulty to the ability of students in multiple choice tests [21]. Only relatively recently has Rasch analysis been used to validate learning inventories [12], motivation [22-24] and learning preference constructs [25], although its use has been limited.

Rasch analysis tests the fit of responses to a questionnaire to a formal scale model developed by Georg Rasch which gives the expected responses if an interval scale measurement is to be achieved [11]. The model is a probabilistic form of Gutman scaling [26] which assumes strict ordering of the questions, from ones easiest for participants to agree with up to ones that are hardest to agree with, thus differentiating between participants with low and high preference for the latent trait being measured. The model can be applied to measuring instruments with both binary and polythomous items [18] and allows questionnaire items to be summed in order to provide a total score as an overall measure for the latent trait under examination. The binary form of the models takes the form of equation (1);

$$P_i(U_j = 1|\theta) = \frac{e^{(\theta - b_j)}}{1 + e^{(\theta - b_j)}} \qquad (1)$$

which expresses the probability of person i whose propensity to agree with the statements of the questionnaire is parameterised by latent trait θ agreeing with item j $(U_j = 1)$. The parameter b_j refers to item difficulty/ease of agreement.

2. Methodology

The VARK questionnaire was administered to 107 postgraduate students studying for a master's degree in business management or closely related subjects e.g. marketing and human resource management. The majority of students who completed the survey were female (68.2%). The average age of the participants was 26.8 years with a standard deviation 5.24 years. 81% of the students in the sample were international and came mainly from Europe, Asia and Africa. The frequency with which the students chose each of the four modes is shown in table 1. The most prevalent learning preferences for this sample are Auditory and Kinaesthetic (t=1.32, df=100, two tailed p=0.187). Read/Write preference received significantly fewer ticks (t=-2.36, df=101, two tailed p=0.004) with Visual being the least popular (t=-6.22, df=97, two tailed p=0.000). The learning style preferences of business students classified by the VARK assessment tool shows that a large proportion of students (38.1%) did not have a particular preference and use all four modes. Over half of the students (58.1%) use a unimodal approach to learning, and a small minority used bimodal. The sample contained no students who used three modes. Of the 61 students who preferred a unimodal approach, the majority (46%) used Auditory, while the remaining three modes, Kinaesthetic, Visual and Read/Write, were spread equally among students (with approximately 18% expressing preference for each mode).

The analysis was carried out using IBM SPSS Statistics version 21.0 using the extended Rasch Model add-on eRm R package (http://www.01.ibm.com/support/docview.wss?uid=swg21488442). The testlet nature of the questionnaire suggests that to assess the questionnaire's overall structure validity a multilevel Rasch model needs to be considered [27]. However, multilevel models require a minimum of 200 observations to ensure that the parameter estimates are reliable [28]. Given the limitation of the small sample size, this paper reports on preliminary analysis that only considers the model fit of the four separate sub-scales of the VARK questionnaire.

Table 1. Number of ticks for VARK preference modes

	Mean	**Standard deviation**
Visual	4.44	2.51
Auditory	6.56	3.03
Read/Write	5.42	2.79
Kinaesthetic	6.17	2.72

3. Results

The results from the Rasch analysis on each of the four sub-scales of the VARK questionnaire are summarised below.

3.1 Visual

Overall the responses to Visual preference questions fit the Rasch model well, although the fit of VQ11 is affected by some outliers indicated by relatively high Outfit MSQ value and a statistical significance of p=0.003. The infit and outfit mean square values for all Visual items are in the range of 0.73-1.43, which is within the acceptable region of 0.5 - 1.5. The person fit responses are acceptable at 5% level of significance on all reported statistics. The person/item map (Figure 1(a)) shows that the items are mainly clustered to the right of the scale, suggesting that only people with stronger Visual modal preference are likely to agree with the items.

3.2 Auditory

Overall the responses to Auditory type items fit the Rasch model well, although the fit of AQ05 is affected by outliers indicated by a relatively high Outfit MSQ value and a statistical significance of p=0.013. The Auditory item infit and outfit mean square values are in the range of 0.64 - 1.32 which is acceptable. The person fit statistics are only acceptable at 1% level of significance for the Casewise Deviance, although this statistic is found to frequently make type 1 error [29]. The person/item map (Figure 1(b)) shows that the items are mainly clustered to the centre and the right of the scale, suggesting that only persons with moderate and stronger Auditory modal preference are likely to agree with the items.

3.3 Read/Write

Overall the responses to Read/Write items fit the Rasch model well, although the fit of RQ01 and RQ08 are affected by some outliers as indicated by a relatively high Outfit MSQ value and a statistical significance of p=0.036 and p=0.045 respectively. The Read/Write item infit and outfit mean square values are in the range of 0.74 – 1.25 and are acceptable. The person fit responses are only acceptable at the 1% level of significance for the Hosmer-Lemershow statistic, although the reliability of this test is affected by small sample sizes [29]. The person/item map (Figure 1 (c)) shows that the items are mainly clustered to the right of the scale, suggesting that only persons with stronger Read/Write modal preference are likely to agree with the items.

3.4 Kinaesthetic

Overall the responses to Kinaesthetic items fit the Rasch model well, although question KQ08 has higher MSQ and fails the significance testing at 5% with p=0.31. Overall the Read/Write item infit and outfit mean square values range between 0.78 – 1.26 and are acceptable. Although the person fit responses are rejected at 1% level of significance when using the Casewise Deviance, this statistic frequently makes type 1 error [29]. The person/item map show that the items are mainly clustered to the right of the scale, suggesting that only persons with stronger Kinaesthetic modal preference are likely to agree with the items.

4. Discussion

The results from the analysis confirm that overall the Rasch model fits the VARK questionnaire responses, and the data supports the internal validity of the four sub-constructs. The majority of items in the four sub-constructs, Visual, Auditory, Read/Write and Kinaesthetic preferences for learning match the data at 5% level of significance, and could be used as predictors for a person's learning preference orientation. The analysis confirmed that the instrument responses match the Rasch model and are, therefore, suitable for parametric statistical analysis. In addition, the binary nature of the responses would not have a negative impact on the reliability of any recommendations made to learners regarding their learning style preference. However, one Visual question had significance level less than 1% and four other questions had significance levels of less than 5% (see table 2).

Table 2. Scenario responses that do not fit the Rasch model

Scenario:	Problem response:	Preference:
Other than price, what would most influence your decision to buy a new non-fiction book?	The way it looks is appealing.	Visual
A group of tourists wants to learn about the parks...	Talk about, or arrange a talk for them about parks or wildlife reserves.	Auditory
You are helping someone who wants to go to your airport, town centre or railway station…	Write down the directions.	Read/Write
You have a problem with your heart. You would prefer that the doctor…	Gave you something to read to explain what was wrong.	Read/Write
	Used a plastic model to show what was wrong	Kinaesthetic

As mentioned, the questionnaire consists of 16 testlet scenarios each with four options, one for each type of learning preference. Previous research has indicated that the scenarios may introduce a bias in the answers within each testlet [16, 27]. Noting the specific scenarios that contain responses that do not fit the Rasch model, one could see that the Auditory and Read/Write options require the respondent to go to either considerably more effort to organise a talk or be in possession of a pen. Similarly, the Visual questions may clash with users' preference for value for money which is another option within the scenario, or external factors such as preference in younger users to read online rather than purchase books. Two of the answers in scenario 8 do not match the expected item fit and this may be due to people choosing answers based on expectations from the scenario (visiting a doctor) rather than reflecting their true learning preference. The testlet nature of the model may also explain the small degree of overfit existing across some items as the responses to questions may be influenced by the particular scenario overarching the question.

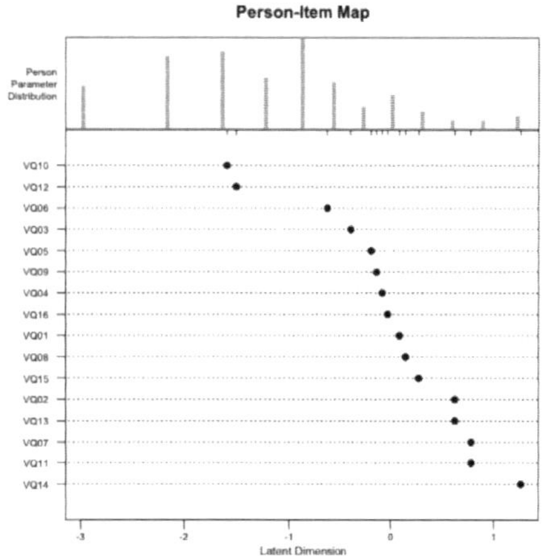

(a) Visual Person-Item

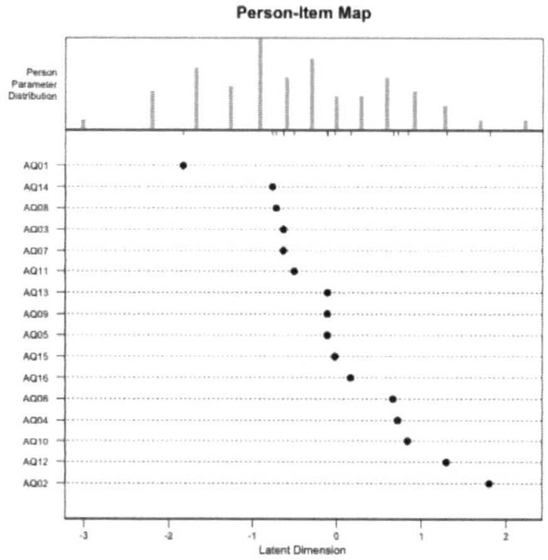

(b) Auditory Person-Item

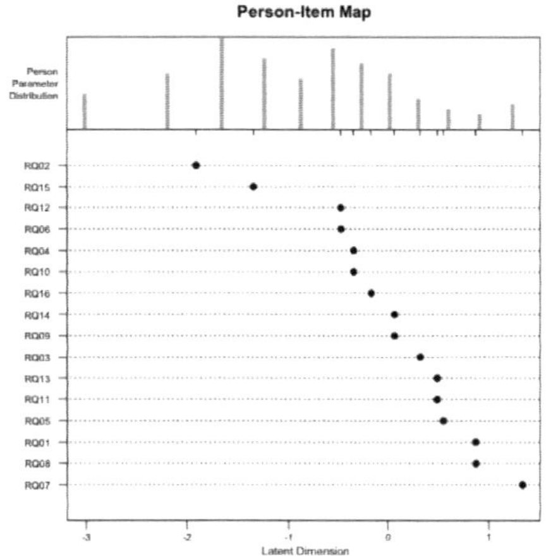

(c) Read/Write Person-Item

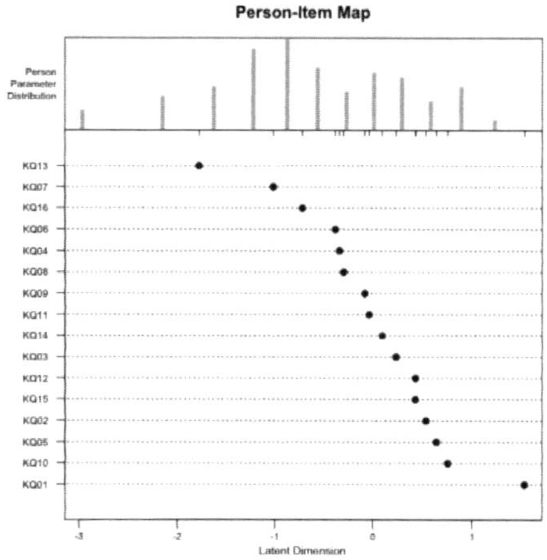

(d) Kinaesthetic Person-Item

Figure 1. Person-Item map results

Some person fit statistics also did not satisfy the 5% peel of statistical significance although this applied only for the Hosmer-Lemeshow and Casewise Deviance measures which are very sensitive to small sample size or much more likely to reject the null hypothesis, when it is in fact true [29]. The most robust of all item fit statistics, Collapsed Deviance, confirms item fit for all four sub-scales. The person-item figures confirm that in most cases people with moderate strength in a particular learning preference are likely to agree with the statements. Very few questions are likely to be agreed upon by people with low preference and, similarly, there are no questions that are likely to be agreed upon only by people with very strong learning preference. This is not entirely surprising for people with low learning preference as the questionnaire is designed to identify a positive preference for a learning approach, rather than dislike. However, the limited number of questions that identify strong preference may be limiting the discriminatory power of the instrument. Due to the very small sample size it is not possible to guarantee its representativeness across all learning preference categories or to check the invariability of the instrument across different of groups of users across factors such as gender, age and background.

5. Conclusions and Recommendations

This pilot study assessed the internal validity of the four sub-scales of the VARK learning preference questionnaire and showed that the response patterns fit the Rasch model. The findings lend support to the questionnaire's suitability and reliability as an instrument for measuring learners' preferences for receiving and processing information in Visual, Auditory, Read/Write and Kinaesthetic ways. However, the limited sample size precluded the examination of the questionnaire's multi-level structure and thus further analysis using a significantly larger sample is required, before any recommendations are made for amending/removing items. In addition, a larger sample size will allow research into the invariability of the instrument across different groups of users grouped by gender, age and background, and provide basis for drawing conclusions about the overall suitability of the VARK measurement instrument.

References

[1] Credé M and Kuncel N R 2008 *Perspect Psychol Sci* **3** 425–53
[2] Popescu E 2009 *Advances in Web Based Learning–ICWL 2009* 343–52
[3] Hsieh C T, Mache M and Knudson D 2012 *Sport Biomech* **11** 108–19
[4] Haq S M, Yasmeen S, Ali S and Gallam F I 2012 *JRMC* **16** 191–3
[5] Dobson J L 2010 *Adv Physiol Educ* **34** 197–204
[6] Fitkov-Norris E D and Yeghiazarian A 2013 *12th ECRM for Business and Management Studies* (Guimaraes, Portugal) pp 144–52
[7] Coffield F, Moseley D, Hall E and Ecclestone K 2004 *Learning Styles and Pedagogy in Post-16 Learning* (Learning and Skills Research Centre)
[8] Brown W F and Holtzman W H 1955 *J Educ Psychol* **46** 75
[9] Entwistle N J, Thompson J and Wilson J D 1974 *High Educ* **3** 379–96
[10] Nonis S A and Hudson G I 2010 *J Educ Bus* **85** 229–38
[11] Tennant A and Conaghan P G 2007 *Arthritis Rheum* **57** 1358–62
[12] Waugh R F and Addison P A 1998 *Brit J Educ Psychol* **68** 95–112
[13] Pallant J F and Tennant A 2010 *Brit J Clin Psychol* **46** 1–18
[14] Fleming N D 1995 Proceedings of the 1995 *Annual HERSA Conference* **18** pp 308–13
[15] Boverman D M 1962 *Psychol Rev* 69 295–305
[16] Leite W L, Svinicki M and Shi Y 2010 *Educ Psychol Meas* **70** 323–39
[17] Belvedere S L and de Morton N A 2010 *J Clin Epidemiol* **63** 1287–97
[18] Kahler C, Strong D, Read J, Palfai T and Wood M 2004 *Psychol Addict Behav* **18** 322–33
[19] Hagquist C and Andrich D 2004 *Pers Indiv Differ* **36** 955–68
[20] Raudenbush S W, Johnson C and Sampson R J 2003 *Sociol Methodol* **33** 169–211
[21] Divgi D R 1986 *J Educ Meas* **23** 283–98
[22] Waugh R F 2002 *Brit J Educ Psychol* **72** 65–86
[23] Waugh R F 2003 *J Appl Meas* **4** 164–80
[24] Waugh R F 2002 *Brit J Educ Psychol* **72** 573–604
[25] Di Nisio R 2010 *Procedia Soc Behav Sci* **9** 373–7
[26] Andrich D 1985 *Sociol Methodol* **15** 33–80
[27] Wang W-C, Chng Y-Y and Wilson M 2005 *Educ Psychol Meas* **65** 5–27
[28] Brune K D 2011 *An Evaluation of Item Difficulty and Person Ability Estimation using the Multilevel Measurement Model with Short Tests and Small Sample Sizes,* PhD Thesis.
[29] Reise S P, Bentler P M and Mair P 2008 IRT Goodness-of-Fit Using Approaches from Logistic Regression *available at: http://escholarship.org/uc/item/1m46j62q*

Characterization of Layer–Delaminations by Ultrasonic Speckles

Silvester Sadjina, Patrick Hölzl and Bernhard G Zagar
Institute for Measurement Technology, Johannes Kepler University Linz, Austria
E-mail: Bernhard.Zagar@jku.at

Abstract. Combined electrical mechanical and thermal loads influence highly structured specimen over time and might cause a variety of failure modes. Some of these failures can be such as to only gradually diminish the nominal behaviour of the components so one cannot conclude there is no evolving problem if nominal function is observed. Standard non–destructive testing methods are only partly able to detect the developing delamination or progressive generation of voids due to stress. These delaminations can cause voids and cracks in the still operative specimen that might never be of any problem in the nominal operating performance but on the other hand they might further the degradation of the component to ultimately result in a catastrophic failure. We investigate the gradual emergence of so–called ultrasonic (US) speckles as a hint to degradation processes deep in the volume of the specimen caused by scatterers (small voids or small cracks) that are too small to be directly imaged by ultrasound techniques. However, due to their density evolving over time they expose themselvs by forming US–speckles that increase in contrast over time. In this paper we are applying the well–known theory on optical laser speckles caused by micro roughness on a specimens surface to the formation of ultrasound speckles caused by randomly distributed sub–wavelength scatterers within the volume of a specimen. Both effects are due to random phase variations in the former case operating on laser light caused by height variations of the scattering profile and in the latter case by sound velocity variations along the propagation path within a specimens volume. We can show that the speckle contrast is a good measure of the average total volume of scattering voids and thus of the onset and evolution of delaminations as defined by Goodman of partly developed speckles.

keywords: ultrasound speckles, non–destructive testing, layer delamination, randomly distributed scatterers

1. Introduction

There are a variety of non–destructive testing schemes allowing to tomographically scan a specimen for voids or cracks deep in the volume — going from fine to coarse resolution — like X–ray diffraction [1], acoustic near–field techniques (SNAM) [2], micro computed tomography scans, scanning acoustic microscopy [3, 4] and thermography techniques [1]. All mentioned techniques are essentially aiming for the detection of each and every (developing) void and thus require an appropriate spatial resolution and significant scanning and processing time. In our contribution we are not aiming for the detection of single failures but are modeling the degradation as randomly scattered acoustic phase objects representing a volume density of voids that is tomographically analyzed using an acoustic microscope. In the acoustic scanning microscopy technique [5] these sub–resolution random scatterers are causing ultrasound speckles

Content from this work may be used under the terms of the Creative Commons Attribution 3.0 licence. Any further distribution of this work must maintain attribution to the author(s) and the title of the work, journal citation and DOI.
Published under licence by IOP Publishing Ltd

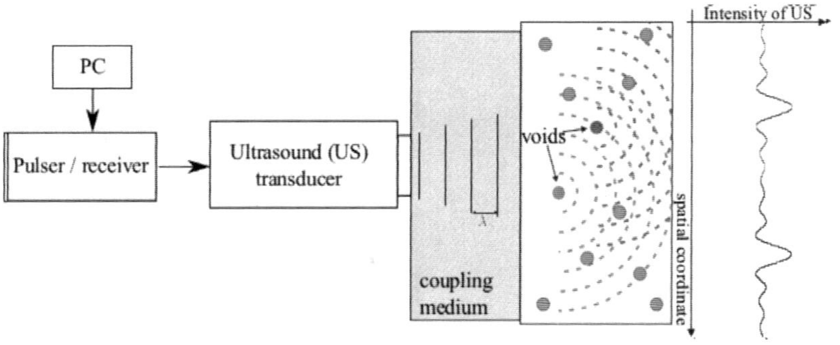

Figure 1. Experimental setup. Voids in the specimen cause diffraction effects. The superposition of resulting waves cause speckles in the acoustic microscope image of Fig. 3.

[6]. These speckles are developing over stress cycles with respect to the initial state of the homogeneous specimen beginning with zero phase deviations over partly developed speckles for specimen in the void initiation phase to fully developed speckles [7] for specimen experiencing sufficiently many randomly scattered voids [8, 9].

Coherent imaging systems be it synthetic aperture radar imaging, laser scanner based imaging systems [10, 11, 12] or be it ultrasonic imaging experience the presence of speckle noise overlaid onto the range image, the laser scan image or the echogenicity map in ultrasound images. Speckles themselvs are a random but still deterministic interference pattern in an image that most of the time has a negative impact on the measurement problem at hand. Speckles, however, convey sufficient information for some other measurement problems since e.g. in laser applications they act like fingerprints of an analyzed surface element whose motion can thus be tracked with sufficient resolution [13].

In ultrasonic imaging speckles are typically reducing the ability to detect fine structures in a specimen by masking its echogenicity. They are formed with coherent radiation of a specimen presumably containing many sub–resolution scatterers [14].

In some fields the development over time of the density and average size of voids can be an indicator to the overall thermal or mechanical stress the specimen was to bear. Most of the time the specimens are keeping their nominal operating performance even if they exhibit a moderate density of cracks and voids with sizes below the detection limit of acoustic microscopes. The idea conveyed in this paper is to utilize the temporal development of speckle noise overlaid onto the B–mode scan of a specimens structure to estimate the average density and possibly size of sub–resolution scatterers (voids or cracks). It is organized as follows: in Section 2 the experimental set–up is described. Section 3 details the theory of speckle noise in US images and derives the speckle contrast as a measure of void density. Section 4 presents first results of an acoustic microscope adapted to the detection of voids and cracks. The outlook concludes the paper.

2. Experimental set–up

Our experimental set–up consists of a Panametrics NDT Mod. 5900 PR pulser receiver, driving a Panametrics 20 MHz transducer V390 with a 0.25 inch element size and a 0.5 inch focal length. An Agilent Infiniium DSO 9254A oscilloscope is used to digitize the US–return with a preset sample rate of 1 Gsample/s (depth resolution approximately 1.5μm for the typical material of the specimen). The mechanical scanner is a Linos x–act LT 100–1 three–dimensional translation stage (resolution 1μm). All the processing is done in Matlab as is the visualization. Volume data is rendered using ParaView open source software. Figure 1 shows our set–up schematically.

Figure 2 shows the temporal impulse response of the transducer with a relative bandwidth of

70% of center frequency in the left panel and the spatial impulse response — the so–called point spread function — schematically in the right panel.

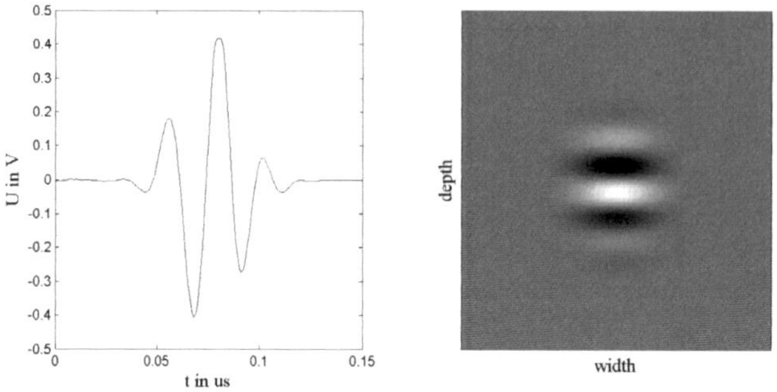

Figure 2. Left diagram: temporal impulse response of a Panametrics 20 MHz US transducer. Right panel: spatial impulse response (point spread function) of the same transducer in the focal plane.

3. Theory of US–speckles

This section is modelling the statistics of US–speckles based upon some simplifying assumptions. (A) that the carrier frequency ν (in our case 20 MHz) is sufficiently monochromatic to exhibit a coherence time that allows the constructive or destructive interference of contributions from scatterers located in the resolution cell of the set–up. For our case this can be interpreted in the temporal domain to have a pulse length of approximately $0.1\mu s$ as can be seen from Fig. 2 left panel equating to approximately $120\mu m$ in depth dimension. (B) in the spatial domain the transducer focuses to approximately $10\mu m$ (FWHM) as is schematically depicted in Fig. 2 right panel. (C) Thus the resolution cell is cigar–shaped with said dimensions.
The sound pressure is modeled as a complex phasor

$$\underbrace{u(x,y,z,t)}_{\text{temporal \& spatial}} = \underbrace{A(x,y,z)}_{\text{spatial}} \cdot \underbrace{e^{j2\pi\nu t}}_{\text{temp.}}$$

with u representing spatial and temporal variations, A, the complex phasor magnitude, and ν the center frequency. The complex phasor A can be decomposed into its magnitude $\mid A \mid$ and phase term $e^{j\Theta(x,y,z)}$.

The complex phasor A results from the summation over all (index n) contributions of single scatterers in the resolution cell with the following assumptions: (A) amplitude a_n and phase Θ_n of the nth scatterer are statistically independent of each other and of all other scatterers in the resolution cell. (B) the phases of the scatterers are uniformly distributed in $[0, 2\pi)$, which means that their spatial distribution is such as to allow many wavelength λ in between them either in depth or lateral direction. Now further applying the central limit theorem allows to express A as the coherent sum of scaled contributions a_n, where the scale factor $\sqrt{N}$ takes into account the decreasing magnitude of single contributions if their number N increases.

$$A(x,y,z) = \frac{1}{\sqrt{N}} \sum_{n=1}^{N} |a_n| \cdot e^{\Theta_n}$$

Analyzing the probability density function for A one can show [11], that a circula Gaussian PDF results.

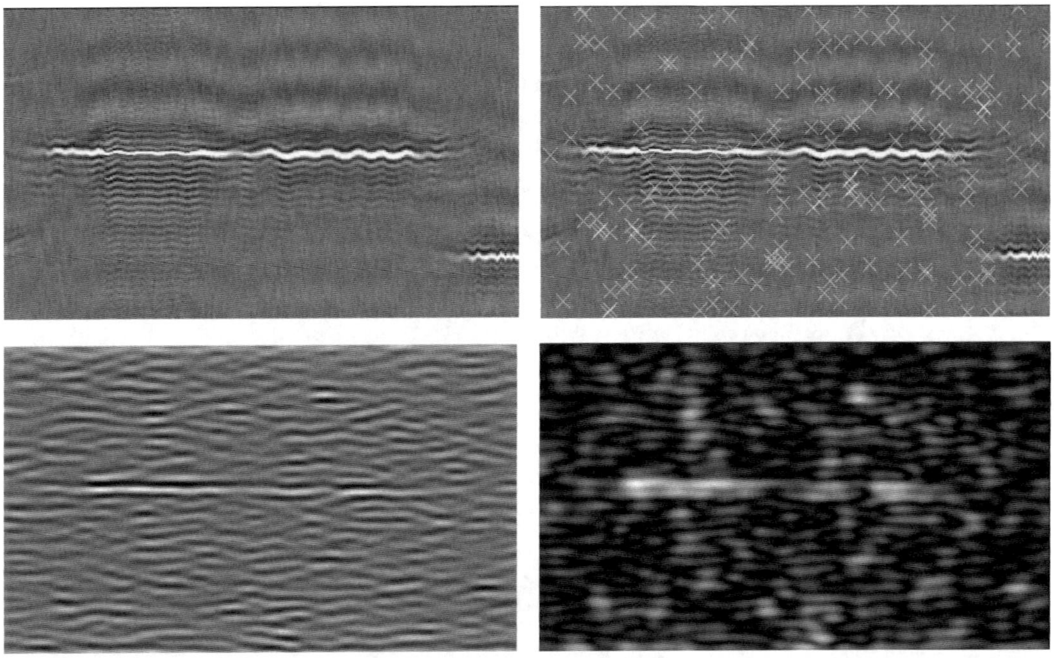

Figure 3. Ultrasound image of the cross–section of a specimen with some pronounced structure top left (remark: the waviness of the imaged structure very likely is caused by the non–ideal mechanical guide of the 3D translation stage.); Top right: the same global structure with corresponding scattering function. This scattering function is derived from a randomly distributed population of sub–resolution scatterers (yellow ×'s). The scattering function is spatially convolved with the point spread function of the ultrasonic transducer (bottom left); Envelope demodulation of the RF–carrier (here a 20 MHz center frequency, 70% relative bandwidth transducer) exhibiting strong speckles bottom right.

Going a step further and assuming that in addition to the random weakly scattering scatterers there is a deterministic structure present that would add a rather strong coherent contribution to A, it can be shown (again [8, 11]) that a Riccian PDF will result.

$$p_A(a) = \frac{a}{\sigma^2} \cdot e^{\left(-\frac{a^2+s^2}{2\sigma^2}\right)} \cdot I_0\left(\frac{as}{\sigma^2}\right) \; ; \quad a \geq 0 \tag{1}$$

Here I_0 is the Bessel function of the first kind with order zero, σ is the standard deviation of the circular Gaussian PDF alone, and s is the strong coherent scattering of the deterministic structure. In analyzing Eqn. 1 one can conclude that the image contrast as defined by the variation of intensity in a flat region of the image in relation to the mean brightness (caused by the deterministic structure) is a statistical parameter measuring the impact of density and average size of the sub–resolution scatterers sought.

4. Measurement results

In order to demonstrate the applicability of the method one would have to resort to the acquisition of a rather lengthy experimental procedure where the specimen needs to be stressed for many thousands of loading cycles without being spatially displaced within an acoustic microscope. The necessary experimental equipment is currently under development and is thus not available yet to acquire real data. For that reason we did in fact acquire true US volume data which was appropriately demodulated using the Hilbert transform method to show in 3D the internal structure of the specimen. Subsequently randomly distributed scatterers of various echogenicity were added and the volume data again convolved with the point spread function shown in Fig. 2 to exhibit US speckles. This process is described in more detail next. Speckle are present both in the RF data and the envelope – detect data as is shown in Fig. 3 for example. The top left panel shows the echogenicity map without random sub–resolution scatterers. The top right panel shows the scattering function by adding artificially some hundred sub–resolution scatterers. The bottom left panel shows the simulated ultrasound RF data obtained from scanning spatially the echogenicity. The simulation is done by convolving in lateral dimension and in the time (depth) dimension with the assumed point spread function of the transducer as depicted in Fig. 2.

The bottom right image is formed by demodulating the RF data to yield the complex envelope whose magnitude is then imaged in grayscale resulting in this very obvious noisy (speckled) image of the original echogenicity (top left).

These speckled data set was then statistically analyzed giving the result shown in Fig. 4. One can clearly observe the increase in the variance σ_I^2 of the image intensity within the presumed flat area boxed, which is a measure of speckle contrast.

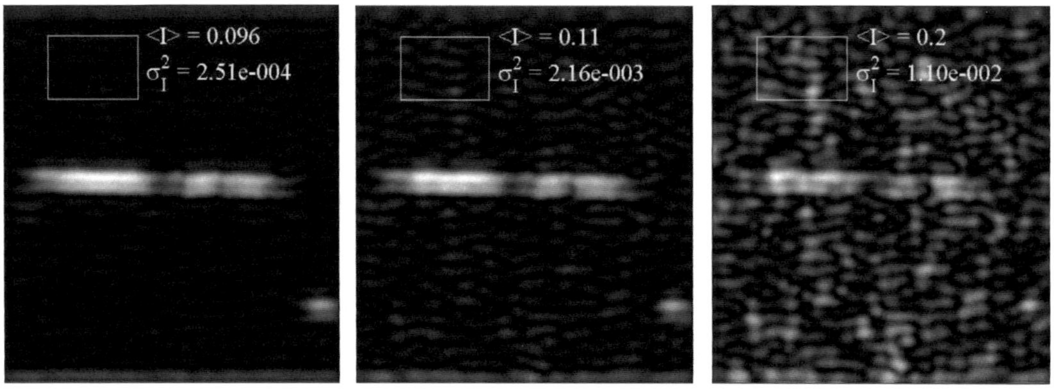

Figure 4. Three B–mode scans exhibiting US speckles in increasing magnitude. All frames were generated adding 5000 sub–resolution scatterers (size $\approx \lambda/50$) each with relative scattering strength of -60 dB, -54 dB and -40 dB relative to the deterministic component (the bright horizontal bar) going from left to right. One can clearly observe the increase in the variance of the intensity σ_I^2.

In Fig. 5 the complete data set is 3D rendered to show the internal structure of the specimen.

5. Conclusions

We have demonstrated that US speckles typically considered a nuisance to be avoided in ultrasonic imaging can be exploited nicely to indicate the emergence of sub–resolution scatterers which might be caused by micro–cracks or small sized delaminations still small enough not to impair the nominal functionality of the specimen but possibly being indicative of an upcoming failure.

Acknowledgments

The authors gratefully acknowledge the partial financial support of this work by the Austrian Research Promotion Agency under grant 838515 (InMet).

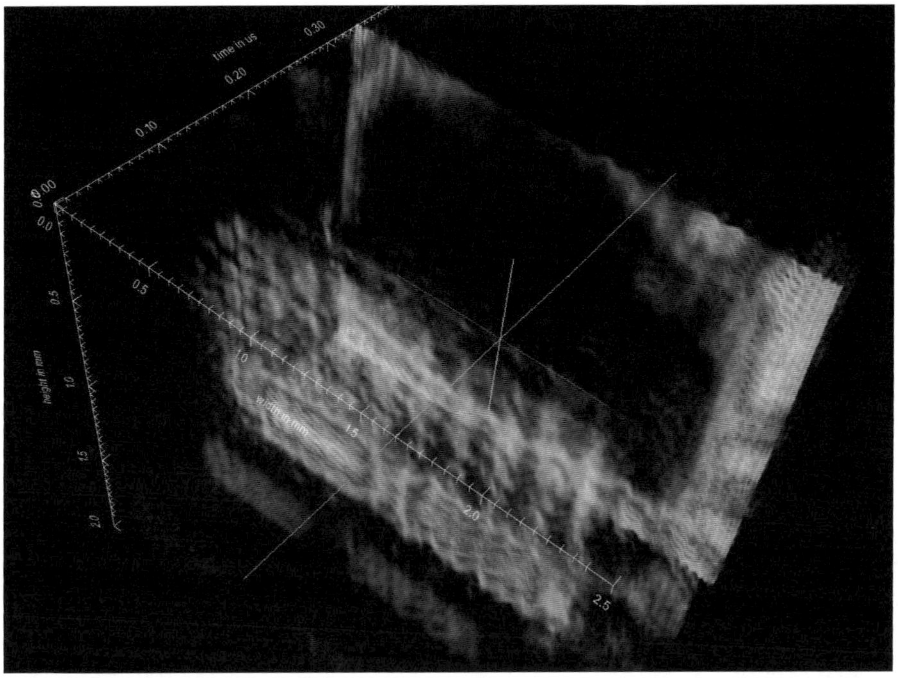

Figure 5. Ultrasound volume data rendered using the ParaView software [15].

References

[1] M. Oppermann et al., "Nano Evaluation in Electronics Packaging", 2nd Electronics Systemintegration Technology Conference, Greenwich UK, 2008

[2] Francesco Marinello (ed.), Daniele Passeri (ed.), Enrico Savio (ed.), Acoustic Scanning Probe Microscopy, Springer, 2013

[3] A.G. Every, M. Pluta, W. Grill, Interference of non–aligned Airy diffraction patterns in transmission acoustic microscopy images of crystals, Proc. R. Soc. A (2005), **461**, pp. 3847 – 3862

[4] Roman Gr. Maev, Acoustic Microscopy — Fundamentals and Applications, Wiley – VCH, 2008

[5] Rupitsch S., Zagar B., Acoustic Microscopy Technique to Precisely Locate Layer Delamination, in: IEEE Trans.on Instr. & Meas., Vol. 56, No. 4, pp. 1429–1434, 2007.

[6] Kothbauer R., Rupitsch S.J., Zagar B.G., Characterization of micro–materials using laser speckles, Proc. of the SPIE — The International Society for Optical Engineering, vol. 6371, pp. 63710Q- 1–9, 2006

[7] Michael Halliwell ed., Acoustical Imaging, Springer, 2013

[8] Gregg E. Trahey, Stephen W. Smith, Properties of acoustical speckle in the presence of phase aberration part I: first order statistics. Ultrasonic Imaging, 10:12-28, 1988.

[9] Stephen W. Smith, Gregg E. Trahey, Sylvia M. Hubbard, Robert F. Wagner, Properties of acoustical speckle in the presence of phase

[10] Lettner J., Zagar B., Two-wavelengths laser-speckle technique for thickness determination of transparent layers on rough surfaces, in: Measurement Science and Technology, Volume 24, Number 11, Page(s) 115204, 2013.

[11] Joseph W. Goodman, Speckle Phenomena in Optics: Theory and Applications, Roberts & Co. Publ., 2010

[12] J.C. Dainty ed., Laser Speckle and Related Phenomena (Topics in Applied Physics), Springer, 1984

[13] S.C. Schneider, S.J. Rupitsch, B.G. Zagar, Signal Processing for Laser–Speckle Strain–Measurement Techniques, IEEE Trans. on Instr. & Meas.; vol. 56, no. 6, pp. 2681–2687, Dec. 2007

[14] Saralees Nadarajah, Statistical distributions of potential interest in ultrasound speckle analysis, Phys. Med. Biol. 52 (2007) N213–N227 aberration part II: correlation lengths.Ultrasonic Imaging, 10:29-51, 1988.

[15] Kitware open source software ParaView, Kitware Inc., Clifton Park, NY 12065 USA

Irreversibility of a measurement procedure

S V Muravyov
Department of Computer-aided Measurement Systems and Metrology
National Research Tomsk Polytechnic University
Tomsk, 634050, Russia

E-mail: muravyov@camsam.tpu.ru

Abstract – It is shown in the paper that any measurement procedure possesses inherent generic property of irreversibility. The property consists in impossibility of a process where a transition is possible from a finite state of a system to its initial state through the same states but in reverse order. The irreversibility is accompanied by property of sequentiality that means multistage character of a measurement. Awareness of these properties can provide some useful ways and tools for development of new measurement methods.

1. Introduction

It is well known that any measurement process is based on operation of a *comparison*. This is not a single elementary operation defining the measurement, however its explicit role is generally recognized, see, for example definition of *measurement* in the 3-rd edition of International Vocabulary of Metrology (VIM):

"process of experimentally obtaining one or more quantity values that can reasonably be attributed to a quantity",

which is provided with the note:

"measurement implies *comparison* of quantities and includes counting of entities".

When to try to go deeply into the sense of term 'comparison', one may arrive to outcomes appearing to be useful for understanding in more detail a nature of measurement.

Earlier in this research direction, a discrete-mathematical model of measurement procedure had been proposed and investigated by the paper author [1,2]. The formal structure, called *PSIQ-model*, consists of four elementary operations as follows, see Fig. 1: **p**artitioning of the initial (before measurement) uncertainty interval S (to that the sought quantity value x belongs) into subintervals; **s**electing one of the subintervals and an element $y_i \in S$ in this subinterval as its representative; **i**nverse mapping the image y_i of the value x to the *empirical set*, i.e. $x_i = f^{-1}(y_i)$; and **q**uestioning whether the result of the inverse mapping x_i is similar to the element x, fixed in the empirical system, using appropriate measure of similarity. The PSIQ-model allows to describe in concise and formal way different measurement processes implemented in various scales. In this paper we restrict a consideration by only ratio scale measurements.

This paper will extend the treatment of a measurement procedure in form of the PSIQ-model involving the notation of irreversibility.

2. Irreversibility

Reversible is a process where a transition is possible from a finite state of a system to its initial state through the same states but in reverse order. If this transition is not possible, the corresponding process is *irreversible* [3]. Ice melting and metal fusing can be examples of reversible processes in some extent. Burning, diffusion, thermal conduction and viscous flow are typical examples of irreversible processes.

The concepts of reversibility and irreversibility arise most frequently in thermodynamics, where a reversible process can be characterized as follows: if in forward stroke, the system, at some elementary por-

Content from this work may be used under the terms of the Creative Commons Attribution 3.0 licence. Any further distribution of this work must maintain attribution to the author(s) and the title of the work, journal citation and DOI.
Published under licence by IOP Publishing Ltd

tion of its state transition, absorbs heat and does work, then, in back stroke, at the same portion, the system releases heat and work is done over it. However, the described situation is ideal one as upon termination of the cycle (that is, the forward and backward strokes) the system returns into the initial state but in its surroundings some changes always take place due to inevitable loss of heat energy. That is why all natural processes are said to be irreversible [3].

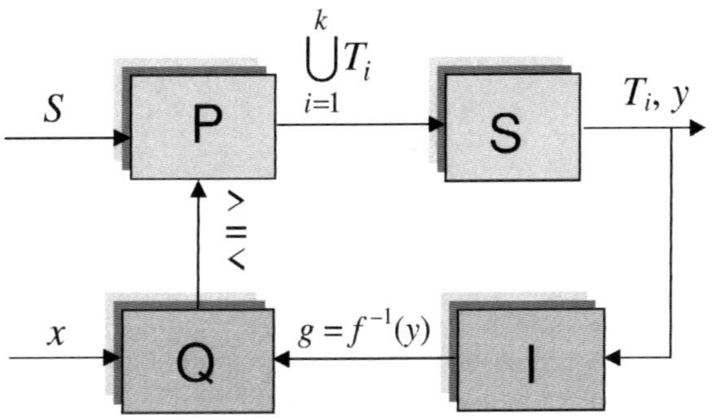

Figure 1. Four-element PSIQ-model of a measurement procedure.

In quantum measurement theory [4], one of principal setting is that a dequantization process is irreversible. Dequantization is defined as a transformation of a quantum sensing system input signal in such a way that the signal could be considered to be classical at an instrument output, see also [5-7]. Thus, overall quantum measurement process consisting of dequantization and classical parts is irreversible.

We claim now that *any* measurement procedure is irreversible and will show this in terms of PSIQ-model using a mental experiment described below.

2.1. Necessary notations and measurement procedure

In the course of the measurement procedure a *measurement problem* is solving that is

$$\mathbf{min}\ \{D(f^{-1}(y), x) \mid y = f(x) \in B,\ B \subseteq S\}, \tag{1}$$

where $y = f(x) \in B$ is a form of mapping of an *empirical set* A to a *numerical set* B [1],
$D(f^{-1}(y), x) = |f^{-1}(y) - x| \leq u$ is a distance between preimage of numerical value y and (empirical) measurand x, and
u is a given measurement uncertainty.

Denote the *operation of partitioning* B into k subintervals T_i through

$$B = \bigcup_{i=1}^{k} T_i\ . \tag{2}$$

where k can take different values in dependence on particular kind of partitioning.

We will *encode* the numerical value $y = f(x)$ in the following way:

$$y = f(x) = < q_1, q_2, \ldots, q_m >, \tag{3}$$

where m takes values in dependence on particular kind of partitioning and

$$q_i = \begin{cases} 0 \text{ if } y \in T_i \\ 1 \text{ if } y \notin T_i \end{cases}. \tag{4}$$

In order to validate expression (4) we suppose that there is a possibility to judge of a membership of y in T_i by result of comparing x and g, where $g = f^{-1}(y)$ is a current preimage of y. That is if $x \geq g$ then $y \notin T_i$ and if $x < g$ then $y \in T_i$. It means that expression (4) is equivalent to

$$q_i = \begin{cases} 0 \text{ if } x < g \\ 1 \text{ if } x \geq g \end{cases}. \quad (5)$$

Finally, we need a method of implementing transformation of the current numerical value y into its current preimage g, i.e. $g = f^{-1}(y)$. This can be done accounting the length (or cardinality) $|T_i|$ of the current interval T_i, that is,

$$g_i = \begin{cases} g_{i-1} - |T_i| \text{ if } x < g \\ g_{i-1} + |T_i| \text{ if } x \geq g \end{cases}. \quad (6)$$

Now we can define a measurement procedure that will find a numerical representation for the given property manifestation x using the following simple recursive dichotomy process (see Fig. 2): at the stage of initialization all q_i are set to 0 and the initial uncertainty interval $T_o = B = \{0, 1, ..., y_n\}$ is formed where the reference point y_n is known to be conformed to $a_n = |A|$; at the first step the initial uncertainty interval T_o is broken up into two equal non-intersected subintervals (that is, in this case, $k = 2$); the subinterval, including y, is selected as T_1; at the next steps every time the previously selected interval is broken up into two parts and the part including y is selected as T_i.

Measurement procedure:

$y \leftarrow$ <0,0,...,0>; $T_0 \leftarrow \{0, 1, ..., y_n\}$ // initialization
DICHOTOMY (1, 0)
procedure *DICHOTOMY* (i, g):
P: $T_{i-1} \leftarrow T_i^1 \cup T_i^2$, $|T_i^1| = |T_i^2|$;
S: $T_i \leftarrow T_i^1$
I: $g \leftarrow g + |T_i|$
Q: **if** $x \geq g$ **then** $q_i \leftarrow 1, T_i \leftarrow T_i^2$
else $g \leftarrow g - |T_i|$
if $|T_i| > 1$ **then** *DICHOTOMY* (i + 1, g)

Figure 2. Formal view of the mesurement procedure in terms of PSIQ-model.

2.2. *Mental experiments with the measurement procedure*

Let the empirical system is modeled by set $A \subset \mathbf{N_0}$, $A = \{0, 1, 2, ..., 7\}$, and an element $x \in A$, $x = 4$, is fixed (see Fig. 3). Measurement procedure shown in Fig. 2 should identify the element $x = 4$ among other elements of A and assign for it an appropriate numerical value $y = \langle q_1, q_2, q_3 \rangle$.

Particular steps and corresponding current solutions of the procedure are reduced to Table 1 and additionally illustrated by Fig. 3. One can observe that the procedure correctly identifies the empirical element $x = 4$ and determines the right numerical value y = <1, 0, 0>.

In order to make sure that the considered procedure is reversible one can fulfill a backward transition from its last step to its initial step. In fact, it is possible to move in inverse order through the same current solutions as shown in Table 1 taking into account that expression (6) should be transformed to its inverse form

$$g_{i-1} = \begin{cases} g_i + |T_i| \text{ if } x < g \\ g_i - |T_i| \text{ if } x \geq g \end{cases}. \tag{7}$$

Table 1
Ideal measurement procedure steps

i-th step	T_i	$y \in / \notin T_i$	q_i	Current solutions
0	{0, 1, 2, 3, 4, 5, 6, 7}			initial interval $B = T_o$; $g = 0$
1	{0, 1,2, 3} ∪ {4, 5, 6, 7}	$\notin$	1	partition and selection of T_1^1 = {1, 2, 3, 4}; $g = 4$; $x = g$; $q_1 = 1 \Rightarrow$ selection of T_1^2 = {4, 5, 6, 7};
2	{4, 5} ∪ {6, 7}	$\in$	0	partition and selection of T_2^1 = {4, 5}; $g = 4 + 2 = 6$; $x < g$; $q_2 = 0 \Rightarrow g = 6 - 2 = 4$
3	{4} ∪ {5}	$\in$	0	partition and selection of T_3^1 = {4}, $g = 4 + 1 = 5$; $x < g$; $q_3 = 0 \Rightarrow g = 5 - 1 = 4$ **y = <1,0,0>**

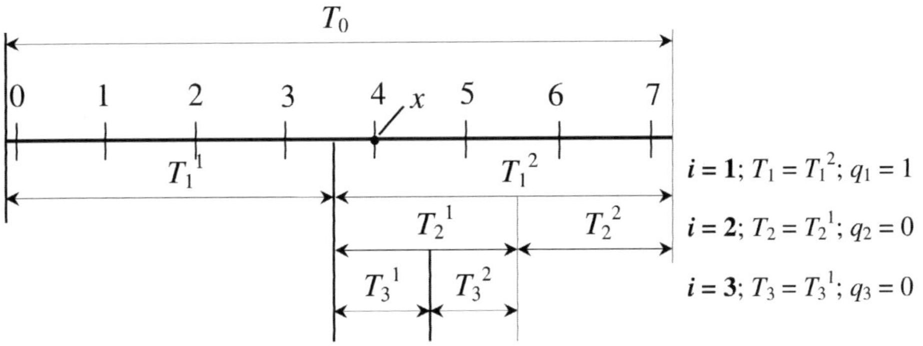

Figure 3. Current partitions and solutions of the mesurement procedure.

Clear that the considered procedure is ideal one as it was implicitly supposed that comparison of values x and g in expressions (5), (6) and (7) is fulfilled without errors. Now let us suppose g have an additive error ε that accidentally takes values ±1. Table 2 shows that presence of ε catastrophically changes operation of the measurement procedure which leads, in this case, to the wrong result y = <0,1,1>.

It is evident that the presence of ε makes the considered measurement procedure to be irreversible as, in this case, both forward and backward transitions between PSIQ-model steps are unpredictable.

Generally speaking, irreversibility of this kind of measurement procedure immediately follows from the fact that inverse mapping $g' = f^{-1}(y)$ never coincides with preimage g since if g is erroneous there are no empirical conditions able to guarantee validity of the hypothesis $g' = g$ [8].

Table 2
Steps of the measurement procedure with erroneous g

i-th step	T_i	$y \overset{\in}{\notin} T_i$	q_i	ε	Current solutions
0	{0, 1, 2, 3, 4, 5, 6, 7}				initial interval $B = T_0$
1	$\underline{\{0, 1, 2, 3\}} \cup \{4, 5, 6, 7\}$	∉	1	+1	partition and selection of $T_1^1 = \{0, 1, 2, 3\}$; $g = 5$; $x < g$; $q_1 = 0 \Rightarrow g = 5 - 4 = 1$
2	$\{0, 1\} \cup \underline{\{2, 3\}}$	∈	0		partition and selection of $T_2^1 = \{0, 1\}$; $g = 1 + 2 = 3$; $x \geq g$; $q_2 = 1 \Rightarrow$ selection of $T_2^2 = \{2, 3\}$
3	$\{2\} \cup \underline{\{3\}}$	∈	0	−1	partition and selection of $T_3^1 = \{2\}$, $g = 3$; $x \geq g$; $q_3 = 1 \Rightarrow$ selection of $T_3^2 = \{3\}$, $\boldsymbol{y}$ **= <0,1,1>**

3. Sequentiality

Sequentiality of measurement procedure means its multistage character which can become apparent as iterativity or recursivity.

This follows from presence of the necessary elementary operation of partition in a measurement procedure and is evident from consideration of corresponding PSIQ-model.

The number of partitioning intervals can be infinite if a measurement accuracy is infinite. However, under infinite accuracy of some particular measurement result, general description of the empirical system becomes rougher, see, e.g. [9].

4. Conclusion

Irreversibility and sequentiality are deserving attention properties of any kind of measurement procedures. Awareness of them can provide some useful ways and tools for development of new measurement methods, see, e.g. [10].

Acknowledgment

This work was partly supported by the Russian Science Foundation, Project # 14-19-00926.

References

[1] Muravyov S V and Savolainen V 1996 Discrete-mathematical approach to formal description of measurement procedure *Measurement* **18** 71-80.
[2] Muravyov S V and Savolainen V 1996 Some structural properties of formal model of measurement procedure *Measurement* **18** 81-7.
[3] Atkins P W 1997 *The Second Law* W.H. Freeman & Co. New York.
[4] Vorontsov Yu I 1989 *Theory and methods of macroscopic measurements* Nauka Publishers Moscow
[5] *The Measurement Problem.* Retrieved February 15, 2014, from Information Philosopher Web site http://www.informationphilosopher.com/problems/measurement.
[6] Kadomtsev B B 2003 Classical and quantum irreversibility *Physics-Uspekhi* **46** 1183-201.
[7] Antonov V and Kondratyev B 2011 Entropy and irreversibility in classical and quantum mechanics *Journal of Modern Physics* **2** 519-32.

[8] Muravyov S V 2012 Chaotic results of multidimensional ordinal measurements *Proc XX IMEKO World Congress (September 9-14, 2012, Busan, Republic of Korea)* **3** 2071-74.

[9] Knorring V G 2013 *Istoriya i metodologiya nauki i tekhiki* Saint-Petersburg Polytechnic University Press (in Rissian).

[10] Srivastav A, Ray A and Gupta S 2007 Irreversibility-based measure of slowly evolving anomalies *American Control Conference (9-13 July 2007, New York)* 3228-33.

The role of adequate reference materials in density measurements in hemodialysis

A Furtado[1], J Moutinho[2], S Moura[1], F Oliveira[2], E Filipe[1]
[1]Instituto Português da Qualidade, Rua António Gião, 2, 2820-010 Caparica, Portugal
[2]Fresenius Medical Care Portugal
E-mail: afurtado@ipq.pt; jorge.moutinho@fmc-ag.com

Abstract. In hemodialysis, oscillation-type density meters are used to measure the density of the acid component of the dialysate solutions used in the treatment of kidney patients. An incorrect density determination of this solution used in hemodialysis treatments can cause several and adverse events to patients. Therefore, despite the Fresenius Medical Care (FME) tight control of the density meters calibration results, this study shows the benefits of mimic the matrix usually measured to produce suitable reference materials for the density meter calibrations.

1. Introduction

Hemodialysis is a renal function replacement therapy. During hemodialysis, the patient's blood is passed through a filter outside the body and then reintroduced to the patient. Tiny pores in the filter membrane filter out toxins, while vital components, such as proteins, are left in the blood. Excess water can also be removed through these tiny pores. The process is controlled by a dialysis machine that is equipped with a blood pump and monitoring systems that ensure safety. The machine can also administer drugs, e.g. Heparin, to prohibit blood clotting during the treatment [1].

1.1. The role of density measurements in hemodialysis

Dialysate consists of purified water and various substances dissolved in it. With the exception of glucose, the substances dissolved in the dialysate are all electrolytes. Their concentration (besides potassium and the buffer substance) closely resembles the concentration of the electrolytes occurring naturally in the blood. Dialysate regulates the electrolyte and acid-base balance of the dialysis patient and removes waste products. Dialysate is produced by the hemodialysis machine combining purified water, acid concentrated solution and bicarbonate concentrated solution. Acid concentrated solution can be obtained in each Hemodialysis Center through a first dilution certified process using certified mixtures of needed electrolytes in ultra-pure water. The final product, i.e. the dilution, must be validated by a calibrated portable oscillation-type density meter. A wrong dilution means a product that is not conform to be used in patient treatments. If the oscillation-type density meters are not properly calibrated, then a wrong mixture of electrolytes will be used for patients' treatments causing several and adverse events to them. A second dilution of these concentrated electrolytes will take place in the dialysis machine. Normally ratios 1:35 or 1:45 are used [2]. To be sure that the final dialysate is adequate for the treatment based on the medical prescription, the maximum permissible error given by the calibration certificate of the density meter, on the first dilution process, is set to 0,001 g/cm^3.

Content from this work may be used under the terms of the Creative Commons Attribution 3.0 licence. Any further distribution of this work must maintain attribution to the author(s) and the title of the work, journal citation and DOI.
Published under licence by IOP Publishing Ltd

1.2. Oscillation-type density meters

The working principle of an oscillation-type density meter is based on the law of harmonic oscillation, in which a U-shaped tube, i.e. the measuring cell, is completely filled with the sample to be analysed, and subjected to an electromagnetic force. The measurement of the frequency and duration of vibration of the tube filled with the sample allows the determination of the density value of the sample. This measuring principle is based on the Mass-Spring Model [3].

Like all measuring instruments, the results obtained by an oscillation-type density meter may vary in time. Errors may be due to: instrumental changes due to physical changes in the U-tube (mass, volume or elasticity coefficient); changes in the electronic operation of the instrument; damage due to mishandling; instrument movement during measurement, especially if at a different angle to the horizontal; effects of liquid on the surface of the tube, such as deposition of material, or erosion by the sample or by the cleaning method. Therefore the calibration is an essential key to understand and take into account the measuring behaviour of the measuring instruments.

1.3. The National Metrology Institute and the calibration of the FME density meters

Density, ρ, is defined as the mass per unit volume of a fluid or a solid [4], and depends, in general, on both temperature and pressure. It is a property of extreme importance, as it is routinely applied in the control of industrial processes, but also used in fields such as biomedical diagnostics, fiscal control and basic research. The uncertainty requirement in a density measurement depends on its specific application and can vary from better than 0,1 % to 1 % [5].

Calibration is an operation that, under specified conditions, in a first step, establishes a relation between the quantity values with measurement uncertainties provided by measurement standards and corresponding indications with associated measurement uncertainties and, in a second step, uses this information to establish a relation for obtaining a measurement result from an indication [6]. Reference materials are often used in calibration operation as measurand generator. According to ISO Guide 34 [7], a reference material can be described by a material, sufficiently homogeneous and stable with respect to one or more specified properties, which has been established to be fit for its intended use in a measurement process.

In the Laboratory of Properties of Liquids (LPL) of the National Metrology Laboratory of Portuguese Institute for Quality (IPQ), density determination of liquid samples can be performed with an oscillation-type density meter DMA 5 000 (Anton Paar). An internal procedure based on the ISO 15212-1 [8] is used for these measurements.The uncertainty budget according to GUM methodology [9], previously established [10], comprises four major uncertainty components: the uncertainty associated with the measurements repeatability, the uncertainty due to the density meters resolution, the uncertainty related to the standard density meter calibration, which depends on the uncertainty of the Certified Reference Materials used, and the uncertainty of the temperature and pressure measurement of the sample inside the measuring cell. The calibration of portables oscillation-type density meters in IPQ is performed by comparative method using reference materials that are measurand generators (Figure 1). These reference materials are characterized with IPQ standard density meter (DMA 5 000, Anton Paar). The density and temperature measurement results of IPQ's standard density meter are traceable to SI units by means of calibration with certified reference materials. The reference solutions are then used in the Fresenius Medical Care (FME) density meters and its indication are compared with the reference values. The measurement error results from the difference between the measured quantity value, i.e. the density value indicated in the FME density meter for a reference solution, and the reference, i.e. the density of the reference solutions determined in IPQ standard density meter. The measurement uncertainty is the non-negative parameter characterizing the dispersion of the quantity values being attributed to a measurand, based on the information used [10]. FME acceptance criterion is given by the modulus of the sum of the measurement error with the expanded measurement uncertainty obtained in the calibration. For a calibration to be acceptable for FME, this sum should be lower or equal to 0,001 g/cm^3.

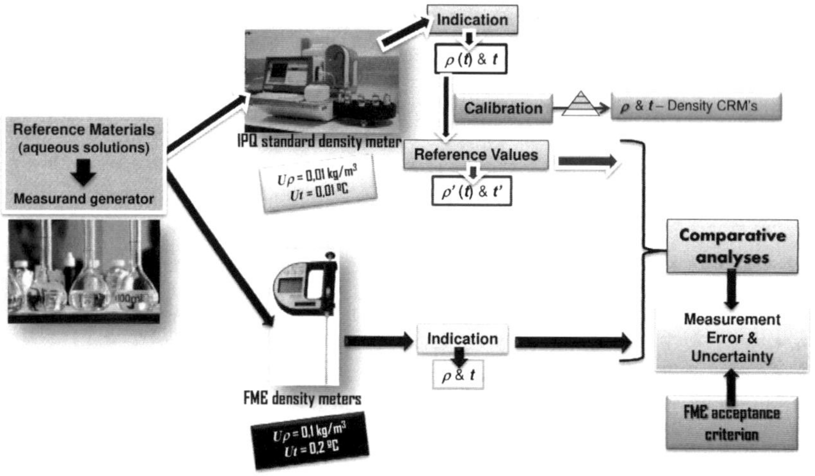

Figure 1. Schematic representation of FME oscillation-type density meters calibration by comparative method in IPQ.

1.4. Calibration of FME oscillation-type density meters by IPQ

Since 2006, IPQ has been performing the calibration of the FME density meters at 20 ºC, for two different densities (1,141 g/cm^3 and 1,163 g/cm^3). Earlier IPQ used standard solutions of sucrose, similar to those used in the Metrological Control of refractometers (OIML R124: 1997), whose stability and homogeneity had been evidenced. But, in 2011, due to the high calibration rejection rate of the density meters, IPQ along with FME began trying to unravel what was the origin of these high failure rates. Then, FME provided to IPQ a sample of acid concentrated solution for further tests.

2. Experimental procedure

2.1. Characterization of acid concentrated solution solution

According to the label the major components of the acid concentrated solution were sodium chloride (~78 % of the solid residue) and anhydrous glucose (~13 %). The coefficient of thermal expansion within the [10, 30] C temperature interval was determined in order to be compared with the temperature correction used by FME.

2.2. Development of new reference material

As the major components of the acid concentrated solution were sodium chloride, two new aqueous solutions of sodium chloride were prepared (mass fraction of 19,2 cg/g and 21,9 cg/g) in order to produce the two points of density desired by FME (1,141 g/cm^3 and 1,163 g/cm^3, respectively).

2.3. Oscillation-type density meters calibration by comparative method in IPQ

The FME density meters were tested with the old reference material (sucrose aqueous solutions with 33,5 cg/g and 38 cg/g), with the new reference materials (sodium chloride aqueous solutions with 19,2 cg/g and 21,9 cg/g) and with the acid concentrated solution.

3. Results

The coefficient of thermal expansion of the acid concentrated solution used by FME ((0,50 ± 0,10) kg m^{-3} K^{-1}) was confirmed by IPQ ((0,48 ± 0,06) kg m^{-3} K^{-1}). The results obtained in the tests with the different solutions showed the similarity of the measuring error obtained with the acid concentrated solution together with the sodium chloride aqueous solutions. Indeed the errors displayed using the previous reference solutions are one order of magnitude lower than the one using the acid concentrated solution or the new reference materials (Table 1). The viscosity (η) of the two kinds of reference solutions shown to be different (Table 1), and this fact could cause errors due the oscillation damping of measuring cell that occurs for viscosities higher than 2 mPa·s. This type of error in the portable density meters cannot be quantified.

Table 1. Calibration results of FME density meters.

Solution tested			**Calibration results / kg/m^3**	
ρ **/ kg/m^3**	X **/ cg/g**	η **/ mPa·s**	**Error**	U **(*k*=2,00)**
sucrose aqueous solutions				
1141	33,5	3,89	-0,4	0,2
1163	38,0	5,89	-0,6	0,2
sodium chloride aqueous solutions				
1141	19,2	1,52	-2,2	0,2
1163	21,9	1,67	-2,4	0,2
acid concentrated solution				
1159	-	-	-2,5	0,2

With the use of the sodium chloride aqueous solutions as reference material for the calibration of the FME density meters, the calibration rejection rate was reduced from 23 % to 4 %.

4. Conclusions

The present study allowed us to demonstrate the correct adaptation of the reference material used in calibration methods in order to reproduce the real operational conditions of the measuring instrument. It was showed that is not enough to guarantee the basic requirements of the reference material, such as homogeneity and stability and specified property [7] (in this case the required density) to obtain acceptable calibration results, as it is also necessary to know the characteristics, such as composition and viscosity of the liquids with which the density meters are intended to be calibrated.

For FME Portugal, the adaptation of the reference material used in the calibration of their density meters allowed an increasing of the stability of the dilution and treatment support process. An important reduction on cost was verified on transport, re-calibration and repairing. The total number of needed density meters was reduced and consequently IPQ response time was shorter. All these performance enabled a more stable global process.

5. References

[1] http://www.fmc-ag.com/36.htm, consultado a 2014-02-22.
[2] http://www.fmc-ag.com/108.htm, consultado a 2014-02-22.
[3] Stabinger, H 1994 Density Measurement using modern oscillating transducers, South Yorkshire Trading Standards Unit, Sheffield.
[4] ISO 80 000-4:2006(E) Quantities and units — Part 4: Mechanics.
[5] Gupta, S V 2002 Practical Density Measurement and Hydrometry, IOP Publishing.
[6] JCGM 200:2008 - International vocabulary of metrology — Basic and general concepts and associated terms (VIM).
[7] ISO Guide 34:2009(E) General requirements for the competence of reference material producers.
[8] ISO 15212-1:1998 Oscillation-type density meters - Part 1: Laboratory instruments.
[9] JCGM 100:2008 - Evaluation of Measurement Data - Guide to the expression of uncertainty in measurement, (GUM 1995 with minor corrections).
[10] Furtado A, Batista E, Spohr I, Filipe E 2009 Measurement of density using oscillation-type density meters. Calibration, traceability and Uncertainties. 14eme Congrès International de Métrologie.

Review of properties of magnetic shape memory (MSM) alloys and MSM actuator designs

N Gabdullin and S H Khan

City University London, School of Engineering and Mathematical Sciences, Northampton Square, London EC1V 0HB, UK

E-mail: Nikita.Gabdullin.1@city.ac.uk

Abstract. Magnetic shape memory alloys are a new group of "smart" materials that exhibit large strain of 6-12% when subjected to magnetic fields. This indicates their enormous potential to be used in different electromagnetic (EM) devices such as actuators, sensors, energy harvesters and dampers. Shape change in MSM materials is controlled by magnetic field and doesn't involve phase transformation, allowing it to overcome a number of disadvantages of conventional shape memory alloys (SMAs). MSM devices are capable of producing large force and stroke output in considerably small dimensions. At the same time they can have fast response and potentially very long lifetime. This paper discusses different modern designs and approaches to MSM actuator design with their advantages and disadvantages. An overview on characteristics of MSM alloys is also presented in order to highlight how different properties of the material influence the total output of a device.

1. Introduction

Magnetic shape memory alloys (MSMAs) or ferromagnetic shape memory alloys (FSMAs) form a new group of shape memory materials in which a shape change can be controlled by the application of a magnetic field or an external stress. Ni-Mn-Ga alloy is the most studied representative of this alloy family and it has been shown to have enormous potential for real applications. Although the material is still at R&D stage of its development and applications, its properties have been significantly improved in recent years since the low-field magnetic shape memory effect was discovered in 1996 [1]. MSM alloys are well-known for their enormous magnetic field induced (MFI) strain, maximum theoretical value of which varies depending on material's microstructure from 6% for five-layered modulated (10M) to 12% for non-modulated (NM) alloys [2,3]. These strain values are higher than maximum strains achievable by conventional magnetostrictive and piezoelectric materials and are of the same order as strains obtained from conventional shape memory alloys (SMAs). However, MSM alloys respond to excitation much faster because the shape change occurs with no phase transformation unlike in conventional SMA materials. The main potential applications of MSM materials could be sensing, damping, energy harvesting and especially high-performance electromagnetic actuation at a frequency range up to a kHz. This paper overviews MSM actuator designs proposed so far and the properties of MSM alloys that are relevant for this application.

Content from this work may be used under the terms of the Creative Commons Attribution 3.0 licence. Any further distribution of this work must maintain attribution to the author(s) and the title of the work, journal citation and DOI.
Published under licence by IOP Publishing Ltd

2. Effects of twinning stress on performance and efficiency

The nature of MSM phenomenon is extensively discussed in literature [4]. When MSM material is subjected to a magnetic field its unit cell can be reoriented producing a strain the maximum value of which is determined by lattice distortion $\varepsilon_0=1-c/a$, where c and a are "easy" and "hard" magnetization axes respectively. Maximum value of MFI stress is limited by crystal's anisotropy energy and is usually considered to be 3 MPa [4] for 10M microstructure though 3.5 MPa was also reported [5]. One of the most important parameters of MSM alloys for their use in actuators is twinning stress, which is a stress to be overcome in order to initiate the shape change. Due to magneto-mechanical coupling, this parameter determines switching field of the crystal (the lowest field at which the shape change occurs). The twinning stress acts as internal friction which decreases the maximum output stress of an MSM element (the so called MSM 'stick') and, hence of an actuator based on MSM alloy. Conventional MSM crystals with type I (twin I) boundaries have a twinning stress of the order of 0.6MPa [6,7]. However, new crystals with type II (twin II) twin boundaries have been reported recently [8,9] with twinning stress less than 0.1 MPa. Such a low twinning stress can lead to corresponding increase in output force coupled with a decrease in required power input enhancing at the same time efficiency of an actuator. Unfortunately, this microstructure is still very unstable and tends to accommodate other twin boundaries which increase twinning stress and dramatically decrease fatigue life of a crystal [6].

The magneto-mechanical characteristics of MSM alloys are very sensitive to temperature. The twinning stress of type I twin boundaries tends to decrease as temperature increases [10]. This may have a tremendous effect on performance parameters, such as increase in strain or output stress with temperature at a constant magnetic field [11] and a decrease in switching field [12]. This may imply that MSM alloys can be used most effectively at high temperatures. However, the current thermal operational range of these crystals is limited by a temperature range of 60-80°C [13, 14] which is rather low for most high-frequency and long-lifetime applications. Due to electric and magnetic losses in various parts of an actuator including the MSM element, excessive temperature rise can lead to the disappearance of MSM effect. The increase of this transition temperature is very important since this temperature essentially defines the operational magneto-thermal limitation of an MSM alloy. Hence much research is being conducted at present by materials researchers to increase this temperature. Also it should be noted that even if the upper temperature limit for twin II crystals is the same as for twin I, the twinning stress of these crystals doesn't depend on temperature [15]. This essentially implies that the characteristics of actuators based on twin II crystals are more stable. However, magneto/electro-mechanical fatigue life of MSM alloys should also be considered.

Theoretically MSM alloys are capable of delivering very long fatigue life. However, this is very much dependent upon its microstructure and operational regimes. Over the years there has been steady improvement in fatigue life from the mere 200×10^6 cycles in 2002 [28] to 2×10^9 in 2010 [6]. This means that it is possible to design MSM devices with very long lifetime. However, two important considerations need to be taken into account to achieve this. Firstly, MSM crystals with low twinning stress were found to be much more brittle with a fatigue life less than 10^7 cycles. This means that twin I crystals with higher twinning stress are more promising for long lifetime devices. Secondly, the operational regimes need to be designed in such a way as to prevent the MSM element transiting into a single-variant state (fully elongated or fully contracted). Although this slightly decreases available stain but it makes very long fatigue life possible.

Overall, 10M MSM alloys with type I twin boundaries can be used to design long lifetime devices with average efficiency, whereas crystals with type II boundaries can be used for very compact and efficient devices with a limited fatigue life. At present crystals with 10M microstructure are considered to be the best for actuation devices as they have high MFI strain and stress accompanied by low twinning stress. This microstructure is the most studied and all designs discussed are based on the use of these crystals.

3. MSM actuators

All MSM actuator designs are based on design elements used as sources of excitation and restoring forces. Almost all possible combinations suitable for actuators are discussed by Holz *et al* [16-18] where each combination is called an "operation mode". Two of them, corresponding to magnetic elongation along with mechanical contraction and only magnetically controlled shape change are discussed in details. The former represents the most common design feature of MSM actuators: a magnetic field produced by coils and/or permanent magnets (PM) controls the elongation of the MSM element in one part of a cycle whereas a mechanical force, which is usually produced by a spring, contracts it in the other part of the cycle. Actuators of this type are the most studied. The main disadvantage of this design feature is reduced maximum force of an actuator as mechanical spring should be strong enough to overcome the twinning stress of a material to contract it [4]. Therefore, MFI stress is opposed by both twinning stress of the material and compressive stress of the spring used. Moreover, the spring represents a weak point from the point of fatigue life. In an alternative design a field generated, by two coils is used for both contraction and elongation depending on the direction of currents. In this case there is no uniform compressive force, thus the output force is higher. Such a design was also discussed in [19], where two coils were used to operate a thin film MSM bridge. However, this design leads to additional power demands and higher electric power losses. In a conventional system, only half of the operating cycle requires electrical excitation while the other half is operated by a spring with no electromagnetic losses and additional power demands. In a system with no spring, both parts of the cycle require current leading to extra losses and extra power consumption. Replacing the spring by permanent magnet(s) can solve the problem described above. If uniform magnetic field of permanent magnet(s) is applied along the direction of the shape change, MFI stress produced by this field will compress the MSM stick. Therefore, no additional excitation is required for contraction. However, magnetic flied inside an MSM stick produced by permanent magnets will depend on MSM permeability which is a function of strain and, hence the magnetic stress produced will not be constant. This means that during elongation the compressive magnetic stress will be excessively high which will lead to a decrease in maximum output force of an actuator by even higher amount than a spring does.

Permanent magnets can also be used to reduce supply current [20]. In this case, magnetic field of magnets is aligned with the coil field which is transversal to the MSM stick. The strength of PM field in the air gap can vary between zero and the switching field. In this case the supply current can be significantly reduced leading to smaller size and higher efficiency of an actuator. However, the PM field will produce magnetic stress which doesn't disappear when the excitation current is removed which means an extra stress to be overcome by the spring. Therefore, stronger spring is required which in turn will decrease the output stress of an actuator during elongation. This operation should be carefully considered as a possible trade-off between efficiency and output force of a device.

As mentioned above, linear actuators with electromagnetic excitation and a spring are the most common design option. The simplest design comprises an MSM stick of rectangular cross-section positioned between the poles of a suitably designed magnetic circuit and connected to a push rod of some sort. Various design variations of this kind can be found in the literature [21] including linear actuator with MSM element of cylindrical shape. Actuators of this type can be fast and small in size delivering considerable output forces. Unfortunately, the excitation coils of these actuators are still relatively large and require extra space limiting the design of very small-size devices. As twin II alloys are shown to have very low twinning stress, actuators based on these alloys can be much smaller due to reduced coil dimensions. Nevertheless, switching field of the material also depends on the total load. Therefore, actuators designed for large force output would require a higher excitation current compared to actuators designed for large stroke output under a small load. This leads to increased number of coil turns and hence, again, to larger dimensions. For actuators with twin I MSM elements, a considerable twinning stress complicates control of intermediate strains due to magneto-mechanical hysteresis [22]. However, there are a number of ways of overcoming this unfavourable property and even to turn in into an advantage.

One design option to use two MSM sticks in order to improve the accuracy of position control has led to a "push-push" actuator design [23], which is receiving a considerable attention nowadays. In this case, the moveable part is placed between two MSM sticks each of which is capable of further straining or contraction. Excitation of the coil corresponding to one of the two sticks leads to elongation of that stick with simultaneous contraction of the other, with the mechanical link moving in the direction of elongation of the first stick. If the current is removed, the actuator maintains its position due to internal holding forces of MSM elements. Therefore, excitation currents can be applied as short pulses in corresponding windings. However, the size of such an actuator tends to be much larger due to the need for two MSM sticks and two excitation coils. Also, even if no restoring force is required, twinning stress of each MSM stick lowers the output stress of the other and thus, decreases the total output force of the actuator. The "push-push" concept of MSM actuator design was behind the design of an MSM-based clamping device [24] where two MSM sticks were used to open or close a single clamping jaw.

Due to variations of length, cross-section and magnetisation during actuation, the resistivity and permeability of MSM crystals also change with the shape change depending on the current strain. This opens up ways to design a device capable of sensing its own position via the estimation of current values of varying parameters. This can be done by adding a circuitry that measures either the resistance [25] or the inductance [26] to calculate the corresponding strain. Such a self-sensing actuator possesses a closed-loop control system which significantly improves position control capability [27]. However, the size of an actuator increases due to the need to accommodate a sensing element e.g. an extra coil for measurements.

In general, the output force of an MSM stick is proportional to its cross-sectional area and the stroke is proportional to its length. An actuator producing a 1kN blocking force by a cylindrical MSM stick of 290mm diameter was presented in [28]. Also an actuator design containing five MSM sticks connected in series producing 4.5mm total stroke was presented in [29]. This illustrates design variations and the capability of MSM alloys to produce large force and large stroke outputs.

At present the main applications of MSM alloys include actuators for valves, linear motors, pumps for macro- and micro-scale applications. The design of a rotating actuator is presented in [30], where movement of a linking rod connected to two MSM sticks of a differential actuator was converted into rotation of a rotor. Application of inverse characteristics was proposed for use in sensing, energy harvesting and damping.

4. Modelling and design of MSM actuators

The structure of a basic MSM actuator is rather simple, consisting of coils, flux guide (magnetic circuit) and a MSM stick with a push-rod connected to it. Therefore, the principles used to design the magnetic circuit of electromagnetic solenoid actuators can be applied here. The only difference is that magnetic properties of MSM elements vary in magnetic field in a different way than those in conventional ferromagnetic materials. These variations turn out to be considerable affecting the entire operation of an actuator. Initially, it was suggested to linearise the magnetisation curves [31] and use the effective magnetisation curve which essentially means taking the average properties of MSM material. This resulted in an equivalent circuit model with lumped parameters which uses average MSM permeability [20]. However, some of the recent research results show that variant distribution plays a significant role in affecting the magnetic field distribution inside an MSM stick. This can be taken into account by finite element (FE) modelling and characterisation of magnetic field distribution in the magnetic circuit [7, 32]. If the average permeability is used the modelling results give good agreement with that obtained from analytical solutions. If the variant distribution is taken into account the MSM properties are represented much more accurately and local non-uniform field distribution can be modelled. It is also possible to take into account the entire magnetic circuit of an MSM actuator into one model enabling optimisation. Overall, the modelling of magnetic field of an actuator leads to significant increase in accuracy due to more accurate representation of piece-wise homogenous nature of MSM properties. The most comprehensive approach to modelling along with experimental verification is presented in [7].

5. Conclusions

A brief overview of modern MSM actuator designs have been discussed with their advantages and disadvantages highlighted. The effects of twinning stress on actuator output force and fatigue life together with its dependence on twin boundary type have been also discussed. The variations of magnetic properties of MSM alloys have been discussed and FE modelling is proposed as a valuable design tool for taking those into account.

References

[1] Ullakko K, Huang J K, Kantner C and O'Handley R C 1996 *Appl. Phys. Lett.* **69** 1966
[2] Murray SJ, Marioni MA, Allen SM, O'Handley RC, Lograsso TA 2000 *Appl. Phys. Lett.* **77** 886
[3] Sozinov A, Lanska N, Soroka A and Zou W 2013 *Appl. Phys. Lett.* **10** **2**021902
[4] Straka L, Hanninen H, Soroka A and Sozinov A 2011 *J. Phys.: Conf. Ser.* **303** 012079
[5] Aaltio I 2011 *Doctoral Dissertation*
[6] Aaltio I, Ge Y, Pulkkinen H, Sjöberg A, Söderberg O, Liu X W and Hannula S-P 2010 *Phys. Procedia* **10** 87-93
[7] Schiepp T, Maier M, Pagounis E, Schlüter A, Laufenberg M 2014 *IEEE Trans. Mag.* **50** 7024504
[8] Sozinov A, Lanska N, Soroka A and Straka L 2011 *Appl. Phys. Lett* **99** 124103
[9] Kellis D, Smith A, Ullakko K, Müllner P 2012 *J. Cryst. Growth* **359** 64-68
[10] Straka L, Hänninen H, Heczko O 2011 *Appl. Phys. Lett.* **98** 141902
[11] Qu W and Lu J 2011 *Third International Conference on Measuring Technology and Mechatronics Automation* pp. 591-594
[12] Heczko O and Ullakko K 2001 *IEEE Trans. Mag.* **37** 2672-2674
[13] Vallal Peruman K, Mahendran M, Seenithurai S 2010 *Physica B* **405** 1770–1774
[14] Pagounis E, Chulist R, Szczerba M J, Laufenberg M 2014 *Scripta Mater* http://dx.doi.org/10.1016/j.scriptamat.2014.04.001
[15] Heczko O, Straka L, Seiner H 2013 *Acta Mater.* **61** 622-631
[16] Holz B and Janocha H 2010 *Proc. Actuator* 307
[17] Holz B, Janocha H and Riccardi L 2012 *Proc. Actuator* 663
[18] Holz B, Riccardi L, Janocha H and Naso D 2012 *Adv. Eng. Mat.* **14** 668
[19] Krevet B and Kohl M 2010 *Phys. Proc.* **10** 154
[20] Suorsa I 2005 *Doctoral Dissertation*
[21] Schmidt H 2011 *J. Phys.: Conf. Ser.* **303** 012078
[22] Riccardi L, Ciaccia G, Naso D, Janocha H, Turchiano B 2010 *IFAC International Symposium on Mechatronic Systems* 478
[23] Gauthier J Y, Hubert A, Abadie J, Chaillet N and Lexcellent C 2006 *Proceedings of the 5th International Workshop on Microfactories*
[24] Wegener K, Blumenthal P and Raatz A 2013 *J. Int. Mat. Syst. and Struct.* **0** 1-7
[25] Niskanen A J and Soroka A 2012 *Proc. Actuator*
[26] Lin C, Wei Y and Dong F 2010 *Proc. ICMTMA* 812
[27] Qingxin Z, Hongmei Z, Yibo L, Jing Z 2008 *Proceedings of the 27th Chinese Control Conference* 257
[28] Tellinen J, Suorsa I, Jääskeläinen A, Aaltio I, Ullakko K 2002 *Proc. Actuator* 566
[29] Zhao Xiang, Wang Sheliang, Zhao Xicheng 2011 *Int. Conf. Electr. Technol. Civ. Eng.* 6730–4
[30] Jun Lu, Zhihong Liang and Wanchun Qu 2009 *Proc. HIS* 95
[31] Suorsa I, Pagounis E, Ullakko K 2005 *Sensors and Actuators A* **121** 136–141
[32] Smith A, Tellinen J, Müllner P, Ullakko K 2014 *Scripta Mater.* **77** 68-70

Calibration of infusion pumps using liquids whose physical properties differ from those of water

[1]E. Batista, [1]A. Furtado, [1]N. Almeida, [1]S. Moura, [2]R. Martins, [2]L.Sousa, [1]E. Filipe
[1] Instituto Português da Qualidade, Rua António Gião, 2, 2829-513 Caparica, Portugal
[2] FCT/UNL - Faculdade de Ciências e Tecnologia, Universidade Nova de Lisboa, UNIDEMI - 2829-516 Caparica, Portugal
E-mail: ebatista@ipq.pt

Abstract. Infusion medical devices are used in field applications, namely in clinical environments, here are used several types of liquids, according to the therapeutic to be administrated into the patient. In order to determine the influence of the fluids physical properties, such as viscosity and density and to produce an adequate reference liquid, tests were performed with a syringe pump, using the gravimetric method as reference calibration method.

1. Introduction

Infusion pumps are measuring instruments that are widely used in clinical environment for nutrition and hydration of patients and for controlled drug delivery. There is a large variability of used liquids in these pumps, namely parenteral nutrition, insulin, cardiogenic amines, analgesics and antibiotics. These liquids have different physical properties that can influence the total delivered volume and the flow rate.

Usually the calibration of flow measuring devices is performed with water as reference liquid [1]. Despite common infusion solutions used in the infusion pumps are low viscous aqueous solutions, some are approximately four times more viscous than water. This fact suggests that the infusion pumps behaviour should be tested with more viscous reference liquids in order to reproduce the real operational conditions.

2. Microflow measurements and calibration of infusion devices

The Volume Laboratory (LVO) of National Metrology Laboratory (LNM) of the Portuguese Institute for Quality (IPQ), in collaboration with FCT/UNL-DEMI, developed a gravimetric system for microflow measurements and calibration of infusion devices [2] in the frame of the European Metrology Research Programme (EMRP), through the participation in the Joint Research Programme - Metrology for Drug Delivery (MeDD) [3].

3. Liquids physical properties

The Laboratory of Properties of Liquids (LPL) of LNM-IPQ, together with LVO, has been performing studies related with the influence of the liquids physical properties in the volumetric instruments accuracy [4]. Therefore, for dynamic measurements of volume, physical properties such as viscosity and density have particular importance in the flow behaviour.

Content from this work may be used under the terms of the Creative Commons Attribution 3.0 licence. Any further distribution of this work must maintain attribution to the author(s) and the title of the work, journal citation and DOI.
Published under licence by IOP Publishing Ltd

3.1 Density

Density, ρ, is defined as the mass (m) per unit volume (V) of a fluid or a solid [4], and depends on both temperature and pressure. It is a property of extreme importance, applied in the control of industrial processes, biomedical diagnostics, legal control and research. In the LPL, the determination of liquid's density is performed with an oscillation-type density meter DMA 5 000 from Anton Paar, with a best expanded uncertainty of 0,010 kg/m^3 (k = 2). An internal procedure based on the ISO 15212-1 [5] is used for these measurements.

3.2 Dynamic viscosity

The dynamic viscosity, η, is a measure of the internal resistance of a fluid to the flow and is obtained from the ratio of the shear force (τ) and the velocity gradient or shear rate ($\dot{\gamma}$) and it is also the product of fluid density by its kinematic viscosity (ν).The measurement of the kinematic viscosity of a liquid through a capillary with a viscometer is based on the measurement of the time that a fixed volume of liquid takes to flow, in laminar flow regime and only, due to gravity, through the capillary at controlled temperature. The kinematic viscosity of the liquid tested is then obtained from the product of the measured flow time and by the constant of the viscometer used [6, 7].

4. Characteristics of the commonly infusion solutions

The common infusion solutions used in hospital are aqueous solutions of glucose ($C_6H_{12}O_6$) with a mass fraction of 5 cg/g and sodium chloride (NaCl) at 0,9 cg/g. When there is an incompatibility with the drugs to be administered or in case of hypertensive patients, sodium chloride at 0,45 cg/g is then used (Table 1).

Table 1. Physical properties of the hospital infusion solutions and water.

	dynamic viscosity @ 20 °C / mPa·s	density @ 20 °C / kg/m^3
Infusion solutions		
$C_6H_{12}O_6$ (5 cg/g)	1,145	1017,5
0,9 cg/g NaCl	1,020	1005,3
0,45 cg/g NaCl	1,011	1001,8
Water	1,002	998,2

The viscosity and the density of these solutions (Table 1) are similar to the ones for water, which is used as reference liquid in the calibration of the syringe pumps. However, there are some commonly infusion solutions, like Hespan® and Dextran 40®, that are four times more viscous than water. This fact suggests that the infusion pumps should be tested with more viscous reference liquids in order to reproduce the real operational conditions.

5. Experimental procedure

5.1 Standard calibration of infusion pumps

The calibration of infusion devices was performed by gravimetric method [1]. A syringe is filled with degas ultra-pure water [8] and a purge is done by passing a sufficient amount of water through the extension line to ensure that there is no bubble in the system. The intended rate flow is programmed in the syringe pump and the water is then collected in a flask inside the evaporation trap in the balance plate (Mettler, AX26). Through the difference between the apparent mass of the liquid displaced into the flask (I_L) and the empty flask (I_E) in the measuring time interval (t_f - t_i) the volumetric flow (Q) is then calculated by a LabVIEW® program using Equation 1 [2].

$$?? = \frac{?}{?_{?} - ?_{?}}\left[\left((?_{?} - ?_{?}) - (???_{???})\right) \times \frac{?}{?_{?} - ?_{?}} \times \left(? - \frac{?_{?}}{?_{?}}\right) \times [? - ?(??? - ?_{?})]\right] + ???_{?????} \quad (1)$$

Where: ρ_W - water density; ρ_A - air density; ρ_B - mass standard density; γ - expansion coefficient; T_0 - measured water temperature; δV_{evap} – evaporation; δm_{imp} - correction due to tube immersion.

5.2 Calibration of the infusion pumps with viscous liquids

Five aqueous solutions with different viscosities (1,5 mPa·s, 2 mPa·s, 2,5 mPa·s, 5 mPa·s and 10 mPa·s) were prepared gravimetrically using two different compounds - glycerol and glucose. The density of these aqueous solutions was determined as described in 3.1, and the nominal viscosity was established through the measured density according to the literature data [9].

These five solutions, along with pure water, were then tested with a Nexus 3000 pump at three different flow rates of 1 mL/h, 5 mL/h and 10 mL/h. Each test was carried out 3 times and the uncertainty of the average was determined based on the pooled standard deviation and the uncertainty method. The error for each test solution was obtained in relation to the difference from the pure water flow results.

A Perfusor® Space infusion pump was also tested at three different flow rates with water and glycerol solutions with a viscosity of 5 mPa·s and 10 mPa·s.

The calibration results were analyzed according one-way ANOVA (Excel®) assuming a Gaussian distribution for the data.

5.3 Calibration uncertainty budget

The calibration uncertainty budget was established according to GUM methodology [10] and comprises the uncertainty components described in Table 2.

Table 2. Uncertainty components for flow determination by gravimetric method.

Uncertainty source	Standard uncertainty	Evaluation process
final mass	$u(I_L)$	1/2 mse (mean square error)
initial mass	$u(I_E)$	1/2 mse
water density	$u(\rho_w)$	Literature [10]
air density	$u(\rho_A)$	Literature [11]
mass pieces density of the	$u(\rho_B)$	Calibration certificate
temperature	$u(T)$	Calibration certificate
expansion coefficient	$u(\gamma)$	Literature [1]
evaporation	$u(\delta V_{evap})$	Polynomial adjustment
final time	$u(t_f)$	Estimation (1 µs)
initial time	$u(t_f)$	Estimation (1 µs)
buoyancy	$u(m_{imp})$	Calibration certificate
repeatability	$u(rep)$	Standard deviation of the measurements

6. Results

From the data presented in Table 3 and Figure 1 one can observe that there is no evident trend in the general results either in flow and viscosity range tested. For the glycerol solutions with a higher viscosity an increasing uncertainty was observed and generally at least two times higher than the ones obtained for the glucose solutions.

Table 3. Resume of the calibration results of the Nexus 3 000 infusion pump for the different flows and aqueous solutions tested assuming water as reference.

Nominal Flow / ml.h^{-1}	η / mPa·s	Sol.	Error / %	U / %	Nominal Flow / ml.h^{-1}	η / mPa·s	Sol.	Error / %	U / %	Nominal Flow / ml.h^{-1}	η / mPa·s	Sol.	Error / %	U / %
	1,5	Gly	-1,3	1,7		1,5	Gly	-0,1	1,1		1,5	Gly	-0,08	0,84
		Glu	0,1	1,3			Glu	0,06	0,97			Glu	0,0	1,4
	2	Gly	-0,6	2,1		2	Gly	0,5	1,2		2	Gly	0,44	0,82
		Glu	-0,6	1,4			Glu	0,13	0,91			Glu	0,1	1,1
1	2,5	Gly	-0,7	4,4	5	2,5	Gly	0,2	1,2	10	2,5	Gly	-0,1	1,0
		Glu	-0,1	1,4			Glu	0,27	0,91			Glu	0,1	1,4
	5	Gly	-3,2	4,4		5	Gly	0,1	1,3		5	Gly	0,4	1,3
		Glu	-0,5	1,5			Glu	0,23	1,1			Glu	0,3	1,2
	10	Gly	-2,6	9,1		10	Gly	0,1	2,4		10	Gly	-0,5	2,2
		Glu	-0,2	1,3			Glu	0,0	1,1			Glu	0,1	1,3

Legend: Flow – nominal flow tested; Sol. – Solutions; Gly – Glycerol; Glu – Glucose; η - Nominal dynamic viscosity @ 20 °C; Error – Relative difference between tested solution and water; U – Relative expanded uncertainty of the measured flow.

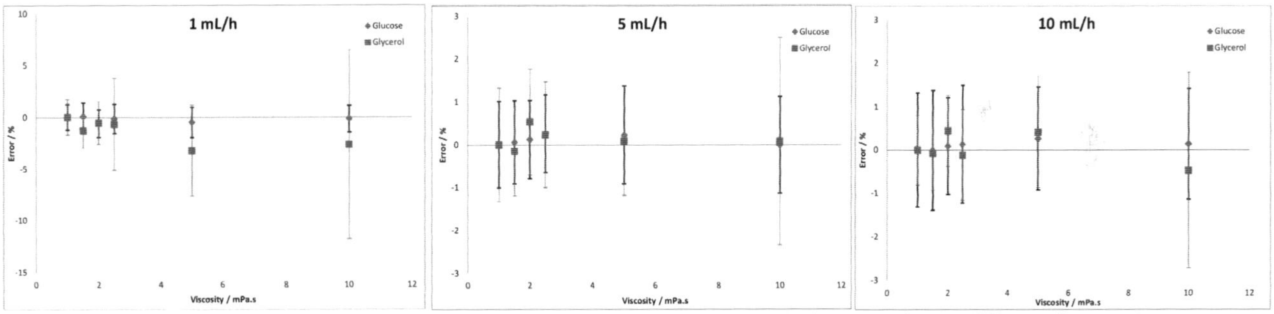

Figure 1. Flow measuring error from water results of the Nexus 3 000 infusion pump for the different flows and for the aqueous solutions tested. (the vertical bars represent measurement uncertainty)

The results of the one-way ANOVA for the effect of the flow and viscosity on flow error of the Nexus 3 000 infusion pump (Table 3) shows that the viscosity of the samples tested (from 1,5 mPa·s to10 mPa·s) has no effect neither on the error of the resulting flow (p-value > 0,05 %). For the flow, the null hypothesis is reject for a significance level of 0,05 %, therefore we can concluded that the flow produces a significant effect on the error of the calibration results.

Table 3. Resume of the one-way ANOVA results for the effect of the flow and the viscosity on the flow error of the Nexus 3 000 infusion pump.

Liquids / Effect	*p*-value (α = 0,05 %)
Water	
Flow effect	0,03
Glycerol	
Flow effect	0,00
Viscosity effect	0,74
Glucose	
Flow effect	0,00
Viscosity effect	0,92

The flow measuring errors from the nominal flow and measuring uncertainty of the Perfusor® Space infusion pump for the different flows and viscosities also shown no evident trend in both (Figure 2).

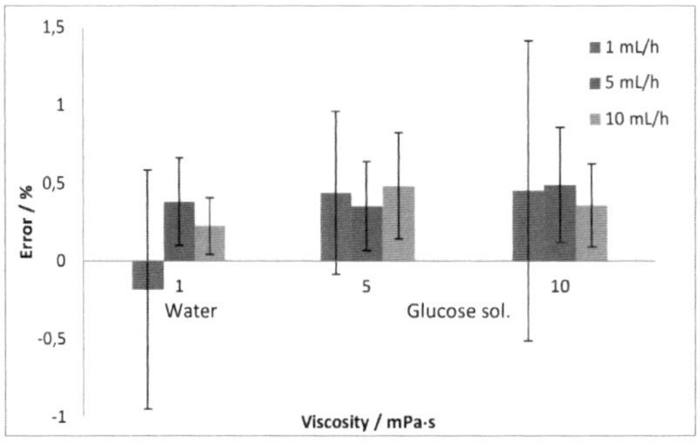

Figure 2. Flow measuring error from the nominal flow of the Perfusor® Space infusion pump for the different flows and for the aqueous solutions tested (the vertical bars represent measurement uncertainty).

The results of the one-way ANOVA for the effect of the flow and the viscosity on the flow error of the Perfusor® Space infusion pump (Table 4) shows again that the viscosity of the samples tested has no effect on the error of the resulting flow (*p*-value > 0,05 %).

Table 4. Resume of the one-way ANOVA results for the effect of the flow and the viscosity on the flow error of the Perfusor® Space infusion pump.

Effects	*p*-value (α = 0,05 %)
Flow	0,65
Viscosity	0,15

7. Conclusions

Based on the results obtained during the calibration of the two syringe pumps tested it can be verified that there are no significant differences between the measured flow for the different liquids tested and no influence of the viscosity in the final flow, within the interval of 1 mPa·s to 10 mPa·s.

The increasing uncertainty observed for the glycerol solutions with the increasing viscosity in the Nexus 3 000 infusion pump can be related with the low vapour pressure of glycerol (0,001 hPa at 20 °C [10]) compared with water (2 300 hPa at 20 °C [10]). This fact should lead to a higher evaporation rate in the more viscous glycerol solutions. It can also be concluded that glucose solutions seems more suitable for these kinds of tests leading to a better repeatability and therefore a lower measuring uncertainty.

The statistical analysis of the flow measuring error by means of ANOVA has shown that the viscosity of the samples tested (from 1,5 mPa·s to 10 mPa·s) has no effect on the error of the resulting flow for both pumps. On the other hand the flow produces a significant effect on the error of the calibration results as confirmed by the one obtained for water for the Nexus 3 000 infusion pump. For the Perfusor® Space infusion pump no significant effect was observed on the flow errors obtained with the tested liquids. We can assume that these different results in the flow effect must be related with different mechanical behaviors of these infusion pumps.

This work leads to the conclusion that water is suitable reference calibration liquid in the case of this equipment within the flow range of 1 mL/h to 10 mL/h. We can also assume that different types of viscous liquids used in hospital environment have no significant influence in the final delivered flow.

8. References

[1] ISO 4787:2010 [1] ISO 4787:2010 – Laboratory Glassware – Volumetric Glassware – Methods for use and Testing of Capacity.

[2] Batista, E., Gala, J., Ribeiro, L., Almeida, N., Filipe, E., Martins, R., *Desenvolvimento de um padrão de medição de caudal de fluidos* 2014 Medições e Ensaios, nº 7.

[3] Lucas, P., Batista, E., Melvad, C. et al Metrology for drug delivery, research project 2012 – 2015, www.drugmetrology.com.

[4] ISO 80 000-4:2006(E). Quantities and units — Part 4: Mechanics.

[5] ISO 15212-1:1998.Oscillation-type density meters - Part 1: Laboratory instruments.

[6] ISO 3104:1994 (E): Petroleum products – Transparent and opaque liquids – Determination of kinematic viscosity and calculation of dynamic viscosity.

[7] DIN 51562-1:1999: Viscometry. Determination of kinematic viscosity using Ubbelohde viscometer, Part *1: Apparatus and measurement procedure.*

[8] ISO 3696:1987 - Water for analytical laboratory use – Specification and test methods.

[9] Lide D 1996 Handbook of Chemistry and Physics 77th edition CRC Press.

[10] JCGM 100:2008 - Evaluation of Measurement Data - Guide to the expression of uncertainty in measurement, (GUM 1995 with minor corrections).

Utilization of Software Tools for Uncertainty Calculation in Measurement Science Education

Hubert Zangl[1], Mariam Zine-Zine[1], and Klaus Hoermaier[2]

[1] Institute of Smart Systems Technologies, Sensors and Actuators, Alpen-Adria-Universitaet Klagenfurt, Universitaetstrasse 65-67, 9020 Klagenfurt, Austria

[2] Infineon Technologies Austria AG, Siemensstrasse 2 9500 Villach, Austria

E-mail: hubert.zangl@aau.at

Abstract. Despite its importance, uncertainty is often neglected by practitioners in the design of system even in safety critical applications. Thus, problems arising from uncertainty may only be identified late in the design process and thus lead to additional costs. Although there exists numerous tools to support uncertainty calculation, reasons for limited usage in early design phases may be low awareness of the existence of the tools and insufficient training in the practical application. We present a teaching philosophy that addresses uncertainty from the very beginning of teaching measurement science, in particular with respect to the utilization of software tools. The developed teaching material is based on the GUM method and makes use of uncertainty toolboxes in the simulation environment. Based on examples in measurement science education we discuss advantages and disadvantages of the proposed teaching philosophy and include feedback from students.

1. Introduction and Motivation

Measurements are required for many tasks and the quality of measurement results has a major impact on the overall outcome. Surprisingly, the concept of uncertainty is comparatively young and was proposed for the first time in 1984 [1]. The idea was to overcome some of the limitations that are associated with the previously used term error [2]. With respect to metrology, the uncertainty reflects the fact that measurements can only provide incomplete knowledge and that a measurement is only useful when the lack of knowledge is somehow quantified. This is particularly true with respect to safety and reliability. Consider, for example, a monitoring system that should validate that a certain parameter lies within a certain interval. However, if the measurement uncertainty of the monitoring system becomes larger than the interval to be monitored, then the monitoring system can *never* be used to validate that the parameter is actually *within* the interval; it can only be used to validate that the parameter (with high probability) resides *outside* of the interval. This may not be apparent for a user or even for a developer of such a system, in particular considering that the engineer may not be an expert in stochastic and uncertainty quantification. Therefore, it seems to be reasonable to provide a method that is commonly accepted by practitioners and experts, can easily be applied for a wide range of problems and still provides good results (yet they may not be optimal in a theoretical sense).

In 1977, as it was recognized that there exists a lack of international consensus on the expression of uncertainty in measurement, the world's highest authority in metrology, the Comit

Content from this work may be used under the terms of the Creative Commons Attribution 3.0 licence. Any further distribution of this work must maintain attribution to the author(s) and the title of the work, journal citation and DOI.
Published under licence by IOP Publishing Ltd

International des Poids et Mesures (CIPM), requested the Bureau International des Poids et Mesures (BIPM) to address the problem in conjunction with the national standards laboratories and to make a recommendation. An effort that finally led to the definition of the Guide to the Expression of Uncertainty in Measurement (GUM) [2]. According to the GUM, the ideal method should be universal (applicable to all kinds of measurements and to all types of input data used in measurements), internally consistent (directly derivable from the components that contribute to it), and transferable (possibility to directly use the uncertainty evaluated for one result as a component in evaluating the uncertainty of another measurement in which the first result is used).

As a result of these requirements, the GUM treats all uncertainty contributions identically, more or less as if the distributions were Gaussian and the relations were linear. This is in accordance with one of the initial requirements for such a recommendation: The approach has to be universal. However, the GUM working group was aware that there are limitations of the original GUM method and supplementary [2, 3, 4] suggested to use Monte Carlo sampling in certain cases. A recent survey [5] on current research activities in the field of measurement uncertainty reports that most recent work addresses the GUM (Guide to the Expression of Uncertainty in Measurement. Consequently, the present paper focuses on this approach, which has a wide acceptance within the field of metrology.

The tools we use for education are Metas.UncLib Matlab toolbox[6], which implements the GUM tree method [7] and also a toolbox for Matlab that is developed in our group. For the basic operations as shown in this paper and as are required for bachelor level measurement science education, both toolboxes are similar in usage and functionality.

2. Teaching Concept

The course in question is on Measurement Science, Sensors and Actuators and is intended for students of information technology in the third year. Consequently, the students are familiar with basic concepts of electrical engineering, have some experience with measurement devices and also have background from a mathematically oriented course on stochastic. Therefore, they are familiar with the concept of random variables, Ohms law and electrical networks. Thus we decided to actually start the course with the concept of measurement uncertainty before we introduce the SI system. Consequently, the discussion on "good" definitions for base units can also be based on the uncertainty concept. Furthermore, traceability is also directly linked to this discussion. This should provide a holistic view of how measurement science work and that knowing the uncertainty is as important as knowing the estimate of some parameter in question.

Our proposed introduction to the concept of uncertainty is illustrated in Table 1. Starting from an interpretation of measurements as realization of random variables that provide some information about the parameter in question we introduce the original GUM including terms such as standard uncertainty, combined uncertainty and determination by means of Taylor series expansion.

The first practical example is the determination of a resistance value and corresponding combined standard uncertainty from the measurement of voltage and current with respective rectangular distribution of uncertainty using Ohm's law

$$R = \frac{U}{I} \tag{1}$$

as the measurement model and the corresponding equation for uncorrelated input quantities and linearization

$$u_c^2(R) = \left(\frac{\partial R}{\partial I}\right)^2 u^2(I) + \left(\frac{\partial R}{\partial U}\right)^2 u^2(U) \tag{2}$$

Section	Content	Educational Objective
1	Introduction to Uncertainty	Understand measurements as realizations of random variables
2	Introduction to original GUM, Error Propagation by means of Taylor Series	Know GUM and it's application to measurement equations
3	Example Resistor Measurement (Voltage/Current)	Practical experience with the original GUM
4	Consideration of systematic errors	Understand importance of measurement model
5	Example Wheatstone Bridge - Simple	Identify pitfalls in approach according to section 2
6	Example Wheatstone Bride - More Accurate	Practice, learn that a correct consideration of uncertainties may be time consuming if done "by hand"
7	Example Wheatstone Bridge - Tool Based	Understand the concepts of "uncertain" as a datatype
8	Example Wheatstone Bridge - Analysis of Sources of Uncertainty	Understand that tools can ease the analysis of measurement chains
9	Example Wheatstone Bridge - Influence of Parameters	Understand that parameters that do not directly contribute to the uncertainty may do so through other parameters.
10	Discussion	Understand that uncertainty must always be considered in measurement. Awareness that tools are available that simplifies most of the calculations. Awareness that the tools and GUM have limitations

Table 1. Steps of the proposed introduction to the concept of uncertainty.

As a second step, systematic errors due to the inner resistance of the measurement instruments are included in the measurement model.

After this simple example we apply the method to the Wheatstone Bridge circuit as another common method to evaluate resistance. There we also introduce software tools for the evaluation of uncertainty. This is described in the next section.

3. The Wheatstone Bridge with Uncertainties

Bridge circuits for the determination of unknown impedances and as realization of the compensation method are important building blocks in measurement science and thus usually treated in introductory courses. We use the Wheatstone Bridge to emphasize how tools for uncertainty calculation may change the way how the material is presented to students. Figure 1 shows the circuit of a Wheatstone Bridge.

First, we start from the classical result for the equation to determinate the value of the unknown resistor R_1, i.e.

$$R_1 = R_2 \frac{R_3}{R_4} \tag{3}$$

and directly apply the GUM method to this, leading to

$$u_C{}^2(R_1) = \frac{\delta R_1}{\delta R_2}^2 u^2(R_2) + \frac{\delta R_1}{\delta R_3}^2 u^2(R_3) + \frac{\delta R_1}{\delta R_4}^2 u^2(R_4) \tag{4}$$

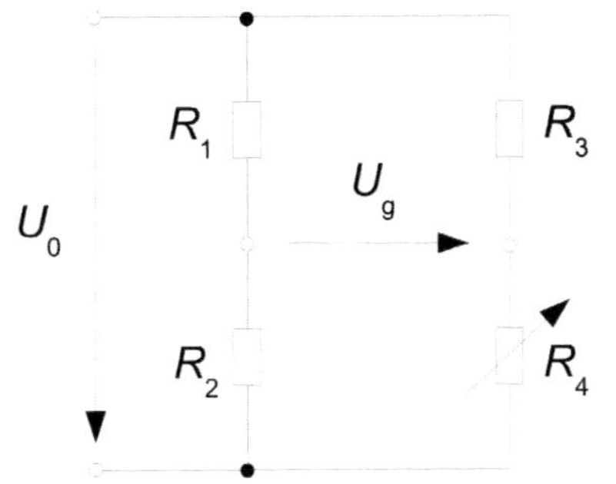

Figure 1. Example for a Wheatstone Bridge.

and let the students do an interpretation of the results. The aim is to point out that apparently some important influences are missing, E.g., it seems that the choice of U_0 and the accuracy of the instrument that measures U_g would not be important with respect to the uncertainty. This is obviously not correct. This leads to a more detailed analysis, showing that it may not be sufficient to only consider a measurement equation to fully determine the standard uncertainty.

The students are instructed to derive a more complete measurement model such as

$$R_1 = \frac{U_0 R_2 (R_3 + R_4)}{U_0 R_4 + U_g (R_3 + R_4)} - R_2 \tag{5}$$

Here we emphasize that a value that is measured to be zero may still be significantly different from zero and must thus not be omitted for uncertainty considerations. In the next step the measurement equation is derived with respect to U_0, U_g, R_2, R_3, R_4, leading to the coefficients $C_{R2} = \frac{\delta R_1}{\delta R_2}$, $C_{R3} = \frac{\delta R_1}{\delta R_3}$ and so on. The resulting uncertainty of the resistor R_1 can be calculated as follows:

$$u_C{}^2(R_1) = C_{U0}^2 u^2(U_0) + C_{Ug}^2 u^2(U_g) + C_{R2}^2 u^2(R_2) + C_{R3}^2 u^2(R_3) + C_{R4}^2 u^2(R_4) \tag{6}$$

with the uncertainty $u(X)$ of the respective input quantities and the coefficients calculated previously.

Following this analysis we introduce toolboxes for Matlab which use a datatype "uncertain". With this we perform the same calculation as above, but numerically and step by step as shown in Figure 2 emphasizing that it does not require any additional effort to obtain the combined standard uncertainty but providing the standard uncertainty for the input quantities. It is also shown that the toolbox could also be applied to equation (3) but leading to an incorrect result (Fig. 3).

The reason - failing to correctly consider the uncertainty of the voltage measurement (the imbalance of the bridge) - is otherwise often not obvious for students. Consequently, several rules for use of uncertain measurement results but also for the derivation of measurement equations (including their simplification)can be derived. This allows obtaining the well known result according to equation (3) but clearly highlights that other parameters that not occur in this equation contribute to the uncertainty.

However, the benefit is not just the automatic calculation of the standard uncertainty. Additionally, toolboxes also provide means to determine the contribution of the uncertain input variables to the combined uncertainty of the result. This is shown in Figure 4. In the present example the main contribution to the uncertainty comes from the uncertainty of the voltage measurement, which is in practice not truly zero.

```
Value            Standard Uncertainty
>> U0=unc(5,0.1);R2=unc(1000,5);R3=unc(1000,5).
>> R4=unc(1001,1);Ug=unc(0,0.01);
>> I4=U/(R3+R4);
>> U4=I4*R4;
>> U2=U4+Ug;
>> I2=U2/R2;
>> R1=(U0-U2)/I2,

R1 =

   (999.001 ± 10.713)
```

Figure 2. Calculation of the measurement result including the standard uncertainty of the unknown resistor R_1 using Metas.UncLib toolbox for Matlab [6].

```
>> R2*R3/R4

ans =

   (999.001 ± 7.13415)
```

Figure 3. Incorrect determination of the standard uncertainty of R_1 due to direct application of the GUM to the classical solution according to equation (3).

```
>> get_jacobi(R1)

ans =

    0.0000     4.9950     4.9950    -0.9980    -7.992(
     (U0)       (R2)       (R3)       (R4)       (Ug)
```

Figure 4. Determination of the contributions of the various sources of uncertainty to the standard uncertainty of R_1 using Matlab and an uncertainty toolbox

In this example the contribution of U_0 to the uncertainty is close to zero. By reducing U_0 to half of its original value we show that although its contribution to the standard uncertainty is still negligible we still see an increase of the combined standard uncertainty as other contributions increase. Here we aim to emphasize such interdependencies and how they are easily studied with the tools.

4. Discussion of Advantages and Disadvantages and Student Feedback

We consider toolboxes for determination of standard uncertainty just as an additional feature of an electronic calculator. Similarly as nonlinear functions such as division or logarithm are difficult to calculate without the help of electronic calculators, the results are readily available without any effort. This implies that most of the time it is fully sufficient to understand what a function e.g. logarithm does but not necessarily how it is actually calculated numerically (although the concept should be known). As with the electronic calculator in general, a frequently observed disadvantage is a loss of the 'feeling' whether a result is feasible or not. We consider toolboxes for determination of standard uncertainty just as an additional feature of an electronic calculator. Similarly as nonlinear functions such as division or logarithm are difficult to calculate without the help of electronic calculators, the results are readily available without any effort. This implies that most of the time it is fully sufficient to understand what a function e.g. logarithm does but not necessarily how it is actually calculated numerically (although the concept should be known). As with the electronic calculator in general, a frequently observed disadvantage is a loss of the 'feeling' whether a result is feasible or not. Our aim with respect to uncertainty is to put the focus on the fact that measurement results are actually realizations of random variables and that the GUM tries to describe the probability density function using the standard uncertainty. This is in contrast e.g. with the electronic evaluation of the logarithm, where the result is highly accurate. This is not necessarily the case for the described tools, in particular when only linearization according to the original GUM is used. In particular revisions of the GUM such as projected in [8]. With the use of the tools students may not be aware of the limitations of these tools as they are used to fully rely on the results from electronic calculators.

The positive students' feedback reasons from the initial introduction of the uncertainties, which are directly coupled to the variables. This allows copying the uncertainties given in the specification directly to the variable itself. Since the students are familiar with Matlab, the barrier of using the tool is pretty low. They can work in the known environment without any additional training. Nevertheless, the most important positive point is that the students' attention can be unglued from the already known mathematical calculations and guided towards the measurement model construction and the analysis of the contribution of the various sources of uncertainty. Especially, the simple access of the contribution of the various sources allows to immediately highlight the main contributors to the uncertainty. The students get a feeling on what shall be tried to improve first. On the other side some minor negative feedback was given, e.g. on naming of commands such as get_jacobi, which may not be intuitive for the students. Additionally, more experienced students objected that the provided number of digits for the results (e.g. 999.001 ± 10.713) is somewhat in contradiction with the actual accuracy. We aim to consider such aspects in the improvement of our tool.

5. Conclusion

We propose an approach for considering uncertainty in measurement science education from the very beginning using special tools. The approach does not offer different or even better results; it is currently simply based on the original GUM method. However, it offers an alternative philosophy for the workflow. We believe that this approach may help to increase the awareness with respect to uncertainty, has a high practical applicability in particular with respect to safety engineering and may thus lead to consideration of uncertainty in early design phases.

References

[1] A. Ferrero and S. Salicone, "Uncertainty: Only one mathematical approach to its evaluation and expression?" *IEEE Transactions on Instrumentation and Measurement*, vol. 61, no. 8, pp. 2167–2178, 2012.
[2] BIPM, "JCGM 100:2008: Guide to the expression of uncertainty in measurement."
[3] ——, "JCGM 101:2008: Evaluation of measurement data - supplement 1 to the "guide to the expression of uncertainty in measurement" - propagation of distributions using a monte carlo method evaluaton of measurement data."
[4] ——, "JCGM 102:2011: Evaluation of measurement data - supplement 2 to the "guide to the expression of uncertainty in measurement" - extension to any number of output quantities."
[5] P. da Silva Hack and C. S. ten Caten, "Measurement uncertainty: Literature review and research trends," *IEEE Transactions on Instrumentation and Measurement*, vol. 61, no. 8, pp. 2116–2124, 2012.
[6] M. Wollensack, "Metas.UncLib - an advanced measurement uncertainty calculator," http://www.metas.ch/metasweb/Fachbereiche/Elektrizitaet/HF/MetasUncLib, September 2012.
[7] B. D. Hall, "Computing uncertainty with uncertain numbers," *Metrologia*, vol. 43, pp. L56 – L61, 2006.
[8] W. Bich, M. G. Cox, R. Dybkaer, C. Elster, W. T. Estler, B. Hibbert, H. Imai, W. Kool, C. Michotte, L. Nielsen, L. Pendrill, S. Sidney, A. M. H. van der Veen, and W. Woeger, "Revision of the "guide to the expression of uncertainty in measurement"," *Metrologia*, vol. 49, no. 6, p. 702, 2012. [Online]. Available: http://stacks.iop.org/0026-1394/49/i=6/a=702

IMEKO IOP Publishing
Journal of Physics: Conference Series **588** (2015) 012055 doi:10.1088/1742-6596/588/1/012055

Body posture measurement in a context of example-based teaching

Eric Benoit, Stephane Perrin, Didier Coquin
LISTIC, Universite de Savoie, FRANCE
E-mail: eric.benoit@univ-savoie.fr

Abstract. This paper presents a measurement process of body postures operated in a context of humanoid robot learning. The basic measured quantities are the angle joints of a human skeleton and the angle joints of a humanoid robot. Due to the differences between the two mechanical structures, the measurement results are expressed into a common representation space by the way of fuzzy scales. This paper shows how the common representation space can be defined, and presents a method to match weakly defined postures with uncertain measurements of a human posture.

1. Introduction

The era of humanoid robots now enters into a new step with the introduction of products for non-professional users. The set of applications of such robots is quickly growing and the conventional systems for the configuration and the robot programming stay dedicated to specific users. A solution to extend the set of potential *robot programmers* is to map education techniques on the context of robots. In particular, we suppose that it is possible to teach postures and gestures to a humanoid robot with a sample-based teaching method. The chosen process to perform such method is given:

- a human called teacher performs a posture,
- this posture is measured then interpreted,
- following the interpretation, the robot tries to perform the same posture.
- the robot posture is measured and compared to the teacher posture.

In order to be compared, the postures of the teacher and of the robot need to be expressed in a common space. Unfortunately, the numerical space in which the angles of the human joints are expressed cannot be the same than the numerical space used to represent the angles of the robots joints. This difference mainly comes from the distribution of the DOF (degrees of freedom) that differs between the teacher and the robot.

The solution we propose is to perform measurements with fuzzy scales [1][2]. This allows to represent a measurement indication as a fuzzy lexical set with dimension 6, each dimension leading to a part of the body : legs, torso, arms and head. For each dimension, i.e. for each body part, the posture of this part is represented with a usual Lexical Fuzzy Subset (LFS). A LFS is then a fuzzy subset of lexical terms, each lexical term being associated to a characteristic posture of this body part. The indication given by the measurement of the full body is then defined as a 6 components vector of LFSs.

Content from this work may be used under the terms of the Creative Commons Attribution 3.0 licence. Any further distribution of this work must maintain attribution to the author(s) and the title of the work, journal citation and DOI.
Published under licence by IOP Publishing Ltd

The main advantage of this approach is that the same set of lexical terms can be used to characterise a part-posture of the teacher or the equivalent part-posture of the robot. This means that the measurements of full body postures of the teacher and of the robot are expressed into the same representation space and can be compared. It has been shown before that fuzzy scales are metrical scales and restrict the set of relations between measurement results to the comparisons of distances. But these relations are enough to perform the comparisons we need and to manage the measurement uncertainty.

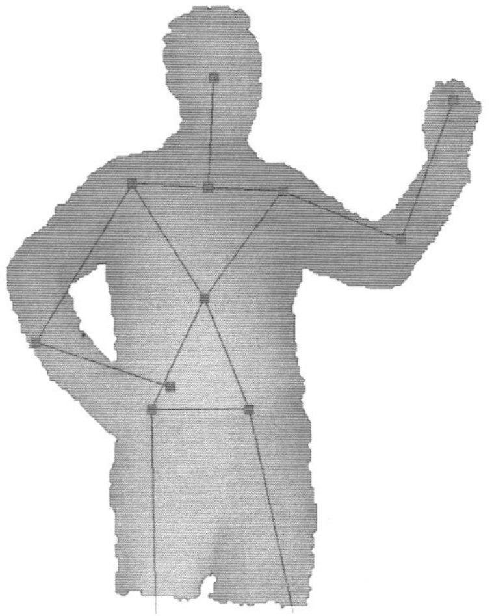

Figure 1. 2D representation of the body perceived by the depth sensor and of the measured joints.

Figure 2. Robot used for the study. The joint states are acquired with proprioceptive sensors.

2. Numerical measurements

The body posture measurements of the teacher and of the robot are performed with 2 different systems. The body of the teacher is acquired with an image depth sensor (a Azuz Xtion pro in our case) [3]. A middle-ware extracts the body shape and computes the 3D coordinates of the joints to give a skeleton approximation as for [4][5]. Each joint is either a 1DOF joint or a 3DOF joint. Each 1DOF joint is characterized by a single angle. Each 3DOF joint is characterized by a normal rotation axe and an angle. The robot used for this study is a Nao robot from Aldebaran Robotics. It holds 13 joints with a total of 25DOF. Each rotation axe has a proprioceptive sensor that is used for the measurement of the body posture.

The teacher body and the robot body are decomposed into 6 parts that are the left and right arms, the left and right legs, the torso, the head. For each part, the information entities used to express indications and measurement results are vectors of numerical data.

Thanks to the proprioceptive sensors, the measurement uncertainty is negligible for the robot posture, but needs to be managed for the teacher posture. Actually the measurement process with an image depth sensor induces two kind of uncertainties. The first uncertainty is an ontic uncertainty [6] and comes from the common statistical dispersion of the 3D coordinates of joints and can be estimated with a type A evaluation [7][8]. The second uncertainty comes from the fact that a joint can be hidden by another part of the body or can simply be confused with another object (see figure 1). A third source of uncertainty named epistemic uncertainty [6] is not directly linked to the measurement but depends on the definition of postures that can be imprecise. In this case a posture is no more defined with a value (actually a vector), but with a set of values. Such set doesn't define a probability distribution but a possibility distribution. As the simultaneous management of the measurement uncertainty and the epistemic uncertainty is not simply performed within a pure probabilistic approach, we use the Dempster-Shafer evidence theory that allows to handle in the same approach probabilities and possibilities [9].

3. The transferable belief model

The Dempster-Shafer evidence theory, also named the transferable belief model (TBM) is a general approach for uncertainty management based on the assignment of belief to sets. In this section, we present a short introduction to this theory.

The main element of the TBM is the *frame of discernment* denoted Ω and presented as a set of elementary events, or as a discourse set for the expression of the belief in a *piece of evidence*. The belief in a piece of evidence E is then expressed with a mapping $m[E]$ (or simply m) called *belief function*, or *basic belief assignment* (BBA). This mapping is defined from the set of subsets of Ω, denoted 2^{Ω}, to the set $[0, 1]$. It represents the distribution of a weight of belief over the event sets. A non null amount of belief assigned to an event set means that the occurrence of the given events are compatible with the *piece of evidence*.

Any belief function verifies:

$$\sum_{A \in 2^{\Omega}} m(A) = 1 \tag{1}$$

The set $\{A \mid m(A) > 0\}$ is called the set of focal elements of m, and the couple $(\{A \mid m(A) > 0\}, m)$ is called a body of evidence of a variable with values on Ω. The link with the probability theory is performed with the body of evidences of the form $(A \in \Omega, m)$. In this case the focal sets are singletons and the BBA m, qualified as a Bayesian BBA, is a probability distribution over Ω. The other categories of BBA used in our studies are the consonant BBAs which focal sets are nested sets. These last BBAs lead to possibility distributions.

The belief in two pieces of evidence E_1 and E_2 issued from 2 independent sources can be computed with a conjunctive combination operator:

$$m[E_1 \cap E_2](A) = \sum_{B \cap C = A} m[E_1](B)m[E_2](C), \forall A \subseteq \Omega \tag{2}$$

This operator can be used if both sources are reliable. In this case, it concentrates the amount of belief on events that are compatible with both pieces of evidence. This operator has the property to assign some belief on the empty set. The value $m[E_1 \cap E_2](\emptyset)$ is actually representative of the degree of conflict between sources. In order to moderate the degree of conflict it is possible to weaken the confidence into a source with a transfer of belief to the set Ω.

4. The fuzzy linguistic representation of the measured values

The method used to build a linguistic representation of measured postures is similar to the method previously used for hand posture measurement in [10]. In the initial study, only measurement indications were considered as measurement results. If we apply this approach to our case, each posture indication is represented by a LFS (Lexical Fuzzy Subset) defined on the cross product Ω of 6 lexical sets, each one corresponding to a body part. Within the scope of evidence theory used in this paper, Ω is interpreted as the exhaustive set of mutually exclusive representations of a posture. As an example, the item $\omega = (vertical, front, down.open, down.open, behind.open, down.open)$ is a synthetic definition of a posture used to check hand. A posture indication is then a fuzzy subset $\{\mu_i/\omega_i\}$ where the membership degree μ_i to each item ω_i is directly interpreted as a Bayesian basic belief function in Ω.

$$\Omega = L_{Body}.L_{Head}.L_{RightLeg}.L_{LeftLeg}.L_{RightArm}.L_{LeftArm} \tag{3}$$

$$L_{Body} = \{vertical, bent, left, right\} \tag{4}$$

$$L_{Head} = \{front, left, right, up, down, leftbent, rightbent\} \tag{5}$$

$$L_{RightLeg} = L_{LeftLeg} = \{down, up, behind, outside\}.\{open, square, closed\} \tag{6}$$

$$L_{RightArm} = L_{LeftArm} = \{down, up, behind, outside, bodyside\}.\{open, square, closed\} \tag{7}$$

5. Building the epistemic knowledge

The first step before any measurement is to define the universe of discourse on which any measurement is expressed. In our approach, the universe of discourse is made of labels that identify postures. So each posture is identified by a unique label but a label fits with a set of postures. For example, the label $chekingHand$ identifies several postures with several left arm locations.

According that, the sets of postures are learned with examples, we propose to model the knowledge on postures with belief functions. Indeed, this knowledge is acquired from various sources and cannot be simply limited to the definition of the labels. It also hold an epistemic uncertainty. The process is the following:

First the meaning of each lexical set L_i is defined as a fuzzy partition of the numerical set such that:

- An observation of the posture is performed and gives an indication expressed as a LFS on each lexical set.
- An aggregation is performed to compute a LFS on the general lexical set Ω.

The membership of the observation to each lexical term is interpreted as a Bayesian basic belief assignment m_1 on Ω^2. The definition of postures can be interpreted as a classification building problem. The idea is to build a BBA that includes the knowledge related to a posture. We suppose that for a given posture to learn, we have a set of samples given by a teacher and represented by LFSs. One possibility is to consider all samples as independent sources and use the conjunctive combination to build a BBA consistent with all samples. This approach is consistent if the samples are close from each other but cannot be used in our case because a posture is imprecisely defined. In [11] Denoeux proposed to weaken the confidence into samples that are not similar with the previous learned samples. But this option stays unsuitable in our case where two dissimilar samples can be strongly representative of the same posture. Another possibility is to aggregate the Bayesian BBA associated to each sample with a simple mean operator. The resulting Bayesian BBA is then translated into a consonant BBA using the discrete version of the Yamada's algorithm [12].

Step 1: Let m_1 be the Bayesian basic belief assignment to translate. Let $j = 1$ be an iterator. Notice that all focal sets are singletons.

Step 2: Let $A = \{\omega_i \mid m_j(\{\omega_i\}) \neq 0\}$ be the set of focal sets of m_j that are singleton. Let a be the smallest belief assignment on a singleton in A.

Step 3: Transfer the amount a of belief assignment from each singleton in A to the set A.

$$\forall \omega \in A, m_{j+1}(\omega) = m_j(\omega) - a \tag{8}$$

$$m_{j+1}(A) = \sum_{\omega \in A} a = a \mid A \mid \tag{9}$$

Step 4: If m_{j+1} has at least one singleton as focal set, increment the iterator j and go to step 2.

At the end of this process, each archetype posture P is weakly defined by a BBA $m[P]$. These BBAs are consonant BBAs and are therefore considered as possibility distributions (see figure 3).

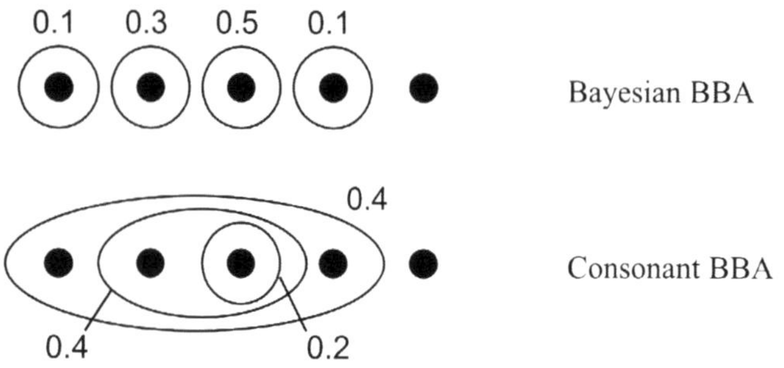

Figure 3. Example of a transformation of a Bayesian BBA into a consonant BBA on a frame of discernment Ω made of 5 symbols. The figure gives for each BBA the focal sets and their belief assignment.

6. measurement of postures

Each observation gives an indication I in the form of a LFS. Such LFS is directly interpreted as a Bayesian BBA $m[I]$ on Ω. For each defined posture P we have a consonant BBA and the conjunctive combination of the 2 BBAs gives the simultaneous belief into the indication value and a posture definition.

The result is a BBA where the focal sets are singletons and the emty set. The amount of belief assigned to the empty set after the combination $m[I \cap P](\emptyset)$ gives the degree of conflict between the indication and the interpretation of the measurement with P. This last information is representative of the impossibility to believe into the assertion *the posture P represents the indication I.*

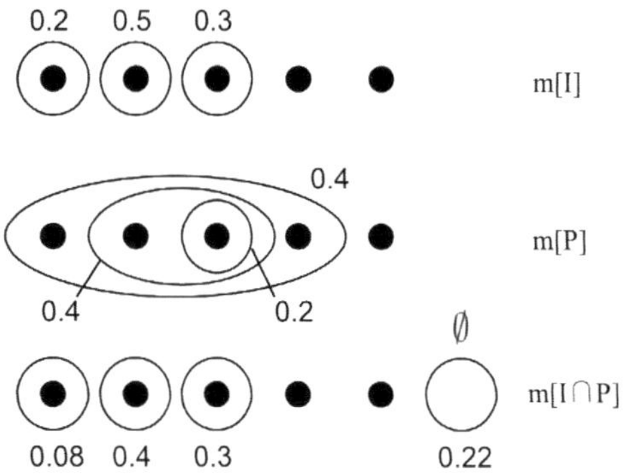

Figure 4. Conjunctive combination of an indication and a posture definition where the indication globally fits with the posture.

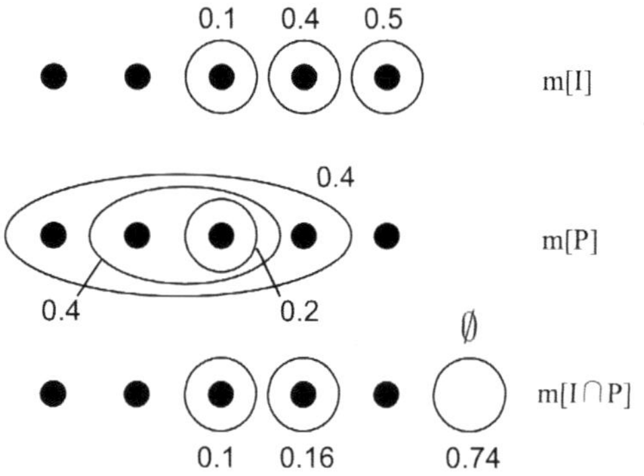

Figure 5. Conjunctive combination of an indication and a posture definition where the indication partially fits with the posture. The empty set holds the majority of the belief assignment.

Conclusion

As for psycho-physical scales, the scale built for human posture recognition depends on a knowledge on the posture definitions: an epistemic knowledge. Thus, the definitional uncertainty known as the epistemic uncertainty is more related to the consistency of the knowledge than related to statistical events. Such uncertainty is then preferably modeled in the frame of possibility theory that allows to model unknown events. The management of ontic uncertainty issued from measurements is based on statistical approaches and are preferably modelled in the frame of the probability theory that gives a large set of mathematical tools. In this paper ontic and epistemic uncertainties interact at two steps. First when the epistemic knowledge is built with a set a measured samples. Then when a measurement indication is compared with the epistemic knowledge. We see in this paper that the immersion of these two kind of uncertainties into the transferable belief model allows to manage them together.

References

[1] Benoit E and Foulloy L 2003 *Measurement* **34** 49–55
[2] Benoit E and Foulloy L 2013 *Measurement* **46** 2921–2926
[3] Bing-ru L, qian S S, min L R, dong Z Z and Yang L 2010 Automatic measurement of scanned human body in fixed posture *Computer-Aided Industrial Design Conceptual Design (CAIDCD), 2010 IEEE 11th International Conference on* vol 1 pp 575–578
[4] Handrich S and Al-Hamadi A 2013 A robust method for human pose estimation based on geodesic distance features *Systems, Man, and Cybernetics (SMC), 2013 IEEE International Conference on* pp 906–911
[5] Xie F, Xu G, Cheng Y and Tian Y 2011 *Image Processing, IET* **5** 420–428 ISSN 1751-9659
[6] Dubois D 2011 Ontic vs. epistemic fuzzy sets in modeling and data processing tasks. *IJCCI (NCTA)* p 13
[7] Dutta T 2012 *Applied ergonomics* **43** 645–649
[8] Clark R A, Pua Y H, Fortin K, Ritchie C, Webster K E, Denehy L and Bryant A L 2012 *Gait Posture* **36** 372 – 377 ISSN 0966-6362
[9] Smets P 2000 Data fusion in the transferable belief model *Proc. of the 3rd conf. on Information Fusion, FUSION2000* vol 1 pp 21–33
[10] Allevard T, Benoit E and Foulloy L 2005 *Measurement* **38** pp. 303–312
[11] Denoeux T 1995 *IEEE Transactions on Systems, Man and Cybernetics* **25** 804–813
[12] Yamada K 2001 Probability-possibility transformation based on evidence theory *IFSA World Congress and 20th NAFIPS International Conference, 2001. Joint 9th* vol 1 (IEEE) pp 70–75

About the Portuguese *VIM3* version

O Pellegrino[1], A Cruz[2], J C Oliveira[3] and E Filipe[4]

[1] IPQ, Rua António Gião, 2, 2829-513 Caparica, Portugal
[2] UNIDO, Maputo, Mozambique
[3] Instituto Nacional de Metrologia, Qualidade e Tecnologia, Rio de Janeiro, Brasil
[4] IPQ, Rua António Gião, 2, 2829-513 Caparica, Portugal

E-mail: opellegrino@ipq.pt

Abstract. For the first time, a unique Portuguese version of the International Vocabulary of Metrology (VIM) was organized and published by the National Metrology Institutes (NMIs) of Portugal and Brazil. This challenge could be met thanks to the experiences of the respective translations of the previous editions of the VIM and to the new Orthographic Agreement (AO) of the Portuguese speaking countries. After a brief historical review of the VIMs and their Portuguese versions, this communication aims to display the main steps that led to the final joint translation. Advantage was taken of this 3rd edition and of the AO to update the Portuguese multiplicative prefix writing "kilo" in coherence with the respective symbol "k". By way of answer to the questions raised by the recent edition of the VIM (VIM3) that stresses on the concepts associated to the terms, some suggestions are proposed and inconsistencies are identified, in order to facilitate the understanding and the dissemination of the document. These few suggestions for the next edition of the VIM also intended to standardize the terminology found in normative texts of different scientific fields which unfortunately does not necessarily tend to be consistent between them.

1. Introduction

In 2008, the Presidents of National Metrology Institutes (NMIs) of Portugal and Brazil, the Portuguese Institute for Quality (IPQ) and the National Institute of Metrology, Quality and Technology (Inmetro) respectively, agreed to publish a unique Portuguese version of 3rd edition of the International Vocabulary of Metrology (VIM3) [1]. This challenge seemed all the more feasible that a Portuguese Language Orthographic Agreement (AO90) between the Portuguese speaking countries, since 1990, was reinforced by a law decreeing its application from 2012 January 1, in all departments, organisms and entities of the Portuguese State. Shortly before, the translation of VIM3 into Spanish, under the auspices of the NMI of Spain, the Spanish Centre of Metrology (CEM), and with the collaboration of various Spanish speaking countries NMIs had been a success story as it turned available the VIM3 for the 2nd largest number of native speaking people on Earth [2]. What could happen with the Portuguese speaking community that is less numerous, although the largest number of native speaking people in South hemisphere? Concerning VIM3, what would be the consequences of more input being provided by new fields in metrology like Chemistry and Medical Science to its elaboration? What evolution followed the traditional concepts of measurement, error and measurement result with the introduction

[1] To whom any correspondence should be addressed

Content from this work may be used under the terms of the Creative Commons Attribution 3.0 licence. Any further distribution of this work must maintain attribution to the author(s) and the title of the work, journal citation and DOI.
Published under licence by IOP Publishing Ltd

of the measurement uncertainty? These are some of the points that are approached in this communication.

2. The VIMs Portuguese translations

2.1. Previous VIMs and their Portuguese translations

The International Bureau of Weights and Measures (BIPM) is an intergovernmental organization that was created by the treaty called "The Convention of the Metre" in 1875. The latter also created the General Conference on Weights and Measures (CGPM), that is made up of delegates of the governments of the Member States and observers from the Associate States and that has a meeting in Paris, one year other four. In its 11th meeting, in 1960, the CGPM adopted the International System of Units, SI, with the rules for the practice of prefixes, the derived units and the base units and for the quantity calculus, setting the language of science, according to the famous I.Mills' expression [3]. This settlement was put further on, in 1969, when the International Organization of Legal Metrology (OIML) published the first edition of the International Vocabulary of Legal Metrology (VIML) and its 2nd edition in 1978. But only in 1984, the first edition of the VIM was published under the responsibility of the BIPM, the OIML, the International Organization for Standardization (ISO) and the International Electrotechnical Commission (IEC). In Portugal, for historical matters, the 1st translation of VIML was released in 1983. It was followed, the year after, with the publication of the Portuguese version of VIM, by the Portuguese Institution responsible for National Organization of Metrology of that time.

After the approximately same time period necessary for the publication of 2nd edition of the VIML, in 1993, the 2nd edition of the VIM (VIM2) released also with the contributions of the International Federation of Clinical Chemistry and Laboratory Medicine (IFCC), the International Union of Pure and Applied Chemistry (IUPAC) and the International Union of Pure and Applied Physics (IUPAP) that joined themselves to the group of international Organizations at the origin of the 1st edition of VIM. The 3rd edition of the VIM (VIM3) was only published in 2007, for which the International Laboratory Accreditation Cooperation (ILAC) collaborated from 2005. In the meanwhile, in 2000, OIML published the most recent edition of VIML, but based on the VIM2. The then recently created NMI of Portugal, the IPQ, published the Portuguese version of the VIM2, in 1994 and, in 2009, the Portuguese version of the 2000 edition of VIML, taking into account the Portuguese version of VIM3, published in 2008. The NMI of Brazil, the Inmetro, published five versions of the Brazilian translation of the VIM2 and, in 2009 November, it published a Brazilian version of VIM3.

2.2. Towards a possible unique Portuguese VIM3

Contrarily to the Spanish language, the European Portuguese language with respect to the American one is similar to the European English language with respect to the one of the American continent. The G B Shaw's aphorism, "England and America are two countries separated by a common language", is also unfortunately valid for Portugal with respect to Brazil. Indeed, as they are built with approximately 400 000 words, English and Portuguese languages can be considered as languages of culture and the percentage of different words between English and American is about 10 %, which is the same difference that exists between Portuguese and Brazilian. However, throughout the twentieth century, some Orthographic Agreements have been made in order to create a unified orthography for the Portuguese Language. The most recent of them was signed in 1990 (AO 90). Considering a sampling of around 140 000 words, studies showed that 3,2 % of them were affected by the AO 90, changing the spelling of 1,0 % of Brazilian Portuguese (BP) words, whereas 1,6 % of the spelling of European Portuguese (EP) words changed. It seemed then to be possible to use the AO 90 to produce a unique Portuguese version of the VIM3.

2.3. The produced VIM3 Portuguese version

In 2012, a new VIM3 version was released that replaced the 2007 version by including the corrections published in 2010. The 2012 VIM3 version was translated in Portuguese by IPQ and Inmetro applying the AO 90 in order to try and solve all possible differences between EP and BP concerning syntactic, morphological (in case of difference, BP was preferred) and lexical matters. For the edition published in Portugal, when the EP and the BP terms are different, first appears the EP term followed by the BP one and a footnote explains in which entry each term is used. For the edition published in Brazil, the BP term is displayed first, followed by the EP one and a footnote has the same role as the European edition. As balance, it can be said that around 6 % of the approximately 410 terms of VIM 2012 were evidenced to be different.

In spite of all, let us now look at how the Portuguese VIM 2012 allowed to solving incoherence in Metrology as practised in countries of Portuguese language for decades. Portugal was one of the first countries to adopt the decimal metric system of units, in 1852. In that time, and following the wording of the decimal metric system, the multiple of unit by a factor one thousand had the prefix "kilo". This lasted until the first orthographic reform, in 1911, when the k letter was put out of the primer, and the writing "quilograma" was admitted instead of "kilograma" for etymologic reason (although the spelling "quilo" comes from the ancient Greek word meaning one thousand, the original decimal metric system adopted symbol was k for the sake of simplicity and dissemination). In 1931, an agreement between the Academy of Sciences of Lisbon and the Brazilian Academy of Letters only maintained the k letter for the symbol of the factor one thousand, until the AO 1990 that reinserts the k letter in the primer, but allowing both the spellings "kilojoule" and "quilograma".. Among the Roman languages, only the Italian with "chilogrammo" and the Catalan with "quilogram" display such discrepancy between the symbol and the writing of the factor for one thousand. Fortunately, the Inmetro not only agreed for adopting the writing "kilogram" in the Portuguese VIM 2012 but also in 2012 edition of the Brazilian SI. These decisions will greatly help that in Portugal also the coherent writing be officially adopted back again.

Before the final publication of the VIM 2012 Portuguese version, the IPQ and the Inmetro launched a 3 months public consultation in the respective websites. Contributions were mainly auditors from the Chemical and Biological fields. The general suggestion of the contributors was for a greater proximity of the translation to the English terms than the French ones.. which was contradictory with the fact that Portuguese is a Romance language and, on the other hand, the fact that French is the official language of VIM3. A typical example of disagreement of the public contributors dealt with the translation of the term precision (*fidélité*, in French). As, in the VIM2, this term had disappeared, the IPQ team translated it by *fidelidade*, which is closer to the French word than *precisão*. Anyway, the great proportion of suggestions from the public consultation for the second word made accept both of them [4].

3. Some challenges for the VIM

3.1. Some basic terms

One of the main criticisms recorded by the VIM3 Online Survey [5], to which comments were able to be sent on the BIPM website during 4 months in 2012, was the difficulty that some readers had to understand and use the VIM3. This was surely not due to the lack of published comments and proposals about the VIM3 terms and concepts. For instance, interesting works have been issued about VIM3 definitions of basic terms like "quantity" and "measurement" [6 - 9]. For instance, VIM3 defines "quantity" as a property having "a magnitude that can be expressed as a number and a reference", in opposition to the "nominal property", which is defined as a property having "no magnitude". Since "measurement does not apply to nominal properties" and the metrology is the "science of measurement and its application", VIM3 does not apply to nominal properties. It can be observed that VIM3 settles a correspondence between quantity and nominal property, quantity value and nominal property value, measurement uncertainty and examination uncertainty and metrological traceability of a quantity value and traceability of a nominal property value [10]. Consequently, *stricto sensus* following the apparent logic of VIM3 might lead to define the term "Examinology", as the science of examination of nominal properties. Projects of Recommendation [11] and Guides [12, 13] were already produced and interestingly dealt with uncertainty and traceability for qualitative analysis or nominal property values issues, i.e. by introducing specific statistics associated to attributes. That is to say, the concepts of quantity value and attribute are clearly supported by the statistics that are fit for the kinds of property, which are not exclusively quantitative [13]. But coherence still needs to be reached among reference documents. For example, as the reference materials are either associated to nominal property value or to quantity value [1], ISO/REMCO [12] is expected to have proposals for the thematic. They interestingly consist on assigning the term "property value" to "quantity value" and the term "property attribute" to "nominal property value", however they still do not have a clear concept associated to quantity and measurement.. A solution of the medical laboratories quality system ISO 15189 [14] is to only consider the term "examination" to deal with both nominal properties and quantities. Due to their evidenced importance in medical and chemical analysis, efforts to clarify and get coherent definitions of the concepts should be continued. Other interesting works that were published about the quantity and the type of quantity [15], about quantity and quantity value in VIM3 [16] and about measurement [17] deserve to be mentioned as fruitful contributions for future version of the VIM.

3.2. Mathematical modelling in Metrology

The growing acknowledged importance of mathematic tools in Metrology turned their use to be "official" by VIM3 publishing associated terms. In particular, there is a general trend of no return in Metrology for mathematical modelling, already initiated with the *Guide to the expression of uncertainty in measurement (GUM)* [18], continued by BIPM Guides about the propagation of distributions using Monte Carlo method [19 - 20] and by other BIPM Guides [21]. As a typical example, a new definition of calibration can be shortly analyzed. According to VIM3, calibration consists of two successive steps: first setting a correspondence between the indication of the measuring instrument (MI) under calibration and the reference values (from standards, for instance), then setting a relationship between the indication of the MI and the measurement result. In the Portuguese translation of VIM3, the new definition of calibration [22] facilitates understanding and implementing this new practice within the accredited laboratories, albeit arisen questions about the choice of the model, its physical meaning, their evaluation and the content of the calibration certificate that have to be solved [23].

When considering measurements involving Chemistry and Biology systems, with great number of entities, the mathematical modelling with analytical function (in the sense of the GUM chapter 8.) are still hard to be solved, even using the propagation of distributions, based on Monte Carlo method. Alternative methods may be used [24] and key terms and associated are referred also in VIM3. However, their definitions need to be coherent with those of ISO 3534-2:2006 and ISO 5725-1:1994 Standards [25].

4. Conclusion

For the sake of better dissemination in the Portuguese-speaking countries, a unique translation of the VIM3 was produced, for the first time, thanks to an Agreement between the Presidents of the NMIs of Portugal and Brazil. A recent officialised Orthographic Agreement between the majority of Portuguese-speaking countries helped its execution, in spite of some still impossible differences in the spelling of 10 % of the terms (as it is the case of the scientific words ending with *ião* in Portugal end with *íon* in Brazil). Anyway, the new logic of VIM in paying more attention to the concepts in VIM3 than the previous editions changed the meaning of measurement (from previously "determining a value of a quantity" to VIM3 "obtaining one or more values that can reasonably be attributed to a quantity") [26] and evidenced the need to clarify concepts associated to the nominal property. The increasing importance of mathematical tools and of Chemistry and Medical Science in Metrology also challenged the applicability of the new definition of calibration and alternative methods to the GUM-based Guides that are present in VIM3, the Portuguese translation of which gave a contribution for more scientists to be involved in.

5. References

[1] JCGM 200:2012 *International vocabulary of metrology — Basic and general concepts and associated terms (VIM) 3rd edition 2008 edition with minor corrections*
[2] CEM:2012 *Vocabulário Internacional de Metrología Coneptos fundamentales y generales, y términos asociados 3ª edicíon 2012 Edicíon del VIM 2008 con inclusíon de pequeñas correciones 3ª edicíon en español Centro Español de Metrología*
[3] Mills I.M. 1997 *Metrologia* **34** 101
[4] IPQ-Inmetro:2012 *Vocabulário Internacional de Metrologia (VIM) 1.ª edição luso-brasileira do VIM 2012*
[5] News from JCGM-WG2 December 2012, accessed on May 2013, on: http://www.bipm.org/utils/en/pdf/news_jcgm-wg2.pdf
[6] Mari L 2003 *Measurement* **34** 17
[7] De Bièvre P 2008 *Metrologia* **45** 335
[8] Ehrlich C, Dybkaer R and Woeger W 2007 *Accred. Qual. Assur.* **12** 201
[9] Rossi G B 2008 *Proc. 13th IMEKO TC1 & TC7 Joint Symposium on Man Science & Measurement (Annecy)* p. 37
[10] Dinu D, Poenaru M-M, Dinu C 2012 *OIML Bulletin* Volume LIII Number 2 18
[11] IFCC-IUPAC Recommendations 201x “Vocabulary for Nominal Properties and Nominal Examinations – Basic and General Concepts and Associated Terms” under development
[12] ISO/REMCO Guide 30:1992/Amd 1:2008
[13] EURACHEM/CITAC Guide “Assessing performance in Qualitative Analysis. The expression of uncertainties in qualitative analysis and testing” under development
[14] ISO 15189:2007 *Medical Laboratories – Particular requirements for quality and competence*
[15] Dybkaer R 2010 *Metrologia* **47** 127
[16] Mari L and Giordani A 2012 *Metrologia* **49** 756
[17] Mari L BEAR Seminar, University of California, Berkeley, 2013
[18] BIPM, IEC, IFCC, ISO, IUPAC, IUPAP and OIML 1993 amended 1995 *Guide to the expression of uncertainty in measurement (GUM)*
[19] JCGM 101:2008 *Evaluation of measurement data — Supplement 1 to the “Guide to the expression of uncertainty in measurement” — Propagation of distributions using a Monte Carlo method*
[20] JCGM 102:2011 *Evaluation of measurement data — Supplement 2 to the “Guide to the expression of uncertainty in measurement” — Extension to any number of output quantities*
[21] Available on: http://www.bipm.org/en/publications/guides/gum.html
[22] Pellegrino O, Furtado A and Filipe E 2013 *Boletim da Sociedade Portuguesa de Matemática, Número Especial, Actas do Encontro Nacional da SPM, Faro, Julho de 2012*, p 217
[23] Groupe de travail CFM “Etalonnage” 2011 *Actes du Congrès International de Métrologie 2011 (Paris)*
[24] Priel M, Amarouche S, Désenfant M and De Graeve J 2010 Using Proficiency Testing Results Latest Alternative Method for Uncertainty Evaluation: Application in Clinical Chemistry *Transverse Disciplines in Metrology* ed John Wiley and Sons chapter 18
[25] Buccianti R, Cibien M, Mari L and Rebaglia B I 2009 *Proc. XIX IMEKO World Congress Fundamental and Applied Metrology (Lisbon)*
[26] Mari L, Carbone P and Petri D 2012 *IEEE Transactions on Instrumentation and Measurement* **61** 2107

AUTHOR INDEX

AUTHOR INDEX